英汉选矿工业词典

陈贤书 编著

北京
冶金工业出版社
2018

图书在版编目(CIP)数据

英汉选矿工业词典/陈贤书编著. —北京：冶金工业出版社，2018.8
ISBN 978-7-5024-7749-3

Ⅰ.①英… Ⅱ.①陈… Ⅲ.①选矿—词典—英、汉 Ⅳ.①TD9-61

中国版本图书馆 CIP 数据核字(2018)第 084598 号

出版　人　谭学余
地　　址　北京市东城区嵩祝院北巷 39 号　邮编 100009　电话 (010)64027926
网　　址　www.cnmip.com.cn　电子信箱　yjcbs@cnmip.com.cn
责任编辑　杨盈园　美术编辑　彭子赫　版式设计　孙跃红
责任校对　石　静　责任印制　李玉山

ISBN 978-7-5024-7749-3
冶金工业出版社出版发行；各地新华书店经销；三河市双峰印刷装订有限公司印刷
2018 年 8 月第 1 版，2018 年 8 月第 1 次印刷
850mm×1168mm　1/32；16.75 印张；614 千字；527 页
196.00 元

冶金工业出版社　投稿电话　(010)64027932　投稿信箱　tougao@cnmip.com.cn
冶金工业出版社营销中心　电话　(010)64044283　传真　(010)64027893
冶金书店　地址　北京市东四西大街 46 号(100010)　电话　(010)65289081(兼传真)
冶金工业出版社天猫旗舰店　yjgycbs.tmall.com

(本书如有印装质量问题，本社营销中心负责退换)

前　言

  这是一本迟来的词典，它本应于20多年前问世。

  词典编著者：陈贤书（生于1912年，故于1997年）1942年毕业于广西大学矿冶系；1942年10月至1957年2月在广西平桂矿务局煤矿、锡矿及广西平桂矿务局总局任职，先后任技师、工程师、基建室主任及设计科科长；1957年2月至1957年12月在北京有色金属选矿研究设计院担任总设计师；1958年1月至1985年9月在北京有色冶金设计研究总院先后担任总设计师、选矿室主任工程师，并成为当时国内有色冶金行业为数不多的教授（研究员）级高级工程师。

  陈贤书先生是我国早期有色金属选矿领域的资深专家，为了适应我国当时采、选专业分家和满足我国选矿工业领域技术水平和行业发展需要，首次创造性地以选矿专业为主，而编写了《英汉选矿工业词典》。

  本词典收编了以选矿技术为核心的英汉词汇20000余条，缩略语约300余条，机构、企业及刊物名称词汇约400余条。其中还包括了与选矿相关的矿物、选矿机械、设备配置、水冶、药剂、土建、总图运输、尾矿、环保、自动控制、水暖通风、计算机技术等技术词汇，并适量收编了采矿、地质、岩石等技术术语，其中涉及部分在当时已出版的冶金方面的英汉词典中尚未收录的新词汇。

  本词典编著者陈贤书先生是我的父亲，他把自己的一生都献给了祖国的冶金采选矿事业，工作中他勤奋努力，认真严谨，一丝不苟，经常到各矿山去搞设计和科研，有时一去就是几十天，具有科技工作者良好的职业道德，"祖国利益高于一切"是他的信条。平常在工作中他非常注意收集、整理冶金选矿方面的英语词汇，并能够结合多年的选矿工作实践经验更贴切和准确地为这些专业英语加以诠释，为本词典的最终定稿付出了大量的心血和精力。即便是在退休后73岁高龄仍然为词典的整理经常工作到深夜。父亲勤勉治学的精神难能可贵并深深感染着我们后辈。遗憾的是本词典在父亲生前由于种种原因未得以出版，但是其技术价值和应用价值在现行的冶金行业依然具有其现实意义。

今天，在社会各方面的大力支持下，本词典才有机会与广大读者见面，这对于我国冶金选矿业是一份迟到的薄礼，同时也实现了我父亲生前的夙愿。

在父亲去世后的这些年，为了完成他出版此词典的心愿，我和姐姐在寻找出版社的同时，和中国恩菲工程技术有限公司的刘国柱翻译一起为完善、丰富本书的内容词汇，使其更加贴近时代发展的需要，查找并增补了很多新词汇，以方便读者阅读及使用。

在此我要感谢为本词典出版做出贡献的人们，其中特别感谢参与后期录入、整理及校对的人员：刘国柱先生（有色院高级英语翻译）；陈幼贤女士（编著者陈贤书的女儿、电气高级工程师），他们为词典的出版做了大量细致的工作。

谨以此词典献给为祖国冶金事业奉献一生的前辈及正在为同一事业奉献的人们。

陈继贤

2017年10月

目　　录

使用说明

词典正文 ………………………………………………… 1~471

附录 ……………………………………………………… 472

　附录 1　缩略语 ………………………………………… 472

　附录 2　机构、企业、刊物名称 ……………………… 506

参考文献 …………………………………………………… 524

后记 ………………………………………………………… 525

作者手稿 …………………………………………………… 526

使 用 说 明

一、英文词条的排列方法

本"词典"中的英文全部词条,均按空格、数字、符号及英文字母顺序排列。

二、标点符号

1. 中、英文的释义或注解及可以省略的词、字、字母均括在圆括号"( )"内;

2. 英文同义词或词条的缩写词括在圆括号"( )"内;

3. 中文译文的同义词括在方括号"[ ]"内;

4. 相近译义的译文之间用逗号","区分,较远译义的译文之间用分号";"区分。

三、关于药剂词条"品名"及"品牌名"的标注

1. "品名"即"商品名"的简称,对单个药剂的词条,中文译文中标注"品名";

2. "品牌名"是同品牌药剂的品牌名称,对列在品牌名称词条下的同品牌系列的具体药剂词条,中文译文中不再标注"品名"。

A

2-arm spider top shell 二臂星形架上罩[壳]

3-arm spider top shell 三臂星形架上罩[壳]

a 30,000TPD concentrator 3万吨日处理量选矿厂

a 6 mesh wire screen 6网目编丝筛面

a 7-digit display direct-reading 7位数字显示直接读数

A fraction starch A级淀粉

a gang sampler 一组[套]取样机(组合成一套的)

a group of conveyors connected in parallel 一组并联运输机

a group of conveyors connected in series 一组串联运输机

a series of pipeline and pump loop system test 一系列管线和泵回路系统试验

a set of base condition 一组基本条件

a set of predetermined value 一组预定值

a set of sequential charts 一套连贯图表

a standard set of testing screens 一套标准试验用筛

AA-rated company AA级公司

AAS 异戊基菲磺酸钠(波兰)

AASHO Road Test AASHO(美国各州公路工作者协会)道路试验

Abbe jar mill 阿贝实验室磨机

Abbe pebble mill 阿贝型砾磨机(实验室用)

abietinol 松香醇(作为疏水性调整剂与胺类捕收剂同时使用,用于钾盐浮选) $C_{19}H_{29}OH$

abnormal ore 非正常[异常]矿石

abrasion characteristic 磨蚀[磨损,研磨]特性

abrasion index testing 磨蚀指数试验

abrasion process 磨蚀过程

abrasion resistant rubber 耐磨橡胶

abrasion resistant steel 耐磨钢材

abrasion-resistant steels 耐磨钢种

abrasive rocks 磨蚀性岩石类

absolute magnitude 绝对量(级)

absolute measurement 绝对测量值

absolute rate of flotation 绝对浮选速度[捕获比]

absolute ratio 绝对比率

absolute value 绝对值

absorbance 消[吸]光度

absorbed radiation 被吸收的辐射线

absorbency 消[吸]光度,吸收度

absorption 吸收[附]
absorption band 吸收谱带
absorption characteristic 吸收特性
absorption column 吸收[附]塔
absorption density 吸附密[浓]度
absorption edge 吸收限
absorption tower 吸收塔
absorptivity 吸收率[性,能力]
AC wound rotor drive motor 交流绕线式转子驱动电动机
AC301(=Aero Xanthate 301) 仲丁基钠黄药(可用作捕收剂)
AC303(=Aero Xanthate 303) 乙基钾黄药(可用作捕收剂)
AC317(=Aero Xanthate 317) 异丁基钠黄药(可用作捕收剂,适于中性矿浆中选黄铁矿)
AC322(=Aero Xanthate 322) 异丙基钾黄药(可用作捕收剂)
AC325(=Aero Xanthate 325) 乙基钠黄药(可用作捕收剂)
AC343(=Aero Xanthate 343) 异丙基钠黄药(可用作捕收剂)
AC350(=Aero Xanthate 350) 戊基钠黄药(可用作捕收剂,适用于浮选氧化铅、铜矿)
academia 学术界
academician 院士
accelerated method 加速法
acceleration due to gravity 重力加速度
acceleration of gravity 重力加速度
acceleration of the earth 地心引力加速度
acceleration time 加速时间
accelerative force 加速力
acceptable concentrate 合格精矿
acceptable dust count 合格矿尘数(标准内矿尘数)
acceptable grade 合格品位
acceptable grade of concentrate 合格的精矿品位
acceptable level 验收[合格]标准,允许程度[含量,范围,水平]
acceptable limit 允许范围
acceptable product 可接受[满意]的产品
acceptance test specification 验收测试规范
acceptor-like state 受电子型状态
acceptor-like surface state 受电子型表面状态
access road 联络道路,便道
accessibility of property 矿区道路,通往矿区的可能性
accessory mineral group 附生矿物族
accident insurance 事故保险
accident-free-day 无事故日,安全日
account 计算,估计,利益,价值,账目[单,户],计算书,报表
account and note payable 应付的账目和票据
account and note receivable 应收的账目和票据

account receivable 应收账目
accountant 会计人员
accounting 会计(学,制度),统计,计算,账目,报表
accounting book 会计账簿
accounting charge 会计用款
accounting department 会计[财务]处[科,室,部门]
accounting earning 会计收益
accounting procedure 会计程序
accrued interest 应计利息
accrued liability 应计负债
accumulated depreciation and depletion 累计的折旧和损耗
accuracy factor 精度因子
accurate to 6 digits 精确到六位数
acetic acid 醋酸,乙酸 CH_3COOH
acetic acid solution analysis 醋酸溶解分析法
acetic acid-ammonium acetate solution 醋酸-醋酸铵溶液
acetone 丙酮 CH_3COCH_3
acetone cyanohydrin 丙酮合氰化氢,2-甲基-2-羟基丙腈(可用作铜-铅-锌矿选择性浮选的捕收剂) $(CH_3)_2C(OH)CN$
acetyl 乙酰(基) CH_3CO-
acetyl acetone 乙酰丙酮,戊间二酮 $(CH_3CO)_2CH_2$
acid 酸,酸的
acid anion 酸性阴离子
acid base equilibrium 酸碱平衡

acid circuit 酸(性)回路,酸(性)流程
acid concentration 酸浓度
acid digestion room 酸分解[溶解,消化]室
acid dissociation constant 酸解离常数
acid dissociation constant K_a 酸解离常数 K_a
acid dissociation pK_a value 酸解离常数 pK_a 值
acid grade 酸级[质]
acid grade fluorspar 酸质[级]萤石
acid group 酸基
acid leach process 酸浸法
acid leached concentrate 酸浸(过)的精矿
acid leaching condition 酸浸条件
acid leaching process 酸浸法
acid media 酸性介质
acid medium condition 酸性介质条件
acid plant 制酸厂
acid process 酸法
acid producing salt 产酸盐
acid production 酸生产(制酸)
acid reagent 酸性药剂
acid resistant rubber 抗[耐]酸橡胶
acid rinsing 酸洗涤
acid rinsing operation 酸冲洗[洗涤]作业
acid salt 酸性盐
acid scrubber 酸洗机
acid sludge 精制石油产品的废酸液、酸性淤渣、酸性油渣(可用于硫化矿的全

浮选,用作起泡剂和捕收剂)
acid soluble copper 酸溶铜
acid solvent 酸溶剂
acid strength 酸强度
acid trap 酸阱,捕酸器
acid treatment 酸处理
acid treatment tank 酸处理槽[罐]
acid-forming constituent 形成酸的成分
acidic 酸性,酸式,酸的
acidic condition 酸性条件
acidic pulp 酸性料(矿)浆
acidic region 酸性区
acidic slurry 酸性矿浆
acidic thiourea 酸性[式]硫脲
acidic type mineral 酸式矿物
acidic water 酸性水
acidol 盐酸甜菜碱(可做锰矿、钾盐、磷酸盐矿的捕收剂)
acidothiourea(tion)(=acidic thiourea) 酸性[式]硫脲
acid-producing potential 生成酸的可能性,生酸潜力
acid-proof flotation cell 防酸浮选槽
acid-soap complex 酸-皂络合物
Acintol 塔尔油制品品牌名
Acintol 2122 塔尔油制品(含 $C_8 \sim C_{10}$ 脂肪酸64.4%,不皂化物35%,松香酸0.5%)
Acintol C 塔尔油制品(含油酸22%,亚油酸20%,松香酸51%)
Acintol D 塔尔油制品(含油酸34%,亚油酸32%,松香酸32%)
Acintol FA$_1$ 塔尔油制品(含油酸46%,亚油酸41%,松香酸4%)
Acintol FA$_2$ 塔尔油制品(含油酸51%,亚油酸45%,松香酸1%)
Aciterge 表面活性剂(烷基恶唑啉脂肪酸盐,可用作湿润剂、乳化剂)
Aciterge OL 阳离子表面活性剂(在酸介质中为非金属矿物的矿泥絮凝剂,还可用作起泡剂、湿润剂)
Aconnon T 分馏的塔尔油(品名,可用作捕收剂)
Acosix T 分馏的塔尔油(品名,可用作捕收剂)
acquisition cost 取得[购置,收购]成本
acrylic polymer 丙烯酸聚合物
Actinol C 粗塔尔油(品名,可用作捕收剂)
Actinol D 精制塔尔油(品名,可用作捕收剂)
action of dissolved oxygen 溶解氧作用
action of nucleating site 成核区作用
action of pulp transfer 传送矿浆作用
action period 动作周期
action relay 动作继电器
activated carbon 活性炭
activated charcoal carbon 活性炭
activated material 活化物质
activating ability 活化能力
activating effect 活化效应[作用]
activation bonding 活化键合
activation effect 活化效应,强化效应

activation reaction 活化反应
activation stage 活化阶段
active catalyst 活性催化剂
active flotation 活化浮选
active gangue slime 活性脉石矿泥
active mineral 活泼矿物
active substance 活化物质
active zone 活动带
actively corroding metal 腐蚀性能活泼(的)金属
activity coefficient 活度系数
activity of negatively charged potential determining ion 阴电荷定位离子活度
activity of positively charged potential determining ion 阳电荷定位离子活度
actual analysis 实际分析
actual calculation 实际计算
actual collector species 实际起捕收剂作用的表面产物
actual concentrate grade 实际精矿品位
actual contact angle 实际接触角
actual cost figure 实际费用数字[值]
actual cost information 实际费用资料
actual data 实际数据
actual experimental result 实际实验结果
actual feasibility constraint 实际可行性约束
actual grinding power 实际磨矿功率
actual measurement 实际测定
actual operating condition 实际操作条件
actual operating data 实际操作[作业]数据
actual operating experience 实际操作经验
actual operating plant 实际生产厂
actual operating work force 实际生产劳动力
actual P. V. level 实际指示值水平
actual performance data 实际性能数据
actual project 实际工程[项目]
actual recovery 实际回收率
actual test 实际试验
actual test data 实际试验数据
actuating unit 动作[执行]装置
actuator 致动[启动,激励,调节]器,传动(装置,机构),拖动装置,操作机构,执行机构
acute oral toxicity 口服剧毒性[物]
adaptive control 自适应控制
added by section 分段添加
added electrolyte 外来[加]电解质
adding water 补加水
addition 添[附]加,补充,扩建部分
addition of reagent 加药
addition range 添加(量)范围
addition rate 添加量,加料速度[率]
additional anodic process 附加阳极过程

A

additional cost 附加费
additional data 补充[额外]资料
additional equipment cost 附加设备费用
additional information 附加[补充]资料
additional input 附加输入量
additional interest 附加利息
additional investment 增加投资
additional item 附加[补充]项目
additional mining equipment 增添的采矿设备
additional operating cost 附加生产费用
additional price 附[追]加价格
additional research 补充研究
additional step 辅助[附加,补充]步骤
additional study 补充研究
additional term 附加项
additional water 额外水分
additive compound 加成化合物
additive requirement 添加剂需要[用]量
address 地址,发言
adherent film 黏附膜
adhesion mechanism 附着[黏结]机理
adhesion work 黏附功
adhesive property 黏性
Adiprene 安迪普林型塑料
adjustable cutter lip 可调截取器切刃[凸缘]
adjustable feed tub 可调给矿盆[槽]

adjustable model parameter 可调(整的)模型参数
adjustable process variable 可调工艺变量
adjustable stroke belt feeder 变速[冲程]带式给矿机
adjustable stroke ore feeder 变速给矿机
adjustable surface 可调表面
adjustable weir 可调溢流堰
adjustable weir overflow 可调的溢流堰[流量]
adjusted data 修正的数据
adjustment cap 调整帽(圆锥破碎机的)
adjustment cap seal 调整盖密封装置,调整环罩密封装置
adjustment collar 调整环
adjustment lock post 调整盖防松螺杆
adjustment lock post washer 调整盖防松螺杆垫圈
adjustment of raw data 原始数据修正
adjustment ring 调整环(圆锥破碎机的)
adjustment ring dust collar 调整环防尘圈
administration cost 管理费
administration office 行政管理办公室
administrative cost 行政管理费
administrative office 行政办公室
administrative personnel 行政管理人员

Adogen 表面活性剂品牌名

Adogen 464 季铵盐,甲基三烷基氯化铵(相当于 Aliquat 336,可做捕收剂、金属的萃取剂)$[R_3N-CH_3]^+Cl^-$, $R=C_9\sim C_{11}$(直链烷基)

Adogen 468 季铵盐,甲基三烷基氯化铵(可做捕收剂、金属的萃取剂)$[R_3N-CH_3]^+Cl^-$, $R=C_9\sim C_{11}$(直链烷基)

adsorbate 吸附体[物],被吸附物

adsorbed ethyl xanthate anion 吸附的乙基黄原酸盐阴离子

adsorbed film 吸附膜

adsorbed layer 吸附层

adsorbed molecule 吸附分子

adsorbed phase 吸附相

adsorbed species 吸附物

adsorbing ion 吸附离子

adsorption behavior 吸附行为

adsorption condition 吸附条件

adsorption curve 吸附曲线

adsorption density 吸附密度

adsorption free energy 吸附自由能

adsorption kinetics 吸附动力学

adsorption layer 吸附层

adsorption mode 吸附模式

adsorption model 吸附模型

adsorption phenomena 吸附现象

adsorption rate 吸附速度

adsorption test 吸附试验

adsorption-dissociation 吸附解离

adsorption-screening circuit 吸附筛分回路

adsorptive bubble technique 气泡吸附法

adsorptive concentration 吸附富集

adsubble technique 气泡吸附法

advanced calculation 高级计算

advanced technological control system 先进的技术控制系统

advancing contact angle 前展接触角

adverse effect 反[副]作用,有害[不利]影响

adverse reaction 不良反应

advertising production manager 广告制作经理

aerated solution 充气溶液

aerating impeller 充气叶轮(浮选机的)

aeration characteristics of machine 浮选机充气特性

aeration effect 充气效应

aeration mechanism 充气机理

aeration rate 充气速度[量]

aeration relationship 充气关系

aeration test 充气试验

aerial seeding 飞机播种

Aero 美国氰胺公司化学药剂品牌名

Aero 130(thiocarbanilide) Aero 130白药(含均一二苯硫脲,乳白粉末,易分散于水中,是方铅矿,特别是含银方铅矿和硫化银矿浮选的重要促进剂和捕收剂)

Aero 238(di-sec butyl thiophosphate) Aero 238(二仲丁基硫代磷酸盐)

Aero 303(potassium ethyl xanthate)
Aero 303(乙基黄原酸钾)

Aero 3477 (alkyl dithiophosphate)
Aero 3477(烷基二硫代磷酸盐)

Aero 3501 (dithiophosphate with strong frothing properties) Aero 3501(具有强起泡性能的二硫代磷酸盐)

Aero 600 series depressant Aero 600系列抑制剂(美国氰胺公司)

Aero brand cyanide 氰化钙(粗制,可用作闪锌矿抑制剂)

Aero depressant 610 Aero 610抑制剂(有机胶质,为吸湿性粉末,可用作滑石、云母类脉石抑制剂)

Aero depressant 615 Aero 615抑制剂(有机胶质,为吸湿性粉末,可用作滑石、绢云母抑制剂)

Aero depressant 620 Aero 620抑制剂(有机胶质,为吸湿性粉末,可用作滑石等抑制剂)

Aero depressant 633 Aero 633抑制剂(有机胶质,为吸湿性粉末,可用作含碳脉石抑制剂)

Aero modifier 158 Aero 158号变态[调整]剂(一种高分子聚合物,可用作矿泥、脉石等的抑制剂)

Aero modifier 162 Aero 162号变态[调整]剂(性质同Aero 158号变态剂)

Aero promoter 美国氰胺公司化学药剂(促进剂)

Aero promoter 135 Aero 135号促进剂(黑药的酰氯衍生物)(RO)$_2$P(S)Cl

Aero promoter 3302 Aero 3302号促进剂(一种黄原酸酯类,不溶于水的油状捕收剂,适于浮选铜、钼)

Aero promoter 3461 Aero 3461号促进剂(为Aero 3302的同系物,黏度比Aero 3302小,可用作捕收剂)

Aero promoter 3477 Aero 3477号促进剂(二硫代磷酸类药剂,白色粉末,易溶于水,浓度可达30%,可用作Cu、Ag、Zn硫代矿的捕收剂)

Aero promoter 3501 Aero 3501号促进剂(二硫代磷酸类药剂,同Aero 3477)

Aero promoter 404 Aero 404号促进剂(2-硫基苯骈噻唑,同Reagent 404,可用于氧化重金属矿和硫化矿浮选(用作捕收剂))

Aero promoter 407 Aero 407号促进剂(同Aero 412号促进剂)

Aero promoter 412 Aero 412号促进剂(可用作浮选金的捕收剂,也用于铜、镍及硫化矿的浮选)

Aero promoter 425 Aero 425号促进剂(硫基苯并噻唑,绿黄色吸湿性粉末状,主要要用于氧化铜矿和硫化氧化混合铜矿的浮选)

Aero promoter 708 Aero 708号促进剂(粗制塔尔油脂肪酸,可用于磷矿等非金属矿浮选,特别是磷的浮选(捕收剂))

Aero promoter 710 Aero 710号促进剂(Aero 708号促进剂的钠皂,除油酸及亚油酸外还含有一定量的松香酸,可用

于非金属矿浮选,特别是磷的浮选(捕收剂))

Aero promoter 712　Aero 712 号促进剂(塔尔油脂肪酸钠皂,可用于非金属矿浮选,特别是磷的浮选(捕收剂))

Aero promoter 723　Aero 723 号促进剂(精制塔尔油,含精制脂肪酸 92% 及松香酸 4%,可用作捕收剂)

Aero promoter 765　Aero 765 号促进剂(精制塔尔油乳剂,含油酸、亚油酸及很少量的树脂酸和非离子乳化剂,属于 Aero 723 号促进剂的乳化物,可用作捕收剂)

Aero promoter 801　Aero 801 号促进剂(水溶性石油磺酸盐,棕黑色糊状物,可用于浮选非金属矿、重晶石、石榴石及铁矿(捕收剂、起泡剂))

Aero promoter 824　Aero 824 号促进剂(水溶性石油磺酸盐,棕黑色糊状物,可用于浮选非金属矿、重晶石、石榴石及铁矿,在酸性矿浆中能浮大于 40 微米的黑钨及白钨矿(用作捕收剂))

Aero promoter 825　Aero 825 号促进剂(油溶性石油磺酸盐,红棕色糊状物,在水中可以分散,起泡性稍次于 Aero 801 号促进剂,捕收性和选择性较强,适用于浮选石榴石、铬铁矿蓝晶石(捕收剂))

Aero promoter 827　Aero 827 号促进剂(石油磺酸盐,用途同 Aero 801 号促进剂)

Aero promoter 830　Aero 830 号促进剂(新的改良的磺酸盐型捕收剂)

Aero promoter 840　Aero 840 号促进剂(阴离子石油磺酸型捕收剂,可用于从玻璃砂中浮选重矿物)

Aero promoter 845　Aero 845 号促进剂(新的改良的磺酸盐型捕收剂,同 Aero 825 号、Aero 830 号促进剂,适用于在酸性矿浆中捕收铁矿)

Aero promoter 899　Aero 899 号促进剂(水溶性石油磺酸盐,可作捕收剂)

Aerodri　美国氰胺公司化学药剂品牌名

Aerodri 100　一种表面张力抑制剂(可用于助滤)

Aerodri 100 dewatering aid　Aerodri 100 脱水剂(属表面活化剂,广泛用于过滤作业)

Aerodri 104 dewatering aid　Aerodri 104 脱水剂(属表面活化剂,广泛用于过滤作业)

Aerodri dewatering aid　Aerodri 脱水剂

Aerofloat　美国氰胺公司化学药剂(黑药)

Aerofloat 15　15 号黑药(甲酚与 15% P_2S_5 作用的产物,液体,可作捕收剂及起泡剂,用于重金属硫化矿、金矿、银矿的浮选)

Aerofloat 203　203 号黑药(二异丙基二硫代磷酸钠,固体,可用于金、银、铜及锌矿的浮选)

Aerofloat 208　208 号黑药(为二乙基与仲丁基黑药的混合中性盐,固体,可用

Aerofloat 211 作捕收剂,用于浮选金、银、铜矿)

Aerofloat 211 211号黑药(为二异丙基二硫代磷酸钠性质类似的黑药,可用作浮选锌矿的选择性捕收剂)

Aerofloat 213 213号黑药,二异丙基铵黑药(二异丙基二硫代磷酸铵,固体,有起泡性,可用作捕收剂)

Aerofloat 226 226号黑药,仲丁基铵黑药(二仲丁基二硫代磷酸铵盐,固体,可用作捕收剂,用于选别金、银、锌和铜矿)

Aerofloat 238 238号黑药,仲丁基钠黑药(二仲丁基二硫代磷酸钠,固体,用于重金属硫化矿浮选(捕收剂))

Aerofloat 241 241号黑药(为25号黑药的水溶性铵盐,液体)

Aerofloat 242 242号黑药(用氨水中和的31号黑药的铵盐,液体,可用作捕收剂,用于浮选方铅矿、银和氧化金矿)

Aerofloat 243 243号黑药(烷基二硫代磷酸盐,固体,可用作捕收剂,用于重金属硫化矿浮选,主要用于闪锌矿浮选)

Aerofloat 249 249号黑药(烷基二硫代磷酸盐,成分同243号黑药,起泡性更强,可用作捕收剂,用于重金属硫化矿浮选,主要用于硫化铜矿浮选)

Aerofloat 25 25号黑药(甲酚与25% P_2S_5 作用的产物,液体,可作捕收剂及起泡剂,用于银、铅以及无泥的铜、铅、锌矿物浮选)

Aerofloat 31 31号黑药(25号黑药加6%白药的混合物,液体,可用作捕收剂及起泡剂,用于方铅矿、金属铜的浮选)

Aerofloat 33 33号黑药(与31号黑药相似,非极性基多一个甲基,液体,可作捕收剂及起泡剂)

Aerofloat B 二异丙基钠黑药(二异丙基二硫代磷酸钠)

Aerofloat sodium 钠黑药(烷基二硫代磷酸钠,可用作捕收剂,用于硫化铜、锌的浮选)

Aerofloc 美国氰胺公司化学药剂(絮凝剂)

Aerofloc 3000 3000号絮凝剂(中分子量的聚丙烯酰胺,作为絮凝剂时(用量5~125克/吨),能造成一种强韧的絮凝物,有效范围pH值为0~12)

Aerofloc 3171 3171号絮凝剂(高分子量的聚丙烯酰胺,用作絮凝剂比3000号絮凝剂更有效,用量5~125克/吨)

Aerofloc 3425 3425号絮凝剂(比550号絮凝剂分子量更高,用于浓缩过滤)

Aerofloc 3453 3453号絮凝剂(高分子量的阴离子型聚丙烯酰胺)

Aerofloc 548 548号絮凝剂(聚电解质型高分子化合物,用为絮凝剂时,适用范围pH值为2~4)

Aerofloc 550 550号絮凝剂(水解了的聚丙烯腈,片状或黄色固体,作为絮凝剂时,用量5~125克/吨,分段添加)

Aerofloc 552 552号絮凝剂(聚电解质型高分子化合物,用为絮凝剂时,适用范围pH值为4~12)

Aerofloc flocculent Aerofloc 絮凝剂

Aerofroth 美国氰胺公司化学药剂(起泡剂)

Aerofroth 63 Aero 63号起泡剂(一种高级醇类起泡剂)

Aerofroth 65 Aero 65号起泡剂(一种水溶性合成起泡剂,可用于硫化矿的浮选)

Aerofroth 70 Aero 70号起泡剂(甲基戊醇或甲基异丁基卡必醇)

Aerofroth 71 Aero 71号起泡剂($C_6 \sim C_9$ 醇混合物)

Aerofroth 73 Aero 73号起泡剂(一种合成起泡剂)

Aerofroth 77 Aero 77号起泡剂(一种合成起泡剂,为直链高级醇混合物)

Aerofroth 80 Aero 80号起泡剂(一种直链醇起泡剂)

Aerofroth frother Aerofroth 起泡剂

Aeromine 美国氰胺公司化学药剂(阳离子型表面活性剂)

Aeromine(R)R-3037 阳离子型表面活性剂(可用作絮凝剂,也可作为石英、氧化锌矿的捕收剂)

Aeromine 2026 阳离子型絮凝剂,用于石英浮选,加速黏土的沉降

Aeromine 3035 阳离子型表面活性剂(胺类),使用时用醋酸中和(可用作捕收剂)

aerophine 二异丁基二硫代次膦酸钠

aerosol 气溶胶;气雾剂;悬浮微粒(如烟、雾等);湿润剂

Aerosol 美国氰胺公司表面活性剂(湿润剂、絮凝剂,中文名称:艾罗素)

Aerosol(R)C-61 絮凝剂(乙醇化的烷基喹尼啶-胺络合物)

Aerosol 18 湿润剂(磺丁二酸;正-十八烷基硫化琥珀酰胺二钠盐,糊状物,含量35%~36%,含水约65%)

Aerosol 22 湿润剂(磺丁二酰胺酸;N-正十八烷基-N-1,2-二羟基乙基-硫化琥珀酰胺四钠盐,可用作重晶石的捕收剂)

Aerosol 404 艾罗素404(即硫醇基苯并噻唑)

Aerosol AS 湿润剂(异丙基萘磺酸钠的水溶液,外加一种溶剂)

Aerosol AY 湿润剂(双戊基磺化琥珀酸钠,作为湿润剂用量为5~100克/吨)

Aerosol C61 一种阳离子胺类润湿剂和絮凝剂(十八烷基氨基甲酸与环氧乙烷的缩合物,十八烷基胍盐)

Aerosol GPG 湿润剂(双-2-乙基己基磺化琥珀酸钠,相对分子质量444),可用于改善过滤

Aerosol IB 湿润剂(双异丁基磺化琥珀酸钠,作为湿润剂用量为5~100克/吨)

Aerosol MA 湿润剂(双-(甲基戊基)-磺化琥珀酸钠,作为湿润剂用量为5~100克/吨)

Aerosol OS 湿润剂(异丙基萘磺酸钠,作为湿润剂用量为5~100克/吨)

Aerosol OSB 湿润剂、捕收剂(丁基萘

磺酸钠)

Aerosol OT 湿润剂、乳化剂(双-2-乙基己基磺化琥珀酸钠,相对分子质量 444,可用于改善过滤)

aerospace industry 航天工业

Aero-thiocarbanilide Aero 白药(二苯硫脲,可用作捕收剂,适用于捕收硫化矿中的铜、铅和银)

Aero-xanthate 301 Aero 301 黄药(仲丁基钠黄药,可用作捕收剂)

Aero-xanthate 303 Aero 303 黄药(乙基钾黄药,可用作捕收剂)

Aero-xanthate 322 Aero 322 黄药(异丙基钾黄药,可用作捕收剂)

Aero-xanthate 325 Aero 325 黄药(乙基钠黄药,可用作捕收剂)

Aero-xanthate 343 Aero 343 黄药(异丙基钠黄药,可用作捕收剂)

Aero-xanthate 350 Aero 350 黄药(戊基钾黄药,可用作捕收剂)

aesthetics 美学

AF 用谷物淀粉所制成的阴离子型絮凝剂

AF-3302 有机硅油类化合物(品名,戊黄烯酯,戊黄原酸丙烯酯,同 S-3302,可用作捕收剂,用于浮选辉钼矿及硫化铜矿)

affinity law 相似规律

Ag amyl xanthogenate 戊基黄原酸银

Ag butyl xanthogenate 丁基黄原酸银

Ag ethyl xanthogenate 乙基黄原酸银

Ag heptyl xanthogenate 庚基黄原酸银

Ag hexyl xanthogenate 己基黄原酸银

Ag isoamyl xanthogenate 异戊基黄原酸银

Ag isobutyl xanthogenate 异丁基黄原酸银

Ag isopropyl xanthogenate 异丙基黄原酸银

Ag octyl xanthogenate 辛基黄原酸银

Ag propyl xanthogenate 丙基黄原酸银

agate vibrating mill 玛瑙振动磨矿机

agate vibratory mill 玛瑙振动式磨矿机

aged PBS sample 老化 PBS 试样

aged sample 老化试样

ageing condition 老化条件

agent bank's prime rate of interest 经办银行优惠利率

agglomerate 团块,团矿,制成团

agglomerate spiraling 团聚螺旋选矿

agglomerated ore 团聚[成团]矿石

agglomerated phosphate 团聚[成块]磷酸盐

agglomerating 烧结,凝结

agglomerating character 烧结特性

agglomeration 团聚[团矿,结块,集块,烧结](作用),加热黏结

agglomeration drum 鼓形团矿机

agglomeration mechanism 团聚机理

aggregate 集合体,组合体

aggregate and mineral information 集料和矿物资料

aggregate industry　集料工业
aggregate plant　集[骨,填充]料厂
aggregate sampling　集料取样
aggregate specification　集[粒]料规格
aggregation of particle　矿[颗]粒团聚作用
aggressive ongoing program　积极进取计划
aggressive process water　腐蚀性工艺用水
Agitair 120A flotation machine　阿基泰尔 120A 浮选机(美国 Galigher Company 生产)
Agitair 120B unit　阿基泰尔 120B 型浮选设备(美国 Galigher Company 生产)
Agitair flotation machine　阿基泰尔浮选机(美国 Galigher Company 生产)
agitated condition　搅拌条件
agitated disc filter　搅拌圆盘[盘式]过滤机
agitated leach　搅拌浸出
agitated mix tank　搅拌混合槽
agitated slurry　已搅拌的矿浆
agitating and holding tank　搅拌储存槽
agitating speed　搅拌速度
agitating type flotation machine　搅拌式浮选机
agitating type machine　搅拌式浮选机
agitation box　搅拌槽
agitation compartment　搅拌室
agitation leach　搅拌浸出

agitation leach tank　搅拌浸出槽
agitation leaching　搅拌浸出
agitation machine　搅拌式浮选机
agitation rate　搅拌速度
agitation tank　搅拌槽
agitation time　搅拌时间
agitation vessel　搅拌器,搅拌槽
agitation zone　搅拌区
agitation-type flotation machine　搅拌式浮选机
agriculture limestone　农业石灰石
air addition　空气添加(量)
air brake　气闸
air bubble - floatable specie (solid) complex　气泡-可浮矿物聚合体
air cannon　空气炮
air capacity number　空气容积数($C_a = Q_a/ND^3$)
air classifier model　空气分级机模型
air clutch　空气离合器
air column　空气柱
air contact　空气接触
air driven　空气驱[传]动的
air dry basis　风干基础
air duct　进气口,空气管道,通风道
air engulfment　空气卷吸
air entry　空气入口
air erosion　空气侵蚀
air evacuation　空气排出
air exhaust outlet　排气口,抽风接口
air flow condition　空气流动条件[状态]

air flow number	空气流量数
air flow rate	气流速度
air gravity concentrating table	风力重选摇床
air hold-up	空气容量
air infiltration	空气渗透,漏风
air ingestion	吸气
air ingestion capability	吸气能力
air ingestion rate	空气吸入量
air inlet duct	空气入口管
air lift carbon in pulp tank	空气提升炭浆槽
air line	风管
air lock valve	气锁安全阀
air motor	气动马达
air operated solenoid valve	气动电磁阀
air oxidation	空气氧化
air path	空气通路
air penetration	空气渗透
air permeation	空气渗透
air pollution control agency	空气[大气]污染控制机构
air pollution control regulation	空气[大气]污染控制规程[条例]
air pressure differential	气压差
air purge	空气吹洗
air quality control	空气质量管理
air rate	充气量
air slide	空气滑塞,气动滑板
air slide conveyer	气滑[风送]式输送机
air sparge	空气喷雾
air transfer	空气吸入[输送]量(浮选槽)
air transfer characteristic	空气吸入[输送]特性(浮选槽)
air-atomizing	空气喷雾
air-avid part	亲气部分
airborne product	气载产品
air-conditioned room	空调室
air-conditioning system	空调系统
air-exposed sample	暴露于空气的试样
air-flow	气流,空气流量,透气量
airflow characteristic	气流特性[特征]
airflow volume	空气流量
air-lift cell	气升式浮选机
air-operated dart valve	气动突进阀
air-operated deflection plate	气动偏斜板
air-powered piston	风动活塞
air-saturated	空气饱和的
air-slide conveyer	压缩空气[气滑式]输送器
air-stream dust loading	气流粉尘载荷[含量,负载]
air-swept milling operation	气吹式磨矿作业
air-swept zone	气扫区(充气作用区域)
air-tight access door	气密入口
air-tight door	气密门
air-tight seal	气密密封

air-tight surge bin　密封[不透气的]缓冲矿仓
air-tight valve　气密阀
air-tight vessel　气密容器
air-to-cloth ratio　空气与滤布比(率)
air-to-concentrate ratio　空气精矿比率
air-type cell　空气式浮选槽
Akins classifier　埃金斯型分级机
Akins double pitch spiral　埃金斯型双节距螺旋
Akins duplex spiral classifier　埃金斯双螺旋分级机
Akins simplex spiral classifier　埃金斯型单螺旋分级机
Akins single pitch spiral　埃金斯型单节距螺旋
Alamac 26　用塔尔油制成的混合脂肪胺醋酸盐(品名,可用作捕收剂)
Alamac 26D　用蒸馏的塔尔油制成的脂肪胺醋酸盐(品名)
Alamac H26D　用加氢的精制塔尔油所制胺的醋酸盐(品名,可用作捕收剂)
Alamine　表面活性剂品牌名(中文名称:阿拉明,脂肪胺或其混合物,可用作腐蚀抑制剂及乳化剂、捕收剂等)
Alamine 21　椰子油胺(含87%伯胺,可用作捕收剂)
Alamine 21D　椰子油胺(含97%伯胺,可用作捕收剂)
Alamine 221　椰子油仲胺(含85%伯胺,可用作捕收剂)
Alamine 26　牛脂胺(含87%伯胺,可用作捕收剂)
Alamine 26D　牛脂胺(含97%伯胺,可用作捕收剂)
Alamine 33　棉籽油胺(含87%伯胺,可用作捕收剂)
Alamine 336(a tertiary amine)　叔胺,三烷基胺
Alamine 33D　棉籽油胺(含97%伯胺,可用作捕收剂)
Alamine 4　正-十二烷胺,月桂胺(含87%伯胺,可用作捕收剂)
Alamine 4D　正-十二烷胺,月桂胺(含97%伯胺,可用作捕收剂)
Alamine 6　正-十六烷胺(含87%伯胺,可用作捕收剂)
Alamine 6D　正-十六烷胺(含97%伯胺,可用作捕收剂)
Alamine 7　正-十八烷胺(含87%伯胺,可用作捕收剂)
Alamine 7D　正-十八烷胺(含97%伯胺,可用作捕收剂)
Alamine H226　氢化牛脂仲胺(含85%伯胺,可用作捕收剂)
Alamine H26　氢化牛脂胺(含87%伯胺,可用作捕收剂)
Alamine H26D　氢化牛脂胺(含97%伯胺,可用作捕收剂)
alarm condition　报警状态
alarm limit　报警极限
alarm logic　报警逻辑
alarm probe　报警探头[针]

alarm signal system 事故警报信号系统
alarm system 报警系统
albumin 蛋白质,白朊(可用作抑制剂)
alcohol 醇 ROH;乙醇,酒精 C_2H_5OH
Alcohol B21(~B30) 高级醇类起泡剂(用量 25~250 克/吨)
alcohol elution 酒精解吸[洗提]
alcohol frother 醇类起泡剂
alcoholate 醇化物
alcoholic 醇的;乙醇的,酒精的
alcoholic frother 醇类起泡剂
alcoholic hydroxyl 醇式羟基
alcohol-type frother 醇类起泡剂
Alcopol R540 二烃基磺化琥珀酰胺钠(同 AP845、Aerosol 18,可作锡石的捕收剂)
Alfa-Laval plate heat exchanger 阿法拉伐板式热交换器(瑞典 Alfa Lava 公司产品)
Alfloc 6701 絮凝剂(一种液态高分子聚合物)
Alframin DCS 磺化高分子脂族醛聚合物(品名)
algae 水藻,藻类
algebraic sum 代数和
alginate 海藻酸盐,藻朊酸盐,藻酸盐(可用作抑制剂、絮凝剂)
alignment 队列,排成直线,对准[齐],校[直]准
Aliphat 塔尔油脂肪酸品牌名
Aliphat 44-A 塔尔油脂肪酸(可用作捕收剂)
Aliphat 44-D 精馏的塔尔油脂肪酸(可用作捕收剂)
Aliphat 44-E 塔尔油中的脂肪酸部分(可用作捕收剂)
aliphatic 脂族的,脂肪质的,脂肪族的
aliphatic acid 脂肪酸 RCOOH
Aliphatic acid No. 100 得自石油的C_8~C_{12}脂肪酸钠皂(品名,可用作捕收剂,用量 125~1500 克/吨)
Aliphatic acid No. 50 得自石油的C_8~C_{12}脂肪酸(品名,可用作捕收剂)
aliphatic alcohol 脂肪醇
aliphatic amine 脂肪胺,脂肪胺族
aliphatic amine molecule 脂肪族胺分子
aliphatic compound 脂(肪)族化合物
aliphatic hydrocarbon 脂肪烃
aliphatic hydrocarbon chain 脂肪族碳氢链
aliphatic monoamine 脂族单胺
aliphatic phosphonic acid 脂族膦酸
aliphatic xanthate 脂肪族黄药
aliphatics 脂肪族化合物
Aliphat-S 塔尔油脂肪酸(含 50%的C_4~C_8脂肪醇,可用作起泡剂)
Alipon A 油酸磺化乙酯钠盐(品名,同 Igepon A,可用作洗涤剂、湿润剂)$C_{17}H_{33}COOCH_2CH_2SO_3Na$
Aliquat 表面活性剂(季[苯]铵盐类,品牌名)
Aliquat 21 季铵盐(可用作捕收剂)

[R-N(CH$_3$)$_3$]$^+$Cl$^-$(RN 基来自椰子油胺)
Aliquat 221 季铵盐(可用作捕收剂)
[R$_2$N(CH$_3$)$_2$]$^+$Cl$^-$(R$_2$N 基来自椰子油仲胺)
Aliquat 26 季铵盐(可用作捕收剂)
[R-N(CH$_3$)$_3$]$^+$Cl$^-$(RN 基来自牛脂胺)
Aliquat 336 季铵盐(三-辛基甲基氯化铵,可用作(磁铁矿)捕收剂、金属萃取剂)[R$_3$N-CH$_3$]$^+$Cl$^-$(R=C$_9$~C$_{11}$)
Aliquat 336-3 季铵盐(三辛基甲基氯化铵,可用作捕收剂、金属萃取剂)(C$_8$H$_{17}$)$_3$N$^+$-CH$_3$Cl$^-$
Aliquat H226 季铵盐(可用作捕收剂)[R$_2$N(CH$_3$)$_2$]$^+$Cl$^-$(R$_2$N 基来自氢化牛脂仲胺)
aliquot 等分试[矿]样
alisonite 闪铜铅矿 2Cu$_2$S·PbS
alizarin 茜素(1,2-二羟基蒽醌,可用作湿润剂、抑制剂及制造染料)C$_6$H$_4$(CO)$_2$C$_6$H$_2$(OH)$_2$
alizarin blue 茜素蓝(可用作抑制剂)C$_6$H$_4$(CO)$_2$C$_9$H$_3$N:(OH)$_2$
alizarin red 茜素红(可用作抑制剂)C$_6$H$_4$(CO)$_2$C$_6$H(OH)$_2$SO$_3$H
alizarin red-S 茜素红-S(可用作(铍矿)抑制剂)
alkalescent 碱性的,弱碱性的
alkali 碱,强碱;(强)碱性的
alkali cation 碱性阳离子
alkali depression of pyrite 黄铁矿碱性抑制
alkali earth 碱土(碱土金属的氧化物)
alkali metal 碱金属
alkali metal ion 碱金属离子
alkali metal phosphate 碱金属磷酸盐(可用作湿润剂、分散剂)
alkali oxide 碱金属氧化物
alkali sulfide 碱性硫化物
alkali sulfide depressant 碱性硫化物抑制剂
alkali-earth minerals 碱土矿物
alkaline 碱的,强碱的,碱性的
alkaline chlorination 碱氯化(消除氰的络合物)
alkaline chlorination cyanide destruction test 用碱氯法分解[破坏,消除]氰化物试验
alkaline circuit 碱性(浮选)流程
alkaline cyanide leaching 碱性氰化浸出
alkaline cyanide solution 碱性氰化(物)溶液
alkaline depressant 碱性抑制剂
alkaline earth 碱土
alkaline earth apatite 碱土磷灰石
alkaline earth bicarbonate 碱土碳酸氢盐,碱土重碳酸盐
alkaline earth soap 碱土皂
alkaline fine textured soil 碱性细粒结构土壤
alkaline pH region 碱性pH值范围
alkaline pulp 碱性料(矿)浆

alkaline pyrogallol solution 焦棓酚碱液

alkaline region 碱性区

alkalinity regulator 碱度调节剂

alkane 链烷,烷烃 C_nH_{2n+2}

Alkanol B 丁基萘磺酸钠(品名,可用作捕收剂、湿润剂)

Alkaterge 阳离子表面活性剂品牌名

Alkaterge-A(-C;-E;-T) 阳离子表面活性剂(烷基噁唑啉类,可用作湿润剂、捕收剂(用于锡石浮选))

化学结构式:$R_3-C\begin{matrix}N-CR_1R_2\\O-CH_2\end{matrix}$

R_1R_2=甲基、乙基或CH_2OH,R_3=长碳链烷基

alkyl 烷基 $R-$,$C_nH_{2n+1}-$

alkyl amine chloride 烷基胺盐酸盐 $RNH_2 \cdot HCl$

alkyl amine fluoride 烷基胺氟氢酸盐 $RNH_2 \cdot HF$

alkyl aminoacid 烷基胺(基)酸 $RCONHCN_2COOH$

alkyl ammonium chloride 烷基氯化铵

alkyl dithiophosphate 烷基二硫代磷酸盐(可用作湿润剂)

alkyl dithiophosphate complex of heavy metal 重金属烷基二硫代磷酸盐络合物

alkyl dithiophosphoric acid salt of K 烷基二硫代磷酸钾盐

alkyl group 烷基

alkyl hydrocarbon chain 烷基碳氢链

alkyl hydroxamic acid 烷基异羟肟酸(可用作捕收剂)$R-C(O)NHOH$

α-alkyl phenoxy carboxylic acid α-烷基酚氧化羧酸

alkyl phosphate 烷基磷酸盐 RPO_4Na_2

alkyl phosphinic acid 烷基次膦酸 $R_2 \cdot PO(OH)$

alkyl phosphite 烷基亚磷酸盐

alkyl phosphonic acid 烷基膦酸 $R \cdot PO(OH)_2$

alkyl pyridinium salt 烷基吡啶盐

alkyl sodium sulfate 烷基硫酸钠,硫酸烷基(酯)钠 $R \cdot SO_4 \cdot Na$

alkyl sodium sulfonate 烷基磺酸钠 $R \cdot SO_3 \cdot Na$

alkyl sulfate(=alkyl sulphate) 烷基硫酸盐[脂] RSO_4Na_2

alkyl sulfonate 烷基磺酸盐 RSO_3M

alkyl sulphonic acid 烷基磺酸 RSO_3H

alkyl-alkyl dithiocarbamate 烷基-烷基二硫代氨基甲酸酯

alkyl-alkyl xanthate 烷基-烷基黄药

alkylamine 脂肪胺,烷基胺 $R \cdot NH_2$

alkyl-ammonium ion 烷基铵离子

alkylarylsulfonate 烷基芳基磺酸盐

alkylated 烷(基)化的,烷基化

alkylated tar 烷基化焦油

alkylhydroxamate 烷基氧肟酸盐

all cyanide flowsheet 全氰化法流程

all departments of mining operation 矿山生产所有部门

all friction clutch 全摩擦离合器
all industrial equipment index 所有工业设备指数
allergy 变态反应,反感,过敏症
Allis-Chalmers catalogue 阿利斯·查尔默斯产品目录(美国)
Allis-Chalmers gyratory crusher 阿利斯·查尔默斯旋回破碎机
Allis-Chalmers low head vibrating feeder 阿利斯·查尔默斯低头振动给矿机
Allis-Chalmers overflow ball-mill 阿利斯·查尔默斯溢流型球磨机
Allis-Chalmers double deck vibrating screen 阿利斯·查尔默斯双层振动筛
Allis-Chalmers modification of equation 阿利斯·查尔默斯改进方程
Allis-Chalmers overflow double scoop fed ball mill 阿利斯·查尔默斯溢流型双勺给矿球磨机
Allis-Chalmers steam dryer 阿利斯·查尔默斯蒸汽干燥机
allocation cost 分摊费
allowable limit 允许限度
allowable tax deduction 允许的税款扣除额
allowance for depreciation 折旧提成
alloy nomenclature 合金名称
all-sliming 全泥化
alluvial overburden 冲积覆盖层
alluvial plain 冲积平地

alphanumeric identification 字母数字识别
Alphasol OT 磺琥辛酯钠
Alrowet 表面活性剂品牌名
Alrowet D 双-2-乙基己基磺化琥珀酸钠(同 Aerosol OT,可用作湿润剂)
alteration 改变,变更,改动,更改
alteration mineral 蚀变[变质]矿物
alteration product 变质[蚀变]矿物[产品]
altered Middle Tertiary granodiorite 变质的中第三纪花岗闪长岩
altered nature 蚀变性质
alternate design 比较设计
alternate means 代替的方法
alternate method 另外[代替,代用]的方法
alternate reagent 代用药剂
alternating bed 交替矿床[层]
alternation 轮流,交替[互,错,换],更[替,迭],交流,间隔
alternative 供选择的,选择性的,交替的,二中择一,供替代的
alternative approach 替代途径[方法]
alternative ball mill model 替代球磨机模型
alternative circuit 流程方案
alternative circuit control strategy 方案性回路控制策略
alternative circuits 流程方案
alternative control strategy 可选择的控制策略

alternative design 方案设计
alternative flowsheet 方案性的流程
alternative frother 替代起泡剂
alternative method of financing 可供选择的筹集资金方法
alternative phase 方案性发展阶段，可替代的相
alternative recovery system 可供选择[用]的回收系统
alternative sample pattern 备选取[试]样布置方案
alternative sequence of construction 方案性的建设程序，建设程序方案
alternative source 替代来源
alternative source of financing 可供选择的资金来源
alternative study 方案研究
alternative system design 交替系统设计
alumina ball, corundum ball 氧化铝[刚玉]球
alumina producer 氧化铝厂
alumina-alkyl sulphonate system 氧化铝-烷基磺酸盐体系
aluminum chloride 氯化铝（可用作抑制剂）
aluminum industry 铝业
aluminum labyrinth member 铝迷宫部件
aluminum mirror 铝镜
aluminum oxide paste 氧化铝膏
aluminum oxide porous barrier 氧化铝多孔隔板
aluminum salt 铝盐
aluminum sulfate 硫酸铝 $Al_2(SO_4)_3 \cdot 18H_2O$（可用作絮凝剂，用量50~2500克/吨）
amalgam retort 汞齐蒸馏罐
amalgamation plant 混汞车间
Amatek string discharge drum filer Amatek型绳索排矿圆筒过滤机
AMB 二烷基甲基苄基溴化铵（品名，可用作捕收剂），化学结构式：
$$CH_3-N^+(R')(R'')-CH_2 \langle \rangle, Br^-$$
Amberlite 离子交换树脂品牌名（美国）
Amberlite IRA-400 碱性离子交换树脂（可用于萃取铀）
Amberlite XE-123 强碱性阴离子交换树脂
ambient air 周围[环境]空气
ambient pH 自然[环境]pH值
ambient temperature 环境[周围]温度，室温
Amdel-Philips in-stream probe 埃姆得尔-菲利浦在线式探头
Amdel-Philips in-stream probe system 埃姆得尔-菲利浦在线探测系统
amenability testing 控制[可行性，可实用性]试验
American Cyanamid 3302 美国氰胺公司捕收剂3302
American dollar 美元
amide 酰胺，氨化物

Amijel 变性玉蜀黍淀粉(可用作絮凝剂、抑制剂,用量 50~1000 克/吨)
amine 胺
Amine 阳离子型表面活性剂品牌名
Amine 220 1-羟乙基-2-十七碳烯基-二氢咪唑(可用作浮选剂,重晶石和氧化铁矿中浮选石英)
　R—$C_3H_4N_2$—C_2H_4OH
　R=$C_{17}H_{33}$(十七碳烯基 heptadecenyl)
Amine 9D-178 十二碳烯胺(可用作萃取剂、液体阴离子交换剂)
Amine F. B. 110 一种阳离子捕收剂
Amine S-24 双异丁二甲辛胺(可用作萃取剂、液体阴离子交换剂)
amine salt 胺盐
amine-ammonium salt 胺-铵类盐
amino- 氨基 —NH_2
Amino 803 醚胺(品名,2-乙基己基氧丙胺,可用作捕收剂)
　$C_4H_9CH(C_2H_5)CH_2OC_3H_6NH_2$
amino-acid 胺酸,氨基酸
　NH_2—R·COOH
amino-alcohol 醇胺,氨基醇
　NH_2R,CH_2OH
amino-aldehyde 氨基醛
amino-benzene 苯胺,氨基苯
　$C_6H_5NH_2$
ammeter 安培计
ammine 氨络(物),氨(络)合物
ammine solvent extraction 氨络物溶剂萃取
ammixable electrolyte 不能混合的电解质
ammonia 氨(水)NH_3
ammonia complex solvent extraction 氨络合物溶剂萃取
ammonia leaching 氨浸(出)
ammonia process 氨浸法
ammonia thiosulfate 硫代硫酸铵
ammoniac 氨草胶,氨树胶,氨的
ammoniacal 氨的,含氨的
ammonium 铵
ammonium acetate 醋酸铵
　$CH_3·COONH_4$
ammonium acetate solution 醋酸铵溶液
ammonium bicarbonate 碳酸氢铵(可用作浮选 Nb(铌)矿活化剂)
　$(NH_4)HCO_3$
ammonium carbamate 氨基甲酸铵
　$NH_2·COONH_4$
ammonium carbonate 碳酸铵
　$(NH_4)_2CO_3$
ammonium fluoride 氟化铵(可用作浮选黄铁矿,磁黄铁矿的活化剂)NH_4F
ammonium hydroxide 氢氧化铵(可用作活化剂,与硫酸铜一起活化闪锌矿)NH_4OH
ammonium ion 铵离子
ammonium metatungstate 偏钨酸铵
ammonium nitrate(AN) 硝酸铵,硝铵(缩写 AN,可作为浮选黄铁矿、磁黄铁矿的活化剂)NH_4NO_3
ammonium paratungstate 仲钨酸铵

ammonium phosphate 磷酸铵
ammonium salt 铵盐
ammonium sulfate 硫酸铵 $(NH_4)_2SO_4$
ammonium sulfide 硫化铵 $(NH_4)_2S$
ammonium sulfite 亚硫酸铵 $(NH_4)_2SO_3$
ammonium sulfocyanate (= ammonium sulfocyanite) 硫氰酸铵 NH_4SCN
ammonium thiocyanate (= ammonium thiocyanite) 硫氰酸铵 NH_4SCN
amorphous component 非晶质[体,形]成分
amorphous nature 非晶形[质]性质
amorphous precipitate 非晶质[无定形]沉淀物
amorphous state 非晶质状态
amorphous structure 无定形结构
amortization 分期偿还,折旧
amortization of term loan 定期贷款分期偿还
amortization period 摊还期,偿还期
amount consumed 消耗量
amount of alteration 蚀变量
amount of grinding 磨矿量
amount of grinding ball 装球量
amount of inflation 通货膨胀率
amount of information 信息量
amount of mine dilution 开采混入废石量
amount of solute adsorbed 溶质吸附量
amount of solute required 所需溶质量
amperage meter 安培计

amplifier indicator 放大指示器
amplitude of motion 运动(的)振幅
Amsco type T pump Amsco 型 T 泵
n-amyl （正）戊基 $CH_3(CH_2)_4—$
amyl alcohol 戊醇(可用作起泡剂、捕收剂) $C_5H_{11}OH$
amyl alcohol sulfate 戊基硫酸盐 $C_5H_{11}SO_3M$
amyl xanthate 戊黄药,戊基黄原酸盐 $C_5H_{11}OCSSM$
amyl xanthate allyl ester 戊基黄原酸烯丙基酯
amyl(=pentyl) 戊基 $C_5H_{11}—$
amylamine 戊胺 $C_5H_{11}NH_2$
amylene 戊烯 C_5H_{10};戊撑、次戊基 $—C_5H_{10}—$
amylene hydrate 叔戊醇,水合戊烯
an engineering project 某工程项目
Analar pure analar 伦琴射线光谱分析纯净的
analar sulfuric acid 纯硫酸
analog and digital I/O 模拟和数字输入/输出接口
analog equipment 模拟量设备,模拟装置
analog simulation 相似模拟
analogous data 相似数据
analogous equation 模拟[类似]方程
analog-to-digit conversion 模拟数字转换
analog-to-digital converter 模(拟)数

(字)转换器

analog-to-pulse duration converter 模拟脉冲持续转换器

analogue application 模拟应用[用途]

analogue backup 模拟备用装置

analogue controlle 模拟控制器

analogue hardware 模拟硬件

analogue input output channel 模拟输入输出通道

analogue panel 模拟盘

analogue pneumatic controller 模拟气动控制[调节]器

analogue signal 模拟信号

analogue system 模拟系统

analogue technique 模拟技术

analysis figures 分析数据[值]

analysis of metal slurry 金属矿浆分析

analysis of response 反应分析

analysis of standards 标准分析

analysis process 分析方法

analytic data 分析资料[数据]

analytic methodology 分析方法

analytical accuracy 分析精(确)度

analytical chemist 分析化学工作者

analytical data 分析资料[数据]

analytical equipment 分析设备

analytical error 分析误差

analytical expression 分析式

analytical grade reagent 分析级试剂

analytical laboratory 分析实验室

analytical model 分析模型

analytical operation 分析作业

analytical reagent 分析纯试剂

analytical technique 分析技术[方法]

analytical variability 分析变化性[率]

analyzer equipment 分析设备

analyzer room 分析室

anchor chain 锚链(气缸定位)

anchored layer 固着层

ancillary specification 辅助设备性能

andalusite flotation curve 红柱石浮选曲线

ANFO truck mixer ANFO卡车式混合器(搅拌机)

angle absolute value 角度绝对值

angle fitting 弯头

angle hole 斜孔

angle of crystal deflection 晶体偏转角

angle of fall 塌落角

angle of installation 安装角度

angle of mantle for crushing head 破碎锥头伞板角度

angle of spatula 刮铲角

angle of wall fraction 壁面摩擦角

angle steel 角钢

angled transfer 角度传送

angstrom unit 埃(长度单位Å,可用于测量波长,1Å=0.1nm(10^{-10}m))

angular bisector 角平分线(法)

animal glue 动物胶

animal life 动物生命

animal tall oil 动物塔尔油

anion 阴离子,负离子

anion adsorption 阴离子吸附

anionic 阴离子的
anionic compound 阴离子化合物
anionic exchange resin 阴离子交换树脂
anionic fatty acid collector 阴离子脂肪酸捕收剂
anionic fatty acid flotation 阴离子脂肪酸浮选
anionic flocculent 阴离子絮凝剂
anionic flotation 阴离子浮选
anionic polyacrylamide flocculent 阴离子聚丙烯(酰)铵絮凝剂
anionic radical 阴离子根
anionic reagent 阴离子式药剂
anionic starch 阴离子淀粉
anisotropic 各向异性[不匀]的,非均质的
anisotropic mineral 各向异性矿物
ankerite 铁白云石
annite 羟铁云母
Annite reagent 胺皂(类)药剂
Annite reagent A 一种胺皂(浮选剂,可用作捕收剂(氧化矿浮选))
annual audited statement 年度审计报表
annual average profit 平均年利润
annual average rate of return method 平均年度收益率法
annual basis 年度基准
annual benefit 年收益
annual capital outlay 年度资本支出
annual cash inflow 年现金流入量
annual cash requirement 年现金需要量
annual copper production 年铜产量[产铜量]
annual depreciation 年折旧率
annual depreciation expense 年折旧费
annual depreciation rate 年折旧率
annual dividend 年股息
annual economic contribution 年度经济贡献
annual economic recovery 全年度经济回收率
annual evaporation 年蒸发量
annual fixed cost 年度固定成本
annual fixed costs 年度固定费用
annual gross value 年总价值
annual indirect cost 年间接费用
annual installment 年分摊费
annual meeting 年会
annual operating cost 全年度生产费
annual output 年产量
annual output value 年产值
annual product value 年产值
annual production 年产量
annual relining cost 年换衬费用
annual report 年度报告
annual sales expense 年销售费用
annual ton 年吨
annual total expense 年度总费用
annual value 年价值
annular orifice 环形孔
annum compound interest return 年复

利还本率
annunciator 信号器,信号装置,信号仪器,卡号器
annunciator message light 警报器消息灯
anode casting 阳极铸造
anode scrap casting 残阳极铸造
anodic charging curve 阳极电荷曲线
anodic current 阳极电流
anodic current component 阳极电流分量
anodic direction 阳极方向
anodic film 阳极薄膜
anodic oxidation 阳极氧化(作用)
anodic oxidation process 阳极氧化过程
anodic oxidation reaction 阳极氧化反应
anodic polarization 阳极极化(作用)
anodic polarization curve 阳极极化曲线
anodic site 阳极区
anodic stripping voltametry 阳极溶出伏安法
anodic sweeping 阳极扫描
anoxic 缺[厌]氧的
anti-blinding action 防止堵塞作用
anticipated operating cost 预计的生产费用
anti-corrosion equipment 防腐设备
anticorrosion pump 耐腐蚀泵
anti-corrosive material 防腐物料

anti-ferrimagnetism 反亚铁磁性
anti-ferromagnetism 反铁磁性
anti-friction bearing crusher 减磨轴承破碎机
antimonial lead 锑铅合金
antiprecipitant 反沉淀剂
AP 845 正十八烷基硫化琥珀酰胺酸二钠盐(品名,成分同 Aerosol 18,可用作(锡石浮选)捕收剂)
apatite lattice 磷灰石晶格
aperture wall 孔墙[壁]
apex capacity 排砂(嘴)能力(水力旋流器)
apex capacity curve 沉砂口能力曲线
apex diameter 排[沉]砂口直径
apex discharge pattern 沉砂[底流]排出形式[形状]
apex selection 沉砂口选择
apex size 沉砂口大小[规格]
apolar solvent 非极性溶液[剂]
apparent density 视[表现,散装,松装]密度
apparent flat band potential 表观平带电位
apparent material density 物料松装[散装,表现,视]密度
apparent pulp rate constant 视在矿浆速度常数
apparent specific gravity 表观[视]比重
apparent velocity constant 视在速度常数

A

appendix 附录[言],附属(物),附加(物),补遗

applicable code 应用代码[符号,代号]

applicable standards 适用的标准

application engineer 应用工程师

application engineer of crusher manufacturer 破碎机制造厂产品操作[应用]工程师

application form 申请书[表格]

application manager 洽谈[接洽]业务[订货]经理,申请[请求]订货经理

application programming 应用程序设计

application software 应用软件

application technique 应用技术

application valve 控制阀

application(operation) program 应用[操作]程序

applied technology 应用技术

appraisal 估[评]价,估计,鉴定

appraisal manual 评估手册

approach 方法,途径,接近,引道[路,槽]

appropriate action 合理行动

appropriate condition 适当条件

appropriate factor 适当因素

appropriate form of mathematic model 数学模型的适当形式

appropriate model response 适当的模型响应

appropriate size 合适[理]粒度

approximate constant 近似常数

approximately constant critical concentration 近似恒定临界浓度

apron feeder 板式给矿机

apron feeder deck 板式给矿机面板层

apron pan feeder 板盘式给矿机

aqueous basic solution 碱性水溶液

aqueous chemistry 水化化学

aqueous electrolyte 含水电解质

aqueous emulsion 水乳浊液

aqueous environment 水(相)环境,液相

aqueous KCl solution 氯化钾水溶液

aqueous medium 水介质

aqueous metal-pregnant solution 含金属水溶液

aqueous phase 水相

aqueous slurry 含水矿浆

aqueous solubility 水溶解度[可溶性]

aqueous solution 水性溶液

aqueous solvent 含水溶剂

aqueous surface 液体[相]表面,水表面

aqueous surfactant 液态表面活性剂

aqueous suspension of apatite 磷灰石水悬浮液

Arabic gum 阿拉伯胶

aragonite type 霰石型

Araldite 环氧(类)树脂,合成树脂黏结剂

Arbiter process 阿比特过程

arbitrary choice 任意[随机]选择

arbitrary classification matrix 任意分

级矩阵
arbitrary classifier matrix 任意分级机矩阵
arbitrary intermediate value 任意中间值
arbitrary shift 任意班
arbitrary unit 任意单位
arching flow factor 结拱流动系数
architectural criteria 建筑标准
architectural drawing 建筑图纸
architectural engineer 建筑工程师
arctic 北极
arctic condition 严寒[寒冷]条件
Arctic syntex A 油酸磺化乙酯钠盐（同 Igepon A）$C_{17}H_{33}COOCH_2CH_2SO_3Na$
Arctic syntex L 椰(子)油单甘油硫酸盐(可作乳化剂)
Arctic syntex M 脂肪酸单甘油硫酸盐（可用作乳化剂浮选辉钼矿）
Arctic syntex T 油酰甲基牛磺酸钠盐（可用作湿润剂、捕收剂）
Arctic Syntex(=Syntex) 表面活性剂品牌名(中文名称:阿蒂克·辛太克斯,简称辛太克斯,可用作选矿药剂)
arcual cut 弧形切[截]取
arcual motion 弧形运动
area conversion constant 面积换算常数
area correction factor 面积修正系数
area facility description list 区域设备描述列表
area for scientific research 科研领域

area of application 应[使]用面积
area of process analysis 流程分析区
area of shortage water 缺水地区
argon 氩
argon gas flow 氩气流
arid region 干旱地区
arithmetic average 算术平均数[值]
arithmetic average flotation recovery 算术平均浮选回收率
arithmetic mean 算术平均值
arithmetic unit 运算器
arkose 长石砂岩
arkosic 长石砂岩的
arkosite 长石砂岩
arm 臂,柄,托臂,武器,装备
arm guard 推杆护板
Armac 脂肪胺盐酸盐阳离子型表面活性剂品牌名
Armac 10D Armeen 10D 的水溶性醋酸盐(正葵胺醋酸盐,可用作捕收剂、湿润剂)
Armac 1120 正十二烷基伯胺醋酸盐（可用作捕收剂）
Armac 12D Armeen 12D 的水溶性醋酸盐(正十二胺醋酸盐,可用作捕收剂、湿润剂)
Armac 14D Armeen 14D 的水溶性醋酸盐(正十四胺醋酸盐,可用作捕收剂、湿润剂)
Armac 16D Armeen 16D 的水溶性醋酸盐(正十六胺醋酸盐,可用作捕收剂、湿润剂)

Armac 18D　Armeen 18D 的水溶性醋酸盐(正十八胺醋酸盐,可用作捕收剂、湿润剂)

Armac 8D　Armeen 8D 的水溶性醋酸盐(正辛胺醋酸盐,可用作捕收剂、湿润剂)

Armac C　Armeen C 的水溶性醋酸盐(椰油胺醋酸盐,可用作捕收剂、湿润剂)

Armac CD　Armeen CD 的水溶性醋酸盐(精馏椰油胺醋酸盐,可用作捕收剂、湿润剂)

Armac CSD　Armeen CSD 的水溶性醋酸盐(精馏棉籽油胺醋酸盐,可用作捕收剂、湿润剂)

Armac HTD　Armeen HTD 的水溶性醋酸盐(精馏氢化牛脂胺醋酸盐,可用作捕收剂、湿润剂)

Armac SD　Armeen SD 的水溶性醋酸盐(精馏大豆油胺醋酸盐,可用作捕收剂、湿润剂)

Armac TD　Armeen TD 的水溶性醋酸盐(精馏牛脂胺醋酸盐,可用作捕收剂、湿润剂)

Armac-Co-Co-B　椰油混合胺(可用作捕收剂)

Armac-flot SD　大豆油制得的胺及腈混合物(可用作捕收剂)

armature　电枢,衔铁,叠片铁心,(电缆的)铠装,装甲

armature current　电枢电流

Armeen　脂肪胺阳离子型表面活性剂品牌名

Armeen 10D　正-癸胺 $C_{10}H_{21}NH_2$

Armeen 12D　精馏正-十二烷胺(含94%伯胺)

Armeen 14D　精馏正-十四烷胺(含92%伯胺) $C_{14}H_{29}NH_2$

Armeen 16D　精馏正-十六烷胺(含95%伯胺) $C_{16}H_{33}NH_2$

Armeen 18　正-十八烷胺(含85%伯胺) $C_{18}H_{37}NH_2$

Armeen 18D　精馏正十八烷胺(含95%伯胺) $C_{18}H_{37}NH_2$

Armeen 2G　椰子油仲胺(含85%仲胺)

Armeen 2HT　氢化牛脂仲胺(含85%仲胺)

Armeen 8D　正-辛胺 $C_8H_{17}NH_2$

Armeen C　椰子油胺(含85%伯胺)

Armeen CD　精馏椰子油胺(含95%伯胺)

Armeen CSD　得自棉籽油的脂肪胺(精馏棉子油胺)

Armeen DM16　二甲基十六叔胺(含80%叔胺) $RN(CH_3)_2$, R=十六烷基

Armeen DM16D　精馏二甲基十六叔胺(含92%叔胺) $RN(CH_3)_2$, R=十六烷基

Armeen DM18　二甲基十八叔胺(含80%叔胺) $RN(CH_3)_2$, R=十八烷基

Armeen DM18D　精馏二甲基十八叔胺(含92%叔胺) $RN(CH_3)_2$, R=十八烷基

Armeen DMC　二甲基椰油叔胺(含

80%叔胺)RN(CH$_3$)$_2$,R来自椰子油胺

Armeen DMCD　精馏二甲基椰油叔胺(含92%叔胺)RN(CH$_3$)$_2$,R来自椰子油胺

Armeen DMS　二甲基大豆油叔胺(含80%叔胺)RN(CH$_3$)$_2$,R来自大豆油胺

Armeen DMSD　精馏二甲基大豆油叔胺(含92%叔胺)RN(CH$_3$)$_2$,R来自大豆油胺

Armeen HT　氢化牛脂胺(含85%伯胺)

Armeen HTD　精馏氢化牛脂胺(含95%伯胺)

Armeen O　油酸胺(含86%伯胺)

Armeen OD　精馏油酸胺(含95%伯胺)

Armeen S　大豆油胺(含85%伯胺)

Armeen SD　精馏大豆油胺(含95%伯胺)

Armeen T　牛脂胺(含85%伯胺)

Armeen TD　精馏牛脂胺(含95%伯胺)

armored car　装甲汽车

Armosol 16　α-磺化软脂酸(品名)

Armosol 18　α-磺化硬脂酸(品名)

Arojel M　絮凝剂(冷水可泡胀的淀粉)

aromatic　芳香族的,芳香的,芳香植物

aromatic amine　芳香胺,芳香胺类

aromatic carbon atom　芳香族碳原子

aromatic ring　芳环

aromatic solvent　芳香族溶剂

Arquad　表面活性剂(水溶性季铵盐氯化物类)品牌名

Arquad 10　三甲基-正癸基季铵盐氯化物

Arquad 12　三甲基-正十二烷基季铵盐氯化物

Arquad 14　三甲基-正十四烷基季铵盐氯化物

Arquad 16　三甲基-正十六烷基季铵盐氯化物

Arquad 18　三甲基-正十八烷基季铵盐氯化物

Arquad 2C　二甲基-二烷基季铵盐氯化物 $[R_2N(CH_3)_2]^+Cl^-$

Arquad 2HT　二甲基-二氢化牛油基季铵盐氯化物(可用作捕收剂)

Arquad 2S　二甲基-二大豆油基季铵盐氯化物(可用作捕收剂) $[R_2N(CH_3)_2]^+Cl^-$,R=大豆油酰基

Arquad 8　三甲基-正辛基季铵盐氯化物

Arquad C　三甲基-椰子油基季铵盐氯化物

Arquad CS　三甲基-棉籽油基季铵盐氯化物(可用作捕收剂)

Arquad HT　三甲基-氢化牛油基季铵盐氯化物(可用作捕收剂)

Arquad S　三甲基-大豆油基季铵盐氯化物(可用作捕收剂)

Arquad T　三甲基-牛油基季铵盐氯化物(可用作捕收剂)

arrangement　整理,布置,配置,安排

arrangement of plant symmetry　工[选

矿]厂对称配置
array 阵列,数组
arsenate(=arseniate) 砷酸盐[脂]
arsenic 砷(As)
arsenic acid 砷酸
arsenical 砷化合物,砷剂,砷的,含砷的
arsenical antimony 砷锑矿 AsSb
arsenical gold ore 砷金矿石
arsenical nickel 红砷镍,红砷镍矿 NiAs
arsenical ore 砷矿石
arsonate 胂酸盐 $R \cdot AsO(OM)_2$
arsonic acid 胂酸 $RAsO_3H_2$
art of cyanidation 氰化技术
art of separation 分离技术
article 物品[件],制[产,成,商]品,论文,文章,项目,条款,章程
artificial bedding 人工床
artificial covellite 人造铜蓝
artificial sand 人工[造]砂
artificial sea water 人工制造[合成]的海水
aryl dithiophosphate 芳基二硫代磷酸盐
aryl dithiophosphate complex of heavy metal 芳基重金属二硫代磷酸盐络合物
aryl dithiophosphoric acid 芳基二硫代磷酸
ascent rate 上升[浮]速度
ascent velocity 上升[浮]速度
ash constituent 灰成[组]分

ash forming mineral 生成灰分的矿物
aspect 方面[向],形势,状况
aspect ratio 纵横比,高宽比,宽高比,长宽比,展弦比
aspiration cyclone 吸入旋风器
assay 化[试]验,(干,定量)分析,检[验,测]定,试金,试(验用)样(品),矿样,试料,试金物,验定法
assay accuracy factor 分析[化验]准确因素
assay calculation 分析计算
assay data 分析数据
assay determination 分析值的确定
assay error 分析误差
assay facilities 分析装置[设施]
assay information 分析资料
assay monitoring system 分析监控系统
assay monitoring tool 分析监控方法[手段,工具]
assay office Muffle furnace 分析室马弗炉
assay screen analysis 化验筛分分析
assay stream 试样流(分析)
assay value 化验值
assayer 化验员
assaying of product 产品分析
assaying procedure 分析过程
assay-size distribution 品位粒度分布
assembler 装配工,装配器,收集器,汇编程序
assemblier-language code 汇编语言

代码

assembly language level 汇编语言水平

assembly program 汇编程序

assets' physical usefulness 资产实际用途

assistance 帮[援]助,辅助设备

assistant engineer 助理工程师

assistant manager 副经理,协[襄]理

associate 相伴因素[元素,物],共生体,同事[行,伙,级],副[助]手

associate professor 副教授

associate professor of metallurgical engineering 冶金工程副教授

associated ash constituent 伴生灰分组分

associated component 配套部件

associated conveyor 辅助运输机

associated editor 附带的编辑程序

associated element 伴生元素

associated gangue 共[伴]生脉石

associated gangue mineral 共生的脉石矿物

associated immersed mechanism 配套的沉没机构

associated ion 缔合离子

associated material handling operation 辅助物料搬运作业

associated metal 伴[共]生金属

associated mineral group 伴生矿物族

associated patch 缔合体[斑点]

associated relationship 共生关系

associated values 伴生有价成分[金属]

associated vanadium 伴生钒

association 共生体,协会,学会,联合会,团体

association of hydrocarbon chain 烃链缔合

association of precious metal 共生贵金属,贵金属共生体

association's laboratory personnel 协会实验室试验人员

asymmetric load 不对称[平衡]荷载

asymmetric stretching mode 不对称伸缩模式

asymptotic value 渐近值

atmospheric carbon stripping 常压碳解吸

atmospheric contamination 大气污染

atmospheric strip 常压解吸

atmospheric stripping 常压解吸

atmospheric weathering 大气[自然]风化

atomic absorption 原子吸收(法)

atomic absorption analysis 原子吸收分析

atomic absorption analysis method 原子吸收分析法

atomic absorption cold NaCN leach method 原子吸收冷NaCN浸出法

atomic absorption method 原子吸收法

atomic absorption silver assay 原子吸收分析的银含量[品位]

atomic absorption spectrophotometer 原子吸收分光光度计
atomic absorption spectrophotometric technique 原子吸收分光技术
atomic absorption spectrophotometry 原子吸收分光光度法
atomic analysis method 原子分析法
atomic analysis system 原子分析法［装置］
atomic method 原子分析法
atomic spectra method 原子光谱法
atomic spectrometry 原子光谱法
atomized ferrosilicon 喷雾［雾化］硅铁
atomizing unit 雾化设备
attached diagram 附图
attached drawings 附图
attached particle 连生颗粒
attachment mechanism model 附着机理模型
attachment method 固定方法
attachment probability 黏附概率
attendant 服务［招待］员,轮换人员,办事员,参与人员
attenuation of ultrasonic signal 超声波信号衰减
attractive potential energy 引力势能
attractive scheme 有吸引力的方案
attractive secondary minimum potential energy 第二最小吸引位［势］能
attrition machine 擦洗机［槽］
attritioner 摩擦机
ATX S-丙烯基异硫脲氯化物(品名,可用作铜矿的捕收剂)
Au amyl xanthogenate 戊基黄原酸金
Au butyl xanthogenate 丁基黄原酸金
Au ethyl xanthogenate 乙基黄原酸金
Au heptyl xanthogenate 庚基黄原酸金
Au hexyl xanthogenate 己基黄原酸金
Au isoamyl xanthogenate 异戊基黄原酸金
Au isobutyl xanthogenate 异丁基黄原酸金
Au isopropyl xanthogenate 异丙基黄原酸金
Au octyl xanthogenate 辛基黄原酸金
Au propyl xanthogenate 丙基黄原酸金
auditor's report 审计［查账］员报告
auger distributor 螺旋分配器
auriferous gravel 含金砾石
auriferous sulphide 含金硫化物［矿］
author 作者
author's view 作者意见［观点］
autocorrelated variable 自相关变量
autocorrelation 自相关作用
autocorrelation aspect of variable 变量的自相关状况
autocorrelation aspect of variables 变量间的自相关情况
autocorrelation function 自相关函数(变异函数)
autocorrelation of regionalized variables 区域化变量的自相关性
autocorrelation study 自相关作用研究
autogeneous carrier flotation 自动载

体浮选
autogenous basis 自磨基础
autogenous grind 自磨
autogenous grinding characteristic 自磨矿性质[特性]
autogenous grinding plant 自磨厂[车间]
autogenous grinding route 自磨路线
autogenous grinding test 自磨试验
autogenous grinding test work 自磨试验工作
autogenous grinding unit 自磨设备[机组]
autogenous media competency test 自磨介质能力试验(用矿石为介质)
autogenous mill 自磨机
autogenous single-stage grinding 自磨单段磨矿,单段自磨
autogenous unit power requirement 自磨单位功耗(自磨功指数)
autogenous work index 自磨(矿)功指数
autogenous-ball mill flowsheet 自磨-球磨流程
autogenous-ball mill-crushing (ABC) process 自磨-球磨-破碎(英文缩写ABC)流程
autojet filter 自动喷射式过滤器
automated set 自动开口宽度
automated set crusher 自动控制排矿口破碎机
automated variable setting crusher 自动调节排矿口破碎机
automatic backstop 自动逆止器
automatic balanced friction hoist system 自动平衡式摩擦提升(机)系统
automatic batch weighing and load-out system 自动分批称重装载系统
automatic block system 自动联[闭]锁系统
automatic cation exchange unit 自动阳离子交换设备
automatic control program 自动控制程序
automatic control system 自动化控制系统
automatic dart valve 自动突进阀
automatic electromotive force measurement 电动势自动测定
automatic flotation testing 自动浮选试验
automatic flow control 流量自动控制
automatic friction hoist 自动摩擦提升机
automatic froth level control 泡沫液面自动控制
automatic grease lubricator 自动油脂润滑器
automatic hoist control room 自动提升控制室
automatic hoisting system 自动提升系统
automatic indexing 自动换位
automatic level control 液面自动控制

automatic level controller 自动液面控制器
automatic loop control 自动环路控制
automatic lubrication system 自动润滑系统
automatic mercury switch 自动水银触点开关
automatic mixing system 自动混合系统
automatic mode 自(动)控(制)式
automatic particle size monitor 自动测量粒度监测仪
automatic pebble feeding system 砾石自动给矿系统
automatic pH control pH值自动控制
automatic plate shifter 板框自动移动装置(板框压滤机的)
automatic positioning 自动定位
automatic printer 自动打印机
automatic proportioning 自动按比例定量[配合,配量]
automatic regulation 自动调节[整]
automatic scrap iron remover 自动除铁器
automatic sensing device 自动传感装置[设备,仪器]
automatic sequence starting 自动程[顺]序起动
automatic setting control system 排矿口宽[开]度自动控制系统
automatic setting regulation 排矿口宽度自动调节

automatic ship loading transloader 自动装载运输机
automatic sump density control 泵池浓度自动控制
automatic take-up 自动拉紧装置
automatic timed cycle 自动定[计]时周期[循环],自动时控周期[循环]
automatic timer 自动定时器
automatic truck hopper 自动卡车料[漏]斗
automatically controlled reagent system 自动控制药剂系统
automatically controlled tripper car 自动控制卸矿[布料]小车
automation control 自动化控制
automobile allowance 汽车补助[津贴]费
automobile crankshaft 汽车曲轴
automotive vehicle 汽车,机动车
auto-return unit 自返装置
auto-weighing feeder 自动测量给矿机
autoxidation cell 自动氧化槽(制酸)
autoxidation process 自动氧化法(制酸)
autoxidation tank 自动氧化槽
auxiliary building 辅助厂房
auxiliary dust system 辅助收尘系统
auxiliary equation 辅助方程
auxiliary equipment 辅助设备
auxiliary facilities 辅助设施
auxiliary plant 辅助车间
auxiliary plant facilities 辅助设备设施

[装置]
auxiliary platinum electrode 辅助铂电极
auxiliary service 附属服务设备
auxiliary services 辅助设施
availability 可用性,有效性,实用性,可获得性
availability factor 有效性因素,可用系数,利用率系数,使用效率,运转系数[因素]
availability fee 可使用的经费
availability index 可用[有效]指数
availability of ore 矿石的可应[利]用性
availability of water 水的利用率[可利用性]
availability system 系统利用率[可用性],有效工作系统
available cash 可利[使]用的现金
available gas lasers 可用[有效]气体激光
available hours 可利用时数
available mill cross-section area 有效的磨机横断面积,磨机有效横断面[积]
available real-time extension function 现成的实时扩大功能
average 平均(数,值)
average annual profit 平均年利润
average assay 平均成分
average Bond work index 平均邦德功指数

average book value 平均账面价值
average capacity 平均生产能力
average character 平均[一般]特征
average chemical analysis 平均化学分析
average circulating load 平均循环负荷
average cost value 平均费用值
average dimension 平均尺寸
average distance 平均距离
average energy of collision 平均碰撞能
average experience 经验平均值
average feed grade 平均给矿品位
average grade 平均品位
average grain size 平均粒度
average head grade 原矿平均品位
average material 一般[普通]物料
average medium 一般[普通]介质
average method 平均法
average net filtration feed rate 平均净过滤给矿量
average operation 一般作业[生产]
average ore 普通矿石(品位不高不低)
average outstanding balance 平均未偿还借款余额
average plant feed 车间平均给矿量
average pore size 平均孔径
average rate of interest 平均利率
average recommended operating speed 推荐[建议]的平均运转速度

average result 平均结果
average size 平均尺寸
average table value 表中平均(数)值
average tolerance 中等药物耐受性
average variogram value 平均变异函数值
average volumetric proportion 平均体[容]积比
averaging grade 平均品位
aviation fuel 航空燃料
aviation gasoline 航空汽油

aviation jet fuel 航空喷射燃料
avoidable product loss 可避免的产品损失
avoirdupois 常衡制
axial mass flow rate 轴向总流量
axial-flow 轴流量
axis of symmetry 对称轴
axisymmetric stream function 对称轴流函数
azimuth position 方位角(位置)

B

B. E. T. krypton adsorption B. E. T. 氪吸附法
B-23 frother(=Dupout Frother B23) B-23 起泡剂(同 Dupout Frother B23, 含 40%~45%伯醇(主要是二甲基戊醇-1)和 45%~50%仲醇(主要是 2,4-二甲基己醇-3)及 8%~12%的酮类)
B-3 frother B-3 起泡剂(聚氧乙烯丁醚)
back calculation 回归计算,反算法
back reef facies 后礁相
back side 反面
backfill storage silo 回填(矿,尾矿)筒仓
background application program 后台应用程序
background corrosion current 本底腐蚀电流
background current 本底电流
background data 背景资料
background drift 本底漂移区
background information 基础[本]资料
background terminal 后台终端
background utility program 后台应用程序
backing material 基底材料
backing material kit 填料件
back-pressure gradient 反压力梯度
backscattered electron image 反光散

射电子图像
backup data 备份资料[数据]
backward flow 逆向流动,回流
backward reaction 逆向反应
backwash sump pump 反洗液地坑泵
backwashing agent 反萃[洗提]剂
bacterial degradation 细菌分解
bacteriological dewatering 使用细菌脱水
baffle plate 稳流板
baffle ring 挡圈,防护圈
Baffled flat blade turbine mixer 稳流平直叶片透平搅拌器[机]
baffled plastic vessel 有挡板的塑料容器[杯,器皿,瓶]
baffling 稳流[板]
baffling effect 稳流效应
bag binding p 袋黏合问题
bag breaker 拆袋[包]机
bag cleaning machinery 清袋机械
bag filter 布袋收尘器,袋式过滤机
bag filter house 布袋[囊式]收尘室
bag house 布袋[囊式]收尘室
bag lead-out 袋装外运
Bagdad wiggler 巴格达式摇动器(喷洒用)
bagging plant 装袋厂[车间]
bag-house walk-in plenum 布袋室可入式强制通风
Bagolax 甲基纤维素(可用作捕收剂、抑制剂及湿润剂)
balance 差额,余额,剩余部分,平衡(表),对照表,结算差额表
balance cathodic reaction 平衡阴极反应
balance of trade 贸易平衡
balance room 天平室
balance sheet 资产负债表,平衡表,结算差额表
balance sheet limitation 资产负债的限额
balance system 天平系统
balanced center line load 平衡中心线负荷
balanced input and output rate 平衡的给(量和)排量
balanced operation 平衡作业
ball charge 装球量,球添加量
ball charge addition 添加[补加]球荷
ball charge level 装球程度[量]
ball charging 装球
ball mill circulating load controller 球磨机循环负荷控制器
ball mill classifier circuit 球磨机分级机回路
ball mill density 球磨机(矿浆)浓度
ball mill discharge percent solids 球磨机排矿固体百分数
ball mill discharge pulp density 球磨机排矿浆浓度
ball mill feed 球磨给矿量
ball mill feed conveyer 球磨给矿胶带运输机
ball mill feed conveyor 球磨机给矿皮

带运输机

ball mill intake scoop 球磨机给[挖]矿勺

ball mill media voidage 球磨机介质孔隙度[量,率]

ball mill oversize feed inefficiency factor 球磨机过大粒度给矿无效系数

ball mill power draft 球磨机牵引功率

ball mill scoop box 球磨机大杓槽

ball mill work index calculation 球磨机功指数计算

ball milling 球磨

ball size correction factor 钢球尺寸修正系数

ball trap 捕球器

ball usage 球使用量

ball wear rate 钢球磨损率

balling disc 成球盘

ball-pan hardness test 球盘硬度试验

band 带,环,箍,圈,环[带]状物,区[域],范围

band diagram 能带图

band shift 谱带偏移

bank grade 系列[浮选机组]品位

bank gravel 河岸[采石坑]砾石

bank interest rate 银行利率

bank loan 银行借[贷]款

bank note 钞票

Bank of China 中国银行

bank overdraft 银行透支

bank recovery 系列[浮选机组]回收率

bank sand 岸砂

bank's consultant 银行顾问

bank's prime rate 银行优惠利率

banker 银行家

bankruptcy 破产,经营失败

bar pitch (筛)棒[条]距

Barco rotary joint 巴可型旋转接头

barded wire fence 带刺铁丝网围墙

bare drive pulley 无外壳驱动滚筒(胶带机的)

bare pulley drive 光面胶带轮驱动

bar-flight feeder 链杆刮板给矿机

barge mounted pumping system 安装在驳船上的泵系统

barge-mounted pump 浮船泵

barium 钡 Ba

barium chemicals 钡化学药品

barium chloranilate 氯冉酸钡

barium chloride 氯化钡(可用作石英活化剂)

barium ore 钡矿,重晶石 $BaSO_4$;氢氧化钡矿

bark extracts 树皮萃(树皮萃取液,含有单宁及木质素,可用作抑制剂(用量50~250克/吨)及细泥分散剂)

Baroid 巴罗德加重剂(一种重晶石粉)

Baroid S-35 巴罗德 S-35 加重剂(特制重晶石粉)

barrel reclaimer 圆筒形取料机

barren assay 贫液品位[含量]

barren bleed loss 贫液排放损失

barren bleed stream 贫[废]液排出流

量,排出的贫[废]液量
barren ore 贫化矿
barren organic solution 贫有机液
barren pyrite 贫黄铁矿
barren reservoir 贫液贮槽[池]
barren solution tank 贫液槽
barren strip solution 贫(化)萃取[洗提]液
Barrett 煤焦杂酚油品牌名
Barrett flot oil No.4 Barrett 4号浮选油(煤焦甲酚酸,可用作起泡剂及捕收剂(矿化矿浮选))
Barrett No.4 Barrett 4号油(煤焦甲酚酸,为高炉焦油中的甲酚酸)
Barrett No.410 Barrett 410号油(煤焦甲酚酸,泡沫的变态剂,可用作捕收剂及调整剂)
Barrett No.634 Brrett 643号油(煤焦馏油,黏度高于 Barrett No.4,可用作浮选硫化矿时的泡沫稳定剂)
Barrett oil Barrett 油(煤焦甲酚酸,可用作起泡剂、捕收剂、调整剂)
Barrett－Haentjens type CT pump Barrett-Haentjens型CT泵
barrier-layer theory 阻挡层理论
Barsky relationship 巴斯基关系式
Barsky rule 巴斯基定律
Bartles cross-belt separator 巴特莱斯型横流皮带溜槽(英国Bartles公司制造)
base condition 基本条件
base course (路)基层,基础[地基]垫层
base e 底数e(自然对数底数)
base metal beneficiation 贱金属选矿
base metal beneficiation plant 普通[贱]金属选矿厂
base metal mill 贱金属选矿厂
base metal mineral 贱[碱]金属矿物
base mounting spring suspension (筛)底架弹簧悬挂装置
base of oxidation 氧化基[底]面
base-metal sulfide 贱金属硫化物[矿]
basic 基本[础]的,碱性[式]的,含少量硅酸的
basic amenability 基本适应性
basic approach 基本途径
basic capacity 基本(生产)能力
basic compound 碱性化合物
basic conclusion 基本结论
basic condition 基本情况,碱性条件
basic conveyor circuit 基本运输机流程,运输机基本流程
basic crusher control philosophy 破碎机控制基本原理
basic crusher control system 基本的破碎机控制系统
basic cyclone 基本旋流器
basic data 基础资料
basic design 基础设计
basic design consideration 基础设计依据
basic design criteria 基础设计标准
basic device 基本装置[设备]

basic element	基本元件
basic engineering	基础设计
basic engineering design phase	基础设计阶段
basic engineering project	基础设计项目
basic equipment	基本[主要]设备
basic factor	基本系数
basic flotation	碱性浮选
basic function	基本作[功]用
basic igneous breccia	基性火成角砾岩
basic industry	基础[重]工业
basic information	基础[原始]资料
basic item	基本[础]项目
BASIC language	BASIC 语言(计算机)
basic level	基本水平[层次]
basic operating system design	基本操作系统设计
basic operation	基本作业[工作]
basic operation unit	基本操作单元
basic operator station	基本操作员控制站
basic parameter	基本参数
basic physical-chemical phenomenon	基本物理-化学现象
basic physics	基本物理学
basic principle	基本原理
basic process	基本工艺过程[方法,流程]
basic process mechanism	基本过程机理
basic property	基本性质
basic raw material	碱性生料
basic reaction	碱性反应
basic reagent combination	基本[主要]药剂配方
basic rock type	基性岩类型
basic sampling procedure	基本采样过程[程序]
basic scale-up design parameter	按比例放大的基本设计参数
basic science	基础科学
basic software	基本软件
basic step	基本步骤
basic tonnage flowrate	基本吨位流量
basic type	基本类型[形式]
basic type mineral	碱式矿物
basic type of circuit	基本流程类型,流程基本类型
basic type of sample pattern	取样布置基本类型
basic type of simulator	模拟程序基本类型
basic unit	基层单位,基础设备,基本单元
basic unit operation	基本[主要]单元作业
basic zinc carbonate	碱式碳酸锌
basicity modifier	碱度调节剂
basis of taxation	征税基础
bassanite	烧石膏
bastnaesite	氟碳酸铈镧矿
batch basis	批量基础
batch cell	单槽

batch condition 分批[批量]条件
batch conditioning 分批试验调整
batch cyanidation 分批氰化
batch cyanidation process 分批氰化法
batch experiment 分批试验
batch float 分批浮选
batch flotation experiment 分批浮选试验
batch flotation test 分批[开路]浮选试验
batch grinding data 分批磨矿数据
batch grinding model equation 批料磨矿模型方程式
batch grinding scale-up 分批磨矿按比例扩[增]大
batch investigation 分批研究
batch laboratory grinding 实验室批量磨矿
batch method 批量法
batch process model 分批过程[试验]模型
batch process of cyanidation 分批氰化法
batch production 间歇生产
batch response 分批因变量
batch retorting 分批蒸馏
batch sample 分批试[矿]样,批样
batch stripping study 分批洗提[解吸]研究
batch stripping[elution] study 分批解吸[洗提]研究
batch tank 分批配料槽
batch testing laboratory 分批[单元]实验室
batch type process 分批(式)法
batch weighing 批量称量
batchwise 分批[断续]的
batchwise grinding 分批磨矿
batchwise operation 分批作业[操作]
bath water 洗澡水
battery of cyclone 旋流器组
bauxite digestion plant 铝土矿溶出厂[车间]
bauxite plant 铝土矿厂(包括采选)
bearing bush 轴承衬,轴瓦
bearing clearance 轴承间隙
bearing gap 轴承间隙
bearing loss 轴承损失
bearing ratio test 承载力比对[例]试验
bearing temperature 轴承温度
Beckman glass electrode 贝克曼玻璃电极(美国)
bed depth of material 物料层厚
bed thickness 层厚
bed volume 床容积
bedding bin 分层掺合矿仓,分层配料仓
bedding screen 床筛(跳汰机的)
bedding stacker 分层堆料机
bed-rock surface 基岩面
beechwood tar 山毛榉焦油
beginning book value 初始账面价值
behavior 行为,过程,(运行)状态,(工

作)情况,特性
bell 铃,钟,叶轮盖板
bell joint 承口接头
Belleville spring 盘[碟]形弹簧
below grade 不合格的,标线[地面]以下的
belt (皮,布,钢,胶,引,传动,子弹,地)带,带状物,区(域),环行铁路,电车环行线路
belt break 胶带断裂
belt conveyer scale system 胶带运输机称量装置[系统]
belt conveyor 胶带[带式]运输机
belt conveyor for recovery 回收地平污水中干矿胶带运输机
belt cover 胶带面层
belt edge distance 胶带宽度余量(带上物料与带边缘间距离)
belt feeder 带式给矿机
belt feeder calculation 皮带给料[矿]机计算
belt flex elastic-transient analysis 胶带挠曲弹性瞬时分析
belt flotation 胶[皮]带浮选
belt guard 胶(皮)带罩
belt hardening 胶带硬化
belt line geometry 胶带运输线路几何形状
belt load 胶带荷载(胶带运输机)
belt load control 胶带运输机负荷控制
belt misalignment detector 胶带跑偏探测器

belt repair 胶带修理
belt retraining 胶带再调整
belt sag between idlers 托辊间胶带(下)垂度
belt scale 拉姆齐型皮带秤
belt scale calibration 皮带秤校准
belt scraper 胶带清扫器,带式刮板(清扫器)
belt slip 胶带打滑
belt slip detector 胶带打滑探测器
belt slip on starting 胶带启动打滑,启动时胶带打滑
belt slope angle 胶带倾(斜)角
belt speed detector 皮带速度探测器
belt speed limit 带速极限[界限,范围]
belt speed sensor 胶带(机)速度传感器
belt starvation detector 皮带待料探测器
belt tension 胶带张力
belt training 胶带调正[位置校正]
belt travel signal 胶带行程信号
belt tripper chute 胶带运输机卸料溜槽
belt weigher 皮带秤
belt width 胶带宽度
belt width selection procedure 胶带宽度选择步骤
beltflex 胶带挠[屈]曲
BELTSTAT 一种胶带运输机设计软件的名称
bench grade 阶段水平

bench limit	台阶界限
bench scale	实验室[小]规模的
bench scale level	实验室[小型]规模水平
bench scale test	实验室规模试验
bench test	小型试验
bench testing	实验室[小型]试验
bench-scale	小型[实验室规模]的
bench-scale experiment	小型[实验室]试验
bench-scale flotation	实验室规模浮选
bench-scale test	实验室规模试验
bench-scale testing	小型[实验室]试验
bend	弯(管,头,道),弯曲处
bend line	折[弯曲]线
bend pulley	改向轮[滚筒]
beneficiate	选矿,精选,选[富]集
beneficiation	选矿,精选,富集
beneficiation of calcium mineral	钙矿物选矿[分选,选别]
beneficiation of ore	选矿
beneficiation of phosphate-bearing mineral	含磷酸盐矿物选矿
beneficiation plant product	选矿厂产品
beneficiation procedure	选矿方法[工艺]
beneficiation process	选矿法
beneficiation product	选矿产品
beneficiation program	选矿计划[方案]
beneficiation scheme	选矿方案
beneficiation technique	选矿技术[方法]
beneficiation test	选矿试验
Benifite	淀粉醚化物(可用作絮凝剂)
bentonite liner	膨润土衬里
bentonitic type	膨润土类型
benzene	苯 C_6H_6
benzene ring	苯环
benzol	苯 C_6H_6
benzolethylene phosphonic acid	苯膦酸乙烯
benzotriazole	苯三唑,硅孔雀石(可用作捕收剂)
benzoyl-	苯酰,苯甲酰 $C_6H_5CO—$
benzoyl alcohol	苯酰醇
benzyl	苄基,苯甲基 $C_6H_5CH_2—$
benzyl alcohol	苄醇,苯甲醇 $C_6H_5 \cdot CH_2OH$
benzyl arsonic acid	苄甲基胂酸,苄胂酸
Berco controller unit	贝尔科型控制装置
Bernoulli equation	伯努利方程(流体动力学方程,瑞士数学家丹尼尔·伯努利提出)
Bernoulli theorem	伯努利定理(流体动力学定理,瑞士数学家丹尼尔·伯努利提出)
Bernoulli's principle	伯努利原理(流体动力学原理,瑞士数学家丹尼尔·伯努利提出)

Bessemer converter matte 贝塞麦转炉冰铜
best available control technology 最佳有效控制技术
best close site setting 最佳窄边排矿口宽度
best operating speed 最好[佳]运转速度
best pattern 最佳模式[形式],最好的药剂制度
best setting 最佳整定值
best value of coefficient 系数最佳值
best value of model coefficient 模型系数最佳值
best values of constants 常数最佳值
best values of parameters 参数最佳值
beta-manganese dioxide β二氧化锰
beta-ray transmission measurement 贝塔射线透射测量
BG Amineacetate 混合胺醋酸盐(含5%C_{14},30%C_{16},65%C_{18},捷克制品)
bias 偏差[离,移,置]
bias test 偏差检验[检查]
bias test connection 偏离[移,转]测试[检验]接头
bias test gate 偏差检验闸门
bias testing 偏差检验[查]
biased sampling pattern 偏离[置]的取样布置图
bibliography 书刊目录[提要,评述,介绍],参考书目[资料]书目(提要),文献(目录)

bid analysis 报价分析
bid evaluation 评标,标书评审
bid invitation specification 招标说明
bid price 投标价格
bid review 评标,标书评审
bidders conference 投标人会议
bidders list 投标单位一览表
biguanidine 双胍
bi-layer adsorption 双层吸附
bimodal distribution 双峰分布
bin activator 料[矿]仓振荡[抖动,激活]器
bin discharge 矿仓排矿
bin discharge rate 矿仓排出量[率]
bin feed bucket elevator 矿仓给料斗式提升机
bin feed chute 矿仓给料[矿]溜槽
binary complex 二元络合物
binary ionic lattice 双[二元]离子晶格
binary mixture 二元混合矿
binary signal 二进制信号
binder-clay 胶黏土
bing 材料堆,垛,废料堆
bin-segregation 矿仓[内材料分层]离析作用
biochemical oxygen demand 生化耗[需]氧量
biodegradable additive 生物降解的添加剂
Biolar Rosano surface tensiometer 皮奥莱尔·罗萨诺表面张力计[仪]

biological degradation 生物分解
biological relevance 生物相关性
biological survey 生物[态]学调查[观测,观察,勘查]
biological treatment 生物处理
biotic community 生物群[体]
biotite diorite 黑云母闪长岩
biotite granodiorite 黑云母花岗闪长岩
Birdsboro jaw crusher 伯兹伯勒型颚式破碎机
bisulfite 亚硫酸氢盐
bit 钻头,刀头,锥,二进制单位(计算机)
bit penetration 钻入[穿,透]
bitum(-en) 沥青,柏油
bituminous 沥青的,含沥青的
bituminous whole coal 烟煤全煤(煤层未采部分)
black 黑色,黑颜料,黑黏土,煤灰,煤质页岩,泥岩,黑的
black amorphous wad 黑色非晶质锰土
black liquor 黑液(造纸副产品,可用作捕收剂)
black liquor soap 黑液皂
black ore 黑矿(日本一种典型的复杂硫化矿石,局部分布的黄铁矿 FeS_2,含铜黄铁矿)
black rubber (Duro 65A) 黑橡胶(Duro 65A)
black water 黑水,煤泥水

Black-type machine 布莱克型颚式破碎机(美国人 E. W. Black 设计制造了世界上第一台颚式破碎机)
blade mill 叶片式擦洗机
Blake(jaw)crusher 布雷克型颚式破碎机
Blancol 表面活性剂品名(甲醛与萘磺酸的缩合产物,可用作分散剂)
blank column 空白栏
blank determination 空白测定
blank flange 盲法兰
blank form 空白表格
blank titration 空白滴定
blast furnace 鼓风炉
blast furnace stock house 鼓风炉料槽
blast gate 鼓[通]风门
blasting material 炸药
blasting technique 爆破技术[方法]
bleed stream 排[放]出的水流[废水]
bleeding of air 排[放]气
blend amine 混合胺
blend pile 混矿矿堆
blend sand, doping sand 混合砂,掺杂砂
blending control system 配料控制系统
blending reclaimer 混匀取料机
blending requirement 配矿要求
blending stockpile 混矿堆场
blending stockyard 混料堆场
blend-rate dial 混合比调节控制盘[刻度盘,指示器]
blister copper 粗铜

blister copper reaction 粗铜反应
BLN torque sensor BLN型转矩传感器
block 块段
block construction 大型砌块建筑,部件[单元]结构
block diagram 方框图
block movement 块体移动
block valve 截止阀,切断阀
blockiness 块度
blocky granodiorite 块状花岗闪长岩
blocky magnetite 块状磁铁矿
blow-scraper discharge 吹气-刮板排料
blue lead 金属铅
bluegill sunfish 蓝鳃太阳鱼
board measure 板材量法,板积计[测量]
board of directors 董事会
body weight 体重
body-centered 体心型(结晶)
boiling temperature 沸腾温度
bolted and gasketed flange joint 螺栓和垫圈法兰盘连接
bolted steel ore bin 用螺栓连接的矿仓
bolted steel tank 螺栓固定钢槽
Boltzmann constant 玻耳兹曼常数
bonanza 大矿囊,富矿带[体,脉]
bond 债券
Bond ball mill grinding work index 邦德球磨机磨矿功指数,球磨机磨矿邦德功指数
bond contract 债券合同

bond gap 键能带隙
Bond grindability index 邦德可磨性指数
Bond grindability procedure 邦德磨矿功指数测定程序[方法]
Bond grindability test 邦德可磨性试验
Bond grindability work index 邦德可磨性[磨矿]功指数
Bond grindability work index method 邦德可磨性功指数法
Bond impact work index method 邦德冲击功指数法
Bond number 邦德功指数值
Bond procedure 邦德程序[方法]
Bond theory 邦德理论(关于破碎的)
Bond work index 邦德功指数(评价矿石被磨碎难易程度的一种指标,首先由美国人邦德(F.C. Bond)提出)
Bond work index calculation 邦德功指数计算
Bond work index formula 邦德功指数公式
Bond work index method 邦德功指数法
Bond work index of impact crushibility 邦德冲击破碎功指数(表征矿石可碎性的指标,是美国人邦德(F.C. Bond)于20世纪50年代提出的)
bondholder 证券持有者
bonding structure 键合结构
Bond's energy and particle size rela-

tionship 邦德的能量与(颗粒)粒度关系
Bond's equation 邦德公式
Bond's procedure 邦德方法
bone glue 骨胶(可用作絮凝剂)
bone phosphate 骨质磷酸盐,骨磷矿
bone phosphate of lime 骨质磷酸钙
book of reference 参考书
book value 账面价值
booklet of drawings 图纸小册子,图纸目录(清)单
booster 辅助剂,增压机,助力器
booster collector 辅助捕收剂
booster drive 辅助驱动
booster pump 升压泵
booster tank 升压池
Booth 225-cu-ft machine 布斯型225立方英尺浮选机(美国)
Booth flotation machine 布斯浮选机(联合使用叶轮和螺旋桨的机械搅拌式浮选机,美国制造)
border 界面,边界[缘]
borderline category 边缘类别
Borg Warner variable speed transmission 博格华纳变速器(美国博格华纳集团产品)
borneol 龙脑冰片,莰醇-[2](可用作起泡剂)$C_{10}H_{17}OH$
boron 硼 B
boron minerals 硼矿物
boron nitrite 氮化硼
borrower 借方

borrowers capital stock 借方资本股票
borrowing power 借款权力
borrowing privilege 借款优惠权
bottle roll 瓶滚动[摇晃]器
bottle roll cyanidation test 摇动台烧瓶氰化法试验
bottleneck 瓶颈,狭道,难关,影响生产流程的因素,妨碍,阻塞
bottom cover 底层
bottom horizontal discharge 底部水平排矿
bottom shell 底壳
bottom unloader 底卸(载)机,底部卸载机
bound water 结合水
bowl 碗,碗形体;固定锥,锥壳(破碎机的);浮槽(分级机的);滚球,铲斗;凹穴,洼地
bowl adjustment arm 碗形(固定锥)调整推杆
bowl assembly 圆锥壳[固定锥]组件(圆锥破碎机的)
bowl classification 浮槽分级
bowl hopper 碗形漏斗
bowl installation 浮槽设备(浮槽式分级机)
bowl level 锥[机]壳中料位
bowl level measurement 锥壳料位测定
bowl liner 固定锥[碗形]衬板
bowl liner U bolt 固定锥衬板U形螺栓
bowl liner U bolt washer 固定锥衬板

U形螺栓垫圈[片]
bowl(-type)classifier 浮槽式分级机
box 箱,匣,盒,接线盒,轴承箱,外壳[套,罩],轴瓦[套],盒形小室,(车,包)厢,岗亭,棚车
box level correction 泵池液位校正
brake energy absorbed 制动[刹车]吸收能量
brake service 脚踏闸
brake time 制动[刹车]时间
brake torque 制动[刹车]转[力]矩
brake unit 制动[刹车]装置
branch flow 支路流量
branch joint 支管接头
branch office 分支机构,分公司
branched chain 支链
branching of hydrocarbon chain 烃链分支
branching program 分支程序
brass mill effluent 黄铜厂排出液
Brazilian rock crystal 巴西水晶[石英](纯净石英)
break even point 盈亏平衡点,平均转效点
breakage and energy relationship 破裂与能量(消耗)关系
breakage characteristic 破碎特性
breakage coefficient 破碎系数
breakage energy 破碎能
breakage energy efficiency 破碎能量效率
breakage facilities 破裂设施

breakage function 破碎函数
breakage function parameter 破碎函数参数
breakage kinetic parameter 破碎动力学参数
breakage matrix 破碎矩阵
breakage rate coefficient 破碎速率[度]系数
breakage rate factor 破碎速度系数
breakage rate parameter 破碎率参数
breakaway friction multiplier 断裂[摩擦力,阻力]系数
breakaway torque 起动[步]转矩
break-down rate 破碎率
breaker ram 破碎夯
breaker strip 防断条,垫层
break-even analysis 无亏损分析,盈亏平衡分析
breakeven economic tradeoff factor 无盈亏经济折中方案系数
breaking characteristic 破碎特性
breaking up 打破
breast mill 侧面磨板[铣刀]
breather pipe 通气管
breathing 呼吸
brick construction 砖石结构,砖石工程
brief introduction 简要介绍
brief summary 简要归纳
bright constituent 发亮成分[煤]
brine 盐水,卤水,海水;加盐处理
British Gum 9084 英国胶9084(品名)
British standard weights and measures

英制标准度量衡
British Thermal Unit 英制热量单位
British unit 英制单位
brittle froth 脆性泡沫
broad guideline 广泛指导准则
broadcast system 广播系统
broken ground 爆破堆
broken line 虚线
broken section 破碎地段
broker 代理人,经纪人,掮客
bromine process 溴化法
bromo-cyanide method 氰化溴法
bromophenol blue 溴酚蓝
Brownian motion 布朗运动
brushless digital unit 无电刷数字式装置
BTC 烷基二甲基季铵盐氯化物(品名,可用作湿润剂)
bubble ascent rate 气泡上升速度
bubble attachment mechanism 气泡附着机理
bubble capture rate 气泡捕捉速度
bubble coalescence 气泡兼并
bubble column selection 气泡选择性
bubble contact 气泡黏着
bubble creation 气泡产生
bubble dispersion 气泡分[弥]散
bubble dissemination 气泡弥散
bubble genesis 气泡发生[成因]
bubble loading 泡沫荷载
bubble loading process 泡沫荷载过程
bubble particle interaction 气泡颗粒

互相作用
bubble pick-up test 掳泡试验
bubble probe 气泡探针
bubble residence time 气泡保留时间
bubble size 气泡尺寸
bubble surface area 泡沫表面面积
bubble-migration hypothesis 气泡迁移假说(即捕收剂由气泡表面向固体迁移)
bubble-particle capture 气泡颗粒捕捉
bubble-particle contact 气泡-颗粒接触
Buchanan jaw crusher 布坎南型颚式破碎机
Buchner flask 布氏烧瓶[抽滤瓶,巴克纳瓶](实验室用,发明者为诺贝尔化学奖获得者 Eduard Buchner)
bucket capacity 斗容,铲斗容量
bucket wheel reclaimer 斗轮式(堆场)装载机
Buckman tilting table 自倾式翻床
budget authentication estimate 预算鉴证估算
budget quotation 预算估价单
budgetary cost 预算成本
buffer bin 缓冲[矿]仓
buffered pH 缓冲pH值
buffering 缓冲作用
buffering effect 缓冲作用
buffering-power relationship 缓冲与功率关系
building and site development 建筑物

及场地开拓

building block 标准部件,预制部件,积木式元件

building foundation 厂房[建筑物]基础

building general arrangement 建筑[厂房]总体[总平面]布置图

building limitation 建筑物[厂房]的限制

building material 建筑材料

building materials 建筑材料

building space 建筑场地

building specification 建筑标准

building structure steel 建筑结构钢

build-up 增长[加,大,高],累[堆]积,聚集,结垢,建造,装配,安[拼]装,组合,结构,构造,形成,产[发]生,出现

built-in pump 装入式泵

built-in reduction 内在降低[减少](作用)

bulk cargo 散装货物

bulk carrier 散装货船

bulk chemical analysis 化学全分析

bulk chemical property 总体化学性质

bulk copper containing complex 大量含铜络合物

bulk copper-molybdenum-pyrite concentrate 铜钼黄铁矿混合精矿

bulk crystal property 整体结晶性质

bulk cyanidation test 大容量氰化试验

bulk density 容积[体积,堆积]密度,容重,松散体重,堆比重

bulk density of ball load 钢球装填的容重[体积重量]

bulk electronic property 整体电子性质

bulk flotation mill 混合浮选厂

bulk flow 主矿流

bulk lead-out 散装外运

bulk lime bin 散装石灰仓

bulk liquid 整体溶液

bulk material 散装[松散,散状,大宗,疏松]材料

bulk phase 整体[全]相

bulk property 总体性质

bulk rougher concentrate regrinding mill 混合粗精矿再磨机

bulk sample 混合[大]样

bulk samples 大批试样

bulk sampling 批量取样

bulk sampling plant 大量取样加工车间

bulk sampling procedure 取大样工艺[方法],取大样过程

bulk shipment 散装货物,散装的装船货物

bulk solid 松散固体

bulk solids 松散固体物料

bulk solution 整[本,主]体溶液

bulk solution pH 整体溶液pH值

bulk specific gravity 散[成堆]比重

bulk storage 外[大容量]存储器

bulk storage device 大容量储存装置

bulk sulfide concentrate 混合(浮选)

硫化物精矿
bulk sulphide float 混合硫化物浮选精矿
bulk test 大容量试验
bulk truck 散装卡车
bulk unit weight 松散物料单位重量
bulk water 重力水,体相水,大体积的水,整体水(物)
bulk-selective flotation 混合优先浮选
bulldozer 推土机
bulletin 通报,公告[报]
bulletin committee 会刊[报]委员会
bullion bar 金锭[条]
bumper bearing 冲击[防撞,缓冲]轴承
bunker coal 燃料[船用]煤
Bunker Hill type bubble tube density controller 邦克山型水泡管浓[密]度控制器(美国)
bunker oil 船用油
buoyancy effect 飘浮[浮力]效应
buoyancy medium 漂浮[浮力]介质
buoyancy principle 浮力原理
buoyant effect 漂浮效应
buoyant gold 漂浮金
buoyant stock market 看涨的股票市场
burning tray 燃烧盘
burnt lime 生石灰
Burtec glass tube Burtec玻璃管(感应炉提取熔融金试样用)
Burtonite No. 78 古耳豆科植物种子精

制液(品名,可用作絮凝剂)
bush fire fighting facilities 灌木火灾清防[防止]设施
bush fires board 灌木火灾委员会
bushing 衬套[管],套管,套筒,轴衬[瓦,套]
bushveld 灌木丛生地带
business department 业务部门
business development 业务开发
business firm 实业公司
business license 营业执照
business tax 营业税
butanol 丁醇 C_4H_9OH
butanol activity 丁醇活度
butanol amine 丁醇胺 $HOC_4H_8NH_2$
butene 丁烯 C_4H_8
butterfly valve 蝶形阀
butyl 丁基 C_4H_9—
butyl aerofloat 丁基黑药
butyl alcohol 丁醇 C_4H_9OH
butyl carbitol 丁基卡必醇(可用作起泡剂)
butylamine 丁(基)胺 $C_4H_9NH_2$
buyer 买方
buyer credit loan 买方信贷贷款
by-pass overflow slurry 分路[支]溢流矿浆
bypassed solids 短路固体(绕过分级的固体)
by-product credit 副产品贷方
byproduct mineral 副产矿物
by-product operation 副产品作业

[生产]
by-product recovery 副产品的回收
By-Prox 仲烷基硫酸酯(品名,可用作
湿润剂)
byte 字节,字组,二进位组

C

1-1-2-2-4 cell arrangement 浮选槽 1-1-2-2-4方式排列
5-3-2 cell-to-cell pattern 5-3-2直流式排列(浮选槽)
60 chamber recessed plate pressure filter 60室板框压滤机
C. I. ball(=cast iron ball) 铸铁球
C. P. M. (= critical path method) 关键路线法,统筹方法
C. P. M. schedule 关键路线法程序(表)
C_{18} unsaturated fatty acid C_{18}不饱和脂肪酸
cable 缆绳,电缆,电报
cable address 电报挂号
cable charge 电报费
cable drive 缆驱[传]动装置
cable fitting 电缆装配附件[接头]
cable trench 电缆沟
cable tunnel 电缆沟[隧道]
cable way 架空索道
cable-belt conveyer 钢绳胶带运输机
CAC ball mill 加拿大阿里斯查尔默斯型球磨机
CAC gyratory crusher 加拿大阿里斯查尔默斯型旋回破碎机
CAC hydrocone crusher 加拿大阿里斯查尔默斯液压圆锥破碎机
CAC Model SH screen 加拿大阿里斯查尔默斯 SH 筛
CAC model XXH screen 加拿大阿里斯查尔默斯型 XXH 筛
cadmium 镉 Cd
cadmium butyl dithiophosphate 丁基二硫代磷酸镉
cadmium ethyl dithiophosphate 乙基二硫代磷酸镉
cadmium isoamyl dithiophosphate 异戊基二硫代磷酸镉
cadmium m-cresyl dithiophosphate m-甲苯基二硫代磷酸镉
cadmium methyl dithiophosphate 甲基二硫代磷酸镉
cadmium o-cresyl dithiophosphate o-甲苯基二硫代磷酸镉
cadmium p-cresyl dithiophosphate

p-甲苯基二硫代磷酸镉
cadmium phenyl dithiophosphate
苯基二硫代磷酸镉
cadmium propyl dithiophosphate
丙基二硫代磷酸镉
cage ladder 笼式扶梯
cage mill 笼型破碎机
caking solids 结块(固体颗粒)物料
calaverite 碲金矿 $AuTe_2$
calcareous 石灰的,石灰质的,含钙的,含碳酸钙的
calcareous apatite 钙质磷灰石
calcareous fluorapatite 钙质[含钙]氟磷灰石
calcareous gangue mineral 含钙[灰质]脉石矿物
calcareous material 钙质物料
calcareous ore 石灰质[铁]矿石
calcareous phosphate 钙质磷酸盐
calcareous spar 方解石 $CaCO_3$
calcine 煅烧[矿],焙解[烧,砂]
calcine agitator 焙砂搅拌器[槽]
calcine bin 焙砂[料]仓
calcined gypsum 烧石膏
calcined ore 煅烧矿[石],焙砂
calciner 煅烧[焙解,焙烧]炉
calcining 煅烧
calcining kiln 煅烧窑
calcite 方解石 $CaCO_3$
calcite type 方解石型
calcium 钙(Ca)
calcium aluminate 铝酸钙

$CaO \cdot Al_2O_3$, $5CaO \cdot 3Al_2O_3$
calcium and magnesium carbonate 钙镁碳酸盐
calcium carbonate 碳酸钙,白垩 $CaCO_3$
calcium carboxylate 羧酸钙
calcium laurate 月桂酸钙
calcium lignin sulfonate 木质[素]磺酸钙
calcium oleate 油酸钙
calcium salt 钙盐
calcium sequestering agent 钙多价螯合剂
calcium soap 钙皂
calcium sulfonate 磺酸钙
calcium-activated gangue 钙活化的脉石
calcium-activated quartz 钙-活化石英
calc-silicate rock 钙硅岩,钙矽岩
calculated 计算的
calculated curve 计算[出的]曲线
calculated equilibrium potential 计算的平衡电位
calculated figure 计算数值[数字]
calculated form 计算(所得的)形式
calculated grade 计算品位
calculated head 计算的原矿
calculated payback period 计算的偿还期
calculated recovery 计算回收率
calculated result 计算结果
calculated size 计算尺寸

calculated thickener area 计算的浓密机面积
calculation 计算
calculation accuracy 计算精[确]度
calculation method 计算方法
calculation of basic expression 基本表式的计算
calculation of screen area 筛板面积计算
calculation of screen size 筛规格[大小]计算
calculation of steep ascent 陡峭上坡的计算
calculation routine 计算程序
calculation sequence 计算程序
calculation sheet 计算表(格)
calculus 微积分(学),演算
Calgon 六偏磷酸钠,卡尔冈
calibrated dropper 校准刻度滴管,经标定的滴管
calibration adjustment 校准调整
calibration check sample 标定校验样品
calibration equipment 校准[标定]设备
calibration experiment 标定试验
calibration test 校准[标定]试验
calibration test sample 校准[标定]试验试样
California Bearing Ratio(CBR) 加利福尼亚土壤[加州]承载比(英文缩写CBR)

caliper 圆规,卡尺[规,钳],两脚[外卡,内径,测径,弯脚]规,测径器,用卡规[测径器]测量
Calolan-Na-Sufonate 油溶性石油磺酸盐(可用作捕收剂氧化铁矿浮选)
Calol-Na-Sufonate Calol-磺酸钠(石油磺酸钠,可用作捕收剂)
calomel filling solution 甘汞充填液
calomel reference electrode 甘汞参比电极
calorific value 热值
calorimetric determination 量热计测定
calorimetric heat 所测得的热量
calorimetric heat measurement 量热测定值
camp 营地,工地宿舍
camp site 营地,工地宿舍
Canadian dollar 加元
canteen 小卖部,食堂
capacitance probe 电容探针
capacitor discharge 电容器放电
capacity and size correction table 生产能力与规格校正表
capacity boost (处理)能力提高
capacity chart 能力[容量](图)表
capacity equation 处理能力方程
capacity of handling machinery 运输机械能力
capacity of hoisting machinery 提升机械能力
capacity safety factor 能力安全系数

capacity table	(生产)能力表
capillary binding force	毛细黏合力
capillary flow meter	毛细管流动计
capillary fluid film technique	毛细液膜方法
capillary liquid film	毛细液膜
capital acquisition funds	基建取得资金
capital amortization cost	资本摊销费
capital asset	固定资产,资本资产
capital construction expenditure	基本建设费
capital cost calculation	建设成本计算
capital cost comparison	投资费用比较
capital cost estimate	基建投资[成本]估算
capital cost estimation procedure	建设成本估算程序
capital cost study	投资费用分析[研究]
capital cost summary	建设费用汇总
capital depreciation cost	资本折旧费
capital equipment cost	设备基建费[成本]
capital equipment replacement expenditure	资本[固定]设备更换费用
capital expenditure	资本[基建]投资
capital fund	基建资金
capital inflow	资本流入
capital investment	基建投资
capital investment requirement	基建投资需求
capital lending	资金拆借
capital movement	资本流动
capital outflow	资本流出
capital outlay	基建投资
capital-expenditure-efficient choice	基建投资[经营费]的有效选择
capitalized asset	资本化资产
capped plastic nipple	冠状塑料连[短]接管[喷嘴]
capric acid	癸酸
caprylic acid	辛酸
Captax	巯基苯骈噻唑(品名,可用作捕收剂)
captive bubble type instrument	捕留气泡[掳泡]型仪器
captive distillation	捕俘蒸馏
captive process	回收[捕获]过程[作业]
capture rate	捕捉速度
capture ratio	捕获比率
car unloading facilities	卸车设施
carbohydrate	碳水化合物,糖类
carbon	碳 C;碳的;石墨,炭精;炭精电极
carbon absorption column	碳吸收柱
carbon acid wash eductor	酸洗固定碳喷射提升器
carbon acid wash fan	酸洗固定碳风机
carbon adsorption cone	炭吸附锥
carbon atom	碳原子
carbon chain alkyl oxypropylamine	碳链烷基氧化丙胺

carbon column 碳吸附[萃取]塔,碳交换柱,碳塔[柱](吸附)
carbon desorption 碳解吸
carbon dioxide bubble 二氧化碳气泡
carbon electrode 碳电极
carbon in leach 碳浸(法)
carbon in leach circuit 碳浸回路
carbon in leach plant 碳浸厂
carbon in leach section 碳浸工段
carbon in leach technique 碳浸技术
carbon in pulp 炭浆法
carbon in pulp practice 炭浆法作业[实践]
carbon load and stripping test 碳载荷[金]和解吸试验
carbon loading test 碳载金试验
carbon matrix 碳脉石[基体]
carbon mould 碳铸模
carbon reactivation 碳再生
carbon screen 碳粒检查筛(碳浆法)
carbon steel 碳素钢
carbon steel ball 碳素钢钢球
carbon steel rod 碳素钢钢棒
carbon steel test mill 碳素钢试验磨机
carbon stripping 碳解吸
carbon tetrachloride activation test 四氯化碳活性试验
carbon wash screen 洗炭筛、炭洗涤筛
carbonaceous matter 含碳物质
carbonaceous ore 碳矿石
carbonate fluorapatite 碳酸氟磷灰石
carbonate group 碳酸盐族

carbonate raffinate 碳酸残液
carbonation 碳化作用,碳酸盐法[化],碳酸饱和
carbonatite core 碳酸盐岩岩心
carbon-in-leach setup 碳浸装置
carbon-in-pulp gold plant 炭浆法金厂(车间)
carbon-in-pulp plant 炭浆法厂(车间)
carbon-in-pulp process 炭浆法
carbon-in-pulp system 炭浆法[系统]
carbon-in-pulp treatment 炭浆法处理
carbonyl 羰基,碳酰(:CO)
carbonyl frequency 羰基频数
carboxyl 羧(基)
carboxylate ion 羧酸盐离子
carboxylic class 羧基化合物类
carboxymethyl 羧甲基
carboxymethyl cellulose 羧甲基纤维素(可用作絮凝剂、抑制剂)
carcase construction 骨架(芯子)结构
carcass crack 支架裂缝[破裂],骨架断裂
cargo 货物
cargo boat 货船
cargo compartment 货舱
cargo hold 货舱
cargo-passenger ship 客货船
carried particle 被负载颗粒
carrier 搬运工,运输公司,运载工具,载体

carrier concentration 载流子浓度
carrier current 载流子电流
carrier density 载流密度
carrier flotation 载体浮选
carrier particle 背负颗粒
carrier precipitation 载体沉淀
carrier vehicle 货运车辆
carry side 承载段[边]
carrying belt training idler 承载胶带导辊
carrying capacity 运送[载]能力
carrying idler 承载托辊
carrying idler application factor 承载托辊应用因素[系数]
carrying idler serial number 承载托辊系列号
carry-over effect 后续效应
cascade construction 分段构造,分段结构
cascade counter-current wash classifier 级联逆流选矿分级机
cascade loop 串联[环路]循环
cascade loop controller 串级环路控制器
cascade machine 梯流(式)浮选机
cascade mill 瀑落式磨机
cascade pressure sampler 串级[瀑落]压力式取样器
Case 1 (实)例1,案例1
case history 典型例证
case study 实例研究
casein 酪朊,酪素(可用作絮凝剂、抑制剂)
cash account 现金账目
cash dividend 现金股息
cash flow 现金流动[量]
cash flow analysis 现金流动[量]分析
cash flow lost 现金流通损失
cash flow method 现金流量法
cash inflow 现金流入(量)
cash outflow 现金流出(量)
cash requirement 现金需要量
cashflow 流动现金,现金流通
casing cross sectional area 套筒[外壳]横断面积
cassiterite flotation plant 锡石浮选厂
cast polyurethane impeller 聚氨基甲酸酯铸造叶轮
cast refractory slab 耐火材料(浇注)板,耐火板
cast silicon carbide crucible 碳化硅浇注坩埚
casting furnace 浇铸炉
Castleman function 卡斯尔曼函数
castor oil 蓖麻油
castor oil acid 蓖麻油脂肪酸
Castor oil acids 135 蓖麻油脂肪酸(品名,可用作捕收剂)
Castor oil acids 9-11 蓖麻油脂肪酸(品名,可用作捕收剂)
castor oil sulfonate 磺化蓖麻油(即土耳其红油,可用作捕收剂和乳化剂)
cat ladder 爬梯,墙上竖梯
Cataflot 一种两性捕收剂的商品代号

catalog capacity 样本容量[能力]	cation adsorption 阳离子吸附
catalog data 样本数据,数据一览表	cation amine collector 阳离子胺类捕收剂
catalog horsepower 样本功率,马力(=0.746kW)	cation amine salt 阳离子胺盐
catalogue 商品目录,产品样本(说明书)	cation exchange column 阳离子交换柱
	cationic amine 阳离子胺
catalogue price 目录价格,定价	cationic ammonium ion 铵阳离子
catalytic surface 催化表面	cationic cell 阳离子浮选槽
category of error 错误[误差]范畴	cationic class 阳离子类
catenary ore bin 悬吊式矿仓	cationic depressant 阳离子抑制剂
caterpillar 履带,履带车,履带式拖拉机	cationic depression 阳离子抑制(作用)
	cationic ether diamine 阳离子醚二胺
cathode copper 阴极铜	cationic exchange resin 阳离子交换树脂
cathode melting and casting 阴极熔铸	cationic flocculent 阳离子絮凝剂
cathode product 阴极产品	cationic flotation area 阳离子浮游区
cathode ray polarograph 阴极射线极谱记录器[仪]	cationic flotation technique 阳离子浮选法[技术]
cathode-anode spacing 阴阳极间距	cationic polymer 阳离子聚合物
cathodic charging curve 阴极电荷曲线	cationic promoter 阳离子促进剂
cathodic current 阴极电流	cationic reagent 阳离子式药剂
cathodic direction 阴极方向	cationic starch 阳离子淀粉
cathodic function 阴极作用	cationic synthetic flocculent 阳离子(的)合成絮凝剂
cathodic limit 阴极极限	cationic type reagent 阳离子型药剂
cathodic polarization 阴极极化(作用)	Cattermole action 卡特莫尔作用
cathodic protection 阴极防护	caulking fines 填隙细粒[粉矿]
cathodic reactant 阴极反应物	caustic cyanide 苛性氰化物
cathodic reaction 阴极反应	caustic cyanide method 碱性氰化法
cathodic reduction 阴极还原	caustic cyanide solution 苛性氰化物溶液
cathodic reduction of oxygen 氧(的)阴极还原	
cathodic scan 阴极扫描	caustic dextrine 苛性糊精

caustic leaching 苛[碱]性沥滤
caustic solution 苛性溶液
caustic starch 苛性淀粉
caustic tapioca starch 苛性木薯淀粉
causticized starch 苛性化淀粉
cavity formation 气腔的形成
cavity size 气腔大小[尺寸]
Cd amyl xanthogenate 戊基黄原酸镉
Cd butyl xanthogenate 丁基黄原酸镉
Cd ethyl xanthogenate 乙基黄原酸镉
Cd heptyl xanthogenate 庚基黄原酸镉
Cd hexyl xanthogenate 己基黄原酸镉
Cd isoamyl xanthogenate 异戊基黄原酸镉
Cd isobutyl xanthogenate 异丁基黄原酸镉
Cd isopropyl xanthogenate 异丙基黄原酸镉
Cd octyl xanthogenate 辛基黄原酸镉
Cd propyl xanthogenate 丙基黄原酸镉
CE Raymond double-whizzer air separator CE 雷蒙德双活动叶片式离心风力分选机(美国)
Cedarapids compactor Cedarapids 压实[捣固]机(美国 Terex 跨国公司产品)
cell assembly 浮选槽机组
cell bottom contour 槽底形状
cell configuration 槽子结构(浮选槽的)
cell current 电池电流

cell holding time (在)浮选槽停留的时间
cell level adjustment 浮选槽液面调整
cell lip 浮选槽溢流堰
cell metallurgical performance 浮选槽工艺性能
cell operating characteristic 浮选槽操作特性
cell performance 浮选槽性能
cell repair 浮选槽修理
cell residence time 浮选槽内停留[浮选]时间
cell sanding 槽中沉砂
cell to cell flotation machine 中间室式浮选机
cell to cell machine 中间室式浮选机
cell volume 槽(子)容积
Cellofase B 羧甲基纤维素钠(可用作絮凝剂、抑制剂, 英国产品)
cell-to-cell type 槽对槽式(浮选机的)
cellulose 纤维素$(C_6H_{10}O_5)_x$
Cellulose 表面活性剂品牌名(美国 Union Carbide 公司)
Cellulose CMC 羧甲基纤维素(可用作矿泥脉石抑制剂)
Cellulose CMHEC 羧甲基羟乙基纤维素(可用作抑制剂、分散剂(用于铁矿和非金属矿物的选别))
cellulose gum 纤维素胶
cellulose phenyl thiourethane 纤维素苯基硫代氨基甲酸乙酯
cellulose phosphate 磷酸纤维素(可用

作稀土矿萃取剂)
cement copper 沉积[渗碳]铜,沉淀[置换]铜,海绵铜,泥铜
cement lined pipeline 衬水泥的管线[路,道]
cement manufacture 水泥制造
cement project 水泥项目
cement silo 水仓,水泥库
cementation 粘结[固],转换沉淀,烧结,渗碳处理,渗金属法
cemented sand filling 胶结矿充填
center clamp 中心夹[压]板,中心压紧装置
center drive mechanism 中心传动机构
center peripheral discharge 中间(央)周边排矿
center peripheral rod mill 中心周边排矿式棒磨(矿)机
center point 中心点
center surveillance room 中心监视室
center well 中心井(浓缩机的)
central air core 中央空气柱[芯](水力旋流器的)
central analyzer device 中央分析器装置
central bank 中央银行
central circular distribution chamber 圆形中心分配室(旋流器给矿系统),圆形中央分配室
central computer 中央计算机
central control station 中央控制站
central core 中央核心

central discharge weir 中央排矿堰(离心金刚石选矿盘)
central draw-off 中央排出(跳汰机的)
central dust-collection system 中央集尘系统
central engineering personnel 核心工程人员
central laboratory 中心实验室
central maintenance department 中央维修部门
central metallurgical office 中心选矿冶金办公室
central mixing chamber 中心[央]混合室
central part 中心部分
central plant 中央选矿厂
central point run 中心点试验
central research group 中心研究组
central rotor 中心转子
central shop 总[中心]修理厂
central water system 中心给水系统
centralized analyzer 集中式分析仪
centralized concentrator X-ray analyzer 集中式选[矿]厂X射线分析装置
centralized concentrator X-ray analyzer installation 选矿厂集中式X射线分析装置
centralized control system 集中控制系统
centralized metering system 集中计量系统

centralized on-stream X-ray analyzer system 集中式 X 射线在线分析装置

centralized on-stream X-ray fluorescence analysis 集中式在线 X 射线荧光分析

centralized X-ray system 集中式 X 射线分析系统

centre zero 中心零点

centrifugal ball governor 离心球调节器

centrifugal concentrator 离心选矿机

centrifugal cone 离心圆锥分级机

centrifugal densifier 离心浓缩机[稠化机]

centrifugal device 离心(分选)装置

centrifugal filtrate pump 离心滤液泵

centrifugal force 离心力

centrifugal force vector 离心力矢量

centrifugal impeller 离心式叶轮

centrifugal pump 离心泵

centrifugal rubber lined pump 衬胶离心泵,离心胶泵

centrifugal scrubber 离心式洗涤器

centrifugal sedimentation 离心沉淀[积]作用

centrifuge product 离心(脱水)产品

centrifuging(separation)test 离心分离试验

ceramic industry 陶瓷工业

ceramic material 陶瓷材料

ceramic tile 陶瓷砖

ceramic tile lining 瓷砖衬里

ceramic wear element 陶瓷耐磨元件

ceramic-lined 陶瓷衬

certain size fraction 某粒级(部分),特定粒度

certificate 证(明)书,执照,检验[合格]证,证明,单据

certificate of delivery 交货证明书

certificate of manufacturer 制造厂证明书

certificate of shipment 出口[装船]许可证

certification of fitness 合格证书

cervantite 黄锑矿,锑赭石 Sb_2O_4

cesium 铯 Cs

cesium chloride 氯化铯

Cesium-137 铯 137

cetane acid 十六烷酸

cetyl pyridinium chloride 十六烷基吡啶氯化物(可用作起泡剂、捕收剂)

chain 链,链条,工程测链长度(1 链=66 英尺);山脉,山系

chain association 链缔合

chain compound 链化合物

chain configuration 链结构

chain driven 链传动的

chain scrapper 链条刮板机

chalcocite(=chalcosine) 辉铜矿 Cu_2S

chalcopyrite association 黄铜矿共

生体
change flow unit 换料装置,转换流向装置
change room 更衣室
changing room 更衣室[间]
channel sampling 刻槽取样
character of mineralization 矿化[成矿]特性
characteristic amount of energy 特殊能量
characteristic analysis 特性分析
characteristic band of monothiocarbonate 一硫代碳酸盐特性谱带
characteristic dimension 特征尺寸
characteristic emission line 表征[特性]发射线
characteristic family 特性曲线族
characteristic frequency 特征[固有]频率
characteristic of grain orientation 颗粒定向特性
characteristic of particle orientation 颗粒定向特性
characteristic radiation 特性辐射
characteristic threshold of liberation (矿石)解离特性界限
characteristic xanthate band 黄原酸盐特征谱带
charcoal concentrate 木炭精矿(采用(美国 Chapman 工艺浮选所得的含金木炭精矿)
charcoal process 活性炭法(用活性

炭从溶液中提取有用成分,然后用热苛性氰化物溶液将它从炭上脱除,再用电解法回收)
charge array 电荷阵列
charge carrier 载流子,电荷载体
charge characteristic 电荷特性
charge characteristic of interface 界面电荷特性
charge density 充填物密度
charge distribution 电荷分布
charge for service 服务费
charge generation 电荷产生
charge generation mechanism 电荷产生机理
charge of ionic solid 离子型固体电荷
charge rate 装入量[率]
charge separation 电荷分离
charge site 电荷区
charge transfer 电荷转移
charge transfer electrode process 电荷转移电极过程
charge transfer reaction 电荷转移反应
charged end 带电端
charged head 荷电端
charged ionic polar group 带电离子极性基
charged polar head group 荷电极性端基团
charged surface area 荷电表面区
charged surface attraction 带电表面

的吸引
charging chute 装料溜槽
charging conveyor 装料运输机
charter party 租船[海运]契约方[人]
charter shipping 包租海运,租船运输
charter vessel 包租船只
chassis 底盘(架),底板[座]
chattering 抖动现象
check weighing system 称量校核系统
checker plate floor 花纹铁板地板
checkout 验收,检查[验,测],测试,试验,调整,校正,验算,结账
checkout console 检验台,检测板
chelate 螯合物,螯合的
chelate compound 螯形化合物
chelate group 螯合基
chelating agent 螯合剂
chelating agent-neutral oil 螯合剂中性油
chelating agent-neutral oil method 螯合剂中性油法
chelating reagent 螯合药剂
chelation product 螯合产品
chemical(wet)analysis 化学(湿法)分析
chemical addition 化学药剂添加
chemical analysis 化学分析
chemical analytical procedure 化学分析方法[过程]
chemical and physical condition 理化条件
chemical assay 化学分析
chemical bond 化学键
chemical bonding 化学键合
chemical characteristic 化学特性[征]
chemical classification 化学分级(类)
chemical composition of aqueous phase 液相化学组成
chemical crystal 化学晶体
chemical depressant 化学抑制剂
chemical determination 化学测定
chemical engineering deskbook 化学工程办公手册
chemical engineering plant construction cost index 化工厂建设费用指数
chemical engineering reactor 化学工程反应器
chemical environment 化学环境
chemical extraction method 化学提取法
chemical factor 化学因素
chemical force 化学力
chemical industry 化学工业
chemical industry waste 化工工业废弃物
chemical interaction 化学(相互)作用
chemical kinetic analogy 化学动力模拟
chemical laboratory 化学实验室
chemical metallurgy 化学冶金
chemical method of examination 化学检验方法

chemical name　化学名
chemical oxygen　化学氧
chemical oxygen demand　化学需氧量
chemical plant　化工厂,化工设备
chemical potential　化学势
chemical processing　化学处理
chemical pure reagent　化学纯试剂
chemical reaction　化学反应
chemical reaction theory　化学反应理论
chemical reactivity　化学反应性
chemical regenerant　化学再生剂
chemical species　化学物种(类)
chemical stability　化学稳定性
chemical state　化学状态
chemical substance　化学物质
chemical test　化学试验
chemical treatment　化学处理
chemical treatment plant　化学处理[化工]厂
chemical usage　化学药剂使用(量)
chemical variable　化学变量
chemically stable particle　化学稳定颗粒
chemicals　化学药剂[品]
chemisorbed layer　化学吸附层
chemisorbed metal-thiolate　化学吸附金属硫醇盐
chemistry of uranium extraction　铀萃取化学
chequing account　支票账户
cherty material　燧石[硅质]物料

chevron method　人字形(断面)法(分层掺和堆积法之一)
chevron packing　人字形衬垫
chief chemist　主任药剂师
chief construction engineer　建筑总工程师
chief counsel　首席顾问[律师]
chief design engineer　设计总工程师
chief electrical engineer　电气总工程师
chief grinding mill engineer　磨矿主任工程师
chief metallurgist　选冶总工程师,主任冶金师
Chili X impeller　奇利X形叶轮
chinook salmon　奇努克鲑鱼
chip　碎片,木屑,芯[薄]片
chip sample　拣块试[岩屑]样
chip sampling　碎片取样
chipping action　切削[削尖,錾平]作用
chisel hopper　凿(刃)形漏斗
chloral　氯醛,三氯乙醛,三氯乙二醇 Cl_3CCHO
chloral hydrate　水合三氯乙醛 $Cl_3C \cdot CH(OH)_2$
chloranilate　氯冉酸盐
chloranilate ion　氯冉酸离子
chlorinated polyethylene　氯化聚乙烯
chlorinated rubber　氯化橡胶
chlorinated water　氯化[加氯]水
chlorination　氯化
chlorinator　加氯机,氯化器
chlorine addition　充氯

chlorine oxidation tank 氯氧化槽
chlorine process 氯气法
chlorite group 氯化(矿)物族
chloritic schist 绿泥片岩
chloritized limestone 绿泥石化灰岩
chlorolignin 氯化木质素(可用作抑制剂)
chloroprene rubber 氯丁二烯橡胶
Chlorosilane 23 二甲基二氯化硅与甲基三氯化硅混合物(品名,法国)
chlorosulphonate polyethylene material 氯化磺聚乙烯材料
chlorosulphonated polyethylene 氯磺化的聚乙烯
choice of equipment size 设备规格选择
choice of equipment type 设备型号选择
choice of regenerant chemical 化学再生剂的选择
choke fed operation 挤满给矿操作
choke feed 挤满给矿[料]
choke feed capacity 挤满给矿能力
choke feed mantle-concave design 挤满给矿的定锥罩和定锥设计
choked condition 挤满给矿情况[条件]
choke-fed 挤满[塞]给矿的
choke-fed primary crusher 挤满给矿粗碎机
choke-fed-hopper level control 填满给矿漏斗料位控制

choke-feed 挤满[塞]给矿
choke-feed condition 挤满给矿状态
chord angle 弦角
chrome molybdenum steel 铬钼钢
chrome-magnesia clinker treating plant 铬镁烧结块处理厂
chrome-moly 铬钼合金钢
chrome-moly casting 铬钼铸件
chromic (正)铬的,三价铬的,六价铬的
chromic acid 铬酸 H_2CrO_4
chromic oxide 氧化铬,三氧化二铬 Cr_2O_3
chromic salt 铬盐
chromic-sulfuric acid cleaning solution 铬硫酸洗液
chromium carbide 碳化铬
chromium carbide hard-faced liner 碳化铬硬面衬里
chromium oxide 氧化铬
chub 鲦鱼
churn drill hole 冲击钻钻孔
chute drop distance 溜槽落下距离
chute flow path 溜槽流矿通道
chute plugging 溜槽[口]堵塞
chute work 溜槽作业
CIL contactor 浸出碳吸附接触器
CIP contactor 碳浆法接触器
circuit 电路,回路,环道,系统,工序
circuit arrangement 回路布[配]置
circuit balance 回路平衡
circuit configuration 流程[回路]结构

circuit constraint identification [布置,配置]	应特性

circuit constraint identification 回路限制判[确,鉴]定
circuit constraints 回路约束条件
circuit control 回路控制
circuit control requirement 回路控制要求
circuit control strategy 回路[流程]控制对策
circuit design 流程设计
circuit diagnosis 回路[流程]诊断[检查,调查分析]
circuit diagnostics 回路诊断
circuit diagram 回路图
circuit feed parameter 流程[回路]给矿参数
circuit gain 回路收益
circuit gains 回路收益
circuit modification 流程[回路]修改[改进]
circuit operating condition 回路操作条件(包括临界速度百分率,装球量百分率,粒度分布,给矿中固体含量,循环负荷)
circuit optimization and economic justification 回路最佳化和经济验证
circuit performance 回路性能
circuit problem identification 回路问题确[鉴,判]定
circuit response analysis 回路响应分析
circuit response characteristic 回路响

circuit schematic 流程简[略]图
circuit selection 回路[流程]选择
circuit simulation 回路模拟
circuit stability 回路稳定性
circuit stabilization 回路稳定化
circuit state model 回路状态模型
circuit stream 回路矿(浆)流
circuit throughput 回路处理[通过,生产]能力
circuit tonnage 回路吨数[位]
circuitry 电[网]路(一套设备中电路的总称),线路,流程,电路图[系统],接线图[法],布[架线],电路原理[学]
circular and angle measure 弧度与角度单位制
circular bin 圆形矿仓
circular chart recorder 圆盘图表记录仪
circular measure 弧度单位制
circular outlet 圆形排(矿)口
circular ventilation shaft 圆形通风竖井
circular vibrating pan 圆形振动盘
circulated load 循环负荷
circulated load ratio 循环负荷率
circulated material 循环物料
circulating asset 流动资本
circulating capital 流通资本
circulating fan 循环风机
circulating load 循环负荷
circulating load optimization 循环负

荷最佳化
circulating loading conveyor 循环负荷（胶带）运输机
circulating medium density control system 循环介质浓度控制系统
circulating medium pump 循环介质泵
circulating water 循环水
circulating water pump station 循环水泵站
circulation well 循环井(浮选机)
circumferential area 圆周[环形]面积
citric acid 柠檬酸
civil commotion 内乱
civil construction package 土建施工包
civil engineering drawing 土建图纸
civil strife 内乱,国内冲突
claim 要求,索赔
clamp ring 锁环,夹圈,夹板环
clamp type joint 夹板[管夹]型接头
clarification lagoon （尾矿,污泥）澄清池
clarified pregnant solution 净化[澄清]贵液[含金溶液]
clarified solution tank 澄清液槽(罐)
clarified water tank 澄清水池(罐)
clarifier underflow pump 澄清器底流泵
clarifying filter 澄清过滤器
clarity 澄清[清晰]度
Clarkson feeder 克拉克森型给料[矿]机
Clarkson Type J solenoid operated valve 克拉克森J形螺线管操作阀
classical approach 传统办法
classical chemical method 经典化学方法(分析)
classical configuration 经典配置
classical grinding circuit 典型磨矿回路
classical statistical method 经典统计法
classification constant 分级常数
classification duty 分级任务[工作]
classification factor 分类因子[数]
classification function 分类函数
classification matrix 分级[类]矩阵
classification of particle 颗粒[粒度]分级
classification of reagents 药剂分类
classification of solids 固体物料的分级
classification pool 分级槽[池]
classification step 分级步骤[阶段]
classification stream 分级矿流
classified mill tailing fill 分级的选矿尾砂充填
classified pulp 已分级的矿浆
classifier cyclone 分级旋流器
classifier flight 分级机刮板
classifier flight and shoe 分级机刮板和刃靴
classifier fluctuation 分级波动
classifier installation 分级机设备
classifier matrix 分级机矩阵
classifier overflow screen 分级机溢流筛

classifier performance characteristic 分级机性能特点
classifier pool 分级机沉淀[降]区
classifier sand 分级机返砂
classifier shoe 分级机(刮板)刃靴
classifier tank slope 分级机槽体倾斜度
classifier underflow 分级机底流[返砂]
classifying device 分级装置
clathrate 笼形[包合]物,窗格形的,插合物的
clay content 黏土含量
clay slime 黏土(型)细泥
clayey material 黏质材料,含黏土的材料
clayey ore 黏土质矿石
clay-forming material 泥质矿物
clay-type constituent 黏土型成分[组分]
clay-type minerals 黏土类型矿物类
clean air 清洁[干净]空气
clean air plenum 干净空气压入式通风
clean gravel 纯净砂砾
clean ore 洁净矿石
clean out grate 清扫格条
clean sand 清洁砂
clean separation 精选,完全[彻底]分选
clean step 精选阶段[步骤]
clean up hopper 清扫漏斗
clean zinc sulfide concentrate 纯净硫化锌精矿
cleaned coal 精选煤
cleaned scheelite concentrate 精选(的)白钨精矿
cleaner 精选浮选机
cleaner cationic flotation 阳离子精选
cleaner circuit 精选回路
cleaner concentrate withdrawal rate 精选(段)精矿回收速度
cleaner feed pump 精选给矿泵
cleaner feed pump box 精选给矿泵池[箱]
cleaner flotation 精选浮选
cleaner flotation circuit 精选浮选回路
cleaner flotation density 精(选浮)选浓度
cleaner flotation machine 精选浮选机
cleaner flotation section 精选浮选段
cleaner flotation tailings 精选尾矿
cleaner grade 精选品位
cleaner grade enhancement 精矿[选]品位的提高
cleaner jig middlings 跳汰精选中矿
cleaner middlings 精选中矿
cleaner operating 精选作业
cleaner performance 精选指标
cleaner plant 精选车间
cleaner reject 精选尾矿
cleaner stage 精选段
cleaner tailings 精选尾矿
cleaners 精选槽
cleaning action 精选作用

cleaning stage 精选次[段]数
cleaning stages 精选段[次]数
cleaning step 精选段
cleaning system （重介质）净化[清扫]系统
cleaning technique 精选技术[方法]
clear water 清水
clearance dimension 间隙[净空]尺寸
cleavage experiment 分裂[解理]试验
cleavage position 解理位置
cleavage technique 解理法
clerical staff 文书[事务]人员
client 客户,买方,委托[当事,交易]人
client billing 委托方[买方]记账[编制账单]
client representative 买[甲]方代表
climatic condition 气候条件
climatological information 气候资料
cloakroom 衣帽间
clocklike 准时的,准确的,正确的
clocklike dial 钟表状刻度盘[表盘]
close association 紧密共生
close co-operation 密切协作
close side discharge setting 最小排矿口（宽度）调节,紧边排矿口调节
close side setting 紧边排矿口调节
close-circuit testing 闭路试验
close-circuited autogenous pebble milling 闭路自动砾磨磨矿（法）
closed circuit 闭路
closed circuit analogue calculation 闭路模拟计算

closed circuit continuous (at steady state) behavior 闭路连续流程(稳态)特性
closed circuit crushing 闭路破碎
closed circuit grinding - classification stage 闭路磨矿-分级段
closed circuit operation 闭路作业[操作]
closed circuit pilot scale grinding 半工业性闭路磨矿
closed circuit screen 闭路筛
closed circuit screen efficiency 闭路筛分效率
closed circuit screen productivity 闭路筛分生产能力
closed circuit system 闭路系统
closed circuit television 闭路[工业]电视
closed circuit television camera 闭路电视摄像机
closed circuit TV monitoring 闭路电视监控
closed horizontal type 密闭卧式
closed loop system 闭环系统（药剂输送）
closed mill 密闭磨机
closed mill water system 选矿厂水循环系统
closed pan vibrating conveyer 封闭槽形振动输送机
closed settings 窄[紧]边开口的调节,最小开口调节

closed stockpile 有盖矿堆
closed-circuit ball mill 闭路球磨
closed-circuit grinding cyclone 闭路磨矿旋流器
closed-loop process water 循环生产[工艺]水
closed-packed film 密集膜
closely packed vertically orientated monolayer 密集垂直定向单分子层
closely sized feed 经严格分级的给矿
close-packed monolayer 密集单分子层
close-packed monolayer absorption 密集单分子层吸附
close-side adjustment 最小[紧边]排矿口调节
close-side setting 最小[紧边]排矿口开度[宽度]设定
close-to-the-rated power draw 接近额定功率牵引[汲取,消耗]
closing cost 结算费用
closing remarks 结束语
cloth specification 滤布技术规格[条件]
cloth support bar 筛网支杆
cloud of neutral atom 中性原子色斑
cluster of peaks 波峰群
co- 一起,共同,联合
coagulant 凝聚剂
Coagulant SX SX 絮凝剂(丙烯酰胺衍生物的阳离子型)
coagulating agent 凝聚剂

coagulation 凝结(物),凝聚,凝固(作用),絮[胶]凝,聚集
coal cleaning 选[洗]煤
coal cleaning method 洗煤方法
coal combustion system 煤炭燃烧系统
coal company 煤业公司
coal division 煤炭处[部]
coal fired steam boiler 烧煤蒸汽锅炉
coal flotation 煤浮选
coal flotation concentrate 煤浮选精矿,浮选精煤
coal handling and combustion system 煤炭搬运和燃烧系统
coal handling system 煤炭搬运系统
coal industry 煤炭工业
coal jig 选煤跳汰机
coal mining 煤矿业
coal preparation 选煤
coal processing 选煤
coal seam 煤层
coal separation 选煤
coal tar 煤焦油
coal weigh-belt feeder 称重带式给煤机
coal yield 煤实[回]收率,煤产量,出煤量
coalesce 聚结[合],凝聚[结],结[联,接,组]合,合并
coal-fired boiler 燃煤锅炉
coal-fired kiln 燃煤窑
coalification 煤化(作用)
coarse aggregate 粗集[骨]料,粗粒集

料(不能通过1/4英寸筛的)　排出口

coarse coal　粗粒煤
coarse coal circuit　粗煤流程
coarse crushing cavity　粗(物料)破碎腔
coarse crushing plant　粗碎车间(包括粗中碎)
coarse dense mineral　粗粒重矿物
coarse filtering medium　粗粒过滤介质
coarse fish　杂鱼
coarse fraction of cone crusher product　圆锥破碎机产品的粗部分
coarse grained ore　粗粒矿石
coarse grind　粗磨
coarse less dense mineral　粗粒轻矿物
coarse light particle　粗的轻颗粒
coarse native piece　粗粒自然金块,粗自然金粒
coarse ore bin　粗矿仓
coarse ore bin inventory level　粗矿仓库存矿量[料位]
coarse ore reclaimer　块矿取料机
coarse ore stockpile　块[粗]矿堆场
coarse ore storage　粗矿仓
coarse ore storage area　粗矿堆场
coarse particle separation　粗颗粒分离
coarse plant　粗碎厂[车间]
coarse rake fraction　粗粒返砂部分
coarse red mud residue　粗红泥废渣
coarse screenings　粗(筛)出物料(大于4英寸)
coarse solid discharge lip　粗粒固体料

coarse-grained free-milling　粗粒嵌布易选的
coarse-grained sulphide　粗颗粒硫化矿物
coarsely crystalline component　粗粒结晶成分[组分]
coarser grained ore　较粗粒矿石
coarse-to-fine ratio　粗细比(例)
cobber　磁选机,手选工
cobber wet drum magnetic separator　粗粒[拣大块]湿式圆筒磁选机
co-chairman　联合主席
coco matting　椰子树衬垫(洗矿槽的)
cocoa nut　可可核
coconut fat acid　椰子油脂肪酸(可用作捕收剂)
coconut oil　椰子油
code address　电报挂号,编码地址
coded form　编码形式
coded unit　编码量
coding　编码,编程
coding model　编码模型
coding relation　编码关系
Coe-Clevenger method　科-克莱文杰法(确定浓缩机有效分离面积的实验室试验方法)
co-editor　合作[联合]编辑
coefficient of equation　方程式系数
coefficient of expansion　膨胀系数
coefficient of resistance　阻力系数
coffe mill　小盘磨机

Coherex 一种树脂乳浊液黏结剂（品名,可用于处理尾矿的黏结剂）
cohesive powder 黏结粉料
cohesive resistance 黏结阻力
co-ions 共同离子
coke bin 焦炭仓
coke formation 焦炭生成
coke oven by-product 焦炉副产品
cold climate 寒冷气候
cold cyanide assay test 冷氰化物分析试验
cold-wound spring 冷煨弹簧
collateral 抵押品
colleague 同事
collected dust 收集到的粉尘
collecting belt conveyor 集矿胶带运输机
collecting effect 捕集作用[效应,影响]
collecting power 捕集能力
collection efficiency 集尘效率
collection of process stages 若干工艺阶段集合体
collection of science essays 科学论文集
collection process 捕集过程
collection sump 集水[料]仓[池]
collection tank 收集箱
collective-selective flotation 混合优先浮选
collectivity 集体[合]性,聚集性
collector action 捕收(剂)作用

collector adsorption 捕收剂吸附
collector adsorption process 捕收剂吸附过程
collector anion 捕收剂阴离子
collector belt 集矿皮带运输机
collector chain length 捕收剂链长
collector coated mineral surface 被捕收剂覆盖的矿物表面
collector coated particle 浮选剂覆盖颗粒
collector coating 捕收剂覆盖层[膜]
collector combination 捕收剂联合剂,联合捕收剂
collector concentration 捕收剂浓度
collector consumption 捕收剂消耗
collector cost 捕收剂成本
collector ion 捕收剂离子
collector ionization 捕收剂离子化
collector solution 捕收剂溶液
collector species 捕收剂物质
collector structure 捕收剂结构
collector system 捕收(剂)制度
collector-coated mineral 捕收剂覆盖的矿物
colliding particle 碰撞颗粒
collinite 无结构凝胶体
collision event 碰撞事件[活动]
collision probability 碰撞概率
collision process 碰撞过程
collision rate 碰撞率
collision-adhesion hypothesis 碰撞-黏附假说

colloid chemistry 胶体化学
colloid ferric hydroxide 胶体氢氧化铁
Colloid Nos 1-5 絮凝剂(动物蛋白质衍生物)
Colloid YG-13 絮凝剂(氧化的动物蛋白质衍生物)
colloidal iron 胶体铁
colloidal material 胶质[体]物料
colloidal silica 胶体二氧化硅
colloidial slime 胶细泥
column flotation 浮选柱浮选
column flotation machine 浮选柱
column flotation operation 浮选柱浮选操作[生产]
column heading 栏[列]标题
column leaching test 浸出柱试验
column machine 浮选柱
column method 萃取塔[柱]法
column mounted pump series 装于柱架上的泵系列
column of resin 树脂交换柱,离子交换柱
columnar arrangement 柱体排列
combination 组合,结合,联合,化合
combination amalgamation-flotation method 混汞浮选联合法
combination container/break-bulk vessel 集装箱与杂货合用船
combination mode 联合方式
combination of operating variable 操作变数组合
combination of process parameters 工艺参数组合
combination of reduction 破碎作用的结[组]合
combination of unit operations 单元作业组合
combination of variable level 变量位数组合
combination ship 客货船
combination track-truck scale 火车卡车联合地中衡
combination truck and railroad scale 卡车和火车联合计量装置
combined flotation 联合浮选
combined method 联合选矿法
combined mixer-settler 联合混合机沉淀机(法),联合搅拌机澄清机(法)
combined sample 混合样
combined setting and feed rate system 排矿口宽度和给矿量联合(控制)系统
combined sluicing-jigging process 溜槽跳汰联[组]合工艺[方法]
combined tailings 混合尾矿
combining change 联[综]合变化
combustible dust 可[易]燃粉尘
combustible matter 可燃物质
combustible mineral substance 可燃矿物质
combustion gas 燃烧气体
combustion reaction 燃烧反应
comfortable working condition 舒适工作条件[环境]
commencement of flotation 浮选起点

commencement of production 投产,开始生产
comment 评论,意见,注解
commerce and business administration 商业与企业管理
commerce paper 商业证券
commercial 商业[品,用,务]的,贸易的,工业(用)的,工厂的,(能)大批生产的,商品化的
commercial application 工业应用
commercial autogenous grinding plant 工业自磨磨矿厂
commercial basis 商业基础
commercial cell 工业浮选槽
commercial deposit 工业矿床
commercial design 工业设计
commercial equipment 工业设备
commercial fine crusher 工业细碎机
commercial froth flotation mill 工业泡沫浮选厂
commercial frother 工业[商品]起泡剂
commercial grade 商业[工业]品位
commercial grade hubnerite 商品级钨锰矿
commercial gyratory crusher 工业旋回破碎机
commercial hydrocarbon mixture 工业碳氢化合物
commercial installation 工业设备[设施,装置]
commercial interest 商业利益,工业价值
commercial laboratory 工业实验室
commercial laboratory personnel 工业实验室试验人员
commercial letter of credit 商业信用证
commercial machine 工业设备[浮选机]
commercial manufacture 工业制造
commercial market (商业)市场
commercial media 工业介质
commercial mill 工业磨机
commercial mill design 工业磨机设计
commercial mill selection 工业磨机选择
commercial mill size 工业磨机尺寸[规格]
commercial milling plant 商业性选矿厂
commercial operating data 工业生产[运转]资料[数据]
commercial paper 商业证券
commercial paper dealer 商业证券经纪人
commercial paper operation 商业证券交易
commercial pellet plant 工业团粒[团矿]车间
commercial plant 工业(性)选矿厂
commercial practice 商业惯例[实务]
commercial process 工业选矿法
commercial processing plant 工业加

工厂

commercial reagent 工业用药剂

commercial scale grinding circuit 工业规模磨矿回路

commercial scale process 大规模生产法

commercial shape 商业形态

commercial size 工业规模[尺寸]

commercial stage 工业阶段

commercial testing of mineral and rock 矿物岩石的工业试验

commercial testing organization 工业性试验机构

commercial use 工业使用

commercial vehicle 商用车辆

commercial weight 商业重量

commercially designed unit 工业设计装置

commercially important deposit 重要工业矿床

commercially perfect screening 工业上理想的筛分

commercially-designed unit 工业设计装置

commercial-scale 工业规模的,大规模的

commercial-sized mill 工业规格[尺寸]磨机

comminution circuit design 破磨流程[系统,回路]设计

comminution circuit dust collection system 破磨[碎磨]回路集尘系统

comminution device 破碎[碎磨]装置

comminution energy 碎磨能量

comminution energy utilization efficiency 碎磨[破碎]能量利用效率

comminution engineer 破磨工程师

comminution model 碎磨模型

comminution plant design calculation 碎磨厂设计计算

comminution project 破磨工程项目

comminution step 破磨段[工序]

comminution system 粉磨[碎磨]系统

comminution system design 破磨系统设计

comminution unit 磨矿[研磨]设备,破碎设备

commissioning 试车,试运行

commissioning and startup cost 试车投产费(用)

commissioning cost 试车费用

commissioning date 投产日期

commissioning stage 投产阶段,交工试运转阶段

commissioning test run 投料试生产

commitment fee 承约费

commodity exchange 商品交易所

common 共用区,共同,共有,公共,普通,平常,低贱,通俗

common cost base 普通[共同]成本基值[础]

common cost index value 通用费用指数值

common figure 一般数值

common files of data　数据外存
common ion　共同离子
common mining analysis　常用的矿业分析法
common modifier　普通[常用]调整剂
common sense　常识
common stock　普通股票
commonly used formula　常用公式
communication　通讯,通信,交流
communication of information　资料交流
communication problem　沟通问题
communication service　通讯工作,联系业务
communication system　通讯系统
compact device　紧密装置
compact installation　紧凑安装
compact mill　紧凑型选矿厂(配置)
compact plant site　紧密的选矿厂厂址
compact plant size　紧密选矿厂规模
compacted layer　压[夯]实层
compacted natural bentonitic-type clay bed　压实的天然膨润土床层
compacted soil　压[夯]实土
compaction　压缩[紧],紧密,严密
compaction test　压紧试验
compaction test procedure　压紧试验方法
compactor　压实器[机],夯实机,压土机
company director　公司董事
company engineer　公司工程师

company experience　公司经验
company financial statement　公司财务报表
company policy　公司政策
company research equipment　公司研究设备
company research facilities　公司研究设施
company staff　公司人员
comparable energy level　可比[类似]的能量水平
comparable grinding data　可比磨矿数据
comparable sizing device　类似筛分[分级]设备
comparative bid analysis　报价对比分析,对比的报价分析
comparative cost　比较成本
comparative frothing test　起泡比对[较]试验
comparative test　比较[对]试验
comparison alternatives　方案比较
comparison of data　数据对比
comparison test　比对试验
compartment area　室面积(跳汰机的)
compartmented car　间隔卡车
compensating balance　补偿余额
compensation agreement　抵偿协定
compensation program　补偿程序
compensation trade　抵偿贸易
compensation trade agreement　补偿贸易协议

competency test 磨矿能力试验
competent 有能力[能干,胜任,合格]的
competent department 主管部门
competent geologist 主管地质人员
competent rock 稳固岩石
competitive bidding 竞争投标
competitive paint 有竞争力的涂料
competitive price (投)标价,竞争价格
compiler 自动编码[程序编制]器,编译程序
compiler language 编译程序语言
complaint 投诉,抗议,不满[平],疾病,失调,不适
complete analysis equipment 整套分析设备
complete analytical equipment 整套分析设备
complete batch weighing system 成套批量称重系统
complete chemical analysis 化学全分析
complete circuit 整个系统
complete closed-water preparation plant circuit 全部回水选煤厂流程
complete cyanide plant 全盘氰化厂
complete design 整套设计
complete dust tight machine 全密封防尘设备
complete equipment 成套设备
complete flotation 完整浮选
complete flotation route 全浮选法

complete monolayer coverage 全部单层覆盖
complete on-off cycle 完整开关周期
complete plant optimization 全(选矿)厂最优化
complete process cost analysis 全过程费用分析
complete reduction system 整套破碎系统
complete set 整套
complete software package 整套软件包
completely engineered unit 整套工程设备
completely-automated system 全自动化系统
completion date 完成日期
complex 聚合[复合,集合,螯合]物
complex acid 络酸
complex base metal ore 复杂贱金属矿石
complex circuiting 复杂流程
complex circuitry 复杂流程
complex copper-lead-zinc ore 铜铅锌复合矿
complex deposition 多金属矿沉积作用
complex equilibrium 络合平衡
complex ion 络离子
complex laboratory 综合实验室
complex solution 复杂溶液
complex steel 多元[合金]钢
complex task 复杂工作[任务]

complexant 配位[络合]剂
complexation 络合(作用)
complexed metal species 络合金属化合物类(药剂)
complexity 复杂性
complicated aggregate 复杂集合体
complicated problem 复杂问题
component 成分,组[元]件,分量,分支[图]
component alignment 部件调整[组合,调直,校正]
component arrangement 构件布置
component availability 部[组]件利用率
component combination 构件组合
component fault index 部件故障指数
component flow 分量流
component operating characteristic 部件[组件]操作特性
component reliability 零部件可靠性
component velocity 分速度
composite constant 复合常数
composite Hamaker constant 哈梅克复合常数
composite impeller-propeller design 混合的叶轮螺旋桨设计
composite mineral particles 连生体矿物颗粒
composite particle 连生体[颗粒]
composite sample 综合试样(混合后)
composition analysis 成分分析
composition measurement 组合测量

compound interest 复利
compound water cyclone 复式水力旋流器
compounding 复利计算
comprehensive assessment 综合评价
comprehensive bibliography 综合参考资料[书面评述,文献介绍]
comprehensive companion program 综合伴随程序
comprehensive experiment 综合[全面]试验
comprehensive metallurgical study 综合冶金[选矿]研究
comprehensive physical analysis 综合物理分析
comprehensive physical test 综合[全面的]物理试验
comprehensive publication 综合性刊物
comprehensive sampling theory 综合采[取]样理论
comprehensive study 综合研究
comprehensive term 词义广泛的名词
comprehensive test 综合试验
comprehensive test plan 综合[扩大]试验计划
comprehensive treatment plan 综合处理计划
comprehensive utilization 综合利用
compressed air filter 压风过滤机
compressed air header 压缩空气联箱
compressed air line 压缩空气管线

compressed air system 压缩空气系统
compression 挤压,压缩,压榨,加压
compression crusher 挤压式破碎机
compression crusher type 挤压式破碎机类型
compression induced tensile 挤压感生张力
compression zone 压缩带
compressive strength 压缩强度
compressive strength test （抗）压强（度）试验
comprised control system 构成式控制系统
compromise 妥协,让步,折中(方案,办法)
compromise model 折中模型
computation 计算(技术,结果),估计
computation procedure 计算程序
computation sheet 计算表格
computational algorithm 计算算法
computational detail 计算细节
computational effort 计算工作量
computational procedure 计算程序
computational scheme 计算方案
computational sequence 计算程序
computational structure 计算结构
computational technique 计算方法［技术］
computed circulating load 计算的循环负荷
computed stream function 计算流函数
computer analysis 计算机分析

computer and teletype system 电子计算机与电传打字系统
computer application 计算机应用
computer control 计算机控制
computer control consideration 计算机控制研究
computer control of grinding 磨矿电子计算机控制
computer control strategy 计算机控制对策
computer control system 计算机控制系统
computer data logging and control system 计算机数据记录和控制系统
computer expert 计算机专家
computer facilities 电子计算机设施
computer hardware 计算机硬件
computer manufacturer 计算机制造厂
computer memory requirement 计算机内存要求
computer plant design program 计算机选矿厂设计程序
computer plotter 计算机绘图仪
computer print-out 计算机打印输出
computer selection 计算机选型
computer service group 计算机服务组
computer simulation 计算机模拟
computer software 计算机软件
computer system supplier 计算机系统供货方
computer terminal 计算机终端
computer treatment 计算机处理

computer vender　计算机供方
computer-aided design　计算机辅助设计
computer-aided design system　计算机辅助设计系统
computer-assisted design　计算机辅助设计
computer-assisted modeling tool　计算机辅助模型工具
computer-assisted off-line optimization　计算机辅助离线最优化
computer-assisted on-line optimization　计算机辅助在线最优化
computer-assisted on-line optimization strategy　计算机辅助的在线最佳化对策
computer-based approach　借助于计算机的方法,应用计算机计算法
computer-based process control system　基于计算机的过程控制系统
computerized calculation program　计算机化计算程序
computerized energy management system　计算机化能源[量]管理系统
computerized plant flow sheet analysis program　选矿厂流程分析计算机程序
computerized program　计算机化程序
computer-oriented personnel　专攻计算机的人员
computer-related ore estimation technique　有关计算机估算矿量技术,应用计算机计算矿量技术
computer-related ore reserve estimation technique　有关计算机矿石储量估计法[技术]
concave configuration　固定锥体构造形式[外形],圆锥破碎机腔部结构
concave liner　固定锥内衬
concave support method　固定锥体支撑[承]方法
Concenco 999 table　康森科999型摇床(悬挂式多层摇床,美国)
concentrate analysis　精矿分析
concentrate assay　精矿试样品位
concentrate bin　精矿仓
concentrate collector ring　精矿收集环
concentrate discharge outlet　精矿排出口
concentrate grade　精矿品位
concentrate handling　精矿处理
concentrate inventory　精矿库存
concentrate leaching　精矿浸出
concentrate load-out　精矿装运
concentrate loss　精矿损失[耗]
concentrate marketing　精矿销售
concentrate marketing condition　精矿销售情况
concentrate moisture content　精矿含水量,精矿水分含量
concentrate output table　精矿产量表
concentrate pipeline　精矿管道
concentrate product　精矿产品
concentrate production　精矿生产

[产率]
concentrate production schedule 精矿产量[生产]表
concentrate receiving bin 精矿受矿仓
concentrate recirculation 精矿循环
concentrate recovery 精矿回收率
concentrate reduction 精矿还原
concentrate removal 精矿排除
concentrate removal rate 精矿排除率
concentrate roasting 精矿焙烧
concentrate separation matrix 精矿分离矩阵
concentrate size 精矿粒度
concentrate slurry holding tank 精矿浆槽[池]
concentrate specification 精矿技术规格
concentrate splitter 精矿分流器(螺旋选矿机的)
concentrate storage bin 精矿储存仓
concentrate stream 精矿流
concentrate thickener overflow pump 精矿浓密机溢流泵
concentrate thickener underflow pump 精矿浓密机底流泵
concentrate trough 精矿槽
concentrate vent fraction 精矿出口部分
concentrated automatic control 集中自动控制
concentrated stock solution 浓缩储存溶液
concentrated stream 集中矿流[料流]
concentrate-storage building 精矿(储存)仓[库]
concentrating circuit 选矿流程
concentrating cost 选矿费用
concentrating plant 选矿厂
concentrating process 选别方法[工艺]
concentrating table 淘汰盘,精选摇床
concentrating tray 精选槽
concentration criterion 分选判据
concentration limit 浓度限制
concentration of scheelite 白钨选矿
concentration of solids 固体浓度
concentration of suspension 悬浮液[体]浓度
concentration polarization 浓度[差]极化
concentration procedure 选别方法[作业]
concentration process 分选[选矿]过程
concentration profile 浓度分布[曲线]图
concentration range 浓度范围
concentration rate 选矿[富集]比
concentration ratio 选矿[富集]比
concentration zone 浓缩带
concentrator 选矿厂
concentrator arrangement 选矿厂配置
concentrator building 选矿厂厂房
concentrator facilities flowsheet 选矿

厂设施工艺流程
concentrator monitoring system 选矿厂监控系统
concentrator office 选矿厂办公室
concentrator operating account 选矿厂生产费用账
concentrator operating equipment 选矿厂生产设备
concentrator operating statistics 选矿厂生产统计
concentrator operation 选矿厂操作
concentrator organization 选矿厂组织[编制]
concentrator payroll 选矿厂工资总额
concentrator performance 选矿厂指标[成绩]
concentrator process control 选矿厂工艺过程控制
concentrator process description 选矿厂工艺(过程)描述[说明]
concentrator proper 选矿厂本身
concentrator services 选矿厂维修[护]
concentrator site 选矿厂厂区
concentrator staff 选矿厂员工
concentrator stream 选矿厂物料流
concentrator supervisory cost 选矿厂管理费
concentrator variable 选矿厂变量
concentrator work force 选矿厂劳动力
concentrator's daily throughput 选矿厂日通过[处理]能力
concentrator's normal and routine maintenance 选矿厂正常例行维修
concentrator's operating water expense 选矿厂生产用水费用
concentric cubical shell 同心立方形壳
concept of comminution 碎磨概念
concept phase 初步[草图]设计阶段
conceptual design 方案[概念]设计
conceptual engineering stage 概性设计阶段
conceptual framework 概念结构
conceptual model 概念模型
conceptual phase 概念[构思,方案设计]阶段
conceptual process system 概念性工艺系统
conceptual structure 概念结构
concise specification 简明技术规格[说明书,条例]
conclusions and recommendations 结论与推荐[建议]
conclusive information 结论性资料
concomitant phenomena 伴随现象
concrete aggregate 混凝土骨料
concrete aggregate plant 混凝土集料厂
concrete aqueduct 混凝土导水管
concrete construction 混凝土结构[施工]
concrete lead-out silo 混凝土装载筒仓
concrete pipeline 混凝土管线[道]

concrete tank 混凝土水池
condensation process 冷凝过程
condensed information 精简[简要]数据[资料]
condensed specification 简明技术规格
condensing agent 冷凝[缩合,凝聚]剂
condition of grinding 磨矿条件
condition precedent 先决条件
conditional mass density function 条件重量密度函数
conditional resource 条件性资源
conditioned pulp 搅拌后的矿浆
conditioner cell 搅拌槽
conditioner motor 调浆[搅拌]槽马达
conditioner power consumption 搅拌槽功率消耗
conditioning chamber 调节室
conditioning period 调整[搅拌]期
conditioning process 调整过程
conditioning reagent 调整剂
conditioning stage 调整作业[步骤,阶段]
conditioning tank 调整槽
conditioning temperature 搅拌[调整]温度
conditioning time 调和[搅拌]时间
conditioning unit 调浆装置,搅拌设备
conditioning-power consumption curve 搅拌功耗曲线
conductance probe 电导探针[头]
conductance probe level signal 电导探针[头]液位信号

conducting liquid 导电液
conducting material 导电[体]材料
conducting phase 导电相
conduction band edge 传导带限界
conductive concentrate 导电精矿
conductive dust 导电粉尘
conductive liquid 导电液
conductor re-treat configuration 导体物料再[精]选形式
cone and quartering 堆锥四分缩样法
cone and quartering technique 堆锥四分(缩样)技术
cone angle 锥角
cone bottom tank 锥底槽(罐)
cone crusher capacity chart 圆锥破碎机生产能力表
cone crusher performance 圆锥破碎机性能
cone distributor 分配锥(圆锥选矿机的)
cone separator 圆锥选矿机
cone type secondary crusher 圆锥形二次破碎机
cone wall 锥体壁
coned disk spring 碟簧,碟锥形弹簧
cone-shaped lower part 锥形下部
confidence 置信度,信心[任],自信
confidence interval 置信区间
confidence level 置信水平,(数值)可信度
confidence limit 置信界限[度]
configuration 布局,格局,构形,组态

configuration matrix 组态矩阵
configuration of cell 槽子形状[结构]
configuration of grinding circuit 磨矿流结构,磨矿回路配置
configuration programming 结构程序设计
configuration set 构形[配置]集
configurational entropy 构型熵
configurational structure 构型,外形[形状]结构
confining face 封闭面
confirmatory test 验证试验
confirmatory work 确定的工作
confiscate 没收
conflicting comminution theory 矛盾的破碎[碎磨]理论
conflicting data 矛盾(的)数据
conglomerate 凝聚成团,砾岩,成团的,跨行公司,联合企业
Congo red 刚果红(染料,可用作抑制剂)
conical blade 锥形叶片
conical button 锥形(金银)锭[块]
conical ended mill 锥端[形]磨机
conical flask 锥形烧瓶
conical hopper 锥形漏斗
conical ledge 锥形突出部分,凸耳[缘]
conical pebble mill 锥形砾磨机
conical section 锥形段,锥体部分
coning and quartering 堆锥四分取样
conjugate direction 共轭方向
conjugate direction method 共轭方向法
conjugate direction technique 共轭方向技术[方法]
connected and consumed power 安装和消耗功率
connecting box 连接箱(浮选机的)
connecting jumper 跨接线连结[接]
Conoco C-50 十二烷基苯磺酸钠(品名,可用作起泡剂、捕收剂)
conservation group 水土保持组,森林[自然]保护组
conservation law 守恒定律
conservation of energy 能量守恒[不灭]
conservative estimate 保守估计
conservative figure 保守[险]数字
conservative result 保守结果
consignment 托运,交付,寄售[销],所托运的货物,代销货物
consistency 一致性,稠度,相容性,稳定性
consistency water (调节)浓[稠]度水,稳流水
consistent error 常有[一贯]误差,常差
consistent feed 稳定给矿[料]
consistent naturally-sized material 一致的自然粒度物料
consistent unit 一致部件[单位]
Consol 骨胶衍生物(品名,可用作絮凝剂)
console CRT terminal 控制台 CRT

终端
console keyboard 控制台键盘
consolidated balance sheet 资产负债汇总表
consolidated current asset 汇总的现有资产
consolidated statement of earnings 收益汇总表
consolidating pressure 固结[压实]压力
consolidation 巩固,固结
consolidation of slime 矿泥固结[化]
consortion 结合,联盟,企业联营,财团
consortium 财团,联合,合伙
constant characteristic 特性常数
constant characteristic of particular material 特定物料常数特性
constant coefficient 恒定系数
constant concentration of neutral electrolyte 恒定的中性电解质浓度
constant critical concentration 恒定临界浓度
constant current charge curve 恒定电流电荷曲线
constant current polarization 恒定电流极化
constant current polarization curve 恒定电流极化曲线
constant current reduction and oxidation period 恒定电流还原和氧化周期
constant dimensionless flow number 恒定无维量流量数

constant dollar value 不变的美元价值
constant feed rate control loop 恒定给矿量控制回路
constant feeder 定量给料机
constant feedrate control 恒定给矿量控制
constant ionic strength 恒定离子强度
constant light intensity 恒定光强度
constant load 恒定充填(率)[载荷]
constant of proportionality 比例常数
constant of Van der Waals attraction 范德华引力常数
constant pitch 不变螺距
constant polarizing current 恒定极化电流
constant potential oxidation process 恒定电位氧化过程
constant potential plateaus 恒定电位平缓曲线
constant power intensity 不变功率强度
constant power number 不变功率数
constant pressure permeameter 恒压渗透仪
constant pressure principle 恒压原理
constant ratio 恒定比率
constant slope 恒定斜率
constant specific energy criterion 恒定的比能标准,恒定的单位功耗标准
constant speed 恒速
constant speed conveyor 恒速运输机
constant speed feeder belt 恒速给矿机

皮带
constant temperature 恒温
constant temperature unit 恒温装置
constant term 常数项
constant time 恒定时间
constant value 常数值
constant-current polarization 恒定电流极化
constant-current polarization study 恒定电流极化研究
constant-pressure flow controller 稳压流量控制器
constituent ion 组分离子
constituent mineral 组成[成分]矿物
constituent part 组成部分
constitutional water 结构水
constrained region 约束领域
constraint 约束,限制
construction 建筑,建造,建设,构造,架设,结构
construction aggregate 建筑骨料
construction area 建筑场地
construction contract 施工合同
construction contractor 施工承包商[方]
construction cost 建设[建筑,工程]费
construction equipment 工程设备
construction items 建筑项目
construction joint 施工[建筑]缝
construction management 施工管理
construction manager 施工经理
construction material 建筑材料

construction member 构件
construction of diagram 制图
construction of ventilation hood 通风罩的制作
construction period 建设[施工]期间
construction phase 建设阶段
construction plan 施工布置[平面]图
construction program(me) 建设[施工]计划
construction schedule 建设[施工]计划
construction service 施工服务[业务]
construction site 工地
construction specification 施工[建筑]说明
construction stage 建设[施工]阶段
construction work 基建工作
consultant 顾问,咨询人员[单位],查阅者
consultant organization 咨询机构
consulting 咨询
consulting cost 咨询费
consulting engineer 顾问工程师
consulting fee 咨询费
consulting firm 咨询公司
consulting metallurgist 顾问冶金师
consulting work 顾问工作
consumable material 可消耗[消耗性]材料
consumable parts 消耗部件
consumable stores cost 消耗品费用
consumables 消耗品

consumed power 消耗功率
consumer countries 消费国
consumer of concentrate 精矿用户
consumer price index 消费物价指数
consumer reporting group 消费者通讯组
consuming country 消费国
consumption rate 消耗率[量]
contact angle 接触角
contact angle data 接触角数据
contact angle measurement 接触角测量
contact angle work 接触角研究工作
contact area 接触面[区域]
contact curve 接触曲线
contact device 接触性装置[仪表,设备]
contact duration 接触时间
contact metamorphic rock 接触变质岩
contact pressure 接触压缩[力]
contact probability 接触几率
contact probe 接触式探头
contact solution with ion exchange resin 离子交换树脂接触溶液
contact switch 触点开关
contact time 接触[作用]时间
contact type level control 接触[触点]式水平[液面,料位]控制
contact-metasomatic deposit 接触交代矿床
contactolite 接触变质岩
contactor 接触器

contacts 联系人,联络方式
container 集装箱
container roll way 集装箱滚装方式
container ship 集装箱船
containerization 集装箱化
containerized shipment 集装箱货运[装运]
containment problem 遏制问题
contaminant 沾染[污垢,掺和]物,杂质,污染物质
contaminated water 污水
contaminating metal 杂质金属
Contekt 一种烷基磺酸盐
contemplated strategy 设想策略
contemporary 同时代的人,同辈,同时期的东西,当代的,同时代的
Continental Divide 大陆分水岭
contingency 偶然[意外]事故,意外费用,不可预见费,应急费,临时费
contingency figure 不可预见费数值
continuous analysis 连续分析
continuous assay data 连续分析数据
continuous circulating oil system 连续循环油系统
continuous counter-current decantation calculation 连续逆流倾析计算法
continuous crushing action 连续破碎作用
continuous cylindrical core of rock 连续柱状岩芯
continuous duty service 连续负荷运转[行]

continuous experiment 连续试验
continuous flotation 连续浮选
continuous flotation experiment 连续浮选试验
continuous flotation process 连续浮选过程
continuous flotation test 连续浮选试验
continuous flow cell 连续流通式矿样盒
continuous flow solvent extraction circuit 连续料流溶剂萃取流程
continuous grinding scale-up 连续磨矿按比例扩[增]大
continuous laboratory test 连续实验室试验
continuous mill 连续磨机
continuous mill input-output relationship 连续磨机输入-输出关系式
continuous multi-cell machine 连续多槽浮选机
continuous nature 连续性
continuous operation 连续操作[作业]
continuous phase 连续相
continuous plot 连续曲线图表
continuous process model 连续过程模型
continuous recycle mode 连续返回方法[式]
continuous sampler 连续取样机(两组或三组破碎缩分的取样机组)
continuous smelting 连续熔炼
continuous strake 连续作业溜槽[选矿槽]
continuous sweeping 连续扫描
continuous time of operation 连续运转时间
continuously-operated vacuum filter 连续作业真空过滤机
contour diagram 等高[值]线图
contract change order 合同变更单
contract insurance 合同保险
contract labor 合同工
contract party 合同当事人
contract rate 合同定额[值]
contract requirement 合同要求
contract worker 合同工
contractor's equipment 承包方设备
contractor's estimate 承包者估算
contractual commitment 按合同规定所承担的义务
contributory agent 促进[媒]剂
control action 控制作用
control action diagram 控制作用图
control algorithm 控制算法
control box 控制箱
control circuit 控制回路
control concept 控制概念
control criterion 控制准则
control duty 控制功能
control engineer 控制工程师
control engineering 控制工程
control equipment 控制设备
control feasibility 控制灵活性
control feature 控制特点

control fluctuation 控制波动
control implementation 控制实施[执行]过程
control instrument 控制仪表
control laboratory 检[化]验室
control loop 控制回路
control matrix 控制矩阵
control measurement signal 控制测量信号
control module 控制模数
control objective 控制目标
control of executive routine 执行程序控制
control parameter 控制参数
control philosophy 控制原理
control probe 控制探头[针]
control room design 控制室设计
control scheme 控制方案
control select logic 控制选择逻辑
control service 控制服务
control signal 控制信号
control standpoint 控制观点
control strategy 控制对策[策略]
control structure 控制结构
control system design 控制系统设计
control tank 控制槽
control unit 控制单元[装置]
control valve 控制阀
controllable electrochemical system 可控(的)电化系统
controllable feed rate 可控给矿量
controllable variable 可控变量

controlled condition 控制条件
controlled environment room 具有环境控制的控制室
controlled feeding 控制给矿
controlled variable 受控变量
controller algorithm calculation 控制器的算法运算
controller cycle 控制器周期
controller of interest 有益的控制装置
controller parameter 控制[调节]器参数
controller response 控制器响应
controller unit 控制装置
controller with integral action 带积分作用的控制器
controlling factor 控制因素
controversial issue 引起争论的问题
convection 对流
convenience 方便,便利
convenience of operation 操作方便
convenient unit 适宜[合适]单位,常用单位
conventional backup 传统备份
conventional cleaning 常规[传统]精选
conventional cone crusher 常规[传统]圆锥破碎机
conventional critical path technique 传统统筹[关键线路,主要矛盾线路]技术
conventional crushing and grinding 常规[传统]的破碎磨矿

conventional equipment 常规设备
conventional feed 一般[常规]给矿
conventional gravity separation technology 传统的重选技术
conventional input 常规[传统]输入量
conventional instrument 传统[常规]仪表
conventional means 传统[常规,常用]方法
conventional measurement and control instrument 常规测(量加)控(制)仪表
conventional method 传统方法
conventional porphyry copper flowsheet 常规斑岩铜矿流程
conventional process 传统工艺[方法]
conventional rod mill-ball mill approach 传统的棒-球磨方法[途径]
conventional rougher-scavenger-cleaner configuration 常规的粗-扫-精选流程结构
conventional roughing 常规[传统]粗选
conventional sign 图例,通用符号[标志]
conventional sulfuric acid leaching 常规的硫酸浸出
conventional test 常规试验
conventional three stage crushing circuit 传统三段破碎流程[回路]
conventional three-stage reduction system 传统[常规]三段破碎系统

convergence criteria 收敛判据
convergence criterion 收敛性判别法[标准]
convergence problem 收敛性问题
convergence test 收敛试验
convergent nature 收敛性
converging hopper 收缩漏斗
converging hopper wall 收缩漏斗壁
converging part 收缩部分
converging tunnel 会聚地道
conversational mode 对话方式
conversion 转换,变换,换算
conversion factor 换算系[因]数,变换系[因]数,转换因子
conversion price 兑换价格
converted mill 经改造的选矿厂
converted operating mill 改建的生产[选矿]厂
converter 转炉,吹(风)炉,变流[整流,转化]器
converter slag 转炉渣
convertible shovel 两用铲,正反铲挖土机
converting process 吹炼工艺[过程]
converting reaction 吹炼[转化]反应
conveyer belt run off switch 运输机皮带离轨[跑偏]开关
conveyer belt speed 胶带运输机速度
conveyer cover 胶带运输机盖板
conveying component 运输分力
conveying equipment 输送设备
conveying part 运输部分

conveying section 运输部分
conveying speed 运输速度
conveying system 运输系统
conveyor component availability 运输机部件利用率
conveyor cover 运输机罩
conveyor design program 运输机设计程序
conveyor entry and exit points 运输机进出口处[点]
conveyor equipment 运输机设备
conveyor flowrate 运输机流量
conveyor general arrangement 皮带运输机总体布置
conveyor hardware 胶带运输机金属构件
conveyor hoisting 运输机提升
conveyor loading and transfer operation 运输机装载和转运作业
conveyor loading chute 运输机装载溜槽
conveyor loading operation 运输机装载作业
conveyor pulley center distance 运输机轮中心距
conveyor size selection 运输机规格选择
conveyor system proper 运输机系统本身
conveyor system tonnage throughput 运输机系统吨位通过量
conveyor tail shaft 运输机尾轴

conveyor transfer operation 运输机转运作业
conveyor transfer point 运输机转运点
conveyor tube 运输机隧[通]道(圆形断面)
conveyor's availability index 运输机的[可]利用指数
cooling and scrubbing installation 冷却洗涤装置
cooling tower blowdown 冷却塔排污
coordinate 坐标
coordinate center 坐标中心
coordinate value 坐标值
coordinating agent 配位剂,络合剂
coordination complex 配位络合物
coordination number 配位数
Coppee vacuum cell 科佩型真空浮选机(选煤用)
copper activated 被铜活化的
copper activated sphalerite 铜活化的闪锌矿
copper amine complex 铜胺络合物
copper and molybdenum sulfide extraction 铜钼硫化矿萃取
copper and zinc amine complex 铜和锌的胺络合物
copper and zinc cyano complex 铜和锌的氰络合物
copper base 铜基
copper bearing clay 含铜黏土
copper butyl dithiophosphate 丁基二硫代磷酸铜

copper cement	沉淀[置换]铜(湿法冶炼)
copper contamination	铜污染
copper converting	铜吹炼
copper dithiophosphate phase	二硫代磷酸铜相
copper ethyl dithiophosphate	乙基二硫代磷酸铜
copper ethyl xanthate	乙基黄原酸铜
copper extraction	铜的提取[萃取]
copper extraction plant	铜萃取厂
copper flotation cleaner circuit	铜精选回路
copper hexyl xanthate	己基黄原酸铜
copper isoamyl dithiophosphate	异戊基二硫代磷酸铜
copper manufactures	铜制品
copper m-cresyl dithiophosphate	m-甲苯基二硫代磷酸铜
copper methyl dithiophosphate	甲基二硫代磷酸铜
copper mineralization	铜矿化
copper molybdenum separation process	铜钼分离过程
copper o-cresyl dithiophosphate	o-甲苯基二硫代磷酸铜
copper p-cresyl dithiophosphate	p-甲苯基二硫代磷酸铜
copper phenyl dithiophosphate	苯基二硫代磷酸铜
copper production	铜生产[产量]
copper propyl dithiophosphate	丙基二硫代磷酸铜
copper range	铜矿区
copper reagent	(硫酸)铜药剂
copper refining	铜精炼
copper salt	铜盐
copper scrap	杂铜
copper smelter	炼铜厂
copper surcharge	铜附加费
copper xanthate phase	黄原酸铜相
copper zone	铜矿带
copper-bearing clay	含铜黏土
copper-bearing pyrite	含铜黄铁矿
copper-extraction industry	铜提取工业
copper-iron separation	铜铁分离[选]
copper-lead bulk cleaner concentrate	铜铅混合精选精矿
copper-lead bulk cleaning	铜铅混合精选
copper-lead bulk flotation	铜铅混合浮选
copper-lead bulk-differential flotation	铜铅混合-优先浮选流程
copper-making blow	铜吹炼
copper-nickel ore	铜镍矿
copper-recovery step	铜回收步骤
co-precipitated	共沉淀の
copric ammonium sulfate	硫酸铜铵
coprocyanide complex	铜氰[氰亚铜酸盐]络合物
copy paper	复印纸
corduroy blanket	灯芯绒或条纹厚棉

布溜矿槽衬底
corduroy cloth concentrator 绒布衬垫选矿机
corduroy sluice 绒衬垫溜槽
core 核心,样[岩]芯
core flow bin 核心流动仓
core image program file 磁心映象程序文件[资料,信息]
core of rock 岩心
core resident scanning program 磁心驻留扫描程序
core size 磁芯规格
core-resident operating system 常驻磁心体操作系统
Corey shape factor 科里形状因子[系数]
corn starch 玉米淀粉
corporate annual report 公司年度报告
corporate equity 公司产权[股权,权益]
corporate management 公司[企业]管理
corporate officer 公司高级职员
corporate research 企业调研
corporate taxes 综合税
corporation 公司,团体,企业
correct combining rate 正确混合率
correct combining ratio 正确配合比
correct mixing rate 正确混合率
corrected base 已校正的基础(数据)
corrected data 修正的数据

corrected efficiency equation 修正效率方程式
corrected load capacity 修正载荷能力
corrected recovery 修正回收率
corrected recovery curve 修正回收率曲线
correction factor 修正系数
correction of feed concentration 给矿浓度的修正
correction of solids specific gravity 固体比重的修正
corrective measure 改[修]正办法[措施]
correlation 相关(性),相互(对比)关系,交互[关联]作用
correlation curve 相关曲线
correlation equation 关系式
correlation work 比对工作
correspondence 信件[函],通信,对应
correspondent 通讯员
corresponding coefficient 相[对]应系数
corresponding constant 相[对]应常数
corresponding constraint relationship 相应约束关系
corresponding equation 对应方程
corresponding flowsheet development 相应的流程编制
corresponding information 相应的资料
corresponding parameter 相[对]应参数

corresponding size 相应规格
corrosion-proof equipment 防腐设备
corrosion-proof pump 防腐泵
corrosion-resisting pump 耐腐蚀泵
corrosive effect 腐蚀作用[影响]
corrosive soluble salt 侵蚀性的可溶盐
corrugated board 波纹金属板
corrugated sheet 波纹板材
cosine 余弦
cosine law 余弦定律
cosmetic industry 化妆品工业
cost and freight 成本加运费价格
cost benefit 成本收[效]益
cost benefit analysis 成本利得分析
cost coding 价值编码
cost data 成本数据
cost effectiveness 成本效益
cost engineer 成本[造价,费用]工程师
cost equation 成本方程
cost estimate 成本估算
cost estimate from project contractor 工程承包商作的费用估算
cost estimating procedure 费用估算方法
cost factor 费用因子
cost index 成本指数
cost keeping 成本核算
cost of asset 资产成本
cost of auxiliary operation 辅助作业费
cost of capital 资本成本,资金成本
cost of construction and equipment 建筑及设备费
cost of electricity 电费
cost of installation 设置[安装]费
cost of labor 人工成本
cost of maintenance 养护费
cost of manufacture 制造费
cost of operation 使用费,运转费,管理费
cost of production 生产成本
cost of production service 生产服务费
cost overrun 成本超支[出]
cost per unit of product 单位产品成本
cost record 成本账
cost reduction 成本降低
cost saving 节省费用
cost saving advantage of automation 自动化节省费用优点
cost trade-off 投资折中办法[方案],投资权衡
cost unit 成本单位
cost unit price 成本单价
cost, insurance & freight 到岸价格(包括货价,运费和保险费)
cost-benefit relationship of higher circulating load 较高循环负荷的费用-受益关系
cost-effective plant flow rate 成本效益[有效]工厂流量
cost-effectiveness 成本效益
cost-effectiveness analysis 成本效益分析(法)
cost-efficient 对成本生效的

costing 成本会计[计算]
cost-price squeeze 成本价格压缩
cotangent 余切
cotton twill filter cloth 棉斜纹滤布
cottonseed oil 棉籽油
Cottrell 科特雷尔电收尘器[静电集尘器]
Cottrell process 静电收尘法
coulomb attraction 库仑引力
coulombic interaction 库仑作用
Coulomb's law 库仑定律
Coulter Counter 库尔特计数器
Coulter counter equipment 库尔特型计数设备
Coulter principle 库尔特原理(被行业作为细胞计数金标准的技术,提供了一种对微小颗粒大小[体积]进行测量的方法,由美国库尔特兄弟发现)
counter current air contact tower heat exchanger 逆流空气接触塔热交换器
counter current flow 逆向流动
counter electrode 反[对]电极
counter shaft 副轴
counter shaft box assembly 副轴箱装置[组件](圆锥破碎机的)
counter shaft housing face 副轴盖[套]面(圆锥破碎机的)
counter shaft keyway 副轴键槽(圆锥破碎机的)
counter shaft speed 中间[副]轴速度
counter washing 反洗
counter weight specification 配重特性

counterbalance bottom-dump skip 配重底卸式箕斗
counter-balanced shaking screen 有平衡作用的摇动筛
counter-current carbon adsorption 逆流碳吸附
countercurrent column machine 逆流浮选柱,逆流圆柱形设备
counter-current decantation wash circuit 逆流倾析洗涤流程
countercurrent ejector-type column flotation machine 逆流喷射型浮选柱
counter-current flotation 逆流浮选
counter-current flow 对流
countercurrent leaching 逆流沥滤
counter-current plant operation 逆流厂作业[操作]
counter-current recleaner flotation 逆流再精选
counter-current stage 逆流阶段
counter-electrode 反电极,电容器的极板
counterflow manner 逆流方式
counterpart 副本,对应物,配对物,相应的对象,地位相同的人,极相似的人或物
counter-productive 阻碍生产的
countershaft box 传动轴箱
countershaft box guard 传动轴[副轴]箱保护装置
countershaft box seal 传动轴箱密封装置

counterweight take-up 平衡锤拉紧装置

counter-weighted double drum hoist 带配重的双滚筒提升机

country of manufacture 制造国家

country of origin 原产地

Courier 300 on-stream analysis system 库尼厄300型在线分析系统

Courier 300 on-stream analyzer 库里厄300型在线分析仪

Courier 300X-ray analyzer 库里厄300型X射线分析仪

Courier multiple-stream 库里厄多路物料流

Courier on-line X-ray analyzer 库里厄在线X射线分析仪

Courier on-stream analyzer 库里厄在线分析仪

Courlose 羧甲基纤维素钠(英国产,可用作抑制剂、絮凝剂)

course of mining 开采过程

covalent bond 共价键

covalent bond formation 共价键形成

covalent character 共价特性

covalent force 共价键力

covariance 协方差

covariance term 协方差项

covenant 契[盟]约,契约条款

cover swells in spot (胶带)面层隆起斑点

covered concentrate shed 加盖式精矿库

co-worker 合作者,同事

CPB (= cetyl pyridinium bromide) 十六烷基溴化吡啶(品名,可用作捕收剂) 结构式:$C_{16}H_{33}-{}^+N\langle\bigcirc\rangle Br^-$

CPC (= cetyl pyridinium chlorde) 十六烷基氯化吡啶(品名,可用作捕收剂)

Crawford process 克劳福德法(美国专利,用木炭作吸收剂,以处理低品位矿石)

crawler type forged track 履带式锻造轨道

creative skill 创造性技能

credit line 信用额度,授信额度

credit rating 信贷率

credit term 信用期限

creditor 债权人

creeping flow 缓流

creosote 杂酚,杂酚油,木馏油,焦馏油

Creosote coal tar 烟煤焦馏油(品名,可用作硫化矿浮选时的起泡剂)

Creosote No.1 干馏硬木焦馏油(品名,可用作浮选硫化矿、金矿的起泡剂)

creosote oil 焦馏油,杂酚油

cresol 甲酚 $CH_3C_6H_4OH$

crest of weir 堰口[顶]

criteria 标准,规范,准则,依据

criteria for mechanical design 机械设计准则[标准]

criteria of merit 优点[质量]判据

criteria of noise control 噪声控制要求

criteria of solubility 可溶标准

criteria selection 准则[判据,标准]选择
criterion of floatability 可浮性判据
criterion of profitability 利润判据
critical air flow number 临界空气流量数
critical carrying velocity 临界运送[载]速度
critical dilution point 临界稀释点
critical dimension 关键尺寸
critical factor 关键[临界]因素
critical flotation concentration 临界浮选浓度
critical flotation concentration of collector 捕收剂临界浮选浓度
critical flow factor 临界流动系数
critical heat of emersion 临界浸润热
critical hemimicelle concentration 临界半胶束浓度
critical hydrophobicity plot 疏水性临界线
critical inflow rate 临界流入流量
critical item 关键项目
critical layer 临界层
critical material 重要原料,关键材料,临界物质
critical micelle concentration 临界胶束浓度
critical operation parameter 主要[关键]操作参数
critical outflow rate 临界流出流量
critical oversize 超[过大]临界粒度,关键性的过大粒度
critical path method 统筹方法,关键路线[途径]法,主要矛盾线路法
critical point 临界点
critical problem 严重[关键]问题
critical property 临界[关键性]性质
critical range 临界范围
critical rathole diameter 临界管状腔[鼠洞]直径
critical size material 临界粒级物料
critical stone 临界碎石
critical time period 临界时间周期,临界时间[时限]
critical value 临界值
critical variable 重要变量
cross angle 交叉角
cross belt magnetic separator 交叉皮带磁选机
cross flow interceptor type 横流拦截器式,横流拢流板式(倾斜浓缩箱的)
cross member 横梁[板,桁]
cross product 叉积,交叉乘积,向量积
cross reference index 交叉引用[参照]索引
cross sectional area capacity 横断面积容量[大小]
cross-connected 交叉连接的
cross-cut sampler 横切式取样器
crossover configuration 交叉结构形式
crossover of potential 电位相交
cross-section availability 横断面利用率

cross-the-stream sample cutter 横切[断]矿流试样截取器
crowding 群[密]集
crowding effect 群集效应
Crowe process 克劳(氰化提金)法
Crowe tower 克劳塔(即脱气[氧]塔)
CRT operation station CRT操作站(计算机操作站)
crucible furnace 坩埚炉
crucible grade flake graphite 坩埚级鳞片状石墨
crucible tilting furnace 可倾坩埚炉
crude blister copper 粗铜
crude native sulfur 未经选别的自然硫
crude ore 原矿石
crude tall oil 粗塔尔油(亚硫酸盐纸浆制造过程中产生的树脂状可以皂化的油)
crude tall oil fatty acid 粗塔尔油脂肪酸
crushability test 可碎性试验
crushed-ore-vat-leaching gold plant 破碎矿石槽浸金厂
crusher application 破碎机的应用
crusher bowl 破碎机锥壳
crusher circuit automation 破碎机回路[系统]自动化
crusher component 破碎机部件
crusher control component 破碎机控制元件
crusher control philosophy 破碎机控制原理

crusher control system 破碎机控制系统
crusher department 破碎机室
crusher discharge area 破碎机排料区
crusher discharge belt 破碎机排矿胶带
crusher feed opening 破碎机给矿口
crusher liner wear rate 破碎机衬板磨损率
crusher mantle 破碎机圆锥锰钢壳
crusher manufacturer 破碎机制造厂家
crusher manufacturer's standard application 破碎机厂家标准应用
crusher motor ammeter 破碎机马达[电动机]电流表
crusher motor power draw 破碎机电机功率输入
crusher outlet 破碎机排矿口
crusher performance 破碎机性能
crusher power rate 破碎机单位功率消耗量
crusher power requirement 破碎机功率要求
crusher product size distribution 破碎机产品粒度分布
crusher screenings 破碎机筛下物
crusher sheave 破碎机滑轮
crusher without automatic setting 无自动调节排矿口破碎机
crushing action 破碎作用
crushing and grinding circuit 破碎磨

矿回路
crushing and grinding process 破碎和磨矿过程[流程]
crushing and grinding throughput 破磨处理量
crushing and screening flowsheet 破碎筛分工艺流程
crushing angle 破碎角
crushing bowl 固定破碎锥
crushing breakage technology 破碎的破裂技术
crushing capacity 破碎能力
crushing cavity 破碎腔
crushing cavity design 破碎腔设计
crushing chamber 破碎腔
crushing chamber configuration 破碎腔构造[形状]
crushing characteristic 破碎特性
crushing department 破碎工段
crushing design tonnage 破碎设计吨数
crushing equipment 破碎机
crushing head 破碎锥头
crushing member 破碎构件
crushing operation 破碎作业
crushing pendulum 破碎摆锤
crushing pendulum potential energy 破碎摆锤位能
crushing plant screen 破碎车间使用的筛子
crushing process 破碎过程
crushing rate 破碎速度,破碎量
crushing schematic 破碎简[略]图

crushing section 破碎工段
crushing stroke 破碎冲程
crushing table 破碎产品目录
crushing-grinding equipment supplier 碎磨设备供应厂家[商]
crushing-grinding facilities 碎磨设施
Cryderman mucker 克莱德曼装岩机（美国）
crystal chemistry 结晶化学
crystal deflection 晶体偏转(角)
crystal diffraction spectrometer 晶体衍射能谱仪
crystal face 晶面
crystal geometry 晶体几何学
crystal lattice 晶格
crystal lattice anion 晶格阴离子
crystal lattice cation 晶格阳离子
crystal lattice geometry 晶体晶格几何形状
crystal structure 晶体[结晶]构造
crystal structure modification 晶体结构变异
crystalline 结晶(质)的,晶状(态,体)的,透明的,结晶体[质],水晶体,晶态
crystalline solid 结晶固体
crystalline structure 结晶结构
crystallization mechanism 结晶机理
CSA's purchase money mortgage 美国联邦的购买货币抵押
CTAB (= cetyl-trimethyl ammonium bromide) 十六烷基三甲基溴化铵（品名,可用作捕收剂）

$CH_3(CH_4)_{15}N^+(CH_3)Br^-$
CTAC (= cetyl - trimethyl ammonium chloride) 十六烷基三甲基氯化铵(品名,可用作捕收剂)
Cu amyl xanthogenate 戊基黄原酸铜
Cu butyl xanthogenate 丁基黄原酸铜
Cu ethyl xanthogenate 乙基黄原酸铜
Cu heptyl xanthogenate 庚基黄原酸铜
Cu hexyl xanthogenate 己基黄原酸铜
Cu isoamyl xanthogenate 异戊基黄原酸铜
Cu isobutyl xanthogenate 异丁基黄原酸铜
Cu isopropyl xanthogenate 异丙基黄原酸铜
Cu octyl xanthogenate 辛基黄原酸铜
Cu propyl xanthogenate 丙基黄原酸铜
cubelet 小立方体,小方块
cubic measure 立方单位制
cubic system 立方(晶)系
cubic unit cell 晶胞立方体
cubical natural grain-shaped product 立方体自然粒级产品
cubical product 立方体产品
cullet 碎玻璃,玻璃片[屑]
cumulative basis 累积基础
cumulative chord - length distribution curve 累计弦长分布[配]曲线
cumulative distribution function 累积分布函数
cumulative feature 累计特性
cumulative floats 累积漂浮物
cumulative fraction 累积[计]分数
cumulative frequency percentage 累积频率百分数
cumulative grade 累积品位
cumulative oversize fraction 累计筛上级别
cumulative passing 累积通过量,累积筛下量
cumulative sinks 累积沉降物
Cu-Ni alloy 铜镍合金
Cupferron 铜铁试剂,铜铁灵,N-亚硝基(β-)苯胲铵(是一种重要的化学分析试剂)$C_6H_5N(NO)ONH_4$
cupric (正)铜的,二价铜的
cupric ammonium sulfate 硫酸铜铵,硫酸四氨酮 $Cu(NH_3)_4SO_4 \cdot H_2O$
cupric hydroxamate 氧肟酸铜
cupric ion concentration 铜离子浓度
cupric octyl hydroxamate 辛基氧肟酸铜
cupric salt 铜盐
cupriferrous 含铜的
cupriferrous pyrite 含铜黄铁矿
cuprous ethyl xanthate 乙基黄原酸亚铜
cuprous xanthate 黄原酸亚铜
currency equivalent 等价货币
current assets 流动资产
current density 电流密度
current design procedure 现行[当前]设计程序

current dollar 现行[现值]美元
current dust and visible emission standard 现行粉尘与可见排放标准
current economic condition 当前经济状况[条件]
current expenditure 经费,经常性支出
current index basis 现行指数基准
current labor rate 现时人工费率
current liabilities 流动负债
current literature 现行[代]文献
current market 当前市场,即期行情
current output 电流输出
current payment 当期付款,经常性支付
current period 目前时期,现阶段
current plant design and practice 现代选矿厂设计和实践
current price 现价,现行价格
current price of equipment 设备现价
current rate 当日汇率,气流强度
current reserves 当前储量
current salary rate 现时工资[薪酬]费率
current series 现行系列
current state 现状
current state of technological development 技术发展现状
current test 现行试验
current transformer 电流互感器,变流器
current transient 电流瞬变
current value 现值
current wages structure 现行工资结构
current-potential curve 电流电位曲线
current-time curve 电流-时间曲线
curvature of crushing member 破碎部件曲率
curve diameter 弧线直径
curve fitting technique 曲线求律法
curve of standard crusher discharge 标准的破碎机排矿曲线
curve radius 弧线半径
curved liquid-gas interface 液气曲界面
cushion of material 物料垫层
custom designed 按常规[传统,定做]设计的
customary addition 通常[习惯]添加量
custom-built 定制的
customer 客户,顾客,用户
customized 定制的
customizing reagent 定制药剂
customs 关税,海关
customs regime 关税制度
cut and fill mining method 分层充填采矿法
cut and try method 尝试法,试凑法
cut size equation 分划[割]粒度方程式
cutinite 角质体[素],角质煤素质
cut-off point 分界点
cut-off rate 界限定额
cut-rate price 减[折]价

cutter 切[截]取器
cutter opening 切[截]取器开口
cutter speed 切取器速度
cyan 氰基 —CN
Cyanamid frothing agent 美国氰胺公司起泡剂
Cyanamid R - 765 (= Aero promoter 765) 氰胺 765 号浮选剂(精制脂肪酸,主要含油酸及亚油酸,同 Aero promoter 765)
Cyanamid R - 801 (= Aero promoter 801) 氰胺 801 号浮选剂(水溶性石油磺酸,同 Aero promoter 801)
Cyanamid R - 825 (= Aero promoter 825) 氰胺 825 号浮选剂(油溶性石油磺酸,同 Aero promoter 825)
Cyanamid reagent 氰胺药剂(美国氰胺公司产品)
Cyanamid reagent 712(=Aero promoter 712) 氰胺 712 号浮选剂(一种脂肪酸钠皂,同 Aero promoter 712)
cyanamide 氨基氰 $H_2N \cdot CN$; 氨腈 RNHCN
cyanamide flocculent 氰胺絮凝剂
cyanidation metallurgy 氰化冶金
cyanidation testing 氰化试验
cyanidation-titration 氰化-滴定法
cyanide base 氰化物基
cyanide base compound 氰化物碱性化合物
cyanide circuit 氰化流程
cyanide compound 氰化物混[复]合物
cyanide concentration 氰化物浓度
cyanide consumption 氰化物消耗量
cyanide containing reagent 含氰化物药剂
cyanide depression 氰化物抑制
cyanide destruction 氰化物的消除
cyanide destruction circuit 氰化物消除回路[系统]
cyanide destruction system 消除氰化物系统
cyanide destruction test 氰化物分解[消除]试验
cyanide eluant 氰化物洗提液
cyanide ion 氰离子
cyanide leach plant 氰化浸出车间
cyanide leach residue 氰化物浸出渣,氰化尾矿
cyanide leaching 氰化浸出
cyanide leaching evaluation 氰化物浸出评价
cyanide level 氰化物含量[浓度]
cyanide mill 氰化厂
cyanide plant 氰化厂[车间]
cyanide reactor 氰化反应槽[器]
cyanide strength 氰化物浓[强]度
cyanide treatment 氰化处理
cyanide zinc complex 氰化锌络合物
cyaniding leaching 氰化浸出
cyaniding test work 氰化试验工作
cyanogen radical 氰基,氰根
cycle 循环,周期
cycle close-circuit test 循环闭路试验

cycle test 循环试验
cycle test result 闭路[循环]试验结果
cycle testing 循环试验
cyclic feed rate 周期性给矿速度[量]
cyclic feed rate variation 周期性给矿速度[量]变化
cyclic fluctuation 周期性波动
cyclic surge 周期性波动
cyclic voltammetry 周期伏安测量(法)
cyclic voltammetry behavior 周期伏安测量特性
cyclic voltammetry measurement 循环伏安法测量
cyclic voltammetry peak position 周期伏安测量峰位
cyclic voltammogram 周期伏安图
cyclo-cell 旋流浮选机
cyclohexane 环己烷
cyclohexanol 环己醇
cyclohexene 环己烯 C_6H_{10}
cyclohexyl 环己基 $C_6H_{11}-$
cyclohexyl xanthate 环己基黄药
cyclone axis 旋流器轴线
cyclone bank 旋流器组
cyclone classification unit 旋流器分级设备
cyclone cluster 旋流器组
cyclone construction 旋流器制作[构造]
cyclone cutaway 旋流器剖视(图)
cyclone description 旋流器描述
cyclone diameter 旋流器直径
cyclone equation 旋流器计算方程式

cyclone feed 旋流器给矿
cyclone feed % solids 旋流器给矿固体百分数含量
cyclone feed flow 旋流器给矿流量
cyclone feed flow rate 旋流器给矿流量
cyclone feed inlet 旋流器给料[矿]入口
cyclone feed percent solids by weight 旋流器给料固体重量百分数
cyclone feed pulp density 旋流器给矿矿浆浓度
cyclone feed pump 旋流器给矿泵
cyclone feed pump box 旋流器给矿泵池
cyclone feed pump sump water volume 旋流器给矿泵池水量
cyclone feed solids 旋流器给矿固体量
cyclone feed sump 旋流器给矿泵池
cyclone feed volume 旋流器给矿体积
cyclone feed water 旋流器给矿水量
cyclone fundamental 旋流器基本原理
cyclone geometry 旋流器几何形状
cyclone header 旋流器头部,旋流器蓄[压力]水池
cyclone inlet 旋流器入口
cyclone inlet pressure 旋流器入口压力
cyclone model 旋流器模型
cyclone model equation 旋流器模型方程
cyclone model parameter 旋流器模型参数
cyclone overflow percent solids 旋流器溢流固体百分数

cyclone overflow pulp density 旋流器溢流矿浆浓度
cyclone performance 旋流器性能[操作效能]
cyclone pump 旋流器用泵
cyclone retention time 旋流时间(分级时矿浆在旋流器内停留时间)
cyclone sand 旋流器沉砂
cyclone scrubber 旋流洗涤机
cyclone shape 旋流器形状
cyclone splitting point 旋流分支[流]点
cyclone underflow box 旋流器底流箱[槽]
cyclone underflow percent solids 旋流器排矿[底流]浓度[固体百分数]
cyclone washing 旋流(器)洗选法,旋流器洗煤法
cyclonic separation chamber 旋流分离室
cyclosizer 旋流分级机[水析仪]
cylinder section 圆通剖面,圆筒形部分
cylindrical core of rock 圆柱岩心
cylindrical drum 圆筒形滚筒
cylindrical feed chamber 圆筒形给料室
cylindrical mixing drum 筒形混合桶
cylindrical overflow ball mill 圆筒溢流型球磨机
cylindrical section 圆筒段(旋流器的)
cylindrical section length 筒体段长度
cylindrical-shape collector 筒形收[集]尘器
cylindrical-shaped impeller 圆柱形叶轮
Cyquest 美国氰胺公司化学药剂品牌名
Cyquest 30HE 正羧基乙基乙二胺三醋酸三钠
Cyquest 3223 antiprecipitant Cyquest 3223 反沉淀剂(丙烯聚合物)
Cyquest 50 一种螯合剂
Cyquest 500 乙二胺四乙酸钠(= EDTA Na·2H$_2$O)
Cyquest 9223 是一种聚丙烯酸(粉末状物质,是一种阻止沉淀药剂,可以除去可溶性盐在浮选中的有害物质)
Cyquest acid 乙二胺四乙酸(= EDTA)
Cyquest EDG 一羟基乙基甘氨酸钠

D

3 deck Nordberg GP screen 3层诺德伯格 GP 筛

D. M. S (= dense medium separation) 重介质选矿

D. P. reagent D. P. 浮选剂(美国杜邦公司产品)

D20B Krebs cyclone D20B 克雷布斯旋流器(美国克雷布斯公司制造)

D50 point D50 点(矿用水力旋流器溢流和底流各为50%的点位称为D50点)

d50 size d50 粒度(该粒度50%颗粒进入沉砂,50%进溢流)

d50 value d50 粒度值(该粒度50%颗粒进入沉砂,50%进入溢流)

D50c parameter of cyclone 旋流器 D50c 参数

D50c point D50c 点(在分级中某一粒度的物料50%进入溢流,50%进入沉矿,该粒度即为D50c点)

daily capacity 日处理能力

daily evaporation 日蒸发量

daily expense 每天费用,日费用

daily processed volume 日处理量

daily tons milled 日处理量

daily variation 日变化

dam angle 坝体倾角

dam building 筑坝

dam crest 坝顶[脊]

dam face 堤,坝面

dam return water 尾矿坝回水

dam wall 尾矿坝围堤,堤坝壁

damaged surface region 表面损坏区域

dampening effect 阻尼效应

damper valve 调节阀

daphnia 水蚤

daphnia magna 大型水蚤

dash and dot line 点画线

dashed line 虚[短划]线

data acquisition plan 数据获取计划

data adjustment procedure 数据修改[正]过程[方法]

data band 数据带

data bank 资料库

data base 数据库

data comparison 数据对比

data equation 数据方程

data evaluation 数据评价,数据鉴定

data file 数据文件[资料,信息]

data handling device 数据处理装置

data log file on diskette 软磁盘数据记录文件

data logging 数据记录

data logging control system 数据记录控制系统

data presentation 数据显示

data processing system 数据处理系统

data requirement 数据需要量

data sheet 资料表

data sheet for liner study 衬板调查表

data storage device 数据存储设备[装置]

database 数据库

database program 数据库程序

database programming 数据库程序设计

data-mesh 数据网格

date of completion 完成日期

date of delivery 交货日期

Davcra flotation cell 达夫克勒浮选槽（借喷嘴达到充气和矿浆循环用，澳大利亚亚锌公司研制）

Davcra rougher scalping cell 达夫克勒型粗粒粗选槽

Davis best operating speed 戴维斯最好[佳]运转速度

Davis equation 戴维斯方程

Davis tube 戴维斯管（实验室磁选用）

Daxad 表面活性剂品牌名（美国W. R. Grace & Co. 公司）

Daxad 11 烷基萘磺酸钠（可用作分散剂）

Daxad 23 二萘甲烷二磺酸盐（萘磺酸与甲醛缩合物，可用作分散剂，用作脉石矿泥分散剂时用量 25~500 克/吨，pH 值为 7~9.5）

DBF（=depth at feed end） 给矿[料]端层厚

DC gearhead motor 直流齿轮电动机

DC motor 直流电动机

DC motor-driven belt feeder 直流电动机驱动的带式给矿[料]机

DC silicone fluid 氯苯甲基聚硅油（品名，美国道康宁公司产品，可用作捕收剂）

DDE（=depth at discharge end） 排矿[料]端层厚

DDS（=di-decyl secondary amine） 二癸基仲胺（品名，可用作捕收剂）

deactivty reaction 去活作用

deactivty system 去活体系

deactivty test 去活试验

dead space 死区

de-aeration process 除气[氧]法（氰化法沉淀金前从溶液中除去溶解氧）

deaeration system 脱[排]气系统

deaeration tower 脱[除]气塔

deaeration zone 脱气区

debenture 债券

debt financing 债务资金筹措，举债筹资

debt-equity ratio 债资比率，债务产权比率

debt-to-equity ratio 负债与产权比率

de-bugging 排除故障

Debye（Debye, Peter Joseph Wilhelm） 德拜（荷兰-美国物理化学家）

Debye crystallogram 德拜晶体衍射图

Debye density 德拜密度

Debye density of states 德拜态密度

Debye ring 德拜光环（德拜晶体衍射图）

Debye-Hückel limiting law 德拜-休克尔极限定律（用于计算活度系数）

Debye-Hückel reciprocal double layer thickness 德拜-休克尔对应双电层厚度

Debye-Scherrer method（= Debye-Scherrer X-ray diffraction） 德拜-谢乐法（X射线衍射分析法）

decalin 奈烷，十氢萘（可用作捕收剂）$C_{10}H_{18}$

decane 癸烷 $C_{10}H_{22}$

decanoic acid 癸酸 $C_9H_{19}COOH$
decant basin 沉淀[澄清]池
decant tower 澄清井,溢流井(尾矿池的)
decantation 滗析,倾析,澄清,调迁,水析
deceleration time 减速时间
Decerosol OT(=Aerosol OT) 表面活性剂(双-2-乙基己基磺化琥珀酸钠,同 Aerosol OT,可用作捕收、湿润剂、活化剂)
decibel level 分贝水平
decimal equivalent 十进制等值
decision analysis 决策分析
decision analysis worksheet 判断分析工作表
decision criteria 决策[判断]准则
decision element 判定[断]元件,计[解]算元件,判定元素
decision group 决策小组
decision maker 决策人员
decision process 判定过程,决策程序
decision statement 决策陈述,断语
deck area 床面面积(摇床的)
deck blinding 筛板堵塞
deck covering structure 筛面结构
deck factor 筛面系数
deck location 筛面层位
deck material 筛面物料
deck replacement 筛面更换
deck support member 筛面支件
decline portion 倾斜[下降]部分

declining balance method 余额递降法
decomposition of reagent 药剂分解
decomposition temperature 分解温度
decreasing particle separation distance 递减颗粒间距
decyl 癸基 $CH_3(CH_2)_8CH_2-$
dedicated device 专用装置
deduction 扣除额
dedusting plant 除尘装置
deep impeller design 深槽叶轮设计
deep rotor 深型转子
deep sea nodule 深海矿结核
deep tank 深槽
deep well pump 深井泵
deep well water 深井水
defective part 有缺陷的部件
deferred credit 递延信贷
define duty 确定任务
definite solution 确定解决办法
definition of process 过程定义
definition of terms 术语定义
definitive design 明确的设计
definitive estimate 确切估算
definitive estimation 最终估算
deflection plate 导向板,偏转板,折向板,挡板
deflector liner 导流[转向]衬板
deflector plate 导流板
deflocculation end point 反絮凝终点
deflocculation period 反絮凝阶段
deformation of bubble 气泡变形
defrothed water 已消泡水

defrothing agent 消[去]泡剂
degassed sample (已)脱气样品
degenerate n-type electrode 简并的n型电极
degradable 可降[分]解的
degradable ore 容易蚀变[降解,分解]的矿石
degradation (能量)衰变,裂[分]解,破[变]坏
degradation of air quality 空气质量下降
degradation of product 产品贫化
degreasing pot 脱脂锅[桶],可用于金刚石选矿
degree Celsius 摄氏温度
degree of agitating 搅拌程度
degree of association 伴[共]生程度
degree of automation 自动化程度
degree of covalence 共价度
degree of cross-linkage of resin 树脂交联程度
degree of crowding 群集度
degree of difficulty 难度
degree of freedom 自由度
degree of hydration 水化程度
degree of hydrophobicity 疏水度
degree of hydropholicity 疏水(性程)度
degree of intergrowth 共[连]生程度,连生度
degree of oxidation 氧化程度
degree of precision 精度

degree of resin cross-linkage 树脂交联程度
degree of saponification 皂化(程)度
degree of sophistication 采用先进技术程度
degree of surface oxidation 表面氧化(程)度
dehydroabietyl amine 去氢松香胺(可用作捕收剂)
Deister laboratory classifier 戴斯特型实验室分级机
Deister table 戴斯特型摇床(选矿机的)
Delamine P 塔尔油胺(品名,不含松香胺,可用作捕收剂)
Delamine PD 精制塔尔油胺(品名,不含松香胺,可用作捕收剂)
Delamine X 塔尔油胺(品名,含40%脂肪酸、60%松香胺,可用作捕收剂)
Delavan sonic plugged chute detector 德拉万型声波溜槽堵塞探测器
delay loop 滞后回路
delay time 延迟[滞后]时间
deleading kiln 除[脱]铅窑
delimitation of orebody 确[圈]定矿体边界
delivered cost 交货费
delivered price 交货价格
delivered product 交货产品
delivery conveyor 运载胶带输送机
delivery cost 运送费用
delivery date 交货日期

delivery fitting 输送管
delivery rate 输送量[速度]
delivery term 交货期限
delivery ticket 交货票,发放票
demagnetizing 去磁,退磁
demand charge 需量电费
demand feed rate 要求给矿量
demand forecasting 需要量预报
demand forecasting technique 需要量预报技术
demineralized water 软化水
demister 除雾器
denatured alcohol 变性酒精
dense medium separating cyclone 重介质分选旋流器
dense medium separating drum 重介质分选滚筒
dense phase conveying 致密[稠密]状态运输
densification 密(实)化,致密化,压[击]实,增稠[浓密],稠化,(密)封,封严
densifying and cleaning system 浓缩净化系统
densifying pump 浓缩(介质)泵
densifying pump box 浓缩泵池
densifying separator 浓缩选矿机
density 密[浓]度,稠密性[度],密集性[度]
density control 浓度控制
density control unit 浓度控制设备
density controller 密[浓]度控制器
density distribution 浓度分布
density function 密度函数
density gauge 浓度计

density gradient column 密度梯度排列
density measurement 密度测量
density of liquid medium 液体介质浓度
density of material 材料密度
density of mill content 磨机内物料的密度
density of solution 溶液浓度
density sensitive gauge 密度灵敏计
density set-point 密度(计)给[设]定点
densometer 浓度计
Denver (non-sliming) rod-ball mill 丹佛型(不泥化)棒球磨机
Denver 200H DR cell 丹佛200H DR 浮选槽
Denver adjustable stroke belt feeder 丹佛型可调冲程皮带给矿机
Denver adjustable stroke diaphragm pump 丹佛型可调冲程隔膜泵
Denver agitator disk filter 丹佛型搅拌圆盘过滤机
Denver airlift agitator 丹佛型气升搅拌槽
Denver amalgamation unit 丹佛型混汞机[装置]
Denver Auto-Flot electrical resistance type automatic level controller 丹佛电阻式自动液面控制器
Denver bag type zinc dust precipitation unit 丹佛型袋式锌粉沉淀设备[装置]
Denver belt conveyer 丹佛型胶带运输机

Denver Buckman tilting concentrator 丹佛巴克曼型自动流槽

Denver cell-to-cell flotation machine 丹佛型直流(槽对槽)式浮选机

Denver center airlift rake type agitator 丹佛型中心气升耙式搅拌槽

Denver combination scrubbing and trammel screen 丹佛型洗矿圆筒筛

Denver cone crusher 丹佛型圆锥破碎机

Denver counter-current drier 丹佛型逆流干燥机

Denver crossflow classifier 丹佛型错流分级机

Denver cyclone classifier 丹佛型旋流分级机

Denver D-1 laboratory flotation machine 丹佛 D-1 型实验室浮选机

Denver dewatering cone classifier 丹佛型脱水圆锥分级机

Denver D-R ring 丹佛 D-R 环

Denver drag 丹佛型刮板脱水机

Denver dry reagent feeder 丹佛型干式给药机

Denver duplex mineral jig 丹佛双室矿物跳汰机

Denver filtrate pump 丹佛型滤液泵

Denver gold matting 丹佛捕金溜槽衬底

Denver heavy-duty thickener 丹佛重型浓缩机

Denver improved 4-compartment Harz jig 丹佛改进型 4 室哈茨跳汰机

Denver indirect rotary dryer 丹佛型间接加热旋转干燥机

Denver M machine 丹佛 M 浮选机(使用叶轮和螺旋浆)

Denver motorized distributor 丹佛型电[机]动(矿浆)分配器

Denver office 丹佛办事处

Denver pan filter 丹佛盘式过滤机

Denver precipitation agitator 丹佛型沉淀搅拌槽

Denver pulp distributor 丹佛型矿浆分配器

Denver pulp distributor-sampler 丹佛型分配取样器

Denver self-rotating distributor 丹佛型自动回[旋]转(矿浆)分配器

Denver self-rotating pulp distributor 丹佛型自动旋转矿浆分配器

Denver side airlift agitator 丹佛型侧边气升搅拌槽

Denver soda ash feeder 丹佛型苏打粉给药机

Denver spiral rake thickener 丹佛型螺旋耙式浓密机

Denver spiral screen 丹佛型螺旋筛

Denver spiral trammel screen 丹佛型螺旋圆筒筛

Denver SRL acid pump 丹佛型 SRL (软橡胶衬里)耐酸泵

Denver SRL pump 丹佛软胶衬泵
Denver standard dryer 丹佛型标准干燥机
Denver steel case pump 丹佛钢壳泵
Denver Sub-A(coal)flotation machine (double overflow) 丹佛型液下充气式(煤)浮选机(双溢流)
Denver Sub-A acid proof flotation machine 丹佛型底吹式防酸浮选机
Denver Sub-A coal flotation machine free-flowing type 丹佛型液下充气煤浮选机自由流动式
Denver Sub-A flotation machine 丹佛型液下充气式浮选机
Denver Sub-A super rougher flotation machine 丹佛型液下充气超级粗选浮选机
Denver sulphidizer 丹佛硫化剂(液体多硫化钙,可用作白铅矿的硫化剂)
Denver super-agitator conditioner 丹佛型超速搅拌调整槽
Denver Type J forced feed antifriction jaw crusher 丹佛J型强制给矿抗磨颚式破碎机
Denver type jig 丹佛型跳汰机
Denver vacuum tank 丹佛型真空桶
Denver vertical centrifugal sand pump 丹佛型立式离心砂泵
Denver vertical concentrate pump 丹佛立式精矿泵
Denver washing trommel 丹佛型洗矿圆筒筛
Denver wet reagent feeder 丹佛型湿式给药机
Denver zinc dust feeder 丹佛型锌粉给料机
Denver-Dillion vibrating screen 丹佛·迪利恩振动筛
Denver-Morton cyclone 丹佛·莫顿旋流器
deoxidation 脱氧
deoxo 脱氧
deoxygenated 除[去]氧的
deoxygenated water 脱氧水
department head 系主任
dependent variable 因[应,他]变数,他[因]变量
depletion 损耗,消耗
depletion allowance 耗竭备抵
deposit liability 存款债务
deposit(ed)metal 溶敷[沉积,电积]金属
depositing reservoir 澄清池
depositional environment 成矿[沉积]环境
depositional feature 矿床特征
depreciation allowance 折旧提成
depreciation cost 折旧费
depreciation method 折旧法
depreciation rate 折旧率
depressant 抑制剂
depressant mixture 抑制剂混合物

depressing action 抑制作用
depressing force 抑制力
depressing mechanism 抑制机理
depressor 抑制剂
depth at discharge end 排矿[料]端层厚
depth at feed end 给矿[料]端层厚
depth of material bed 物料层厚
depth of oxidation 氧化深度
deputy general manager 副总经理
deputy president 副总裁,副董事长,副经理
Dequest 有机磷酸盐类螯合[分散、阻垢]剂的品名(中文名称:德奎斯特)
derivation 导[引]出,(公式)推[求]导,演算,推理[论],证明,分支[流,路],引水道,偏转[差],根源,出处,由来,衍生(物)
derivative 导数,衍[派]生物
derived equation 推导(出的)方程式
Derjaguin wettability model 捷尔佳金可湿性模型
Derrick vibrating screen 德里[瑞]克型振动筛(美国)
description 说明,描述
description of equipment 设备描述
descriptive classification 描述性分类(法)
desiccator 干燥器[剂]
design and construction 设计和建设
design and installation of comminution circuit 破磨系统的设计与装备
design approval 设计批准
design assumption 设计假设[定]
design basis 设计基础
design calculation 设计计算
design capacity 设计能力
design complexity 设计复杂性
design concept 设计概念[原理,观点]
design condition 设计条件
design condition data 设计条件数据[资料]
design consideration(s) 设计根[依]据
design considerations 设计(应考虑的)问题[事项,条件]
design constraint 设计制约,设计约束条件
design criteria 设计标准
design engineering 设计工作[程]
design feature 设计特点
design geometry 设计的几何形状[图形]
design information 设计资料
design limitations 设计限制
design mode 设计方式
design mode simulator 设计方式模拟器
design objective 设计目标
design of experiment 试验设计
design of sampling plan 取样方案设计
design organization 设计机构

design parameter 设计参数
design phase 设计阶段
design philosophy 设计原理[准则]
design practice 设计实践
design principle 设计原理
design problem statement 设计问题语句
design procedure 设计程序
design purpose 设计目的
design requirement 设计要求
design residence time 设计停留时间
design run 设计游程
design specification 设计要求[任务]
design stage 设计阶段
design technology 设计技术
design test 鉴定试验
design tonnage 设计吨数[位]
design tonnage rate 设计吨位处理量
design value 设计数值
design variable 设计变量
design velocity 设计速度
designated representative 指[委]派的代表
designed capacity 设计能力
designed flowrate 设计流量
designed performance level 设计性能水平
designed throughput 设计处理[通过]量
designer's preference 设计者的偏爱
designer's expectation 设计者的期[希]望

designing guideline 设计指南[准则,方针]
desirable component 需要[有用]成分
desired circuit throughput 需要的回路生产[处理,通过]能力
desired mineral 目的矿物
desired percent solids 要求固体浓度
desired power 期待[给定]功率
desired value 期待值
desk study 案头研究,初步研究,粗略估计
deslagging screen 脱渣筛
deslimed ore 脱泥矿石
deslimed pulp 脱泥矿浆
desliming plant 脱细泥厂
desolvation 去溶剂化
desorb 解除吸附,解析,使放出(吸收的相反过程)
desorption isotherm 解吸等温线
desorption regeneration circuit 解吸再生回路
desorption test 解吸试验
destabilizing action 使不稳定作用
destabilizing effect 不稳定效应
destabilizing influence 不稳定影响
DETA(diethylenetriamine) 乙二撑三胺 $NH_2C_2H_4NHC_2H_4NH_2$
detachable drill bit 活接钻头,钎子头
detail 详细,细节,大样图,详图
detail design 详细[施工]设计
detail design drawing 施工(设计)详图

detail drawing 样图,大样图
detail of testing 试验细节
detail requirement 详细规格[技术要求]
detail specification 详细规格[说明]
detailed cost breakdown 详细费用[开支]细目
detailed design 施工图[详细]设计
detailed estimate 详细估算
detailed full scale feasibility study 详细全面的可行研究,详细的工业规模可行性研究
detailed head analysis 详细的原矿分析
detailed information 详细资料
detailed logging 详细记录
detailed market evaluation 市场[销路,行情]的详细评估
detailed methodology 详细方法
detailed model 细节模型
detailed summary 详细总结
detailed survey 详细调查
detailed test result 详细试验结果
Detanol 湿润剂(烷基磺酸盐)
detecting device 探测装置
detection electronics 测试电子仪器[设备,线路]
detection limit 探测范围[极限]
detection statistics 检测统计(量)
detector data 探测器数据
Detergent 85 85号洗涤剂(烷基苯磺酸盐,可用作湿润剂)

Detergent MXP MXP洗涤剂(聚氧乙烯硫醚,可用作湿润剂)
detergent-type molecule 洗涤[去垢]剂型分子
determination of variogram 变异函数的测定
determination of work index 功指数测定
determining thickness 确定厚度
detinned cans 除锡罐
deuterium acetone solution 重氢丙酮液
Deutsche mark 德国马克
Deval abrasion test 德瓦尔[台佛尔]磨耗试验,双筒式磨耗试验
developer 显影剂,开发人员[公司]
developing countries 发展中国家
development 发展,进展,展开,开发,进化,研制,培育,开采[发,辟,拓]
development and design engineer 研制和设计工程师
development cost 研制费用
development era 发展时期[阶段]
development ore 开拓矿石
development personnel 研发人员
development phase 发展阶段
development potential 发展潜力
development problem 开拓[发]问题
development sample 开拓矿样
development shaft 开拓竖井
development team 研发团队
development time 研制时间

Devereaux conditioner 德弗罗型调整槽
deviation alarm 偏移警报
device 装置,设备,器械[件,具],机构[械],仪表[器]
dewaterability 脱水能力
dewatering aid 助脱水剂
dewatering cyclone 脱水旋流器
dewatering magnetic separator 脱水磁选机
dewatering pump box 排水泵池
dewatering section 脱水工段
dewatering step 脱水步骤[阶段]
dextrin(e) 糊精,葡聚糖 $(C_6H_{10}O_5)_n$
Dextrin(e) 152 152号糊精(玉蜀黍淀粉部分水解产品,可用作湿润、抑制剂)
Dextrin(e) 4356 4356号糊精(浮选菱铁矿时用作调整剂)
dextrin(e) depression procedure 糊精抑制法
dextrin(e) xanthate 糊精黄药(钠盐,可用作氧化铁矿的捕收剂)
dezincing 脱锌,除锌
DF 250 (=Dow froth 250) 道[Dow]起泡剂250(三聚丙二醇甲醚) $CH_3—(OC_3H_6)_3OH$
di-2-ethyl hexylacid phosphate 双[二]-2-乙基己基磷酸盐(可用作捕收剂)
di-2-ethylhexyl acid phosphate 二-2-乙基己基磷酸(可用作捕收剂)
diagnostic program(me) 诊断程序

diagnostic test 诊断[特征]试验
diagnostics 诊断程序
diagonal 对角
diagonal element 对角元素
diagonal matrix 对角矩阵
dialkyl 二烷基,二烃基
dialkyl disulfide(dixanthogen) 二烷基二硫化物(双黄药)
dialkyl dithiophosphate(aerofloat) 二烷基二硫代磷酸盐[脂](黑药) $(RO)_2PSSM$
dialkyl thionocarbamate 二烃[烷]基硫代氨基甲酸酯 $R'NH·C(S)OR$
Diam 表面活性剂品牌名
Diam 21 烷基丙二胺(80%二胺,可用作捕收剂) $RNHC_3H_6NH_2$,R来自椰油
Diam 26 烷基丙二胺(80%二胺,可用作捕收剂) $RNHC_3H_6NH_2$,R来自椰油
diamagnetic grain 抗[反,逆]磁(性)颗粒
diameter efficiency factor 直径效率系数
diameter efficiency multiplier 直径效率系数
diameter inside mill liners 磨机衬板内径
diameter inside shell 磨机内径
diamine 二(元)胺,联氨
Diamine green B(Direct Green) 双胺绿B[直接绿](品名,可用作染料、脉石的抑制剂) $C_{34}H_{22}N_8Na_2O_{10}S_2$
diaminopropane 二氨基丙烷

diammonium phosphate 磷酸氢二铵
diamond 金刚石,金刚钻,钻石,菱形菱形断面
Diamond black PV 金刚黑PV(可用作染料、脉石的抑制剂)
diamond drill core 金刚钻岩芯
diamond paste 金刚石磨浆
Dianol 11 湿润剂(烷基苯磺硫酸)
diaphragm pulsing valve 隔膜脉冲阀
diaphragm-type level control 隔膜式料位[面]控制
diarrhea 腹泻
diaspore (硬)水铝石,一水硬铝石,水矾土 $Al_2O_3 \cdot H_2O$
diasporite(=diaspore) (硬)水铝石,一水硬铝石,水矾土 $Al_2O_3 \cdot H_2O$
dibutyl 二丁基
dibutyl carbitol 二[双]丁基卡必醇
dibutyl phthalate 邻苯二甲酸二丁酯 $C_6H_4(COOC_4H_9)_2$
dichlorostearic acid 二氯硬脂酸(可用作捕收剂) $C_{17}H_{33}Cl_2 \cdot COOH$
dicyclohexyl dithiocarbamate 双环己烷基二硫代氨基甲酸盐(可用作捕收剂)
dicyclohexylamine 二环己(基)胺,环己仲胺(可用作捕收剂) $(C_6H_{11})_2NH$
dielectric bead 绝缘垫圈,电介质小球
dielectric constant 介电常数
dielectric strength 介电[绝缘]强度
diesel driven portable pump 柴油机驱动的轻便泵

diesel electric generator unit 柴油发电机设备
diesel forklift 柴油叉车
diesel generator set 柴油发电机组
diesel hauler 柴油拖运机
diesel loader 柴油装载机
diesel power plant 柴油发电厂
diesel powered equipment 柴油驱动设备
diesel-powered transporter 柴油驱动[动力]搬运机
dietary element 食物元素
diethyl 二乙基
diethyl dithiophosphate 二乙基二硫代磷酸盐[脂]
diethyl dixanthogen 二乙基双黄药
diethyl ether 乙醚
diethyl homologue 二乙基同系物
diethylnitrosoamine 二乙基亚硝基胺
different arrangement of jaw type machines 颚式破碎机的不同配置
different inflation pattern 不同的膨胀方式
different length grinding test 不同磨矿时间试验
different manufacturers 不同厂家[制造厂]
different metal-bearing compound 含不同金属化合物
different time period 不同时期
different types of ore 不同类型的矿石
differential absorption measurement of

ultrasonic energy 超声波能不同吸收测量值
differential energy 微分能
differential equation 微分方程
differential flotation rate 不同浮选速度
differential movement 差速[动]运动
differential movement of mineral grain 矿粒差速运动
differential pressure 压降[差]
differential pressure orifice plate flowmeter 压差孔板流量计
differential pressure recorder 压差记录器
differential separation circuit 优先分选[离]回路[流程]
differential thermogram 差热谱图
differential-rate cleaning 不等比率[差速]精选
differently sized blocks 大小不同的矿块
difficult ore 难选矿石
difficult species 难选矿物[种]
difficult-to-float mineral 难浮选的矿物
difficult-to-treat ores 难选矿石类
difficult-to-treat waste liquor 难处理的废液
difficulty of screening 筛分难度
difficulty of separation 分选难度
diffraction crystal 衍射晶体
diffraction effect 衍[绕]射效应
diffraction grating 衍射光栅
diffraction pattern 衍射图谱
diffraction spot 衍射斑点
diffraction technique 衍射技术
diffusate 渗出液[物],扩散物
diffuse layer 扩散层
diffuse portion 扩散部分
diffused double layer 扩散双电层
diffuser wearing plate 扩散器耐磨(衬)板
diffusion coefficient 扩散系数
diffusion mechanism 扩散机理
diffusion rate 扩散速度
diffusiophoretic effect 扩散电泳效应
diffusive radiation 散射辐射
difraction grating 衍射光栅
digester 消化槽
digestion agitator 消化搅拌槽
digestion circuit 浸煮[消化]回路
digital analogue 数字模拟
digital control 数字控制
digital control computer 数字控制计算机
digital controller 数字控制器
digital equipment 数字设备
digital filtering 数字滤波
digital flowmeter 数字流量计
digital forecast system 数字预报系统
digital input output channel 数字输入输出通道
digital meter 数字式仪表
digital readout 数字读出

digital speed sensor 数字(式)速度传感器
digital thumbwheel 数字式拇指压轮
digital totalization 数字加法累积[求和]
digital-to-analog output 数字—模拟输出
dihydroabietyl amine 二氢松香胺(可用作捕收剂)
di-isobutyl acetone 双异丁酮(可用作选煤剂)
di-isobutyl carbinol 双异丁基必醇(可用作选煤剂)
dike-like mass 似岩脉[墙]岩体
dilute medium cyclone 稀介质旋流器
dilute sulfuric acid leaching procedure 稀硫酸浸出法
dilute system 稀溶液体系
dilution accuracy 稀释精度
dilution factor 稀释系数
dimensional analysis 因次分析
dimensional group 因次族
dimensional similitude 量纲相似(性)
dimensionality 维[度]数
dimensionless air flow number 无因次空气流量数
dimensionless capture efficiency 无量纲的捕获效率
dimensionless form 无量纲形式
dimensionless number 无维数
dimensionless settling velocity 无量纲沉降速度
dimensionless time variable 无量纲时间变量
dimensionless unit 无维量单位
dimensionless variable 无量纲变数
dimeric sulphur 二聚硫
dimethyl 二甲基
dimethyl isophthalate 间酞酸二甲酯
dimethyl phthalate 酞酸二甲酯
dimethyl polysiloxane 二甲基聚硅烷(可用作捕收剂)
dimethyldodecylamine 二甲基十二烷胺
dimethylglyoxime 丁二酮肟(可用作捕收剂)
dimethylhexylamine 二甲基己烷基胺
diminishing balance basis 缩小差额基本原则
diminishing balance method 余额递减法
Dings cross belt magnetic separator 丁斯型交叉带式磁选机(美国 Dings 公司生产)
Dings magnetic pulley 丁斯型(电)磁滚筒(美国 Dings 公司生产)
Dinoramac SH 捕收剂(十八烷基二胺,浮选氧化锌矿时用作捕收剂)
diode rectifier 二极管整流器
Dionil W 湿润剂,絮凝剂(聚氧乙烯)
di-o-tolylguanidine 二邻甲苯胍(可用作捕收剂)
dipentene 二戊烯,萜二烯(可用作起泡剂、活性剂)$C_{10}H_{16}$

diphenyl 联苯(C_6H_5)$_2$,二苯基(C_6H_5—)$_2$

diphenyl guanidine(=melaniline) 二苯胍,蜜苯胺(可用作捕收剂)(C_6H_5NH)$_2$C：NH

diphenyl picryl hydrazine 二苯苦基联氨

diphenylurea 二苯脲

di-phosphonicacid ester 双膦酸酯

dipolar surface 偶极表面

dipole effect 双极效应

dipole potential 偶极电势

direct acting pneumatic cylinder 直接动作的风动气缸

direct annual operating expense 直接年生产费用

direct assay 直接分析化验(品位,成分)

direct charge reverberatory furnace 直接加料反射[焰]炉

direct climbing method 直接攀登法

direct correlation 直接相关

direct cost item 直接费用项目

direct current arc 直流电弧

direct cyanidation 直接氰化

direct digital control 直接数字控制

direct digital control strategy 直接数字控制策略

direct electric motor drive 电动机直接传动

direct examination 直接检验

direct experimentation 直接试验

direct fired boiler system 直接燃烧锅炉系统

direct flotation 正[直接]浮选

direct head 原生[直接原]矿

direct head assay 直接原矿分析

direct heating 直接加热

direct interception 直接拦截

direct leasing 直接租用

direct mill cost 选矿厂直接成本

direct observation 直接观测

direct operating cost 直接生产费用

direct operator-computer communication 操作者与计算机直接联系

direct payroll 直接工资单

direct production cost 直接生产费用

direct reading gauge 直接读数仪

direct reading panel meter 直接读数屏[面板]仪表

direct search 直接搜索

direct search method 直接寻找法

direct shipping lump ore 直接外运块矿

direct-connected digital belt speed sensor 接连数字带速传感器

direct-dump truck 自卸式载重汽车[卡车]

direction of polarization 极化方向

direction of rotation 旋转方向

directional heterogeneity 定向多相性[非均匀性]

director 主任,主管,理[董]事,经理,首[社,所,局,处,校]长

director cooperation 董事协作
director of beneficiation department 选矿处处长
director of concentrator 选矿厂厂长
director of department 系主任
director of engineering department 工程处处长
director of milling 选矿厂厂长
director staff services 董事会工作人员服务部
direct-vision method 直观法
dirty area 肮脏地区
dirty separation 不洁净[劣质]分离
disc 6 ft. Eimco type filter 圆盘6英尺的埃姆科型过滤器
disc resident averaging and control program 磁盘平均驻留和控制程序
discharge annulus 排矿口(圆锥破碎机的)
discharge box 排矿箱,尾矿[石]箱(浮选机的)
discharge chamber 排矿[矿]室
discharge chute plug-up 排矿[料]溜槽堵塞
discharge chute plug-up detector 卸料槽堵塞探测器
discharge detail 排矿装置详图
discharge diaphragm 排矿隔[孔]板
discharge duct 下料溜槽,排料槽
discharge flange 排出(漏斗,管)口法兰
discharge grate 排矿格子

discharge moisture 排料水分
discharge opening 排矿口(破碎机的)
discharge setting 排矿口调节(圆锥破碎机的)
discharge stack 排气[通风]管,烟囱
discharge tonnage 排矿吨数[量]
discharge trajectory 排料轨迹[抛射线]
discharge trajectory of conveyor 运输机排料轨迹
discharge trunnion opening 排料耳轴孔
discharge water 排水
discharger 排出装置,卸载器,推料机
discipline designation 学科[专业]名称
disclosure agreement 公开协议
discount rate 贴现[折扣]率
discount sequence 贴现顺序
discount yield 贴现收益率
discounted cash flow analysis 贴现的现金流动分析
discounted net cash flow 贴现的净现金流量
discounting technique 折扣[贴现]技术
discrete analog 离散模拟
discrete fraction 离散分数
discrete grain 离[分]散颗粒
discrete kinetic approach 离散动力法
discrete kinetic scheme 离散动力(学)方案
discrete phase 不连续相,分散相

discrete size class 离散粒级
discrete technique 离散技术
discrete-distributed kinetic scheme 离散分布动力(学)方案
discretely distributed parameter flotation model 离散型分布参数浮选模型
discussion of difference 差距讨论
disjoining pressure 分离压力
disk 磁盘,盘
disk memory 磁盘存储器
disk pack 磁盘组[部件]
disk-based 磁盘支持的
diskette 软磁盘,塑料磁盘
disordered precipitate 无定形沉淀物
disordered structure 无定形结构
dispersant-depressant agent 分散-抑制剂
dispersed air flotation cell 分散空气浮选槽
disperser 扩[分]散器
disperser support spider 分散器十字支架
disperser-cell wall distance 分散器与槽壁距离
dispersible gold 乳[悬浮]状金
dispersing effect 分散作用[效应]
dispersing meter 分散计
dispersion 分散(体,相,体系,作用),弥[扩]散(现象)
dispersion force 色[弥]散力
dispersion height 分散高度

dispersion period 分散阶段
dispersion state 分散状态
dispersion variance 离散方差
displaced value 错置值
displacement current 位移电流
displacement oil pump 旋转[排代]油泵,活塞式油泵
display system 显示系统
disposal area （选矿原矿）处理区
disseminated copper deposit 浸染铜矿床
disseminated copper porphyry 浸染状含铜斑岩
disseminated porphyry copper 浸染型斑岩铜矿
dissociative chemisorption 离解化学吸附
dissolubility 溶解性
dissolution reaction 溶解反应
dissolution type phenomena 溶解类现象
dissolution-precipitation reaction 溶解沉淀反应
dissolved metal 溶解金属
dissolved mineral salt 溶解矿物盐
dissolved oxygen 溶解氧
dissolved oxygen concentration 溶解氧浓度
dissolved substance 溶解物质
dissolved valuable substance 已溶解的有价物质
dissolving action 溶解作用

dissolving metal ion 溶解金属离子
dissolving power 溶解力
distillation-colorimetry 蒸馏-比色法
distillation-titration 蒸馏-滴定法
distilled water 蒸馏水
distinctive condition 独特条件
distributed flotation kinetics 分布动力学
distributed flotation model 分布浮选模型
distributed parameter 分布参数
distributed parameter model 分布参数模型
distributed stream 均[分]布料[矿]流
distributing belt conveyor 分配胶带运输机
distribution and marketing cost 分配[发]销售费
distribution coefficient 分布系数
distribution cone 分配锥(赖克特圆锥选矿机的)
distribution diagram 分布图
distribution modulus 分布模数
distribution of floatability 浮游性分布
distribution potential 分布电位
distribution system 分配系统
disturbance variable 扰动变量
disulfide-type collector oxidation product 二硫化物型捕收剂氧化产物
disulphide oxidation product 二硫化物氧化产物
Ditalan 十二烷基硫酸盐(东德产品,可用作重晶石的捕收剂)
dithiocarbamate 二硫代氨基甲酸盐(可用作捕收剂)
Dithion 捕收剂(二乙基二硫代磷酸盐)
dithiophosphate 黑药(二硫代磷酸盐,可用作浮选剂)
dithiophosphate anion 二硫代磷阴离子
dithiophosphate promoter 二硫代磷酸盐促进剂
dithizone 双硫腙,二苯基硫卡巴腙(可用作捕收剂) $C_6H_5N:NCSNHNHC_6H_5$
divalent 二价的
divalent alkaline earth cation 二价碱土阳离子
divalent anion 二价阴离子
divalent metal cation 二价金属阳离子
divalent metal n-butyl xanthate 二价金属(的)正丁基黄原酸盐
diversification 多样化
diversified operation 多种经营
diversion box 转向[换]箱
diversity factor 差异因[系]数
diverter box 分流器箱
diverter gate 转向闸门
diverter plate 转向板
dividend 红利,股息,被除数
dividend limitation 股息限额
dividend payment 股息付款
dividend right 分红权利
divider 分配[隔,切,流]器,圆规

divider plate 分配[切]板

divider sampler 分流[配]式取样器

division 组成部分[单元],部分[门],区,段,片,师,局,处,科,组,分队

Divulsion D 湿润剂(烷基磺酸盐)

dixanthate 二黄原酸盐(俗称双黄药)

dixanthogen 双黄药(黄原酸盐离子氧化而生成的一种中性二聚物)

dixanthogen oxidation product 双黄药氧化产物

DLT-reagent 浮选药剂(二乙醇胺和高级脂肪酸的缩合产物,可用作捕收剂)

DLVO theory DLVO 理论(一种关于胶体稳定性的理论)

DMS cyclone 重介质旋流器

DMS drum 重介质滚筒

DOBM 捕收剂(十二烷基辛基—苄基甲基氯化铵)

docent 讲[教]师,讲解员

dock loading and unloading system 码头装卸系统

document 文件[献],公文,文档,资料,单据,证书

document controller 文件管理人员

document distribution 文件分配

document distribution chart 文件分配图表

dodecane 十二(碳)烷 $C_{12}H_{26}$

dodecanoic acid 十二烷酸,月桂酸

dodecyl 十二(碳)烷基,十二基 $C_{12}H_{25}-$

dodecyl octyl benzyl methyl ammonium chloride 十二烷基辛基苄基甲基氯化铵

dodecyl sulfate 十二烷基硫酸酯[盐](可用作捕收剂、乳化剂、湿润剂) $C_{12}H_{25}OSO_3M$

dodecyl sulfonate 十二烷基磺酸盐(可用作捕收剂、分散剂、乳化剂、抑制剂) $C_{12}H_{25} \cdot SO_3M$

dodecyl sulfonate anion 十二烷基磺酸盐阴离子

dodecyl xanthate 十二烷基黄原酸盐

dodecyl xanthate ion 十二烷基黄原酸盐离子

dodecylamine 十二烷胺,月桂胺(可用作捕收剂) $C_{12}H_{25}NH_2$

dodecylammonium acetate 十二烷基醋酸铵

dodecylammonium chloride 十二烷基氯化铵

dodecylammonium fluosilicate 十二烷基氟硅酸铵盐

Dodge(jaw)crusher 道奇型颚式破碎机(可动颚板摇动轴在下部)

Dodge pulverizer 道奇型粉碎机(六角形转筒球磨机)

Doittau 14SM 捕收剂(一种硫酸化或磺化脂肪酸,法国产品)

dolomitic calcite 白云石质方解石

domestic iron resources 国内铁矿资源

domestic output 国内产量

domestic production 国内产量

domestic purpose 家用目的
domestic smelter 国内[本地]冶炼厂
domestic-water tank 生活用水水池
dominant electrode 主电极
Dominion ball mill 多米尼翁型球磨机
donated electron 析出电子
donation 捐赠(物),捐款
donor-like surface state 施电子型表面状态
doormat 门垫
Doppler(Cheristian Johann Doppler) 多普勒(奥地利物理学家,克里斯琴·约翰·多普勒),多普勒的
Doppler effect 多普勒效应(主要内容：物体辐射的波长因为波源和观测者的相对运动而产生变化)
Doppler effect method 多普勒效应法
Doppler effect ultrasonic flowmeter 多普勒效应超声波流量计
Doppler lidar 多普勒激光雷达
Doppler radar 多普勒天气雷达
Doppler shift 多普勒效应漂移
Doppler sodar 多普勒雷达
Doppler velocimeter 多普勒速度计
Doppler velocity 多普勒速度
Dore bar 多尔合金[金银]棒
Dore bullion 合质金,多尔金银块
Dore button 金银小块[球,珠,粒]
Dore furnace 金银炉
Dore metal 多尔合金,金银合金
Dore silver 多尔银,粗银(含有少量金的银)

dormitory area 宿舍区
Dorr airlift agitator 多尔[道尔]型空气提升搅拌器
Dorr thickener 多尔[道尔]型浓密[稠、缩]机
dorrclone plant 多尔[道尔]水力旋流器分级厂
Dorrco agitator 道尔科型搅拌槽[器]
Dorr-Oliver 12253 thickener 道尔-奥利弗12253型浓缩机(美国Dorr-Oliver公司生产)
Dorry abrasion testing machine 道瑞磨损试验机(英国)
Dorry hardness test 道瑞硬度试验
dosage rate 剂量
dosing point 加药点
dotted line 点[虚]线
double 两[加]倍的,双重的,双(倍,重,幅,联)的,复(式,合)的,两用的,(成)双[对]的,重合的,两种(意义)的,模棱两可的
double 220kg capacity Wabi reverberatory furnace 两[双]室220kg能力的Wabi型反射炉
double bond 双键
double concentrating cone 双锥选矿机(赖克特圆锥选矿机)
double deck screen 双层筛
double display 双显示
double distilled water 二次蒸馏水
double drum ore hoist 双滚筒矿石提升机

double drum service hoist 双滚筒辅助提升机

double eccentric shaft 双偏心轴

double electrical layer concept 双电层概念

double layer 双电层

double layer attraction 双电层引力

double layer characteristics 双电层特性

double layer effect 双电层效应

double layer formation 双电层形[生]成

double layer potential 双电层电位[势]

double overflow spitzkasten 双溢流角锥槽(浮选用)

double pitch spiral 双节距螺旋

double salt 复盐

double scoop feeder 双勺给矿机

double screw fine material washer-classifier dehydrator 双螺旋粉料洗涤-分级脱水装置

double screw Holoflite dryer 双螺旋Holoflite式干燥机

double skirted arrangement 双边拦矿板配置

double spiral 双螺旋

double spiral construction 双螺旋结构

double tapered stepped cast grizzly 双楔形分段式铸造格筛

double wave shell liner 双波纹筒体衬板

double-compartment mill 双室磨机

double-deck scalping screen 双层除[隔]粗筛

double-door 双层门

double-drum magnetic separator 双滚筒磁选机

double-duty machine 两用浮选槽[机]

doublex machine 双螺旋设备(分级机的)

doubtful account 未定账款

doubtful ore zone 有疑问的矿带

Douglas CZ pearl starch Douglas CZ玉米淀粉(品名,可用作絮凝剂)

Douglas No. 502 canary dextrine Douglas 502号玉米糊精(品名,可用作絮凝剂)

Dow chemical SA-1236 道氏起泡剂SA-1236

Dow Corning 200 DC 200号矿泥分散剂(非离子型聚硅油)

Dow Corning Antifoam A DC聚硅油消泡剂(可用作捕收剂、分散剂、调整剂)

Dow Corning Silicone Fluid 苯基甲基聚硅油(DC聚硅油系列产品,可用作捕收剂、分散剂、调整剂)

Dow Corning Silicone Fluid 550 苯基甲基聚硅油

Dow Corning Silicone Fluid F-258 二甲基聚硅油(可用作捕收剂、分散剂、调整剂)

Dow Corning Silicone Fluid F-60 氯苯甲基聚硅油(可用作捕收剂、分散剂、调整剂)

Dow froth 美国 DOW(中文名称:道氏)化学公司生产的起泡剂系列产品

Dow froth 250 DOW 起泡剂 250(三聚丙二醇甲醚) $CH_3-(OC_3H_6)_3OH$

Dow SA 1797 Z-200 的乳化形式(可用作浮选硫化矿的捕收剂)

Dow Z-200(=Z-200) DOW Z-200 捕收剂(O-异丙基-N-乙基硫代氨基甲酸酯,硫代表面活性剂,是硫化矿的主要浮选药剂)

downstream equipment 下游设备

downstream type tailings dam 下游[顺流]式尾矿坝

downtime prediction 停机[车]预测[报,告,料]

dozer 定量器

dozing equipment 堆土设备

DP 243 DP 243 选矿药剂(脂肪胺盐酸盐的混合物,含 55%月桂胺、18%癸胺、18%辛胺、17%十四烷胺,可用作钾盐矿的捕收剂)

DPC DPC 捕收剂(十六烷基甜菜碱)

DPG(diphenylguanidine) 二苯胍[促进剂 DPG](可用作橡胶硫化的促进剂)

DPG material balance test 二苯基胍物料平衡试验

DPLA DPLA 捕收剂(十八烷基三甲基氯化铵)

DPN DPN 捕收剂(十八烷基甜菜碱)

DPQ DPQ 捕收剂(十二烷基三甲基氯化铵)

D-R ring assembly D-R 环部件(美国 D-R 公司专利产品)

draft 草稿,草案,草图

draft prospectus 募款[提款(汇票,支票)]计划书

draft tube 引流管

draft tube agitator 通[吸]风管搅拌槽

draft tube mechanical agitated tank 通风[吸入,轴流]管式机械搅拌槽

draft tube-rotor engagement 竖管-转子啮合

drafting 起草,制图,拖拽

draft-tube agitator 带通气管搅拌器

drag coefficient 阻力系数

drag force 拖曳力

drag line 绳斗(电)铲,索斗[拉铲]挖掘机

drag line pulley 绳斗铲滑轮

dragline mining 索斗铲采矿

drain cycle 脱[排]水(周)期[程序]

drain port 排水[液]口

drainage 排水

drainage characteristic 排水性

drainage deck 排[脱]水板

drainage deck length 脱水段[台面]长度(分级机)

drainage section 排[脱]除区,脱介区(介质回收筛的)

dramatic relationship 戏剧性[引人注

draw down	引来,招致,拉[扯,放]下
draw draught	牵引功率
draw power	抽运[牵引]功率
drawback	缺点,不利条件
drawdown	放矿
drawdown angle	放矿角,矿石排角
drawing category	图纸类别[种类]
drawing file	图纸档案
drawing identification	图纸辨[识,鉴]别
drawing list	图纸目录
drawing office	绘图[设计]室
drawing schedule	制[绘]图计划
drawing sequence	图纸顺序
drawing status report	制图[图纸]情况报告
drawing status summary	图纸情况概述
drawing title	图(纸)名(称)
Drene	捕收剂(十二烷基硫酸钠)
Dresinate	表面活性剂品牌名
Dresinate 7V6	松脂酸钠皂(可用作捕收剂)
Dresinate 7XI	松脂酸钠皂(可用作捕收剂)
Dresinate 7XIK	松脂酸钾皂(可用作捕收剂)
dribble chute	集矿[接漏]溜槽
dribbles belt	集[滴]矿胶带
dried concentrate	干燥的精矿
dried product	干燥产品
dried-out area	已干涸[燥]区域
Dri-Film	湿润剂、分散剂(非离子型聚硅油)
drift timbering	水平巷道支护
drill core sample	钻孔岩芯样
drill-based sampling method	钻孔[机]取样法
drill-based sampling technique	钻孔[机]取样技术
drilling activity	钻探活动
drilling approach	钻孔方法[手段]
drilling arrangement	钻孔布置
drilling grid	钻孔网度
drilling mud	钻探用的泥浆
drilling pattern	钻孔布置
drilling program	钻孔程序
drip irrigation system	滴灌系统
drip proof	防水,防滴
drip tight	防滴的,不透水的
drive	传动(装置)
drive arrangement	驱动装置
drive component	传动部[组,构]件
drive coupling	驱动联轴节
drive design	驱动设计
drive efficiency	传[驱]动效率
drive friction factor	传动摩擦系数
drive inefficiency	驱[传]动效率低下
drive inertia	传动惯性
drive loss	驱[传]动损失
drive requirement	驱动(功率)要求
drive sprocket	驱动链轮
drive station	驱动站

drive unit 驱动装置

drive unit with overload control 带过负荷控制的拖[驱]动装置

drive wrap angle 传[驱]动包角

driven sheave 从动轮

driven sprocket 从动链轮

driver sheave 驱动轮

driving force 驱动力

driving tension ratio 驱动张力比

droplet of solution 液滴

drop-wise 滴状,液滴方式

drossing and modeling kettle 造渣建模锅

Drucol CH 磺盐酸(品名,可用作捕收剂)R—COOC$_2$H$_4$·SO$_3$Na(R=十二烷基)

drum dumper 圆筒翻车机

drum reclaimer 筒型取料机

drum-scoop feeder 联合给矿机

drum-type wet magnetic separator 鼓型湿式磁选机

dry aerofall mill 干式气落式磨机

dry air-swept mill 干式气吹式磨机

dry autogenous mill 干式自磨机

dry basis 干组分

dry benzene 纯苯

dry bulk density 干容重[体积重量]

dry cement grinding circuit 水泥干磨系统,干式水泥磨矿系统

dry depressed air 干压缩空气

dry dust collector 干式集[吸,除]尘器

dry fiber collector 干式纤维收集[尘]器

dry grate ball mill 干式格子球磨机

dry gravity concentration 干法[式]重选

dry grinding alternative 干磨方案

dry grinding installation 干磨装置[设备]

dry ground product 干磨产品

dry lip contact seal 干式唇形接触密封

dry magnetic cobbing 干式磁选机

dry measure 干量单位制

dry mineral-matter-free (DMMF) carbon content 干燥[无水]的无矿物质的固定碳含量

dry operation 干式作业,干法生产

dry peripheral rod mill 干式周边排矿棒磨机

dry pre-concentration 干法[式]预富集

dry preparation 干式预备,干式处理(筛分)

dry preparation screen 干式预筛

dry process 干法过程[加工]

dry processing 干式工艺[加工,选矿]

dry reagent 干药剂

dry separation technique 干式分选技术

dry state 干态

dry tons of concentrate 干吨精矿

dry weight basis 干重基

dryer combustion burner 干燥机燃烧喷嘴

dryer exhaust gas 干燥机排气
dryer exhaust wet scrubber 干燥机排气湿式洗涤器
dryer-cooler complex 干燥冷却联合机,干燥机冷却器成套设备
dryer-cooler system 干燥机冷却机系统
drying and sizing stage 干燥筛分阶段
drying pad 干燥盘[板]
dry-type gas blowing grinding mill 干式气吹磨机
DT 120 选矿药剂(聚丙烯酰胺,德国药剂名,同Separan 2610,可用作絮凝剂)
DTM 捕收剂(羟基十二烷基三甲基溴化铵) $HO(CH_2)_{12}N^+(CH_3)_3Br^-$
DTP-reagent 二硫代磷酸盐药剂
dual cleaning 二重精选
dual drive 双重传动
dual function 对偶函数
dual impeller agitator 双叶轮搅拌器
dual pulley 双(重传动)滚筒
dual-purpose machine 双作用设备
duct lining 管道衬里
duct sizing 管径的确定,管道定径
duct support 管道支架
duct system balancing calculation 输送管道系统平衡计算
duct system pressure drop calculation 输送管道系统压力降计算
duct work 管道敷设[系统]
ducting system 风管系统
dull constituent 发暗成分(煤)

dump and return cycle time 卸矿复位周期(时间)
dump cycle 卸矿循环
dump hopper 卸矿漏斗仓[漏斗式小仓]
dump leaching 废石堆浸
dump signal light 卸载信号灯
dump truck 自(动倾)卸卡车,矿山自卸车
dumper operator 放矿工,翻车机操作人员
dumping angle 卸料角,卸车角
Duomac S 大豆油脂肪胺醋酸盐(品名,可用作捕收剂)
Duomac T 牛油脂肪二胺醋酸盐(品名,可用作捕收剂、絮凝剂)
Duomeen 表面活性剂品牌名
Duomeen C 烷基丙二胺(含二胺80%) $RNHC_3H_6NH_2$,R来自椰油
Duomeen CD 烷基丙二胺(含二胺84%) $RNHC_3H_6NH_2$,R来自椰油
Duomeen S 大豆油脂肪二胺(可用作捕收剂、絮凝剂)
Duomeen T 牛油脂肪二胺(可用作捕收剂、絮凝剂)
duplex adjustable stroke diaphragm pump 双联可调冲程隔膜泵
duplex double pitch submerged flared spiral classifier 沉没式,双头,双螺旋加宽螺旋分级机
duplex jig 双室跳汰机
duplex mineral jig 双室矿物跳汰机

duplex mineral rougher jig 双室矿物粗选用跳汰机

duplex quadrant gate 双扇形闸门

duplex screw classifier 双螺旋分级机

duplicate 复本,复制品[物],副[复,底]本,副[复,抄]件,双联[重,份]的

duplicate equipment motor 备用设备电动机

duplicate part 备品[件]

duplicate pump 备用泵

duplicate sampling 重复取样

duplicate split 复分样

Duponol 阴离子表面活性剂(烷基硫酸盐类药剂)品牌名

Duponol 100 正-辛烷基硫酸钠(干燥水溶性白色粉末,可用作捕收剂)

Duponol 80 正-辛烷基硫酸钠(可用作捕收剂、乳化剂、分散剂湿润剂)

Duponol C 十二烷基硫酸盐(可用作氧化铁矿的捕收剂) $C_{12}H_{25}OSO_3M$

Duponol CA(=Na-Oleinsulfate) 十八烯醇硫酸钠(同 Na-Oleinsulfate,可用作氧化铁矿的捕收剂) $C_{18}H_{35}OSO_3Na$

Duponol D 混合高级醇硫酸钠(可用作湿润剂、乳化剂) $R-OSO_3Na$

Duponol LS-paste 十八烷基硫酸钠(可用作重晶石的捕收剂、分散剂、乳化剂) $C_{18}H_{37}OSO_3Na$

Duponol ME 十二烷基硫酸钠(可用作氧化铁矿的捕收剂) $C_{12}H_{25}OSO_3Na$

Duponol MP 纯十二烷基硫酸钠(可用作分散剂、湿润剂、乳化剂)

Duponol OS 脂肪醇胺硫酸盐混合物(可用作捕收剂)

Duponol WA 十二烷基硫酸盐(或钠)(可用作捕收剂、分散剂、湿润剂、乳化剂)

Dupont flocculant EXR-102A Dupont EXR-102A 絮凝剂(羧甲基纤维素钠)

Dupont frother B22 Dupont B22 起泡剂

Dupont frother B23(=B23 frother) Dupont B23 起泡剂(高级醇和酮的混合物)

Duro 丢洛(表示硬度的标度,近似于国际橡胶硬度单位 IRHD)

dust (灰,微,粉,烟)尘,尘埃[土],粉(末,剂),药(粉,末)

dust air 含尘[高粉尘]空气

dust and fume control schematic diagram 粉尘烟雾控制简[示意]图

dust capture velocity 捕[集]尘速度

dust chute 集[收]尘溜槽

dust collection and ventilation criteria 收尘通风标准

dust collection equipment 集[收]尘设备

dust collection system 收尘系统

dust collection system total pressure 收尘系统总压力

dust collection tube sheet 收尘器管材薄板

dust collector details 收尘器细部

dust collector hopper 收尘器漏斗

dust collector housing 集[收]尘器罩
dust collector structure steel 收尘器结构钢
dust containment system 矿尘抑制系统
dust control practice 粉尘控制实践[施]
dust cover 粉尘盖
dust emission 粉尘排放[出](物)
dust emission control regulation 粉尘排放控制条例
dust enclosure 防尘罩
dust entrainment 带走粉尘量
dust handling equipment 粉尘处理[搬运]设备
dust handling system 粉尘处理系统
dust ignition proof 防粉尘着火
dust loading 粉尘荷载[装载]
dust loss 灰尘损失
dust nature 粉尘性质
dust pick-up hood 吸尘罩
dust seal 防尘密封
dust suppressant spray 防[抑]尘喷雾器
dust suppression 粉尘抑制
dust suppression equipment 粉尘抑制设备
dust suppression location 粉尘抑制地点[地段,场所]
dust suppressor 粉尘抑制器
dust system 集[收,防]尘系统
dust system design 粉尘系统设计

dust tight 防尘
dust tight enclosure 防尘密封罩
dust tight surge bin 防尘缓冲仓
dust-air quantity 含尘空气量
dust-collecting cyclone 集尘旋流器
dust-collecting system 集尘系统
dust-conveying worm 集尘螺旋运输机
dustiness 含尘量
dust-proof box 防尘罩[盒形小室]
dusty operation 高粉尘[多尘]作业
Dutch State Mine Screen 荷兰国家矿业制筛
duty 任务,工作(制度,状态),功能[用]
duty and responsibility 职责
duty of pump 泵的效率
duty specification 任务规范
Duval patent 杜瓦尔专利
DX-882 絮凝剂(阴离子型高分子聚羧酸)
dye house effluent 染坊废液
dynamic angle 动态(接触)角
dynamic angle of repose 动态休止角
dynamic behavior 动态
dynamic compensation 动态补偿(作用)
dynamic condition of flotation 浮选动态条件
dynamic contact angle 动态接触角
dynamic diffusiophoretic effect 动力(学)扩散电泳效应[作用]

dynamic factor 动力[态]因素
dynamic froth 动态泡沫
dynamic load 动载[负]荷
dynamic model 动态模型
dynamic optimization 动态最优化
dynamic parameter 动态参数
dynamic performance 动态特性
dynamic programming 动态规划
dynamic response 动态反应
dynamic signal 动态信号
dynamic simulation 动态模拟
dynamic state model 动态模型
Dynesol F20 湿润剂(烷基纯聚酯磺酸盐)

E

1384 EHD hydrocone 1384 EHD 液压圆锥破碎机
E. P. 487 捕收剂(十二烷基胺盐酸盐,可用作蔷薇辉石($MnSiO_3$)的捕收剂) $C_{12}H_{25}NH_2 \cdot HCl$
early days 早期
early evaluation 早期评价
early evaluation of orebody 矿体早期评价
early production phase 生产初期
early stage of exploration 勘探初期
earning power 赢利能力
ear-protection device 保护听力装置
earth acceleration of gravity 地心引力加速度
earth gravity 地心引力
earth moving equipment 推[搬]土设备
earth's crust 地壳
earth-fill dam 填土坝
earth-fill tailing dam 土尾矿坝
earthing protection 接地保护
earthing wire 接地线
easily handled product 易处理[搬运]的产品
eastern hemisphere 东半球
easy axis 易磁化轴
easy operability (容)易操作性
easy ore 易选矿石
easy-to-crush material 易于破碎的物料
eccentric bushing 偏心轮轴衬
eccentric drive belt 偏心轮传动皮带
eccentric shaft 偏心轴
eccentric wheel assembly 偏心轮装置[组件]
eccentricity 偏心距(离)
economic advancement 经济进展
economic advantage 经济优势

economic analysis 经济分析
economic and financial feasibility 经济财政可行性
economic and technical standing point 经济技术立场
economic aspect 经济观点[情况,方面]
economic assessment 经济评价
economic benefit 经济利[效]益
economic benefit calculation 经济利益计算
economic comparison 经济比较
economic concentration 经济富集
economic condition 经济条件[情况,形势,状况]
economic consideration 经济问题
economic contribution 经济贡献
economic criterion 经济标准
economic equation 经济方程式
economic evaluation 经济评价
economic exploitation 经济开发
economic factor 经济因素
economic feasibility 经济可行性
economic index 经济指数
economic indicator 经济指标
economic indices 经济指标
economic justification 经济合理性
economic liberation 经济解离
economic life 经济寿命
economic limit 经济效果界限
economic metallurgy 经济冶金
economic mineral 经济[有用]矿物

economic mineral deposit 有经济价值的矿床,经济矿床(可开采矿床)
economic objective function 经济目标函数
economic obligation 经济义务
economic optimum operation 最佳经济作业
economic ore deposit 工业[有经济价值的]矿床
economic orebody 经济[工业]矿体
economic parameter 经济参数
economic performance 经济效能
economic pit limit 露天矿经济(开采)境界,经济露天矿极[界]限
economic power requirement 经济功率需要量
economic problem 经济问题
economic process 经济方法[工艺]
economic processing method 经济加工方法
economic production 经济生产
economic proportion 经济比例
economic recovery 经济回收率
economic research (技术)经济研究
economic result 经济后果
economic return 经济回收[收益,回报]
economic standpoint 经济立场
economic study 经济研究
economic value 经济价值
economic viability 经济生存能力,经济可行性

economic worth 经济价值
economical capital 经济投资
economical comminution technique 经济粉[磨]碎技术[方法]
economical justification 经济合理性,经验效益的验证[证明]
economical justification study 经济合理性研究
economical method 经济方法
economical power consumption 经济功耗
economical treatment 经济处理
economically acceptable route 经济上可取的方法
economically optimum circulating load 经济上的最佳循环负荷
economy and financial condition 经济财政状况
economy of scale 规模经济
eddy current braking torque 涡流制动转[扭]矩
eddy current coupling 涡流联轴器
edge distance selection （胶带）边缘距离选择（的带宽余量）
edge of molybdenite layer 辉钼层棱边
Editas B 选矿药剂（羧甲基纤维素钠,英国产,可用作絮凝剂、抑制剂）
editor 编辑,编辑器
editor program 编辑程序
editorial 社论[编辑]的
editorial assistant 编辑助理
EDTA（= ethylene diamine tetraacetic acid） 乙二胺四乙酸（螯合试剂,可用作捕收剂）
educator 教育工作者
effect of basicity 碱度影响
effect of bubble size 气泡大小影响
effect of conditioning 搅拌影响[效应]
effect of fines 细粒[粉矿]的影响
effect of frothing agent 起泡剂的影响
effect of hydrocarbon chain length 碳氢链长度效应
effect of hydrostatic pressure 静水[液]压的影响
effect of ionomolecular complex 离子分子络合物效应
effect of mass 质量影响
effect of material transport 物料转移效应
effect of neutral molecule 中性分子效应
effect of particle size 粒度影响
effect of pH pH值的影响
effect of radiation 辐射作用[效应]
effect of solid state change 固态变化的影响
effect of tank size 槽[罐]尺寸的影响
effect of temperature 温度影响
effect of thickness 厚度效应[作用]
effect of ultraviolet light 紫外线的影响
effective concentration 有效浓度
effective density 有效密度
effective guidance 有效的指导

effective hours 有效时数
effective quadrature method 有效的求积方法
effective quadrature routine 有效的求积方法
effective screen mesh 有效筛孔
effective screening 有效筛分
effective separation 有效分选
effective utilization rate 有效利用率
effective vertical chute drop distance 溜槽有效垂直下降高度
effective volume 有效容积
effectiveness of particle size separation 粒度分离有效性
efficiency equation 效率方程
efficiency factor 效率系[因]数
efficiency of bubble capture 气泡捕集效率
efficiency of collection 捕收效率
efficiency of power application 功率利用率
efficiency of transfer 转换效率
efficiency of undersize recovery 筛下粒级回收效率
efficiency of undersize removal 筛下级别排[筛]除(效)率
efficiency percentage 效率百分比
efficient grinding 有效磨矿
effluent 污水,废水,流出物,废气
effluent limitation 污[废]水排放限度标准
effluent water quality 排水水质

Eh 氧化还原电位[平衡电位](单位为mV,是溶液氧化性或还原性强弱的衡量指标)
Eh measurement 氧化还原电位[平衡电位]测定
Eh value 氧化还原电位[平衡电位]值
eight equal semi-annual installment 八个等值的半年分期付款
eight-hour shift 八小时一班(制)
Eimco Agidisc filter 艾姆科爱杰迪斯克型过滤机
Eimco belt filter 艾姆科带式过滤机
Eimco disc filter 艾姆科盘式过滤机
Eimco drum filter 艾姆科圆筒过滤机
Eimco high-capacity thickener 艾姆科型高效浓缩机
Eimco thickener 艾姆科型浓缩机
Eimco traction thickener 艾姆科周边传动式浓缩机
ejection-dispersion type of column 喷射扩散型浮选柱
elaidic acid 异油酸(可用作捕收剂) $C_{17}H_{33}COOH$
elaidinate 反油酸盐
elastic rubber flap 弹性橡胶铰链板[挡板]
Elastiol LL 浮选剂(生产亚麻仁油的副产物,黑钨矿浮选时作为捕收剂)
elasto-mechanic behavior 弹性机械状态
elasto-mechanical transient behavior 弹性机械瞬时状态[瞬变过程]

elastomer lining 合成橡胶衬里
elastomer material 弹性材料
electric attribute 电气属性,电气特性
electric attribute of separator 分选机的电气属性
electric control panel 电控制盘
electric dryer 电干燥器[机]
electric equipment room 电气设备室
electric field of cathodic polarization 阴极极化电场
electric furnace 电炉
electric furnace method 电炉法
electric gate valve 电动闸阀
electric iron making process 电炼[制]铁法
electric line 电气线路
electric power generation 发电
electric powered walking dragline 电动步式索斗铲
electric resistance air-froth interface probe 电阻空气-泡沫界面探测器[探针]
electric station 配电站,电气站
electric undertaking 电业
electric vibrator 电振(动)器
electrical capacitor 电容器
electrical component 电气配[组,部]件
electrical contact 电接触[点],电触头,电气接触
electrical control 电气控制
electrical distribution system 配电系统
electrical disturbance 电干扰
electrical double layer 双电层
electrical double layer concept 双电层概念
electrical drawing 电气图纸
electrical effect 电效应
electrical energy 电能
electrical energy input 电力[能]输入
electrical engineering 电工技术,电机工程,电气工程
electrical equipment room 电气设备室
electrical interaction phenomenon 电的相互作用现象
electrical loss 电力损失
electrical property of surface 表面电性
electrical room 配电室,电气室
electrical sensing 电测
electrical separation 静电分选,电选
electrical substation 变电站
electrically conducting solution 导电液
electrically heated retort 电热蒸馏器[罐]
electrically-operated solenoid valve 电动电磁阀
electrical-resistance helix 电阻螺旋线
electric-hydraulic control system 电动液压控制系统
electro switch station 电开关站,电力配电室
electrocatalyst 电催化剂
electrocatalytic property 电催化性

electrochemical approach 电化学方法
electrochemical cell 电化学电池
electrochemical characteristic 电化学性质
electrochemical condition 电化(学)条件
electro-chemical deposition 电化学沉积
electro-chemical dissolution 电化学溶解
electrochemical flotation 电化浮选
electrochemical flotation separation 电化浮选分离
electrochemical heterogeneity 电化学不均匀度[性]
electrochemical history 电化学历程
electrochemical investigation 电化学的研究
electrochemical measurement 电化学测量
electrochemical mechanism 电化学机理
electrochemical observation 电化学观察(结果)
electrochemical oxidation 电化学氧化
electrochemical parameter 电化学参数
electrochemical phase diagram 电化学相位图
electrochemical phenomena 电化学现象
electrochemical phenomenon 电化学现象
electrochemical polarization 电化学极化(作用)
electrochemical polarization technique 电化学极化技术
electrochemical potential 电化学势
electrochemical process 电化过程
electrochemical property 电化学性质
electrochemical reaction 电化学反应
electrochemical theory 电化学理论
electro-cybernetics 电子控制论
electrode assembly 电极装置[组合]
electrode process 电极过程
electrode process mechanism 电极过程机理
electrode reaction 电极作用
electrode surface 电极表面
electroformed micromesh sieve 电铸微网目筛
electroformed sieve 电铸筛
electro-hydraulic power pack 电动液力部件
electrokinetic behavior 动电学特性
electrokinetic condition 动电状态
electrokinetic data 动电数据
electrokinetic experiment 动电试验
electrokinetic measurement 动电测定
electrokinetic phenomena 动电现象
electrokinetic process 动电过程
electrokinetic result 动电结果
electrokinetic study 动电学研究
electrolysis 电解

electrolyte surfactant 电解质表面活性剂
electrolytic cell 电解槽
electrolytic deposition 电解沉淀
electrolytic recovery 电解回收
electrolytic zinc plant 电解锌厂
electromagnet 电磁铁
electromagnetic measurement method 电磁测定法
electromagnetic recorder 电磁记录器
electromagnetism 电磁(性)
electromechanical device 机电装置
electro-mechanical drive 电动机械驱动(装置)
electromechanical sounding probe 机电测深计
electro-metal plant 电冶工厂
electromotive force 电动势
electron beam micro-analysis 电子束微量分析
electron current 电子流
electron diffraction 电子衍射
electron diffraction experiment 电子衍射试验
electron diffraction pattern 电子衍射谱图
electron doning ability 电子施主能力
electron energy band 电子能带
electron excitation 电子激发
electron image 电子图像
electron microprobe 电子显微探针
electron micro-probe technique 电子显微探针[测]技术
electron microscope 电子显微镜
electron microscopy 电子显微镜
electron spectrometry 电子光谱法
electron spin resonance spectroscopy 电子自旋共振分光学[测定法]
electron transfer 电子转移
electron transfer step 电子传递阶段[步骤]
electron volt 电子伏特
electron-deficient space charge region 乏电子空间电荷区
electron-depleted space charge region 贫电子空间电荷区
electronegativity 负[阴]电性[度]
electronegativity difference 负电性差
electroneutral plane 电中性平面
electroneutrality 电中性
electron-hole pair 电子-穴对
electronic analogue signal 电子模拟信号
electronic analysis 电子分析
electronic batch weigh system 电子分批称量系统
electronic belt scale 电子皮带秤
electronic calibration 电子校准
electronic central processing unit 电子中央处理机
electronic charge transfer 电子电荷传送
electronic component 电子元件

electronic conduction 电子传导	电泳浊度测定
electronic conductivity 电子传导性	electrophoretic property 电泳性质
electronic conductor 电子导体	electrophysical parameter 电物理参数
electronic controller 电子控制器	electrophysical property 电物理性质
electronic drift 电子漂移	electropolished 电抛[磨]光的
electronic force 电子力	electrorefined and electrowon cathode 电解电积阴极
electronic scale controller 电子地磅控制器	electro-refined copper 电解精炼铜
electronic signal 电子信号	electrorefining 电解精炼
electronic simulation 电子模拟	electrostatic adsorption potential 静电吸附势能
electronic spectroscopy of chemical analysis 化学分析电子[光]谱学	electrostatic attraction 静电吸引[力]
electronic stream scanning 电子流[束]扫描	electrostatic circuit 静电(选矿)回路[流程,系统]
electronic structure 电子结构	electrostatic contribution 静电分量
electronic void proportion 电子空穴比例	electrostatic double layer 静电双电层
	electrostatic energy 静电能
electronic weigh-scale 电子秤	electrostatic energy of repulsion 静电排斥能
electronics 电子学,电子仪器[设备,线路]	electrostatic force of repulsion 静电(排)斥力
electron-to-hole rate 电子空穴比例	electrostatic interaction 静电相互作用
electro-optical scanning 光电扫描	electrostatic potential energy 静电势能
electro-osmotic dewatering 电渗透脱水	electrostatic precipitator 静电收尘器
electro-osmotic method 电渗透法	electrostatic repulsion 静电排斥
electro-osmotic technique 电渗透技术	electrostatic repulsive force 静电排斥力
electro-oxidation 电氧化(作用)	
electrophoresis curve 电泳曲线	electrostatic response 静电反应
electrophoretic mobility 电泳浊度	electrostatic term 静电项
electrophoretic mobility determination 电泳浊度测定	electrostatic testing 静电(选矿)试验
electrophoretic mobility measurement	electrostatic theory 静电原理

electrostatic theory of flotation 浮选静电理论

electrostatically charged water 带静电的水

electrowinning 电(解沉)积,电解冶金(法)

electrowinning cathode 电解阴极

electrowinning cell 电积槽

electrowinning plant 电积厂[车间]

electrowon copper 电积铜

element of interest 感兴趣[有价值,所关心]的元素,目的元素

element of progress 进步环节[(组成)部分]

elemental composition 元素组成

elemental sulphur 元素硫

elementary analysis 元素分析

elevated roadway 高架道路

elimination method 消元[去]法

Elliott automation 艾略特型自动器

ellipsometry 椭圆对称法[光度法,测量术,偏振技术]

elliptic motion screen 椭圆运动筛

Elmore bulk-oil machine 埃尔莫尔全油浮选机

Elmore cell 埃尔莫尔浮选机[槽]

eluate 洗出[释,提]液,洗出物

Eluex process(=Fluex process of uranium) 埃留克斯法(铀淋萃法流程)

elution column 洗提塔

elution curve 洗提曲线

elution plant 洗提车间

elution process 洗提过程[工序,方法]

elution sequence 洗提顺序

elution system 洗提系统

elution time 洗提时间

elution vessel 洗提器

elutriation 淘选,淘析,水析

elutriation technique 水[淘]析技术

elutriator 淘[水]析器

Emcol 表面活性剂品牌名(美国威科(Witco 或 Witco/Oganics)公司产品)

Emcol 4150 脂肪胺硫酸盐,可用作锰矿的捕收剂

Emcol 5100 烷基醇胺与脂肪酸的缩合物,可用作锰矿浮选湿润剂

Emcol 607-40 烷基氯化吡啶(可用作捕收剂)

$RCOOCH_2CH_2NHCOCH_2 \cdot C_5H_5N^+, Cl^-$

Emcol 660B(=Emulsol 660B) 十二烷基吡啶(碘化物,同 Emulsol 660B,可用作捕收剂)

Emcol 888 聚烷基萘甲基吡啶(氯化物,可用作捕收剂)

Emcol X1 硫酸铵基乙基月桂酸脂(可用作乳化剂、捕收剂、湿润剂)

$C_{11}H_{23}COOCH_2CH_2OSO_3NH_4$

Emcol X25 烷基硫酸的乙醇胺盐(可用作油酸与煤油的乳化剂)

emergency power system 事故[备用,应急]供电系统

emergency stop cord 事故停止线

emergency usage 紧急使用

Emersol 表面活性剂品牌名(德国

Henkel/Emery 公司产品）

Emersol 201　油酸（可用作捕收剂）

Emersol 300　精馏植物油脂肪酸（可用作捕收剂）

Emfac 1202　捕收剂（壬酸）$C_8H_{17}COOH$

Emigol　浮选用非离子型乳化剂（德国赫司特化工厂）

emission of dust　粉尘排放[出]

emission spectrograph method　发射摄谱仪分析法

emission spectrometer　发射光谱仪

emission spectroscopy　发射光谱法

emissivity　放射率

emitted energy　发射能

Empicol　表面活性剂品牌名

Empicol CHC　十八烯基/十八烷基硫酸盐（可用作湿润剂）

Empicol CST　十六烷基/十八烷基硫酸盐（可用作湿润剂）

Empicol L　十二烷基硫酸盐（可用作湿润剂）$C_{12}H_{25}OSO_3M$

empirical allowance　经验提成率

empirical capacity　经验能力[处理量]

empirical classification　经验分级法

empirical classification of locked particle　连生颗粒经验分级法

empirical coefficient　经验系数

empirical dynamic revision　经验的动态修正

empirical equation　经验方程式

empirical formula　经验公式

empirical information　经验数据

empirical limit　经验界限[范围,极限]

empirical method　经验法

empirical mill power draft relationship　磨机汲取功率经验关系式

empirical model　经验模型

empirical nature of test　试验的经验性质

empirical pattern　经验图形

empirical relationship　经验关系式

empirical rule　经验法则

empirical technique　经验法

empirically derived efficiency parameter　经验推导的效率参数

empiricism　经验(主义)

employee relation　雇工[雇用人员]关系

employee working condition　雇用人员的工作条件

empty bin　空（矿,料）仓

empty pallet storage　空托盘储存处[储藏库]

empty rathole　空心斜管

empty-load　空负荷

Emulphor　表面活性剂品牌名（美国GAF公司产品）

Emulphor AG　脂肪酸聚乙醇醚（可用作分散剂、乳化剂、矿泥的分散剂）

Emulphor EL-719　脂肪酸聚氧乙烯醚（可用作湿润剂、乳化剂）

Emulphor O　水溶性脂肪醇衍生物（可用作湿润剂、乳化剂）

Emulphor P　脂肪醇聚乙醇醚（可用作

选矿药剂)
emulsibility 乳化性
Emulsifier 表面活性剂品牌名
Emulsifier 610A 脂肪酸聚乙二醇醚(可用作分散剂)
emulsifying action 乳化作用
emulsifying property 乳化性
emulsion flotation mechanism 乳浊液浮选机理
emulsion promoting power 乳化促进力
emulsion tendency 乳化趋势
Emulsol 表面活性剂品牌名
Emulsol K-1243 脂肪酸类的季铵衍生物(可用作选矿药剂)
Emulsol K-1339 脂肪酸类的季铵衍生物(可用作选矿药剂)
Emulsol K-1340 脂肪酸类的季铵衍生物(可用作选矿药剂)
Emulsol X25(=Emcol X25) 烷基硫酸的乙醇胺盐(同 Emcol X25,可用作捕收剂、湿润剂、乳化剂)
Emulsol XI(=Emcol XI) 十二烷基二聚乙二醇硫酸盐(同 Emcol XI,可用作捕收剂、抑制剂)
encapsulate 封装,密封,装入胶囊,压缩
encapsulating sulfo-minerals 密封的磺基矿物类
enclosed 封闭的,密封的,封装的,包装的,附上
enclosed dust tight elevator 封闭式防尘提升机
enclosed screw feeder 封闭式螺旋给矿[料]机
enclosed storage pile 封闭式堆场
enclosure 封闭[密封]罩,集[收]尘罩
enclosure housing 防护罩
end bearing 端轴承
end book value 终了账面[注册]价值
end disk 端盖
end loader 后端装载机
end of pipe technology 管末技术
end peripheral discharge 端部周边排矿
end plate 端面[末端]板
end suction 端[侧]吸入
end use form 使用格式[结构]
end use test 使用(期)试验
endless belt concentrator 无极皮带选矿机
endothermic physical process 吸热物理过程
energetics (力,动,水)能学,动力(学)
energetics of floatability 可浮性动能学
energy barrier 能垒
energy charge 电费
energy conservation 能量守恒
energy consuming effect 能耗效应
energy consumption 能(量消)耗
energy efficiency 能量效率[输出],电能效率
energy figure 能量数值[图形]

energy information 能量数据
energy input level 能量输入等级
energy intensity 能量强度
energy level 能级
energy level detection 能量等级探测（法）
energy level diagram 能级图
energy level discrimination technique 能级分析[辨]技术
energy normalized form 能量规格化形式
energy of collision 碰撞能
energy of hydrophobic association 疏水缔合能
energy pool 能谷
energy shortage 能源短缺
energy size relationship 能量[耗]与粒度关系
energy utilization 能量利用
energy valley 能谷
energy/ton proportionality constant 能耗/吨（矿石）比例常数
energy-absorbing 吸收能量的
energy-size relationship 能量[耗]-粒度关系（破碎过程的）
energy-size relationship for crushing process 破碎过程的能量[耗]-粒度关系
engineer's thinking 工程师的想法[思考,见解]
engineered equipment 设计的设备
engineered unit 工程设备[装置]

engineering 工程(学)设计
engineering and construction schedule 设计建设计划
engineering and procurement progress （工程）设计,采购进度
engineering and project management 工程项目管理
engineering and supervision manpower 工程及管理用人力
engineering and technical matters 工程技术问题
engineering and technical personnel 工程技术人员
engineering assignment 设计任务
engineering calculation 工程[设计]计算
engineering college 工学院
engineering company 工程公司
engineering construction 工程建筑[结构,建设]
engineering conveying system 工程[业]运输系统
engineering cost 工程费
engineering council 工程委员会
engineering data 技术[设计]数据
engineering department 工程部门
engineering design 工程设计
engineering design calculation 工程设计计算
engineering design phase 工程设计阶段
engineering detail 工程细节

engineering division 工程部
engineering drawing 工程图纸
engineering economics 工程经济学
engineering facts 工程实际情况
engineering file index 工程档案索引
engineering firm 工程公司
engineering formula 技术公式,设计公式
engineering information 设计资料
engineering instruction 技术说明书[细则]
engineering judgment 工程评价
Engineering News-Record Construction Cost Index 工程新闻记录建设费用指数
engineering office 工程设计室
engineering practice 工程实践
engineering project 工程项目
engineering science 技术学科
engineering specification 工程说明书[细则]
engineering study 工程研究
engineering supervision 设计监督
engineering unit 工程单位
engineering worker 技工
English composition 英语措辞[写作]
enhancement factor 增强因子
enhancement factor matrix 增强因子矩阵
enhancement ratio 增强比
enol 乙酰丙醇,烯醇,可用作抑制剂
enriched zone 富集带

enrichment ratio 富集比
enrichment zone 富集带
enthalpy change 焓变化
entire pH range 整个pH值范围
entity 实体[物]
entrained gangue 夹带的脉石
entrainment 带走,夹带,夹杂,带气物
entrainment separator 分离[捕集]器,吸取液体分离器
entrainment trap 雾沫分离器
entrainment-drainage mechanism 机械夹杂排流机理
entrapped undersize 裹挟的细粒[筛下产品]
entropic effect 熵效应
entry 输[引,进]入,通路[道],引[记]入,记录,登记,项[条]目,计入表中的事项[数值],报关,报关手续
enviroclear thickener 环保浓密机
environment consideration 环境[保]问题
environment problem 环境(保护)问题
environmental analysis program 环境保护分析规划
environmental and occupational safety measures 环保和职业安全措施
environmental condition 环境条件[情况]
environmental consideration 环保问题,环保(需要)考虑的事项
environmental control 环境控制

environmental disposal problems 环境保护的处理问题

environmental engineering 环境工程学

environmental expense 环境保护费

environmental factor 环境因素

environmental greening 环境绿化

environmental guideline 环境准则

environmental hazard 环境的危害

environmental impact 环境影响

environmental landscaping 环境绿化

environmental policy 环境保护政策

environmental problem 环境问题

Environmental Protection Agency 环境保护署

environmental regulation 环保规程[条例]

environmental regulations 环境保护规程[条例]

environmental research 环境(保护)研究

environmental suit 环境诉讼

environmental technology division 环境技术部[室,科]

environmentally sensitive area 环境敏感地区

Envirotech thickener 环境保护技术公司产浓缩机[增稠剂]

envisaged capital expenditure 预计的基建投资

eosin 曙红,四溴荧光素 $C_{20}H_8O_5Br_4$

eosin blue 荧光蓝(染料,浮选铜钼矿时作为辉铜矿抑制剂)

eosin yellow 荧光黄(染料,浮选铜钼矿时作为辉铜矿抑制剂)

epoxy 环氧,表氧,氧撑,桥氧基

epoxy encapsulation 环氧树脂密封

epoxy floor surface 环氧树脂楼板面

epoxy resin 环氧树胶

epoxy resin plug 环氧树脂塞

epoxy sealing 环氧树脂密封

EPPA(=ethyl phenyl phosphonic acid) 乙基苯基膦酸(可用作锡石的捕收剂)

equal amine adsorption line 等胺吸附线

equal amount 等量

equality constrained model 等式约束模型

equality constraint 等式约束

equality-constrained model 等式约束模型

equi probability theory 等概率论

equilibrating condition 平衡条件

equilibrium bulk concentration 平衡整体浓度

equilibrium cell 平衡浮选槽[机]

equilibrium concentration 平衡浓度

equilibrium condition 平衡条件

equilibrium contact angle 平衡接触角

equilibrium loading 平衡荷载

equilibrium solution surface 平衡溶液表面

equipment appraisal 设备估价

equipment availability 设备利用[完

好,可用]率
equipment capacity 设备处理能力
equipment category 设备类型
equipment characteristic 设备特性
equipment content 设备内容
equipment cooling 设备冷却
equipment cost ratio 设备费用比率
equipment cost ratio method 设备费用比推法
equipment cover 设备外罩[防护盖]
equipment diameter 设备直径
equipment evaluation 设备估价
equipment for mineral processing 选矿设备
equipment foundation 设备基础
equipment installation 设备安装
equipment installation cost 设备安装费
equipment items 设备项目
equipment life 设备寿命
equipment list 设备清单[表]
equipment manufacturer 设备制造厂家
equipment number 设备编号
equipment opening 设备开口
equipment operating parameter 设备操作参数
equipment operator 设备操作人员
equipment parameter 设备参数
equipment parameter data 设备参数数据
equipment performance 设备性能

equipment selection 设备选型
equipment size 设备规格
equipment specification 设备规格
equipment suite 成套设备
equipment supplier 设备供应商[厂家]
equipment supplier starting aid 设备供应商启动援助
equipment type 设备类型[型号]
equipment ventilation criteria 设备通风标准
equiprobability theory 等概率理论
equivalent bubble radius 气泡当量半径
equivalent compound discount criterion 等值复利贴现标准
equivalent factor 当量系数
equivalent length 当量长度
equivalent metallurgy 等效选矿
equivalent radius 当量半径
equivalent stoichiometric concentration 当量化学计算浓度
equivalent weight 当量
erection and installation 架设[装配]与安装
erection supervision 安装监督
erratic flow 反常[无定]流动
error analysis residual 误差分析残差
error component 误差分量
error estimate 误差估计[算]
error in surface estimation 表面估算误差

error limit 误差范围
error monitoring 误差监控
error rate 误差率
error variance 误差方差
error variance of volume 体积误差的方差
error volume 误差量
erucic acid 一种二十二碳不饱和脂肪酸,参酸,芥酸(可用作捕收剂)
escalating rate of revenue from sales 销售额[销售收入]上升率
escape tunnel 太平地道,逃生隧道,放空管
esoteric device 机密装置
essential part 基本部分
establishing price 制[确]定价格
establishment 建[设]立,创办,开设,机关,企业,部门,组织
establishment charge 开办费
estimate 估计[算,价],预算,概算,判断
estimate accuracy 估计[算]精确度
estimated annual net revenue 估算的年度净收益
estimated cost 估算成本
estimated recovery 估计回收率
estimated scrap value 预计残值
estimated shaft diameter 估计轴径
estimated useful life 预计使用寿命
estimated variance 估计方差
estimating department 预算部门
estimation of error 误差估计

estimation program 估算程序
estimator 估算人员
ESTIMILL program flowsheet ESTIMILL 程序流程图(一种线性模型研磨仿真与参数估计的程序)
ether 醚,乙醚$(C_2H_5)_2O$
ether amine 醚胺(阴离子捕收剂)
ether carboxylic acid 乙醚羧基酸,醚酸
ether diamine 醚二胺
ether diamine monoacetate 醚二胺一醋酸盐
Ethomeen 表面活性剂品牌名($C_8 \sim C_{12}$脂肪胺与环氧乙烷的反应物,可用作浮选剂)
Ethomeen 18/15 1克分子硬脂胺与5克分子环氧乙烷反应产物(可用作捕收剂、絮凝剂)
Ethomeen 18/60 1克分子硬脂胺与50克分子环氧乙烷反应产物(可用作捕收剂、絮凝剂)
Ethomeen C/20 1克分子椰油胺与2克分子环氧乙烷反应产物(可用作捕收剂、絮凝剂)
EthomeenS/10 叔胺(可用作捕收剂、絮凝剂)
EthomeenS/12 1克分子大豆油胺与2克分子环氧乙烷反应产物(可用作捕收剂、絮凝剂)
EthomeenS/15 1克分子大豆油胺与5克分子环氧乙烷反应产物(可用作捕收剂、絮凝剂)

EthomeenS/20　1克分子大豆油胺与10克分子环氧乙烷反应产物(可用作捕收剂、絮凝剂)

EthomeenS/25　1克分子大豆油胺与15克分子环氧乙烷反应产物(可用作捕收剂、絮凝剂)

Ethomeen T/12　1克分子牛油胺与2克分子环氧乙烷反应产物(可用作捕收剂、絮凝剂)

Ethomeen T/15　1克分子牛油胺与5克分子环氧乙烷反应产物(可用作捕收剂、絮凝剂)

Ethomeen T/25　1克分子牛油胺与15克分子环氧乙烷反应产物(可用作捕收剂、絮凝剂)

ethyl　乙基,乙烷基 CH_3CH_2-

ethyl compound　乙基化合物

ethyl phenyl phosphonic acid　乙苯基膦酸乙基苯基膦酸

ethyl silicate　乙基硅酸盐(硅酸乙酯,在酸性或中性矿浆中,用作矿泥脉石的抑制剂)$(C_2H_5)SiO_4$

ethyl sodium　乙基钠 C_2H_5Na

ethyl xanthic acid　乙基黄原酸

ethylene　乙烯 $CH_2:CH_2$;乙烯[基] $CH_2:CH-$;乙撑,次乙基 $-CH_2CH_2-$

ethylene diamine　乙(撑)二胺,乙二胺-(1,2) $NH_2CH_2CH_2NH_2$

ethylenediamine tetraacetic acid　乙二胺四乙[醋]酸,乙底酸[依地酸]

ethylenediamine tetraacetic acid (=EDTA)　乙二胺四乙酸(即EDTA)

eugenol　丁子香酚(可用作起泡剂) $C_{10}H_{12}O_2$

Euler number (Eu)　欧拉数(Eu)(流体力学中:$Eu=\Delta P/\rho u^2$,反映流体流动时压力影响的无因次数群)

Euro currency　欧洲通货[货币]

Eurodollar bond　欧洲美元证券

Eurodollar note　欧洲美元票据

Eurodollar term loan　欧洲美元定期贷款

Euro-dollars　欧洲美元,美国国外持有的美元

evacuation　搬空,抽(成真)空,抽气,排空[泄,气],消除,清除

evaluation　评价[估],估价,求[赋,计]值

evaluation of data　数据评价[评定]

evaluation of error　误差评价

evaluation of flowsheet alternatives　流程方案的评价

evaluation of procedure　过程评价

evaluation of variogram　变异函数的求值

evaluation program　评价方案

evaporated film　蒸发膜

evaporation loss　蒸发损失

evaporation system　蒸发系统

evaporative environment　蒸发环境

evaporative pond　蒸发池

event of default　违章事件

eventual design capacity　最终设计生

产能力

every level surface 各[所有]标高平[地]面

evidence 证[论]据,佐证,形[痕]迹,迹象

evolutionary operation 调优运算

exact circuit condition 准确的回路条件

exact science 精确科学

examination 检验[查,测]

examination method (矿物)检定[检验]方法

examination of mineral specimen 矿物标本的检定[检验,分析,研究]

example application 实例应用

example of crusher application engineering information 破碎机应用技术[管理]资料实例

example problem 例题

example run 实例运算

excavating equipment 采掘设备

excess cash flow 超额现金流量

excess income 超额收益

excess liquidity 过剩流动性,额外流动资金

excess revenue 超额收入

excess surface concentration 表面过量浓度

excessive belt stretch 胶带过度伸长[伸张]

excessive bottom cover wear 底层过度磨损(胶带的)

excessive coarseness 过粗粒度

excessive dusting 超粉尘化

excessive expense 过多费用

excessive fineness 过细粒度

excessive time 过多时间

excessive time and expense 过多的时间和费用

exchange of information 情报交换[流]

exchanging ball mill liner machine 更换球磨机衬板机械

excited atom 受激[激活]原子

exciter (筛)主控振荡器

exciting radiation 激发辐射

exciting radiation source 激发辐射源

execution control 执行控制

execution cycle 执行周期

executive 行政的,经营的,执行者,总经理

executive assistance 行政协助

executive committee 执行[常务]委员会

executive component 操作元件,执行部分

executive editorial committee 执行编辑委员会

executive office 行政办公室

executive portion 执行部分

executive real-time program 执行实时程序

executive report 施工(图)报告书

executive structure 执行机构

executive summary 行政性摘要
executive vice president 执行副总裁
exhaust fan 抽风机
exhibiting spectrum 显示光谱
existing condition 现状
existing equipment 现[原]有设备
existing facilities 现有设施
existing flowsheet 现有流程(图)
existing mining equipment 现有采矿设备
existing plants and mines 现有厂矿
existing situation 当前形势
existing testing facilities 现有试验设施
exothermic chemical process 放热化学过程
exothermic heat 放热
exothermic nature 放热性
exothermic process 放热过程
exothermicity 放热性
expanded flow bin 扩张流动仓
expanded performance evaluation test 扩大性能评价试验(浮选机的)
expansion 扩建[展,大,张,充],膨胀
expansion investment 扩建投资
expansion possibility 扩建可能性
expansion program 扩建计划
expansion project 扩建工程
expectation value 期望值
expected delivery date 预计交货日期
expected life expectancy 预期寿命
expected operating life 预计使用寿命
expected surcharge angle 预[估]计[期,料]的超载角
expected system 预期[期望]的系统
expected time period 预定期间[周期,时间间隔]
expected value 期待[望]值
expediting report 催交报告
expenditure 支出(额),经费,费用
expenditure of capital 投资费用
expenditure of energy 能量消[损]耗
expense 经费,费用,开支,消费
expense of holiday and vacation 节假日费用
expensive equipment 昂贵设备
expensive reagent 昂贵药剂
experience 经验
experience coefficient 经验系数
experience factor 经验[因]系数
experienced chemist 有经验的化学家
experienced flotation engineer 有经验的浮选工程师
experienced mill maintenance personnel 有经验[熟练]的选矿厂维修人员
experienced mill repair men 熟练[有经验]的选矿厂维修人员
experienced personnel 有经验的人员
experiences 经历
experiment error 试[实]验误差
experiment variogram 实验变异函数
experimental anodic polarization curve 阳极极化试验曲线
experimental condition 试[实]验条件
experimental data 试验数据

experimental design 试验设计
experimental detail 试验细节
experimental determination 试验测定
experimental error 实[试]验误差
experimental evidence 实验数据[资料,证据]
experimental form 试验形式
experimental investigation 试验研究
experimental measurement 实[试]验测定值
experimental method 实[试]验方法
experimental photovoltage behavior 实验光电压特性
experimental procedure 试验方法[步骤,程序,作业过程]
experimental program 试验程序[计划]
experimental recovery 试验回收率
experimental region 试验区
experimental response 试验响应
experimental result 试验结果
experimental setup 试验装置
experimental study 试验研究
experimental system 试验系统
experimental variogram 实验性变异函数
experimentally determinable function 通过实验可确定的函数
expert operation 专家操作
expertise 专门技术
explicit 显式,显的
explicit constraint 显式约束

explicit fashion 显式
explicit function 显函数
exploitation stage 开采阶段
exploration 探索[测,究],踏勘
exploration activity 勘探[找矿]活动
exploration arm 探[找]矿杠杆[指针]
exploration cost 勘探成本
exploration expense 勘探费
exploration means 勘探手段
exploration stage 勘探阶段
exploratory 探索[勘探,探究]的,考察的,试探性的
exploratory move 探索步骤
exploratory survey 探测,踏勘测量
exploratory underground 地下勘探
explosion relief vent 防爆安全口,泄爆口
explosion suppression system 爆炸消除系统
explosive breakage 爆破破碎
explosives 炸药
exponential absorption process 指数吸收过程
exponential function 指数函数
exponential model 指数模型
exponential probability density function 指数概率[密度]函数
exponential relationship between power drawn and close setting 传动[抽运,牵引]功率和紧边排口宽度间有幂指数关系
export license 出口许可证

export market 出口市场
export tax 出口税
export trade 出口贸易
exporter 出口商[方],输出国
exporter commercial bank 出口商商业银行
expression 表示,表达式,公式,符号
expropriation 没收,征用
extended geostatistical study 扩大的地质统计研究
extended test 扩大试验
extended testing 扩大试验
extender 补充[辅助,增量]剂,扩展[延长]器
extension error 外延误差
extension flotation 延长[扩大]浮选
extension variance 外延方差
extensive bench scale metallurgical testing 扩大[广泛]的实验室规模工艺[选矿]试验
extensive cyanidation investigation 扩大的氰化试验研究
extensive drilling program 扩大钻探计划
extensive geostatistical argument 外延地质统计论证
extensive geostatistical theory 扩大的地质统计学理论
extensive grinding and flotation trial 扩大的磨浮试验
extensive laboratory and pilot plant test work 扩大实验室和半工业性试验(工作)
extensive laboratory grindability test 扩大[广泛]的实验室可磨性试验
extensive milling and flotation trial 扩大的磨浮试验
extensive operating experience 广泛的操作经验
extensive prior oxidation 大面积预先氧化
extensive research 扩大试验研究
extensive research program 扩大研究计划
extensive sampling 扩大抽样
extensive study 广泛研究
extensive test program 扩大试验项目
extensive test work 扩大[广泛]试验工作
extensive theory of regionalized variable 区域化变量的广延理论
extensive variance 外延方差
extent of deactivity 去活程度
extent of disseminate 浸染程度
extent of dissemination 浸染[嵌布,散布]程度
extent of reaction 反应程度
external air pressurization 外部压气
external air supply pump 外部供气泵
external blower 外部鼓风机
external current flow 外电路电流
external electrical circuit 外部电路
external galler-type plate and frame clarification filter press 明流型板框

式净化压滤机
external main frame pin 外部[侧]主机座螺栓[钉]
external main frame pin bushing 外部[侧]主机座销[螺]栓衬套
external service 对外服务,外部检修
external source 外部来源
external storage of data 数据外存
external timing operation 外部定时操作
external vibrator 外部振动器
external walkway 外部人行道
extinction coefficient 消光系数
extra coarse crushing cavity 特粗(物料)破碎腔
extra cost 额外费用[成本]
Extra heavy duty truss 超重型桁架
extra power 多余功率
extra trained operator 经特别训练的操作人员
extractant 萃取剂
extracting gold 提取黄金
extracting value 提取价值
extraction 提取,萃取,浸出,榨出,榨取,采掘,开采,摘取
extraction and stripping step 萃取和解吸[洗提]作业
extraction circuit 萃取回路
extraction condition 萃取条件
extraction facilities 萃取设施
extraction facility 萃取装置[设备]
extraction fan 抽风机

extraction machinery 采掘设备
extraction of uranium 铀(的)萃取
extraction phase 萃取相
extraction plant 萃取厂[车间]
extraction rate 萃取率
extraction ratio 回采比
extraction reagent 萃取剂
extraction stage 萃取段
extraction step 萃取工序[步骤]
extraction-esterification-chromatography 萃取-酯化-色层法
extractive metallurgical research department 萃取冶金研究室
extractive operation 回[开]采作业
extra-heavy bolt-on balance weight 特重型螺栓固定平衡配重
extra-heavy service line 特重型(负荷)作业线
extra-heavy-duty conditioner 特重型调整[搅拌]槽
extrapolated sample data 外推样品数据
extrapolated value 外推值
extrapolation method 外推法
extreme case 极端[极罕见]的例子
extreme efficiency figure 效率极限数值[字]
extreme example 极罕见[极端]的例子
extreme grade-recovery curve 极限品位与回收率曲线
extreme left position 左端极限位置

extreme range 极限范围
extreme right position 右端极限位置
extreme type 极限类型
extreme value 极值,最大[小]值

extremely close association 极其紧密共生
eye nut 吊环螺母

F

16 foot diameter 10-hearth roaster 16英尺直径10膛焙烧炉
F-126 全氟化的$C_4 \sim C_{10}$的混合脂肪酸铵(品名,可用作捕收剂)
F-258 二甲基聚硅酮(品名,可用作氧化及硅酸盐矿的捕收剂)
F-550 苯基甲基聚硅酮(品名,可用作捕收剂)
fabric decay (纤维)织物衰变
fabric filter 织物滤尘器
fabric filter dust collector 织物过滤(机)集尘器
fabric media 织物介质
fabric reinforced rubber 纤维增强橡胶
fabrication 制[建,构]造,制作[备],装配,生产,加工,建造[制作]物
fabrication cost 造价,生产成本,制造费用
fabrication metallurgy 制造冶金
face center 面中心,面心
face par 票面价格

face value 面值
face-edge ratio 面棱比
fact sheet 记事表,记录表,登记表
factored capital cost estimate guide 投资费用分项估算准则
factored estimate table 析因估算表
factored estimation method 分项估算方法
factorial design test 析因设计实验
factorial experiment 析因试验
factorial experiment design 析因试验设计
factorially designed flotation test 析因设计浮选试验
factorially designed laboratory flotation test 析因设计实验室浮选试验
factory price 工厂价格
factory representative 厂方代表
faculty of engineering 工学院
Fagergren(=FAG) 法格古伦(浮选机械设备品牌名,简称FAG,美国威姆科公司生产)

Fagergren flotation cell 法格古伦浮选槽

Fagergren step-down flotation cell 法格古伦型阶梯式浮选机[槽]

Fagergren trough-type flotation cell 法格古伦型槽式浮选机

failure rate 故障率

fair cost 公平成本

fall throu 漏下量(露天或井下爆破结果所形成的破碎机给矿中比其排口还小的细粒物料)

fall throu factor 漏下量系数(瞬时生产能力与基本生产能力的比率)

falling stream 落下的矿[料]流

false conclusion 错误结论

false declaration 虚报

false information 错误资料,假资料

family 家庭,亲属,种类,系列

family curve relationship 亲族曲线关系

family of curves 曲线族

family of spiral classifier 螺旋分级机系列

family parameter 族的参数

family relationship 亲族关系

fan design criteria 风机设计标准

fan specification 风机技术说明(书)

fan total pressure 风机总压力

Faraday constant 法拉第常数(符号:F,单位:C/mol,代表每摩尔(mol)电子所携带的电量)

fast floatable component 快速浮游成分

fast floating mineral 速浮矿物

fast floating species 快速浮游矿物[种],速浮矿物

fast initial response 快速起始响应

fast photovoltage 快速光电压

fast response photovoltage 瞬时响应光电压

fastening 连接(法),紧固,夹紧,固定(零件),紧固接头,扣件

fastening bearing 紧固轴承

fate of flotation reagent 浮选药剂去向

fatigue life 疲劳寿命

fatigue loading 疲劳负荷

fatty acid 脂肪酸 $C_nH_{2n+1}COOH$

fatty acid concentrate 脂肪酸浮选精矿

fatty acid flotation system 脂肪酸浮选体系

fatty acid promoter 脂肪酸促进剂

fatty acid salt 脂肪酸盐

fatty acid soap 脂肪酸皂

fatty acid-fuel collector 油酸-燃料油捕收剂

fatty acid-primary amine alternative flotation system 脂肪酸伯胺交替浮选系统

fault gouge mud seam 断层泥的泥夹层

feasibility engineering 可行性设计

feasibility of autogenous grinding 自磨的可行性

feasibility of control 控制灵活性
feasibility of gravity separation 重选可行性
feasibility study cost 可行性研究费
feasibility study estimate 可行性研究估算
feasibility study phase 可行性研究阶段
feasible alternative 可行方案[途径]
feature article 特写,专题文章
feature of equipment 设备特性
fed crushing cavity 给矿破碎腔
fee 费用,酬金
feeble bubble contact 弱(的)气泡接触
feeble reaction 轻微[微弱]反应
feed adjustment nut 给料平台[装置]调整螺母
feed adjustment post 给料平台[装置]调整丝杆
feed alteration 给矿[料](量)变化
feed arrangement 给矿方案[制度]
feed assay 给矿分析品位
feed belt conveyer particle size analyzer 给矿皮带粒度分析仪
feed chamber 给矿室
feed characteristic 给矿[料]特性
feed concentration 给料[矿]浓度
feed concentration zone 给矿浓缩带
feed condition 给矿条件
feed cone 给矿[料]锥
feed consistency 给矿稳定性,稳定给矿

feed conveyor discharge 给矿运输机排矿[料]
feed corrosion 给料的腐蚀
feed density 给矿浓度
feed dilution 给矿稀释(度)
feed distributing and regulating plate 给矿分配调节板(圆锥破碎机的)
feed distribution 给矿分配
feed distribution system 给矿分配系统
feed distributor 给矿分配器[槽],分料器
feed distributor bolt 分料器[给矿分布器]螺栓
feed end bed depth 给矿端层厚(给矿筛的)
feed entrance 给料[矿]口
feed entry 给料[矿]口
feed extraction 抽[排]料
feed flow rate 给矿流量
feed forward control 前馈控制
feed grade 给矿品位
feed grading 给矿粒度分级[组成],给矿级配
feed hopper cone 给料漏斗锥体
feed jiggability 给矿跳汰能力
feed level 给矿料位
feed line 给矿[料]管线
feed liquid 给料液体
feed material 给料
feed material moisture 给料水分
feed opening 给料[矿]口
feed pH 给料pH值

feed piping　给矿管道铺设[布置]
feed plate　给矿盘(圆锥破碎机的)
feed platform　给料平台
feed post　给矿[器]支柱(圆锥破碎机的)
feed preparation　备料
feed pressure　给矿压力
feed process requirement　给矿的工艺要求
feed pump　给矿泵
feed rate　给矿量
feed rate automatic control system　给矿量自动控制系统
feed rate control　给矿量控制
feed rate control sub-loop　给矿量控制子回路
feed rate meter　给矿量计量计[表]
feed scalping　给矿筛除大块
feed screw　螺旋进[给]料器
feed segregation　给料偏析
feed side　进[喂]料端
feed size distribution rate　给矿粒度分配率
feed size distribution vector　给矿粒度分布向量
feed size factor　给矿粒度系数
feed size range　给料[矿]粒度范围
feed slurry　给料矿浆
feed slurry distributor　给矿矿浆分配器
feed slurry liquid　给矿矿浆液
feed slurry temperature　给矿矿浆温度

feed slurry volumetric flowrate　给料矿浆体积流量
feed solid particle size distribution　给矿固体粒度分布
feed solid size　给矿固体粒度
feed solid tonnage　给矿[料]固体吨数[吨位]
feed solids　给料固体
feed solids concentration　给料固体浓度
feed stock　给矿原料,原材料
feed sump　给矿泵池
feed temperature　给料[矿]温度
feed tonnage　给矿吨数
feed variation　给料[矿]变化
feed volume fluctuation　给料体[容]积波动
feedback control　反馈控制
feedback control loop　反馈控制回路
feedback control strategy　反馈控制对策
feedback strategy　反馈对策
feedback system　反馈系统
feeder adjustment dial　给矿[料]机调整转盘
feeder box　给料箱
feeder breaker　馈电断路器
feeder guideline　给矿机(设计,使用)准则[指南]
feeder head box　给矿[料]机头部漏斗
feeder hopper volume chart　给矿机漏斗容积(计算)表

feeder main frame assembly 给矿[料]机主架机组
feeder opening 给料[矿]机排口
feeder speed automatic adjustment 给矿机速度自动调节
feeder speed controller 给矿机速度控制器
feeder speed manual adjustment 给矿机速度手动调节
feedforward system 前馈系统
feedforward-feedback control 前馈-反馈控制
feeding system 给矿系统
feedrate controller 给矿量控制器
feed-rate controller 给矿速度[量]控制器
feed-water flow 给水流量
felted fabric 毡制织物
fenchol 葑醇 $C_{10}H_{17}OH$
fenchyl 葑基 $C_{10}H_{17}$
fenchyl alcohol (= fenchol) 葑醇(同 fenchol,可用作起泡剂) $C_{10}H_{17}OH$
Fenopon A (= Igepon A) 油酸磺化乙酯钠盐(品名,同 Igepon A,可用作洗涤剂、湿润剂) $C_{17}H_{33}COOCH_2CH_2SO_3Na$
Fenopon T (= Igepon T) 油酰甲氨基乙基磺酸钠(品名,同 Igepon T,可用作洗涤剂、捕收剂)
$RCON(R')C_2H_4SO_3Na$, R=油烯基 $C_{17}H_{33}$, $R'=CH_3$
Feret's diameter 费雷特直径(一种常用的粒径表示方法)

Fermi level 费米能级(对于金属,绝对零度下电子占据的最高能级就是费米能级)
ferric (正)铁的,三价铁的
ferric hydroxamate 高铁羟氧肟酸盐,氧肟酸铁
ferric hydroxide 氢氧化铁
ferric iron 三价铁,高价铁
ferric oxide mineral 氧化铁矿物
ferric sulfate 硫酸铁 $Fe_2(SO_4)_3 \cdot nH_2O$
ferric sulfide 硫化铁 Fe_2S_3
ferric xanthate 黄原酸铁
ferricyanide ion 铁氰离子
Ferri-Floc 絮凝剂(硫酸铁)
ferrocyanide 氰亚铁酸盐,亚铁氰化物
ferrocyanide ion 亚铁氰离子
ferrodolomite 铁白云石
ferromagnetic mineral 铁磁性矿物
ferro-magnetic mineral 铁磁性矿物,磁性铁矿物
ferronickel of commercial value 有商品价值的镍铁合金
ferrotungsten 钨铁(合金)
ferrous iron 二价铁,低价铁
ferrous silicate 硅酸亚铁
fiber glass construction 纤维玻璃制造
fiber glass rotary distributor 玻璃纤维旋转分配器
fiber-reinforced plastic electrolytic cell 玻璃钢电解槽
field density 场[实地]密度
field engineer 现场[安装,野外]工

程师

field expense 现场费用(施工管理)

field experience 现场经验

field force 场力

field maintenance supervisor 现场维修管理[检查]人员

field modification 现场改造

field of ion 离子场

field office 现场办公室

field operation 现场[野外]施工[作业]

field plant 矿区[矿场,矿田]选矿厂

field screw 野外工作队

field test 现场试验

field test result 现场试验结果

field-configurable digital process controller 现场配置的数字过程控制器

figure-of-merit parameter 品质因数参数

file 文件,档案,文件夹,锉刀

file allocation table 文件分配表

file index 档案索引

file number 档案[文件]号

filing system 档案系统,归档制度

film deposition 薄膜沉淀

film sizing device 流膜分级设备

filter aid tank 助滤剂搅拌槽

filter bag 过滤机滤袋,滤袋

filter bag changing 滤袋更换

filter bag cleaning 滤袋清理

filter bag cleaning cycle 滤袋清理周期

filter bag fabric 过滤机滤布

filter candle 过滤棒

filter discharge 过滤机排料[矿]

filter feed tank 过滤机给矿槽[池,罐]

filter product 过滤(机)产品

filter table 平面过滤机,筛选表

filter test 过滤机试验

filter tub 过滤机槽[桶]

filter unit 过滤机机组[设备,装置]

filtered product 过滤产品

filtered solution 过滤溶液

filtered water pump 过滤水泵

filtering media 过滤介质

filtering rate 过滤速度

filtering vacuum 过滤真空度

filtrate receiver 汽水分离器,滤液接水器,滤液罐

filtration rate 过滤速度

filtration test 过滤试验

final acceptance test 最终(的)验收试验

final bench 最终台阶

final carbon screen 载金碳筛

final cleaner concentrate 最终精选精矿

final computer program 终端计算机程序

final control component 终端控制元件

final control element 终端控制元件

final cost 最终成本

final design parameters 最终设计参数

final drawing 最终图纸

final feasibility study 最终可行性研究

final flotation concentrate 最终浮选精矿
final flowsheet 最终流程图
final flowsheet selection 最终流程选择
final grade product 最终品位产品
final grinding index 最终磨矿指数
final judgment 最终裁决
final maturity 最终期限
final price 最后价格
final process flowsheet 最终工艺流程
final product 最终产品
final product assay 最终产品分析
final repayment date 最终偿还日期
final report 最终报告
final selection 最终选择
final site testing 最终现场试验
final size 最终粒度
final tendering 最后投标
final testing 最终试验
final upgrading 最终精选
finance and accounting 财务会计,财会
finance and feasibility study 财政和可行性研究
finance company 金融公司
finance cost 财政费(用)
Finance Post 财政邮报
financial ability 财力
financial affairs 财务
financial analysis 财务分析
financial analyst 财务分析人员
financial backer 财政支持方
financial calculation 财务计算
financial condition 财务状况
financial disaster 财政灾难
financial evaluation 财务评价
financial expert 金融[财务]专家
financial institution 金融机构
financial lease 金融性出租
financial management 财务管理
financial manager 财务管理人员
financial market 金融市场
financial people 财会人员
financial point of view 财政观点
financial position 财务状况
financial ratio 财务比率
financial record 财务记录
financial report 财务报告
financial resources 财力
financial return 财务[政]收益
financial risk 财政风险
financial statement 财务报告,决算表,借贷对照表
financial structure 财务结构
financial study 财务[政]研究
financial support 财政支持[援助]
financial year 财政[会计]年度
financier 财务人员
financing decision 筹资决策
financing function 筹集资金功能
financing issue 筹措资金问题
fine aggregate 细粒集料(通过1/4英寸筛)
fine ash biotite granite 细粒灰色黑云

母花岗岩
fine clay binder 细黏土胶[粘]结料
fine coal circuit 细煤流程
fine comminution 细磨碎
fine cone crusher 细碎[粒]圆锥破碎机
fine crush product 细碎产品
fine crushing cavity 细(物料)破碎腔
fine crushing facility 细碎设施
fine crushing machine 细碎机械
fine crushing performance 细碎作业[工作,生产,操作效能]
fine crushing plant 细碎车间
fine dense mineral 细粒重矿物
fine fraction 细粒级
fine gold 细粒金
fine grinding concept 细磨概念
fine heavy particle 细的重颗粒
fine mill tailing 细粒[泥]选矿厂尾矿
fine ore crushing building 细碎厂房
fine ore reclaim feeder 粉矿给矿机
fine ore storage area 粉矿堆场
fine plant 细碎厂[车间]
fine red mud residue 细红泥废渣
fine screenings 筛除[出]的细料(小于3/8英寸)
fine sieve 细筛
fine sizes 细粒级
fine slimy material 细粒泥浆物料
fine splinter 细裂片
fine stringer 薄夹层,细脉
fine textured froth 细密结构泡沫

fine tuning 微调
fine vein filling 细脉充填
fine-coal cleaning 选粉煤
fine-grained nature 细料嵌布性质
finely dispersed nature 微细分散性
finely disseminated 细粒浸染的
finely disseminated form 细粒嵌布形式
finely ground heavy mineral 细磨重矿物
finely ground material 细磨物料
fineness of grind 磨矿细度
fineness of grind factor 磨矿细度系数
finer fraction 更细[较细]粒级
fines content 细粒含量
fines factor 细粒[粉矿]系数
finger splitter 指状分流器
fingerling channel catfish 海峡小鲴鱼
fingerling trout 小鳟鱼
finish grinding mill 最终磨矿机
finished flowsheet 最终流程图
finished grade elevation 平整后场地标高
finished ground product 最终磨矿产品
finished particle size 最终(磨矿)粒度
finished plate 抛光板
finished product 制成品,最终产品(破碎筛分后达到预期粒度的物料)
finished product output 最终产品产量
finished product size range 最终产品粒度范围

finished size 成品尺寸
finished tailings 最终尾矿
finished-size material 最终粒度的物料[矿石]
finisher flotation 精选
finishing plant 最终[精选]选矿厂
finite contact angle 限定接触角
finite slope 有限斜率
fire damage 火灾损失
fire procedure 火焰分析法
fire protection system 消防系统
fire suppression 防火抑制,火灾排除
fire water pump 消防水泵
fire water tank 消防水池
fire-polishing 加火抛光
fire-refining technique 火法精炼技术
firm 公司,商号,坚定[牢固,严格,结实]的
firm accounting statement 企业会计表
firm contract for future delivery 企业的期货交割合同
firm mass 坚固[实]块
firm of consultants 咨询公司[商号]
firm price 确切[可靠,稳定]价格
Firm's annual report 企业年度报告
Firm's common share 企业普通股票
Firm's debt-paying capacity 企业偿债能力
Firm's financial position 企业财务状况[态]
Firm's weighted average cost of capital 企业的加权平均投资费

first aid 急救(人员,药品,方法)
first aid appliance 急救用品
first aid facilities 急救设施
first aid supply 急救用品
first cleaner 第一段[次]精选
first generation system 第一代系统
first hydroxy complex 早期氢氧络合物
first mineral flotation 第一种矿物浮选
first order derivative 一阶导数
first order equation 一阶方程
first order kinetic model 一阶动态模型
first order kinetics equation 一阶动力学方程
first order rate law 一阶速度定律
first order term 一阶项
first recleaner 第一次再精选
first recleaner concentrate 第一再精选精矿
first recleaner reject 第一再精选尾矿
first shift 第一班
first stage reduction crusher 第一破碎段破碎机,第一段破碎机
first-in first-out cost 先进先出成本
first-in-first-out flow sequence 先入先出流动顺序
first-order rate model 一阶速度模型
first-stage development program 第一阶段发展计划
first-stage financing 第一阶段资金筹集

first-stage program 第一阶段计划
fiscal year 财政[会计]年度
Fischer double diaphragm pinch valve 费舍尔双隔膜夹紧[节流]阀
Fischer-Porter magnetic flowmeter 费舍尔·波特磁流量计
fish food 鱼食[饵]
fish glue 鱼胶(浮选钾盐矿时作为矿泥抑制剂)
fish liver oil fatty acid 鱼肝油脂肪酸(可用作萤石的捕收剂)
fish oil fatty acid 鱼油脂肪酸(可用作铀、铜、萤石、重晶石、磷矿、石英的捕收剂)
fish oil soap 鱼油皂(可用作石英的捕收剂)
fish oil sulfonate 磺化鱼油(可用作氧化铁矿的捕收剂)
Fisher criterion 费希尔准则
Fisher sub-sieve analyzer 费歇尔亚筛级粒度分析仪(美国 Fisher Scientific Co. 研制)
Fisher sub-sieve sizer 费歇尔亚筛级粉末粒度测定仪(美国 Fisher Scientific Co. 研制)
fitted function 拟合后函数
fitted model 拟合模型
fitted planar equation 拟合平面方程
fitting arrangement drawing 附件装配图
fitting assembling 零件装配
fitting metal 配[附]件

fitting strategy 拟合对策[方法,策略]
five-year period 五年期
five-year plan 五年计划
fixation method 固定方法
fixed assets 固定资产
fixed belt conveyer 固定式胶带运输机
fixed capital 固定资本
fixed capital cost portion 固定投资费用部分
fixed capital portion 固定资金[资本]部分
fixed charge debenture 固定费用债券
fixed hammer 固定锤
fixed load characteristic 固定载荷特性
fixed peripheral screen 周边固定筛
fixed pillow block 固定轴台
fixed price contract 固定价格合同
fixed product size 固定的产品粒度
fixed pump speed 固定泵速
fixed rate 固定利率
fixed sequence 固定顺序
fixed setting mode 固定排口宽度形式
fixed speed pump 固定速度泵
fixed value 固定值
fixture 夹具,卡具,固定物,定位器,安装用具
flake portion 片状部分
flake-like particle shape 薄片状颗粒形态
flanged chute connection 法兰盘溜槽连接[接头]

flanged inlet 带法兰的入口管
flanged outlet 带法兰的出口管
flap gate 翻板[舌瓣]闸门
flare 闪光[耀],耀斑,扩散管
flared tank spiral classifier 喇叭[槽]型螺旋分级机(槽侧向外扩张)
flash filtering 急骤[速]过滤
flash flotation 闪速浮选
flash point 闪点
flash smelting 闪速熔炼
flash smelting furnace 闪速熔炼炉
flash-furnace process 闪速炉工艺[过程]
flash-furnace reaction 闪速炉反应
flat angle 平缓角
flat band condition 平带条件
flat belt 水平胶带(运输机的)
flat bladed turbine 平叶片涡轮
flat bottom with full slot 全长排矿口的矿仓平底
flat cost 直接费(预算)
flat filter bag 扁平式滤袋(织物收尘器的)
flat grizzly 平面棒条筛
flat spring 扁簧
flat test screen 平面[水平]试验筛
flat testing screen 平底[板]试验筛
flat-lying clay layer 平伏黏土层
flat-lying sedimentary deposit 平伏沉积矿床
flat-topped curve 平顶曲线
flexible connection 柔性连接

flexible connector 活动接头
flexible design 灵活[通用]设计
flexible unit 通用设备[装置]
Flexol plasticizer TOF Flexol TOF 增塑剂(三乙基已基磷酸),可用作分散剂
Flexricin 表面活性剂品牌名
Flexricin 9 丙二醇单蓖麻酸酯(可用作捕收剂)
flight pitch (叶片)螺距(螺旋分级机的)
flight width 螺旋叶片宽度(螺旋分级机的)
Flip-Flow(screen) 弛张筛
float 浮选,浮选精矿,浮体[筒,子,球],无边台车,运煤车,冲击层[土]
float cell 浮选槽,浮选机
float coal (可)浮煤
float configuration 浮物[泡沫]形状[构型],漂浮图形
float decantation device 浮标倾析装置
float feed alkalinity 浮选给矿碱度
float product 浮品(重介质选矿的)
float rock 轻产品,废石(重介质选矿的)
float screen product 浮物筛分产品
floatability data 可浮性数据
floatability distribution 浮游性分布
floatable condition 可浮(游)条件
floatable constituent 可浮成分
floatable form 可浮形[状]态
floatable mineral 可浮矿物
floatable pyrite 易[可]浮黄铁矿

floatable specie concentration	可浮矿物的富集
float-and-sink process	浮沉法,重介质[重液]选矿法
floating charge	浮动充电
floating charge debenture	浮动金额债券
floating holiday	浮动假日
floating material	漂浮物
floating oil	漂浮油
floc(=flock)	絮凝物[体],絮团,絮状沉淀,蓬松物质
Floc aid 1038	絮凝助剂(改性的阳离子淀粉衍生物)
Floc aid 1063	絮凝助剂(改性的阳离子淀粉衍生物)
Floc gel	法国产淀粉类絮凝剂
flocculation peak	絮凝高峰
flocculation period	絮凝阶段
flocculation-defloculation characteristic	絮凝反絮凝特性
flocculent bed	絮凝层(浓密机的)
flocculent feed pipe	絮凝剂给入管
flocculent screening test	絮凝剂筛选试验
Flockal 101	絮凝剂(水解聚丙烯腈)
Flockal 152	絮凝剂(水解的甲基丙烯酸与甲基苯烯酸甲酯共聚物)
Flockal 202	絮凝剂(部分水解的淀粉)
Flockal 5	絮凝剂(部分水解的淀粉)
Floculys	法国产淀粉衍生物絮凝剂
floor	地板[面],楼板[面],底[部,面,板],地[楼,层,台,桥]面,(楼)层,底价
floor area	房屋面积,楼面面积,占地面积
floor drain	地沟
floor grating	格子楼板
floor space	建筑面积,占地面积
floor sump	泵池,地坑
floor sump pump	地面污水泵
floor sump pumpage	[地坑]泵(池,砂泵)抽运量
floorwash	地板冲洗
flop gate	跌落[导向]闸门
flop valve	跌落闸门,瓣阀
Flotagen	浮选剂(巯基苯骈噻唑衍生物,可用作捕收剂)
Flotal	起泡剂(醇与萜类的合成混合物)
Flotal B	起泡剂(含 α-萜烯醇及其他萜醇至少92%,法国赫斯特化工厂产品)
Flotanol	起泡剂(乙基甲基吡唑衍生物,可用作捕收剂)
Flotanol F	起泡剂(高分子醇类,法国赫斯特化工厂产品)
Flotanol G	起泡剂(高分子醇类,法国赫斯特化工厂产品)
flotation activity	浮选活度
flotation air	浮选空气
flotation air blower	浮选空气鼓风机
flotation area	浮选段[区]
flotation bank	浮选机组

flotation bank level 浮选机液面[位]
flotation behavior 浮选(变化)过程,浮选性能[功效,效应]
flotation cell criteria 浮选槽设计准则[标准]
flotation cell model 浮选机模型
flotation cell of automatic laboratory 自动化实验室浮选槽
flotation characteristic 浮游[选]性质
flotation chemistry 浮选化学
flotation chemistry research 浮选化学研究
flotation circuit metallurgical performance 浮选回路工艺性能
flotation circuit simulator 浮选回路模拟程序
flotation circuitry 浮选流程[回路]
flotation column 浮选柱
flotation concentration process 浮选精选[富集]法
flotation control system 浮选控制系统
flotation cycle test 闭路浮选试验
flotation density 浮选浓度
flotation edge 浮选边界
flotation engineer 浮选工程师
flotation facilities 浮选设施
flotation feed 浮选给矿
flotation feed density 浮选给矿浓度
flotation feed distributor 浮选给矿分配器
flotation feed line 浮选给矿管线
flotation feed pump 浮选给矿泵

flotation feed size 浮选给矿粒度
flotation field 浮选现场[领域,区域]
flotation floor 浮选段
flotation flowsheet 浮选流程
flotation froth 浮选泡沫
flotation fundamentals 浮选(基本)原理
flotation headings 浮选原矿
flotation helper 浮选工助手
flotation hypothesis 浮选假说[设]
flotation kinetic behavior 浮选动力学特征
flotation kinetic effect 浮选动力学作用[效应]
flotation kinetics 浮选动力学
flotation limit 浮选极限
flotation liquor 浮选溶液
flotation machine 浮选机
flotation machine family 浮选机种类[系列]
flotation machine part 浮选机部件
flotation mechanism 浮选机理
flotation metallurgical balance 浮选冶金平衡
flotation metallurgist 浮选专家
flotation micromechanism 浮选微观机理
flotation model 浮选模型
flotation model provided 所提供的浮选模型
flotation modifying agent 浮选调整剂
flotation network 浮选网络

flotation of non-metallics 非金属物质[矿物]浮选

flotation operation 浮选作业

flotation parameter 浮选参数

flotation particle size 浮选粒级

flotation peak 浮选峰值

flotation performance 浮选性能

flotation performance criteria 浮选性能指标[准则]

flotation personnel 浮选人员

flotation plant 浮选车间[工厂]

flotation plant performance 浮选厂[车间]性能

flotation plant simulation 浮选厂[车间]模拟

flotation procedure 浮选方法[作业]

flotation process 浮选工艺,浮选法

flotation process circuitry 浮选工艺流程

flotation process variable 浮选过程变量

flotation pulp alkalinity 浮选矿浆碱度

flotation range 浮选范围

flotation rate 浮选速度

flotation rate constant 浮选速度常数

flotation reagent behavior 浮选药剂行为

flotation region 泡选区

flotation response 浮选反应

flotation retention time 浮选时间

flotation scientist 浮选科学家

flotation section 浮选工段

flotation selectivity 浮选选择性

flotation separation 浮选(分离)

flotation separation process 浮选分离过程

flotation stage 浮选阶段[步骤]

flotation stream 浮选流程

flotation summary 浮选总结

flotation technique 浮选技术[法]

flotation test 浮选试验

flotation testing technique 浮选试验技术[法]

flotation theory 浮选理论

flotation time 浮选时间

flotation unit 浮选机

flotation unit cell 单槽浮选机

flotation unit operation 浮选单元操作

flotation variability 浮选变化性

flotation waste 浮选废水

flotation yield 浮选效率[回收率,生产能力]

flotation-magnetic-electrostatic separation 浮选-磁选-静电分离

flotation-no flotation edge 浮-不浮边界线

flotation-regrind-flotation flowsheet 浮选-再磨-浮选流程

Flotbel AM20 两性捕收剂(十八氨基磺酸盐类,英国 Float-Ore 公司生产) $R-NH_2-R_1-SO_3M(R=C_{18}H_{37}, R_1$ 在3个碳以内)

Flotbel AM21 两性捕收剂(油烯氨基烷基磺酸盐类,英国 Float-Ore 公

司生产) R—NH$_2$—R$_1$—SO$_3$ M (R = 油烯基)

Flotbel R107 T-A 一种重晶石捕收剂

Flotigam OA 捕收剂(油烯基伯胺醋酸盐,含量95%~97%)

Flotigam PA 捕收剂(十六烷基伯胺醋酸盐)

Flotigam SA 捕收剂(十八烷基伯胺醋酸盐)

Flotigan 捕收剂(德国染料公司生产,可用于浮选硅酸盐)

Flotigol CS 二甲苯酚混合物(德国赫斯特化工厂生产,可用作起泡剂)

Flotol 蒋醇(品名,工业纯,可用作起泡剂)

Flotol A 合成起泡剂

Flotol B 合成起泡剂

flow arrangement 流程配置[结构]

flow bin 流动仓

flow cell 流通[动]式样品盒

flow channel 流动渠道[沟]

flow chart (工艺)流程图,操作程序图,设备形象系统图

flow coefficient 流动系数

flow control 流量控制

flow conveyor 连续流(刮板)运输机

flow criterion 流动准则

flow data 流程(图)数据

flow discharge 泄流

flow factor 流动系数

flow factor for plane convergence 平面收缩流动系数

flow factor for two-dimensional convergence 二维空间收缩的流动系数

flow follower 流动跟踪物

flow function 流动函数

flow function of a solid 固体(物料)流动函数

flow mechanics 流动力学

flow meter 流量计

flow path 物料流动路线,流迹,流动道

flow pattern 流型,流动形态,流线图

flow property 流动性

flow proportional counter 流气式正比计数器

flow rate 流量(率,速)

flow rate correction 流量校正

flow rate service factor 流量工作系数

flow reactor 连续反应器

flow reducing equipment 矿流缩分设备

flow regime 流动状态

flow sensor 流量传感器

flow sequence 流动顺序

flow transmitter 流量变送器

flow velocity 流速

flowability test 流动性试验

flow-algorithm for design process 设计计算法流程

flowing bed 流动料层

flowing criterion 流动准则

flowing film concentrator 流膜选矿机

flowing solids 流动固体颗粒[物料]

flowmeter 流量计

flowmeter panel 流量计组合盘	fluidized bed upflow ion exchange system 流态化床上升气流离子交换系统
flowrate of feed 给料流速[量]	
flowrator 流速计,流量表,转子流量计,浮标式流量计	fluidized electrode 流态化电极
flowsheet alternatives 流程方案	fluidized flow 液化流
flowsheet change 流程变化	fluidized stream 液化流
flowsheet configuration 流程结构	fluidizing system 流态化系统
flowsheet design 流程设计	Fluor Standard Procedure 福禄标准程序(美国福禄公司)
flowsheet development 流程编制	
flowsheet development engineer 工艺流程编制工程师	fluorescent measuring channel 荧光测量道
flowsheet development stage 流程编制阶段	fluorescent reaction 荧光反应
	fluorescent spectrophotometry 荧光分光光度计法
flowsheet development team 流程编制组	fluorescent tagging 荧光示踪
flowsheet investigation 流程考查[研究]	fluorite-calcite separation 荧石-方解石分选
flowsheet of processing plant 选矿厂流程图	fluosolid roasting process 流态化[沸腾]焙烧法
fluctuating motor power signal 电动机功率波动信号	Fluo-Solids drying 沸腾焙烧干燥
	Fluo-solids roaster 沸腾焙烧炉
fluid bed roaster 沸腾焙烧炉	Fluo-solids system 流态化固体燃烧系统
fluid density 流体浓度	
fluid drive pump 液压驱动泵	Fluo-solids unit 沸腾焙烧炉[设备]
fluid film 液体薄膜,薄液膜	flush line 冲水管线
fluid medium 流体介质	flyash precipitator 烟[飞]灰集尘器
fluid path 流体通道	foaming characteristics 起泡特性[性能]
fluid process plant 流体处理设备	
fluid velocity 流体速度	fogging equipment 喷雾设备
fluidity 流动性	folding 褶皱(作用)
fluidization for conveying 输送流态化	following page 下页
fluidized bed roaster 沸腾焙烧炉	follow-up concentration process 后面

的选别过程

food processing waste 食品加工废弃物

foot brake 脚踏闸

foot pulley 尾轮

footage drilled 钻孔[探]进尺

footnote 脚注,补充说明

force field 力场

force of gravity 重力

force of magnetic field 磁场力

forced amalgamation treatment 强制混汞处理

forced vortex 强力涡[旋]流

foreground background capabilities 前台后台作业能力(计算机)

foreign business organization 外国商业团体

foreign currency 外国货币

foreign financial institution 外国金融机构

foreign government 外国政府

foreign lending 对外贷款

foreign material 杂质,外来物质,异物

foreign metal 杂质金属

forest service 森林管理部门[机关]

form 表格,格式

form of model 模型形式

formal test 正式试验

formal test period 正常的试验周期

formaldehyde 甲醛

format 格[版]式

format used 使用格式

formation of ionomolecular complex 离子分子络合物的生成

formed steel plate 成形钢板

former value 原值

formula 公式,配[药,处]方

formula translator 公式翻译程序语言

Forrester air machine 福雷斯特型充气浮选机

Fortran compiler Fortran 语言编译程序

Fortran program Fortran 程序

Fortran(=FORmula TRANslator) 公式翻译程式语言(计算机高级程序设计语言,即 FORmula TRANslator)

forward flow 向前矿流,顺矿流

forward reaction 正向反应

foundation course 基础垫[砌]层

foundation sketch 基础草图

foundry industry 铸造工业

foundry sand 铸造用砂

four bearing screen 四轴承筛

four compartment pulp distributor 四室矿浆分配器

four double-single cone in series 串联四阶段双单锥

four stage attritioner 四段摩擦机

four-cell OK 16 machine 4槽OK16型浮选机

four-fold axes 四对称轴

four-membered and six-membered rings of tetrahedral silicate group 四联和六联环状四面体硅酸盐基团

four-storey service building 四层操作[业务,服务]楼

fourth power 四次方

fourth stage crusher 第四段破碎机

Foxboro differential pressure orifice plate flowmeter 福克斯波罗型压差孔板流量计(美国 Foxboro 公司产品)

Foxboro level control 福克斯波罗型液面控制仪(美国 Foxboro 公司产品)

fraction 分数,份额,粒级

fraction critical speed 临界速率[速度系数],临界转数率

fraction of critical speed 临界速度百分率[数]

fractional ball load 钢球装载百分数[率]

fractional column 分馏塔

fractional critical speed 部分临界速度

fractionate 分级[离,别],分馏

fractionating column 分馏柱[塔]

fractionation 分馏(作用)

fracture pattern 破[断]裂形式

fragmental material 碎块[屑]状物料

fragmentation (晶粒的)碎化,碎分,碎裂,残片

fragmented rock 碎片状岩石

frame (机,底)座,框架,破碎机固定锥

frame clarifying press 净化[澄清]板框压滤机

frame depth 支架[构架]深度

frame precipitation press 置换[沉淀]板框压滤机

framework silicate 骨[框]架式硅酸盐

franckeite 辉锑锡铅矿

francolite 细晶磷灰石

Fransil silica Fransil 硅石粉

free acid form 游离酸形态

free bubble surface 气泡自由表面

free chlorine 游离氯

free crushing rock 易破碎岩石

free cupric ion 自由铜离子

free cyanide 游离氰化物

free cyanide level 游离氰化物浓度[程度,含量]

free electron concentration 自由电子密度

free energy 自由能

free energy of formation 成矿自由能

free enterprise system 自由企业制度

free flow design 直流式设计(浮选机的)

free metallic gold 单体金属金

free milling gold ore 易汞齐金矿石

free oil 游离油

free oxygen 游离氧

free particle 单体颗粒

free settling velocity 自由沉降速度

free settling [falling] speed 自由沉降[下落]速度

free sulphur 单体硫

free swinging hammer 自由摆动[旋转,回转]锤

free vortex 自由涡[旋]流

free water 游离水(分)

free-floating 自由浮动
free-flow type 直流型,自由流动式
free-flowing hydrophobic powder 自由流动的疏水性粉末
free-flowing powder 自由流动粉末,流态粉末
free-flowing solids 自由流动固体物料
freely circulating surface 自由环流表面
freely cleaved surface 自由解理表面
free-milling coarse-grained chalcopyrite 易选粗粒黄铜矿
free-swinging two bearing screen 自由摆动双轴承筛
freezing problem 水冻问题
freezing unit 冷冻设备
freight cost 运费,货运成本
freight rate 运价
freight rate per ton-mile 吨英里运费
freight vehicle 货运车辆
French curve 曲线板(规)
French curve technique 云形曲线技术
French franc 法国法郎
frequency function 频率函数
frequency tuneable 可调频式
fresh carbon return screen 新鲜碳返回筛
fresh feed 新给料[矿]
fresh feed rate 新给料[矿]量
fresh make-up water 新补加水
fresh metallic face 新鲜金属面
fresh mineral surface 新鲜的矿物表面

fresh ore 新鲜矿石
fresh sample 新鲜试[矿]样
fresh stock 新货
fresh water aquatic life 淡水水栖生物
fresh water fish 淡水鱼
fresh water flotation 淡水浮选
fresh water line 新水管线[道]
fresh water reservoir 新鲜水水库
freshly broken surface 新鲜解理面
freshly cleaved surface 新解理表面
friction brake 摩擦制动[刹车]器
friction head loss 摩擦压头损失
friction loss rate 摩擦损失率
frictional loss rate factor 摩擦损失率系数
fritted glass 烧结[多孔]玻璃
fritted glass disc 熔结[多孔]玻璃盘
front panel of control unit 控制单元[装置]面板
front side 正面
frontispiece 卷头插图,标题页
froth base 泡沫底部
froth bed 泡沫层
froth behavior 泡沫特点[性],泡沫变化过程
froth breakage 泡沫破裂
froth buildup 泡沫堵塞(跑槽)
froth cleaning action 泡沫精选作用
froth color 泡沫颜色
froth column 泡沫柱[层]
froth column formation 泡沫层形成
froth cross section area 泡沫截面积

froth depth 泡沫深[厚]度	froth stability 泡沫稳定性
froth effect 泡沫作用[效应]	froth stability index 泡沫稳定性指数
froth factor 泡沫因素	froth surface area 泡沫表面积
froth flotation cell mechanism 泡沫浮选槽机构	froth transmission coefficient 泡沫传送系数
froth flotation mill 泡沫浮选厂	froth volume 泡沫体积
froth flotation separation process 泡沫浮选分离过程	froth waste 泡沫废品,泡沫尾矿产品
	froth water 泡沫水
froth flotation technique 泡沫浮选技术	froth weir 泡沫溢流堰
	frothability 起泡能力[性能]
froth formation 泡沫形成[生成,产生]	froth-breaking property 消泡性能
	frother 起泡剂,泡沫剂
froth height 泡沫高度	Frother 52 一种合成的高级醇类起泡剂(色暗红,密度 $0.850 g/cm^3$,含有2号柴油)
froth holding time 泡沫维持时间	
froth lip 泡沫溢流堰(浮选槽)	
froth lip height 泡沫溢流堰高度	Frother 58 一种合成的高级醇类起泡剂(色暗红,密度 $0.865 g/cm^3$,含有2号柴油)
froth meter test 泡沫计试验	
froth model 泡沫模型	
froth paddle 泡沫刮板	Frother 60 一种合成的高级醇类起泡剂(淡草黄色,密度 $0.830 g/cm^3$,含有2号柴油)
froth phase 泡沫层[相],起泡阶段	
froth product 泡沫产品	
froth pump 泡沫泵	Frother 63 一种水溶性合成醇类起泡剂
froth removal 泡沫排除,刮泡	
froth removal rate 泡沫排除速度[率],泡沫排除量	frother addition rate 起泡剂添加量
	Frother B-23 B-23起泡剂(混合高级醇,含2,4-二甲基戊醇-1;2,4-二甲基己醇-3 及酮类)
froth removal zone 泡沫排除区	
froth residence time 泡沫停留时间	
froth scraper 泡沫刮板	frother concentration 起泡剂浓度
froth separation 泡沫分选[离]	frother molecule 起泡剂分子
froth separator 泡沫层分选浮选机,浮选泡沫分选机	frother species 起泡剂种类
	Frother SW4 SW4起泡剂
froth sprinkling 泡沫喷水	frothing chamber 起泡室

frothing characteristic 起泡特性
frothing characteristics 起泡性能[特性]
frothing compartment 起泡室(浮选机的)
frothing effect 起泡效应
frothing performance 起泡性能
frothing power 起泡能力
frothing property 起泡性能
frothing test 起泡试验
frothy middlings 多泡沫的中矿
Froude number 弗劳德数(离心加速度与重力加速度之比)
frozen lump 冻结块
Frumkin-Fowler isotherm 弗郎金-福勒等温线
fuel oil 燃料油
fuel oil storage tank 燃油储罐
fuel rate 燃料消耗率
fuel storage 燃料仓库,燃料贮存
fuel-fired reverberatory furnace 烧燃料反射[焰]炉
full autogenous basis 全自磨基础
full description 全部说明
full diameter 大[主]直径,外径
full feasibility study 完整的可行性研究
full force and effect 充分效力
full load countershaft speed 满负荷副轴速度(圆锥破碎机)
full plant operation 全厂[满负荷]生产

full pre-run procedure 整个预运行过程
full rank matrix 满秩矩阵
full rank matrix notation 满秩矩阵符号
full scale 大[全]规模,全尺寸,全刻度,满量程
full scale commercial placer operation 全规模工业砂矿生产
full scale construction 全面施工
full scale contour 原尺寸外形
full scale filtration rate 实体尺寸器过滤速度(工业生产规模的过滤速度)
full scale gravity separation equipment 工业生产规模的重选设备
full scale mechanism hydraulic test 工艺性机构水力学试验,机构水力学工业试验(浮选槽)
full scale model 足尺模型
full scale operation 大规模生产
full scale pilot test 全负荷[全面]的半工业试验
full scale pilot trial 生产规模的试生产
full scale plant design 全面工厂设计
full scale plant test 全厂规模试验
full scale production 实际生产
full scale production data 实际生产数据
full scale production equipment 工业规模生产设备
full scale production unit 实际生产装置

full scale test 实物(真实条件,工业规模,全尺寸)试验
full scale units 大规模生产机组
full size 实际[全]尺寸
full size machine 正式机器,足尺寸机器
full size plant 正式厂
full skill support 全方位技术支持
full stream sample 全宽截流试样(即对整个矿流宽度切取到的试样)
full weight 全[毛]重
fuller's earth 漂白土(一种黏土,具有吸收油脂的性质)
fuller's earth burnt 烧制的漂白土
fuller's earth raw 生漂白土
full-in-the-block type format 填入空格型的格式
full-scale economic assessment study 全面经济评价研究
full-scale mill 工业生产规模的磨机
full-scale plant 规模生产(选矿)厂,足尺寸设备(相对于小规模试制,实验性设备)
full-scale plant trial 工业试验
full-scale test 工业试验
full-size Deister wet table 工业用戴斯特湿式摇床
full-time employee 全日制雇员
fully autogenous 全自磨的
fully autogenous grinding 全自磨[法]
fully autogenous grinding mill 全自磨机
fully autogenous grinding system 全自磨系统
fully effective outlet 全部有效排口
fully liberated ore 完全解离的矿石
fully riffled deck 全格条床面
fully-automatic sieving machine 全自动筛分机
fume potential 产生烟雾的可能性[潜力]
function of angle 角函数
function of collector 捕收剂的功能
function of logarithm 对数函数
function of particle size 矿[颗]粒粒度函数
function of time 时间函数[作用]
functional division 职能处[部]
functional form 函数形式
functional objective 功能目标
functional specification 功能规范,功用说明书
fund 资[基]金,经费
fund acquisition 资金取得
fund source 资金来源
fundamental 基本[础]的,根本[主要]的,原理,原则
fundamental aspect 基本方面
fundamental basis 基本依据
fundamental characteristic 基本特性
fundamental equation 基本方程式
fundamental flotation chemist 浮选基础理论化学家
fundamental goal 基本目标

fundamental knowledge 基础知识
fundamental law of classification 分级基本定律
fundamental of cyclone 旋流器基本原理
fundamental purpose 基本目的
fundamental right 基本权利
fundamental science 基础科学
fundamental separation characteristics 基本分离[选]特性
fundamental strategy 基本控制策略
fundamental surface chemistry 基础(的)表面化学
fundamental tool 基本工具
fundamental understanding 基本认识
fundamentals of spiral classifier sizing 螺旋分级机分级原理
funding loss 资金损失
funds committed to construction 建筑专用资金
funds statement 资金报表
funnel flow hopper 漏斗式流动漏斗
funnel-flow bin 漏斗流动仓
funnel-flow bin outlet 漏斗流动仓排口
furfural 糠醛 $C_4H_3O \cdot CHO$
furfural xanthate 糠醛黄药(可用作捕收剂) $C_6H_5O_2S_2K$
furnace effluent 炉子溢出[渗漏]的废气
fused alumina ceramic tile 熔凝[成]的氧化铝瓷砖
fused cast basalt lining 熔铸玄武岩衬里
fusel oil 杂酚油,可用作捕收剂、起泡剂
future delivery 期货,期货交割
future reference 未来参考
future units 预留机组

G

gain 增益,增加,利润,收获,获利
Galigher 120 flotation cell 加林格尔 120 浮选槽[机]
Galigher cell 加林格尔浮选机[槽]
Galigher vertical sump pump 加林格尔垂直油池泵
galvanic contact 电流接触
galvanic potential 伽伐尼电势,电偶电位
galvanized iron sheet 镀锌铁皮
galvanized steel sheet 镀锌薄钢板
galvanomagnetic measurement 电磁测

定法
gamma distribution 伽马[γ]分布
gamma function parameter 伽马[γ]函数参数
gamma gauge 伽马[γ]表[辐射计]
gamma ray density controlling device 伽马[γ]射线浓度控制装置
gamma ray irradiation 伽马[γ]射线辐照
gamma-ray isotope 伽马[γ]射线同位素
gang sampler 组合[共轴]取样机
gangue 脉石,矸石,尾矿
gangue constituent 脉石组分[成分]
gangue depressant 脉石抑制剂
gangue material 脉[废]石
gangue removal 脉石排除
gangue sulfide 脉石硫化矿
gantry unloading crane 龙门卸载吊车
garbage 垃圾,废料
garbage disposal plant 垃圾处理厂
Gardinol CA 湿润剂(油烯基硫酸盐) $C_{18}H_{35}OSO_3M$
Gardinol KD 湿润剂(十八烷二醇-9,18-二硫酸酯)
Gardinol WA 湿润剂(十二烷醇硫酸酯,或瓦斯油) $C_{12}H_{25}OSO_3M$
Gardner-Coleman oil-absorption test 加德纳·科尔曼油吸附试验
Gardner-Denver crawler jumbo 加德纳·丹佛履带式钻车
garnetized limestone 石榴石化灰岩

gas adsorption 气体吸附
gas collection system 气体收集系统
gas collector header 气体收集器集气管
gas condition 气体条件
gas distribution device 气体分配装置
gas emission 排放废气
gas fired crucible furnace 烧气坩埚炉
gas inlet 气体入口
gas outlet 排气管[口]
gas phase chromatography 气相色谱分离法
gas train line 气体管道
gas volume 气体体积
gaseous ion 气态离子
gaseous phase 气相
gaseous state 气态
gas-fired Harding rotary dryer 哈丁燃气式回转干燥机
gas-fired tilting reverberatory furnace 燃气倾倒式反射炉
gas-liquid chromatography 气相-液相色谱法
gas-liquid-bubble interface 气-液-气泡界面
gas-phase experiment 气相实验
gas-solution interface tension 气-液界面张力
gas-tight vessel 气密容器
gate controller 大门[闸门]控制装置[器]
gate feeder 闸门给矿[料]机

Gates' gyratory crusher 盖特斯旋回破碎机
Gaudin distribution function 高登粒度分配函数
Gaudin-Schuhmann function 高登-舒曼粒度分配函数
Gaudin's correction 高登(粒度)修正数[值]
Gaudin's equation 高登粒度分布方程
Gaudin's relation 高登(粒度)关系式
gear pinion loss 传动齿轮损失
gearbox ratio 减速箱减速比
geared motor 齿轮马达[发动机]
Geary-Jennings cross-cut sampler 基瑞-詹宁斯型横切式取样器
Gefanol I 一种木焦馏油(品名,可用作捕收剂湿润剂、起泡剂)
Geiger-Muller tube detector 盖革·缪勒管探测器
gel time 凝胶时间
gelatin(e) (白)明胶(可用作抑制剂、絮凝剂)
Genapol-AS 湿润剂(脂肪醇聚乙二醇硫酸酯,"AS"为铵盐英文缩写) $R(OCH_2CH_2)_n \cdot OSO_3NH_4$
general 总则,综述,普通,综合
general appreciation 综合评价
general arrangement plan 总平面布[配]置图
general aspect 一般方面[情况]
general assembly of standard heavy-duty cone crusher 重型标准圆锥破碎机总装配图
general assignment of book debts 预约借款总则
general characteristic 通性,一般特性
general chemical grounds 一般化学原理[根据]
general conclusion 一般性[总]结论
general construction division 综合施工处[部]
general consultant 一般咨询单位
general contractor 总承包人[商]
general description 总说明
general description of construction 施工说明书
general discussion 一般性讨论
general dust system layout 通用管道系统布置
General Electrix X-ray unit 通用电气公司X射线设备
general expense 一般费用
general expression 普通式,通用表达式
general features of construction 施工概要
general flotation cell model 通用浮选槽模型
general flotation concentrate 总浮选精矿
general flow pattern 概略流程图
general flowsheet 原则[一般,简略]流程
general form 通式

general grinding circuit simulator 通用磨矿回路[系统]模拟程序
general guide 一般指导[南]
general idea 一般概念,大意
general layout 总布置[平面]图,总图
general maintenance 一般[普通]修理
general manager 总经理
general manager's expense account 总经理的用度费
general mathematic structure 通用数学结构
general mathematical formula 一般数学公式
general mill tailing 总(选矿)尾矿
general mineralogical study 综合矿物研究
general model 通用模型
general office cost 总办公费
general outline 概要
general plan 总平面图
general principle 原[通,总]则,普通原理
general principle of dust collection 集[收]尘一般原理
general procedure 通用程序
general process 一般工艺过程
general program 综合[通用]程序
general purpose 通用[普通,一般用途]的
general purpose grinding program 通用磨矿程序
general purpose reagent 通用药剂
general purpose simulation program 通用模拟程序
general purpose simulator 通用模拟器
general relationship between input and output 输入与输出间一般关系式
general remark 概要,总论
general service man 一般服务[技术,修理]人员
general trend 总趋势
general view 大纲,概要,全视图
general-instrument engineering 通用仪表工程技术
general-instrument engineering capacity 通用仪表设计能力
generalized flowsheet 一般化的流程图
generalized scheme 通用图式[方案]
generalized structure 通用机构
generator gas tar 发生炉煤气焦油(可用作捕收剂、起泡剂)
gentle agitation 缓慢搅动
geocentric gravitational constant 地心引力常数
geochemistry 地球[质]化学
geologic evidence 地质论据
geologic factor 地质因素
geologic information 地质资料
geologic reserve 地质储量
geologic structure 地质构造
geological and minable reserve 地质与可采储备
geological character 地质特征
geological character of deposit 矿床地

质特征
geological characteristic 地质特性
geological characteristics 地质特性
geological complexity 地质复杂性
geological evaluation 地质评价
geological exploration data 地质勘探资料[数据]
geological information 地质资料
geological ore reserve 地质矿石储量
geological relationship 地质关系
geological reserve 地质储量
geological science 地质科学
geological viewpoint 地质(学)观点
geologists' report 地质(工作者的)报告
geometric construction 几何结构
geometric mean 几何平均值,等比中项[数]
geometric mean size 几何平均尺寸[粒度]
geometric pattern of sample 取样的几何布置形式
geometric programming 几何规划
geometric similitude 几何相似
geometric variable 几何变量
geometrical parameter 几何参数
geometrical relation 几何[图形]关系
geometrical relationship 几何关系
geometrical shape 几何形状
Georgia Iron Works pump 佐治亚钢铁厂泵
geostatistical study 地质统计研究

geostatistical theory 地质统计学理论
geostatistics 地质统计学
geotechnological method 地质[土工,地球科学]技术法
Gibbs(Josiah Willard Gibbs) 吉布斯(约西亚·威拉德·吉布斯,美国物理和化学家,1839~1903年)
Gibbs adsorption equation 吉布斯吸附方程式
Gibbs adsorption isotherm 吉布斯吸附等温式
Gibbs adsorption theorem 吉布斯吸附定理
Gibbs surface free energy 吉布斯表面自由能
gift 礼物,赠品
Gilmore needle 吉尔摩水泥稠度试验针
Gilmore scale 吉尔摩秤
GIM pump GIM 型泵
given deposit 给定矿床
given value 给定值
glacial acetic acid 冰醋酸
gland sealing system 密封盖水封系统
gland water 水封水
glass bead 玻璃珠
glass electrode 玻璃电极
glass jar 玻璃大[广]口瓶
glass measuring cylinder 玻璃量筒
glass sand 石英砂
glass sand flotation 玻璃砂浮选
glass sphere 玻璃球
glass tube 玻璃管

glass tubing 玻璃管	gold telluride 碲化金,二碲化金
glass-sand specification 玻璃砂规格	gold treatment plant 金处理厂
global reserve 总[全局]储量	gold tulluride 碲金矿
global scale 总体规模	gold yield 金生产率
glossary of terms 术语集[汇编]	gold-amalgamation operation 混汞选金作业[操作,生产]
glove box 密闭操作箱,手套箱	gold-bearing gravels 含金砾石
glue 动物胶,骨胶(可用作絮凝剂)	gold-bearing solution 含金溶液
Glue-Beeds No. 22 动物蛋白质[骨胶](品名,可用作絮凝剂)	gold-saving plant 选金车间
gold amalgamation 黄金混汞法	goniometer and microscope system 测角器与显微镜装置
gold and silver assay 金银鉴定	good and workmanlike manner 良好而熟练的方式
gold and silver producer 金银生产厂(家)	good power measurement 完善的功率检测
gold assay 试金法	goodness of fit test 拟合良好性检定
gold bearing leach solution 含金浸出液	goodness of fitting 拟合优[良]度
gold bearing sulfide 含金硫化物[矿]	Good-rite K721S 絮凝剂(水溶性合成高聚物)
gold bearing zone 含金矿带	Good-rite K770S 絮凝剂(水溶性合成高聚物)
gold concentrator 金选矿厂	goods 货物
gold cyanide compound 氰化金化合物	goods vehicle 货运车辆
gold cyanide extraction 金氰化提取[浸出]率	Gortikov apparatus 戈蒂科夫仪
gold extraction 金的提取,提金	goulac 木素磺酸钙(可用作碳质脉石的抑制剂)
gold foil 金箔	Gouy method 古伊方法(用于估算引力势能)
gold leaf 金箔	governing element 控制元件
gold mill 金选厂	governing law 管理法
gold plant 金[氰化]厂	government agency 政府机构[关]
gold porphyry 含金斑岩	government guaranteed security 政府
gold precipitate 金沉淀物,沉淀金泥	
gold recovery area 金回收区[段]	
gold recovery thickener 金回收浓缩机	
gold room 炼金室[车间]	

担保债券
government owned firm 国有企业
government regulations 政府法规
government statistics 政府统计
governmental action 政府的作用
governmental agency 政府机构
governmental bodies 政府主管部门
governmental clearance 政府的出[入]港执照
governmental control 政府控制
governmental health monitoring agency 政府健康监督机构
governmental laboratory 政府实验室
government-controlled firm 国有企业,政府控制企业
grab and run technique 先捞后捕技术
grab sample 抓斗[随机]取样
grade class 品(位等)级
grade constraint 品位约束[限制]
grade control 品位控制
grade distribution 品位分布
grade elevation 平整场地标高
grade estimate 品位估算
grade estimation 品位估算
grade improvement 品位提高
grade of concentrate 精矿品位
grade of diluting material 废石品位,贫化物料品位
grade of flotation tailings 浮选尾矿品位
grade recovery curve 品位回收率曲线
grade separation 等级分类

grade variability 品位可变性
grade variation 品位变化
graded charge 级配加球
graded media filter 分级介质过滤机
grade-recovery curve 品位回收率曲线
grade-recovery matrix 品位回收率矩阵
grade-recovery performance 品位-回收率指标
grade-recovery (curve) scope 品位-回收率(曲线)范围
gradient assay 梯度分析
gradient method 梯度法
graduate thesis 毕业生论文
graduated burette 刻度滴定管
grain 颗[细,晶]粒,格令(英制重量单位:1格令=0.0648克)
grain count assay 颗粒计数分析
grain gradation 粒度分级
grain size 颗粒尺寸[粒级,粒度]
grain size distribution factor 粒度分配系数
grain structure 晶粒构造
grain-size distribution factor 粒度分配系数
granite 花岗岩
granite porphyry phase 花岗斑岩相
granitic rock 花岗岩岩石
granitic stock 花岗岩岩株
granular product 粒状产品
granular rock 粒状岩石

graph comparison of crusher production 破碎机产量图形[解]比较

graphic display 图形[解]显示

graphic representation 图解表示法

graphical cyclone data 旋流器图解数据

graphical data 图解数据

graphical form 图解方式

graphical information 图解资料

graphical plot 图解,图表

graphite 石墨

graphite inserts 石墨衬垫[套,片]

graphite mould 石墨模

graphitic carbon 石墨碳

graphitic gold ore 石墨金矿

grasshopper conveyor (蚱蚂式)振动运输[输送]机

grate basket 箅子板

grate discharge 格子排料

grate discharge rod mill 格子排矿棒磨机

grate size 格[算]条尺寸

grated floor 格子楼板

grating plate 格板(筛分)

grating walkway 格子走道

gravel sandstone 砾石砂岩

gravimeter 比重计

gravimetric feeder (测)重量给料[矿]机

gravimetric flowrate measurement 流量重力分析测量(法)

gravimetric gas or vapor adsorption 重力气体或蒸汽吸附(作用)

gravitational acceleration 重力加速度

gravitational concentration method 重选法

gravitational concentration tool 重选工具

gravitational effect 重力效应

gravitational energy 重力能

gravitational field 重力场

gravitational force 重[引]力

gravitational sedimentation technique 重力沉降[淀]技术

gravitational settling 重力沉降

gravitional hindered settling 重力干扰沉降

gravitional X-ray particle size analysis 重力X射线粒度分析

gravity 重[引]力

gravity beneficiation 重(力)选矿

gravity characteristic 重力特性

gravity circuit 重选回路

gravity concentrate 重选精矿

gravity concentration device 重力选矿装置,重选装置

gravity concentration investigation 重选研究

gravity concentration method 重(力)选矿法

gravity concentration process 重选工艺

gravity concentration processing 重选工艺[作业],重力选矿

gravity concentration table 重选摇床
gravity cone sluice concentrating table 重力圆锥溜槽选矿机
gravity device 重选装置
gravity feed 自流加[给]料[矿],重力加[给]料[矿]
gravity feed line 自流给矿管线
gravity field 重力场
gravity flowrate 自流流量[速度]
gravity force 重力
gravity means 重选法
gravity method 重力选矿法,重选法
gravity plant 重选厂
gravity plant effluent 重选厂污水
gravity process 重选方法
gravity process design 重选设计
gravity processing technique 重选技术
gravity sampler 重力取样器
gravity separation 重(力)选矿
gravity separation equipment 重(力)选矿设备
gravity separation method 重选法
gravity separation product 重选产品
gravity separation technology 重选工艺[技术]
gravity settling 重力沉降[淀]
gravity settling of particle 颗粒重力沉降
gravity technique 重选技术
gravity treatment 重选
gravity unit 重选设备[装置]
gravity unit process 单一重选工艺[方法,作业]
gravity-plane table 重选平板槽
grazing land 畜牧场
grease belt 油脂胶带机(选金刚石用)
grease lubricated vibrator 油脂润滑振动器
grease lubrication 油脂润滑
grease-lubricated 油脂润滑的
grease-lubricated pinion bearing 油脂润滑的小齿轮轴承
grease-lubricated seal 油脂润滑密封
great contribution 伟大贡献
Greek alphabet 希腊字母
Greek letter 希腊字母
Gregory spectrometry 格雷戈里光谱法
grey iron 灰口铁
grid (网)格
grid of coordinate 格子坐标
grid point 网格点
grid search 格点[网格]搜索(法)
grid search technique 格点[网格]搜索法
grid walkway 格[箅]条行人道
grind circuit size distribution 磨矿回路粒度分配[值]
grind index 磨矿(功)指数
grind size 磨矿细度
grind work index 磨矿功指数
grindability 可磨性
grindability of ore 矿石可磨性
grinding action 磨矿作用

grinding and flotation flowsheet 磨浮工艺流程
grinding and pelletizing stage 磨矿团矿阶段
grinding bay 磨矿跨
grinding bay pump sump 磨矿跨泵池
grinding characteristic 磨矿性质[特性]
grinding charge 磨矿给矿
grinding circuit 磨矿回路[循环，流程]
grinding circuit behavior 磨矿回路特性
grinding circuit capacity 磨矿回路能力
grinding circuit constraint 磨矿回路约束条件
grinding circuit performance 磨矿回路性能
grinding circuit simulator 磨矿回路模拟程序
grinding circuit subprocess 磨矿回路子过程
grinding concentration process 磨石选别过程
grinding condition 磨矿条件
grinding data 磨矿数据
grinding equipment 粉磨设备
grinding feed size 磨矿给矿粒度
grinding index 磨矿指数
grinding kinetics 磨矿动力特性[学]
grinding kinetics factor 磨矿动力学因素
grinding line 磨矿系列
grinding media data 磨矿介质数据
grinding media expansion 磨矿介质膨胀
grinding media void ratio 磨矿介质孔隙率
grinding method 磨矿方法
grinding mill dynamics 磨机动力学[动态特性]
grinding mill information sheet 磨机资料表
grinding mill operation 磨机操作
grinding mill performance 磨机性能
grinding mill power calculation 磨机功率计算
grinding mill pressure drop 磨机压降
grinding model development 磨矿模型研制
grinding model parameter 磨矿模型参数
grinding period 磨矿周期
grinding power 磨矿功率
grinding power calculation 磨矿功率计算
grinding power consumption 磨矿功率消耗
grinding power data 磨矿功率数据
grinding power efficiency 磨矿功率效应
grinding process 磨矿过程
grinding rod 磨矿钢棒

grinding section 磨矿工段
grinding stage model 磨矿阶段模型
grinding throughput 磨矿通过[处理]能力[量]
grinding time 磨矿时间
grinding unit 磨矿单元[设备]
grindings housing comminution circuit 破磨[碎磨]系统厂房
grind-recovery relationship 磨矿细度与回收率关系
grind-recovery-throughput relationship 磨矿细度-回收率-处理量间互相关系
grizzly stockpile 格条筛堆矿场
grizzly undersize product 格筛筛下产品
Groch machine 格罗奇浮选机(封闭叶轮,中空轴底吹式)
gross annual benefit 总年收益
gross cash flow 总现金流量
gross earnings expense 总收入[益]费用
gross margin 毛利(润,率)
gross organic content 有机物总含量
gross output value 总产值
gross pollution 严重污染
gross power 总功率
gross profit 总利润,毛利
gross sales 销售总额
gross tonnage 总吨位[数]
gross volume 总容积
ground area 地面面积
ground grain 已磨颗粒
ground ore 被磨矿石
ground protection 接地保护
ground slurry 磨细的矿浆
ground sulfide 已磨[碎磨]硫化矿
grounding technique 接地技术
grounding wire 地线
group frequency 基团频率[波数]
group frequency shift 基团频率偏移
grouping 分类[组],编组
growing season 植物生长季节[期]
guaiacol 愈疮木酚,邻甲氧[基]苯酚(可用作起泡剂) $CH_3OC_6H_4OH$
guanidine 胍 $NH:C(NH_2)_2$
guanidine hydrochloride 盐酸胍(可用作闪锌矿的捕收剂) $H_2N \cdot C(:NH)NH_2 \cdot HCl$
guanidine nitrate 硝酸胍(可用作闪锌矿的捕收剂) $H_2N \cdot C(:NH)NH_2 \cdot HNO_3$
guar dextrin depressant 古尔糊精抑制剂
guar flour 古尔粉(成分为半乳糖甘露聚糖,可用作抑制剂、絮凝剂、分散剂、活化剂)
guar gum 古尔胶(成分为半乳糖甘露聚糖,可用作抑制剂、絮凝剂、分散剂)
guarantee loan 担保贷款
guaranteed analysis 保证分析
Guartec 古尔胶(品名,半乳醋甘露聚醣,可用作抑制剂、絮凝剂、分散剂)
guesstimate 初步估算值
guesswork 猜想,假设[定],推论

guidance and examination fee 指导与审查费

guide conveyor 导向运输机

guide plate 导向[流]板

guideline 指南,指导路线,方针,准则

guideline for discharge waters 排水标准

guidelines for environmental care 环境保护[管理]准则[方针,指南,政策]

Guidelines for Project Evaluation 工程项目评价指南

gulch gold 砂金

Gum 3502 抑制剂(水溶淀粉制剂,可用作氧化铁矿的抑制剂)

gum Arabic 阿拉伯树胶(可用作云母的抑制剂、脉石的分散剂)

gum rubber 树胶橡胶,天然橡胶,生橡胶

gunite lining 水泥喷射灌浆

Gy basic formula 吉氏基本公式

Gycolate 抑制剂(可用作硫化铜矿的抑制剂)

gypsum scaling 石膏结垢

gyradisc crusher 旋盘式破碎机

gyratory crusher component weight 旋回破碎机部件重量

gyratory crusher operation 旋回破碎机操作

gyratory crusher with hydraulic system 带液压系统的旋回破碎机

gyratory primary crusher 旋回粗碎机,旋回第一段破碎机

gyratory shaft 旋转轴

gyratory-type crusher 旋回型破碎机

H

8-hydroxyquinoline 8-羟基喹啉(可用作捕收剂)HO·C_9H_6N

H-120 一种硅酮油(品名,可用作黄铜矿和方铅矿的捕收剂)

hafnium-free zircon metal 无铪锆金属

Hagan coagulant 11 凝结剂(聚电解质加膨润土,可用作絮凝剂)

Hagan coagulant 18 凝结剂(聚电解质加膨润土,可用作絮凝剂)

Hagan coagulant 2 凝结剂(聚电解质,可用作絮凝剂)

Hagan coagulant 50 凝结剂(聚电解质,可用作絮凝剂)

Hagan coagulant 7 凝结剂(聚电解质加膨润土,可用作絮凝剂)

half cell reaction 半电池反应
half size material 粒度小于筛孔半径的物料
half-size 半尺寸,一半大小的
half-size correction factor 小于筛孔半径的粒度修正系数
Hall effect 霍尔效应(导电材料中的电流与磁场相互作用产生电动势的效应,是美国物理学家霍尔于1879年发现的)
Hall effect measurement 霍尔效应测定
Hall flowmeter 霍尔流量[流速]计,霍尔流动性测量仪
Hall settling detection device 霍尔沉降探测装置
Hallimond microflotation cell 哈里蒙微型浮选槽
Hallimond tube 哈里蒙(单泡浮选试)管(广泛用于浮选理论研究)
Hallimond tube flotation 哈利蒙管浮选
hall-mark (金银的)纯度,品质证明,检验烙印,标志,特点
halocarbon 卤化碳,卤烃
halogen 卤素
halogenated hydrocarbon 卤代烃
halohydrocarbon 卤代烃
Hamaker(Hugo Christiaan Hamaker) 哈梅克(H. C. Hamaker 荷兰科学家,1905~1993年)
Hamaker constant 哈梅克常数(表征物质之间范德华吸引能大小的参数)
hamartite 氟碳酸铈矿 $Ce[CO_3]F$
hammer axle extraction device 锤轴抽[取,拔]出装置
hammer crusher 锤式破碎机
hammer mill product analysis chart 锤磨机产品分析表
hammer replacing device 锤更换装置
hand addition 人工添加
hand calculation 手算
hand computation 笔算,手算
hand control 手控
hand dressing 手选
hand magnifier 手持放大镜
hand sample 手工试样[标本]
hand sorted high grade product 手选高品位产品
hand sorted waste product 手选废石[品]
hand sorting 手选
hand specimen examination 手标本检查[验]
handbook for estimating major equipment cost and preliminary capital cost 估算主要设备费用与初步建设成本手册
hand-held ripper 手持凿岩机[破碎机]
handling cost 搬运费
handling procedure 加工过程
handling process 加工程序(试样)
handling variability 操作变率,加工变

化性

handrail 扶手,栏杆

handwritten state 手写状态

hanger bearing 悬挂轴承

hard ceramics 硬陶瓷

hard coal 硬煤

hard coal sampling 硬煤取[采]样

hard constraint level 严格[硬性]约束水平

hard gangue mineral 硬脉石矿物

hard metal article 硬质合金制品

hard mine water 硬矿井水

hard money 硬通货,硬性货币

hard ore 硬矿石

hard polyurethane 硬质聚氨基甲酸酯

hard sphere model 硬球体模型

hard sphere repulsion model 硬球相斥模型

hard zone 硬(矿)带

harder grinding fraction 较难磨的部分

Harding ear 哈丁电耳

Harding electric ear 哈丁电耳

Hardinge drum heavy media separator 哈丁圆筒重介质分选机(美国哈丁公司产品)

Hardinge feeder 哈丁给料[矿]机(美国哈丁公司产品)

Hardinge semi-autogenous mill 哈丁半自磨机(美国哈丁公司产品)

Hardinge tri-cone mill 哈丁三锥磨机(美国哈丁公司产品)

hard-to-start 难起动的

hardware 硬件

hardware configuration 硬件配[布]置(计算机的)

hardware selection 硬件选择

hard-water cation 硬水阳离子

hardwood creosote 硬木焦馏油

harmful condition 有害条件

harmful effect 有害[不良]后果[效应]

haulage train 运输列车

Haultain infrasizer 豪尔顿微粒空气分级机

Haultain superpanner 豪尔顿淘(砂)矿机,摇动V形淘盘(试验用)

hazardous 有害

hazardous condition 危险条件

Hazen-Williams formula 黑曾-威廉斯公式(计算矿浆管道摩擦损失率系数)

head 动锥(圆锥破碎机的)

head analysis 原矿分析

head assay 原矿分析(品位分析)

head concentration 原矿富集

head diameter 锥体[端部]直径

head displacement 锥头位移

head drive 头部传动(胶带运输机)

head grade 原矿品位

head mineral (入选)原矿矿物

head office 总公司,总部

head office cost 总公司费用

head office staff 公司总部员工

head sample 原矿样
head sample analysis 原矿样分析
head sampler 原矿取样工
head shaft 驱动[首端]轴
head sprocket 头部链轮
header 头[首]部,磁头,顶盖,端板,标题,首长,集(气,水,流)管,联管箱,管座,水箱,蓄[集]水池
heading 标题,题目,平巷
health and life insurance 健康与人寿保险
health authorities 卫生当局
health authority 卫生管理机关
health hazard 健康危害
health of the general public 公众健康
health problem 健康问题
health requirement 保健需要[要求]
heap construction equipment 筑堆设备
heap leaching procedure 堆浸法
heap size 浸堆规格
heaping equipment 筑堆设备
hearing protection 听力保护
hearth-type furnace 床式炉
heat distortion 热变形
heat engineering 热工学
heat exchanger 热交换器
heat interchanger 热交换器
heat of emersion 浸润热
heat of immersion 浸湿[湿润,润温,浸渍]热
heat of micellization 胶粒化热

heat of micellization 胶束化生成热
heat of reaction 反应热
heat transfer ratio 传热比率,比热流
heat treated rolled steel 热轧钢
heat treatment 热处理
heat treatment step 热处理阶段
heat value 热值
heated slurry 已加热的矿浆
heating cost 采暖费
heating cycle 加热(周)期[程序]
heating equipment 采暖设备
heating plant 加热车间[设备]
heating system 加热系统
heat-set 热定型
heat-treated product 热处理产品
heat-wound spring 热煨弹簧
heaviest upper shell 最重的上壳
heavily damped conventional controller 强阻尼作用的常规控制器
heavily marked curve 粗线标出的曲线
heavily milled 过磨
heavily oxidized 严重氧化的
heavily-loaded belt 重负荷胶带运输机
heavy duty apron feeder 重型板式给矿机
heavy duty performance 重负荷工作[操作]
heavy duty truss 重型桁架
heavy element 重元素

heavy gas oil 重瓦斯油
heavy gravity particle 比重大的颗粒,大比重颗粒
heavy impact load 沉重冲击负荷
heavy industry 重工业
heavy liquid analysis 重液分析
heavy liquid analysis of ore 矿石重液分析
heavy liquid media 重液介质
heavy liquid separation 重液分选[离]
heavy liquid separation test 重液分选[离]试验
heavy liquid test result 重液试验结果
heavy maintenance 大(规模)修理
heavy media concentration 重介质选矿
heavy media densifier 重介质浓缩机
heavy media liquid 重介质液[流]体
heavy media separator 重介质分选机
heavy medium circuit 重介质流程
heavy medium cyclone preconcentration plant 重介质旋流器预富集车间
heavy medium drum 重介质滚筒
heavy medium preconcentration plant 重介质预选车间[厂]
heavy metal alkyl xanthate 重金属烷基黄药
heavy metal cation 重金属阳离子
heavy metal cyanide complex 重金属氰化物络合物
heavy metal dithiophosphate 重金属二硫代磷酸盐

heavy metal element 重金属元素
heavy metal hydroxide 重金属氢氧化物
heavy metal ion 重金属离子
heavy metal salt 重金属盐
heavy metal soap 重金属皂
heavy metal sulfide 重金属硫化矿[物]
heavy metal thiolate 重金属硫醇盐
heavy metal xanthate complex 重金属黄原酸盐络合物
heavy metallic ion 重金属离子
heavy mineral 重矿物
heavy mineral contour 重矿物轮廓线
heavy mineral separation 重矿物分选
heavy mineralized froth 厚[粘]矿化泡沫
heavy mud 重泥浆(钻油井用)
heavy organic liquid 重有机液
heavy particle 重颗粒
heavy plant 重型机械设备
heavy service line 重型(负荷)作业线
heavy snowfall 大[暴]雪
heavy sulfide ores of copper 重硫化铜矿石类
heavy-duty 重型的,重载[负]的
heavy-duty centrifugal industrial exhauster 重型离心工业抽风[排气]机
heavy-duty cutter 重型切[截]取器
heavy-duty Denver duplex conditioner 重型丹佛双室搅拌槽

heavy-duty equipment 重型设备
heavy-duty grease lubricated self-aligning ball bearing 重型油脂润滑自动定位滚珠轴承
heavy-duty grease lubricated self-aligning roller bearing 重型油脂润滑自动定位滚柱轴承
heavy-duty jaw crusher 重型颚式破碎机
heavy-duty lubricating grease 重型润滑油
heavy-duty oil 重型油
heavy-duty pipe frame 重型管支架
heavy-duty truck 重型卡车
heavy-duty vibrating feeder 重型振动给矿机
heavy-duty welded steel pedestal 重型焊接钢底座
heavy-gauge pipe 厚壁管
heavy-media coal-washing process 重介质洗[选]煤法
helical starch-oleate clathrate 螺旋形淀粉油酸包合物
helical structure 螺旋结构
helical tube 螺旋管
helium-nitrogen gas 氦氮气体
helper 辅助工
hematitic-martitic jasper 赤铁矿-假象赤铁矿碧玉
hemi-micelle 半胶束
hemi-micelle association 半胶束缔合
hemi-micelle concentration 半微泡[胶束]浓度
hemi-micelle formation 半胶束形[构]成
hendecanoic acid 十一烷酸
heptaldoxime 庚醛肟(可用作石灰石、铜矿的捕收剂)
heptyl 庚基
Hercules CMC 羧甲基纤维素钠(品名,美国 Herculues 公司产品,可用作絮凝剂)
heterogeneous distribution 不均匀分布
heterogeneous distribution of mineral 矿物不均匀分布
heterogeneous liquid phase 不均匀液相
heterogeneous material 不均匀物料
heterogeneous mineral surface 非均质矿物表面
heterogeneous nature 不均匀性
heterogeneous ore 不均匀的矿石
heterogeneous oxidation 多相氧化
heterogeneous reduction 多相还原
heterogeneous solid phase 不均匀固相
heterogeneous surface 不均匀表面
heterogeneous system 多相体系
heteropolar organic compound 异极有机化合物
heteropolar surface-active compound 异极的表面活化化合物
Hewitt Robbins vibrating screen 休伊特·罗宾斯型振动筛
Hewitt Robins vibrating feeder 休伊特·罗宾斯振动给料机

Hewlett Packard computer 惠普计算机

hex head mechanical bolt 六角头机械螺栓

hexadecane 十六(碳)烷 $C_{16}H_{34}$

hexadecanoic acid 十六烷酸,棕榈酸

hexagon nut 六角螺母[帽]

hexagonal structure 六方晶结构

hexagonal unit cell 六方晶胞

hexametaphosphate 六偏磷酸盐(可用作分散剂、湿润剂、调整剂)$(MPO_3)_6$

hexamine 六亚甲基四胺,六胺,乌洛托品,$(CH_2)_6N_4$

hexavalent 六价的

hexavalent chromium 六价铬

hexavalent ion 六价离子

hexyl 己基 C_6H_{13}—

hexyl mercaptan 己基硫醇

hexyl xanthate 己基黄原酸盐

Heyl & Patterson cyclo-cell 海尔和帕特森旋流浮选机(美国 Heyl & Patterson Inc. 生产)

Hg amyl xanthogenate 戊基黄原酸汞

Hg butyl xanthogenate 丁基黄原酸汞

Hg ethyl xanthogenate 乙基黄原酸汞

Hg heptyl xanthogenate 庚基黄原酸汞

Hg hexyl xanthogenate 己基黄原酸汞

Hg isoamyl xanthogenate 异戊基黄原酸汞

Hg isobutyl xanthogenate 异丁基黄原酸汞

Hg isopropyl xanthogenate 异丙基黄原酸汞

Hg octyl xanthogenate 辛基黄原酸汞

Hg propyl xanthogenate 丙基黄原酸汞

HI dry magnet 强磁场[高强度]干式磁选机

HI wet magnet 强磁场湿式磁选机

hi-chrome iron 高铬铁

hide glue 骨胶,皮胶(可用作絮凝剂)

hierarchical analysis of variance 等级方差分析法

high abrasion steel component 耐磨钢部件

high acid grade fluorspar concentrate 高酸级萤石精矿

high alarm relay 高限报警继电器

high brass 优质黄铜

high calcium salt concentration 高钙盐浓度

high chemical grade 高化验品位

high chromium cast iron 高铬铸铁

high chromium content alloy 高铬合金

high degree 高度

high degree of flexibility 高度灵活性

high density polyethylene pipe 高密度聚乙烯管

high dimensionality 高维数

high displacement vacuum pump 高排量真空泵

high dissolution rate 高溶解速度

high efficiency hydrofoil 高效率水叶[翼]

high energy device 高能量装置[设备]
high energy electron diffraction 高能电子衍射(法)
high energy excitation source 高能激发源
high energy twin pendulum tester 高能双摆锤试验机
high energy ultrasonic emulsifier device 高能超声波乳化装置
high energy X-ray source photon 高能量 X 射线源光子
high grade concentrate 高品位精矿
high grade marketable concentrate 高品位商品精矿
high grade product 高品位产品
high grade sand 高品位砂子
high grade strip solution 含金高[高品位]的解吸溶液
high grade unit cell concentrate 高品位单槽(浮选)精矿
high grade zone 高品位矿带
high gravity mineral 高比重矿物
high homologue of xanthate 高级黄药类
high impact backing kit 高冲击填料件
high impact formula 高冲击配[药]方
high intensity magnetic fractionation 强磁选
high intensity magnetic separator 强磁选机
high level calculation 高级运算
high level detector 高料位探测器

high level device 高级设备
high level interpretative language 高级翻译语言
high level language 高级语言
high level multivariable control 高级多变量控制
high level of confidence 较高置信度
high level of protection 防护高限
high lime ore 高钙矿石
high metal volume 高金属(含)量,高金属有用成分
high molecular weight flocculent 高分子(量)絮凝剂
high molecular weight lignin-sulfonate 高分子量木质磺酸盐
high percent solid product 百分数高的固体产品
high performance 高性能(浮选机在低功率下的最大生产能力,最佳品位和回收率)
high power number 高功率数
high power rate crushing procedure 大[高]功率消耗破碎法
high power rate operation 大[高]功率运转
high pressure air equipment 高压[空气]设备
high priority foreground mode 高优先级前台方式
high profile liner 高型衬板[里]
high purity metal 高纯度金属
high quality 高质量

high quality mineral sample 优质矿样
high quality 高质量[优质]的
high rate 高速,高效率
high ratio of reduction factor 高破碎比系数
high reactive metallic iron 高反应性金属铁
high resolution electron microscopic study 高分辨率电子显微镜研究
high specific gravity coarse mineral 大比重粗粒矿物
high specific gravity ore 高比重矿石
high speed analysis 高[快]速分析
high speed mixing unit 高速混合装置[器]
high speed shaft 高速轴
high speed shaft bearing 高速轴轴承
high starting torque 高起动转矩
high steel 高碳钢,硬钢
high sulfide ore 高硫(化物)矿石
high sulphide ore 高硫矿石
high surface activity 高表面活性
high tailings assay 高尾矿品位
high temperature autoclave treatment 高温高压釜处理
high tension roll separator 高压辊式分选机
high tonnage plant 大型[吨位,规模]选矿厂
high turbidity region 高混浊度区
high turbulence flow cell 高度湍流的流动式样品盒
high voltage source 高压电源
high voltage target tube 高压带靶 X 射线管
high voltage X-ray tube 高压 X 射线管
high weir classifier 高堰式分级机
high yield 高回收率,高产(如选煤)
high-carbon steel rod 高碳钢棒
high-chrome iron 高铬铁
high-efficiency classification 高效率分级(法)
high-efficiency crushing circuit 高效率的破碎回路
high-energy X-ray source photon 高能量 X 射线源光子
higher mass particle (较)高质量颗粒
higher order 高阶
higher order term 高阶项
higher specific gravity ore 较高比重矿石
higher term 高次项
high-grade massive sulfide 高品位致密硫化矿
high-intensity agitation 高强度搅拌
high-intensity agitator 高强度搅拌槽
high-intensity crushing procedure 高强度破碎过程
high-intensity excitation source 高强度激发源
high-intensity magnetic separator 高强度磁选机

high-intensity plastic 高强度塑料
high-intensity separating technique 高强度分选技术
high-level circulating water pond 高位循环[回]水池
high-level macroinstruction 高水平宏指令
high-level profile 高标高断面
high-low water level alarms 高低水位警报信号
highly conducting and noble metal sulfide 高导性的惰性金属硫化矿
highly marketable inventory 畅销库存
highly mineralized froth with low thickness 低厚度高矿化泡沫
highly non-polar reaction product 强非极性反应产物
highly polar anion 强极性阴离子
highly refined white oil 高纯度白油
highly selective collector 高选择性捕收剂
highly specialized procedure 高度专业化方法
highly speculative mineral exploration field 高度投机性探矿领域
highly toxic 剧毒,高毒性的
highly-floatable carbonaceous pyrrhotite 有高度可浮性的碳质[含碳]磁黄铁矿
highly-refined fatty acid 高纯脂肪酸
high-pressure air 高压[压缩]空气
high-pressure air distribution system 高压空气分配系统
high-purity water 高纯水
high-speed paper tape unit 高速纸带机
high-speed screen 高速筛
high-speed stereoscopic cinematography 高速立体摄影技术
high-speed, dedicated, fixed-program system 高速,专用的固定程序系统
high-temperature and high-pressure stripping technique 高温高压解吸技术[工艺,方法]
high-temperature belt 高温胶带(机)
high-temperature form 高温形式
high-velocity duct 高速输送管(道)
high-velocity duct system 高速管道系统
high-voltage distribution room 高压配电室
high-voltage distributor 高压分配器
high-voltage equipment 高压设备
hill climbing method 登山法
hill climbing technique 登山法[技术]
hindered settling rate 受阻[干涉]沉降速度[率]
hinged chute gate 悬吊式槽形闸门
historic method 过去[历史]的方法
historic strategy 历史性策略
historical data 历史资料[数据]
historical development 历史沿革,历史发展过程,发展史
history data 历史资料[数据]

Hitachi – Perkin – Elmer model R24A spectrometer 希达奇-珀金-埃尔默 R24A 型分光计

HMP (hexametaphosphate) 选矿用药剂(六偏磷酸钠,可用作分散剂、湿润剂、调整剂)$(NaPO_3)_6$

HMS-flotation method 重介质浮选法

HMS process 重介质选矿法

Hodag flocs 絮凝剂(一种高分子聚合物)

hog-trough flotation unit 猪槽式浮选机

hoisting rate 提升速度

hold up tank 收集箱

hold-back 暂时停顿,抑制,保持[拉紧]装置

holdback payment 暂扣留付款

holding area 储存区

holding tank 储存槽

holding time 停留时间

hold-up 支持,举起,提出,阻塞[滞],障碍,停止[顿,车],滞留量[液],塔储量,容纳量,容器体积

hole size factor 筛孔规格[大小]系数

hole-rich space charge layer 多空穴空间电荷层

hole-rich space charge region 多空穴空间电荷区

hole-rich surface region 多空穴表面区

hollow cathode lamp 中空[空心]阴极灯

hollow glass tube 空心玻璃管

hollow impeller shaft 空心叶轮轴

hollow trunnion 空心[中空]耳轴

Holo-Flite dryer Holo-Flite 型干燥机

home office 总公司,总部

home office cost report 总公司经费[费用]报告

homogeneous sample 均匀[质]样品

homogeneous surface 均匀表面

homogeneous suspension 均质[匀,相]悬浮液[体]

homologous series 同源系列,同族

honeycomb material 蜂窝状物料

Honeywell computer 霍尼韦尔计算机

Honeywell H-316 control computer 霍尼韦尔 H-316 型控制计算机

hook limit 吊钩极限位置(吊车的)

hoop tension 圆周张力,箍[环向]张力

Hoosier pearl 灰色玉米(珍珠米),淀粉(可用作絮凝剂)

hopper 漏斗

hopper flow factor 漏斗流动系数

hopper hole (矿仓)漏斗口

hopper liner 漏斗衬板[里,垫]

hopper slope angle 漏斗(斜)角

hopper-feeder unit 漏斗-给矿机机组

hopper-flow bin 漏斗式流动矿仓

horizontal belt conveyer 水平胶带运输机

horizontal belt filter 水平带式过滤机

horizontal bending moment 水平弯矩

horizontal conditioner 卧式调整槽

horizontal cross-section 水平横断面
horizontal end-suction slurry pump 卧式端吸砂[泥浆]泵
horizontal gate 水平闸门
horizontal labyrinth clearance type seal 水平[卧式]迷宫间隙型密封
horizontal labyrinth grease seal 水平迷宫式油脂密封
horizontal labyrinth seal with wiper 带清除器水平迷宫式密封
horizontal line 水平线
horizontal pipe 水平管
horizontal slurry pump 卧式矿[泥]浆泵
horizontal table filter 水平圆盘[平面]过滤机
horizontal vacuum filter 水平真空过滤机
horizontal velocity component 水平分速度
horse power of transmission 传动马力
horse power requirement 马力[功率]需要量
horse power unit 功率[马力]单位
horsepower column 功率栏
host 主机,主人,主持人
host country 所在国[主办国,东道主]
host country government 所在国政府
host country government's agencies 所在国政府代理机构
host country withholding 所在国扣除费

Hostapal 表面活性剂品牌名(德国 Hoechst 公司产品)
Hostapal B 烷基酚聚乙二醇硫酸盐(可用作湿润剂)
Hostapal C 烷基酚聚乙二醇醚(可用作湿润剂)
Hostapon 表面活性剂品牌名(德国 Hoechst 公司产品)
Hostapon CT 椰子油酰氨基乙基磺酸钠(可用作湿润剂及细粒钨矿的捕收剂) $R-CONHCH_2CH_2SO_3Na$ (R = 椰油酰)
Hostapon MT 牛脂酰氨基乙基磺酸钠(可用作湿润剂及细粒钨矿的捕收剂) $R-CONHCH_2CH_2SO_3Na$ (R = 牛油酰)
Hostapon T 油酰甲氨基乙基磺酸钠(可用作湿润剂及细粒钨矿的捕收剂) $R-CON(CH_3)CH_2CH_2SO_3Na$ (R = 油酰)
Ho-sulfate paste $C_{10} \sim C_{12}$ 烷基(伯)硫酸盐
hot alcohol-caustic cyanide method 热酒精-碱性氰化法(解吸)
hot area 受热面积
hot caustic cyanide solution 热苛性氰化物溶液
hot rolled steel 热轧钢
hourly capacity 小时能力
hourly rate 小时费率
household abrasive 家庭磨料
household detergent 家用洗涤[去污,去垢]剂

housing 壳体
housing wall 壳壁
high starting torque (HST) 高起动转矩
huebnerite 钨锰矿 $MnWO_4$
huebnerite-bearing veinlet 含钨锰矿细脉
Hukki cone classifier Hukki型分泥斗[锥形分级机]
Hukki pneumatic classifier Hukki型空气分级机
human being 人类
human element 人为因素
human engineering 人类[体]工程学
human judgment factor 人为的判断因素
Humboldt impeller 洪堡特型叶轮(不对称型,使矿浆脉动旋转)
humid atmosphere 潮湿大气
Humphries spiral 汉弗莱斯螺旋选矿机
hurdle rate 门坎比率
hurdled ore 经粗筛矿石
hutch concentrate 跳汰机筛下精矿
HX 19 淀粉衍生物(品名,法国产,可用作絮凝剂、活化剂)
Hyamin(e) 一种阳离子表面活性剂(烷基苄基二甲基氯化铵,可用作乳化剂)$(C_6H_5CH_2)RN^+(CH_3)_2Cl^-$
Hydrastroke feeder 液压驱动槽式给矿机
hydrate 水合物

hydrated counter ion 水化配衡离子
hydrated species 水合物
hydration energy 水合(作用)能
hydration free energy 水合自由能
hydration of lattice ion 晶格离子水合作用
hydration sphere 水合
hydraulic adjustment system 液压调整系统
hydraulic bagging press 水压装袋机
hydraulic ball trap 水力捕球器
hydraulic closure 液压闭合器(板框压滤机的)
hydraulic diameter 水力[水压]直径
hydraulic engineering consultant 水利工程顾问
hydraulic equilibrium 水力平衡
hydraulic gold trap 水力捕金器
hydraulic gradient 水力坡[梯]度
hydraulic lock post 液力锁紧栓
hydraulic mantle positioning 液压传动锥体定位,锥体液压定位
hydraulic mechanism 液压装置[机构]
hydraulic performance map 水力(学)特性图
hydraulic power pack 液压动力机组
hydraulic power unit 液力驱动装置
hydraulic ram 水压吸扬机(水撞泵)
hydraulic sand filling 水砂充填
hydraulic settling device 液压沉降装置
hydraulic size 水力分级尺寸[粒度]

hydraulic station 液压站
hydraulic stripping 水力剥离
hydraulic structure 水工结构
hydraulic support 液压支架[撑]
hydraulic support assembly 液压支承装置[组件]
hydraulic test 水力学试验
hydraulic water manifold 压力水歧管[集水管]
hydraulically operated compression testing machine 液压操纵压缩试验机
hydraulically operated device 液压操作[纵]装置
hydrocarbon chain 烃链
hydrocarbon chain association 烃链缔合作用
hydrocarbon chain configuration 烃链结构
hydrocarbon chain length 烃链长度
hydrocarbon collector scheme 烃捕收剂方案[计划]
hydrocarbon fuel 碳氢[烃类]燃料
hydrocarbon layer 烃层
hydrocarbon oil 碳氢[烃]油
hydrocarbon oil emulsification 烃油乳化
hydrocarbon type reagent 烃类药剂
Hydrochem 66 絮凝剂(Ogava silillana 的多羧基衍生物)
hydroclone 水力旋流器
hydrocyclone 水力旋流器
hydrocyclone dimension 水力旋流器尺寸
hydrocyclone model constant 水力旋流器模型常数
hydrocyclone overflow density 水力旋流器溢流浓度
hydrocyclone overflow particle size 水力旋流器溢流粒度
hydrocyclone separation process 水力旋流器分离法
hydrocyclone separation size 水力旋流器分级[离]粒度
hydrocyclone underflow 水利旋流器底流
hydrodynamic collision 流体动力碰撞
hydrodynamic collision rate 流体动力学碰撞率
hydrodynamic condition 流体动力条件
hydrodynamic design 流体动力设计
hydrodynamic effect 流体动力效应
hydrodynamic performance criteria 流体动力学特性准则
hydrodynamic scale-up criteria 流体动力学按比例放大准则
hydrodynamic streamlining 流体动力学流线型
hydrodynamics of collision 碰撞流体动力学
hydroelectric development 水电开发
hydroelectric plant 水力发电厂
hydroelectric power 水电
hydroelectric project 水力发电项目,

水电工程
hydrogen 氢
hydrogen bonding 氢键合
hydrogen form of DPG (DPGH+) 二苯基胍氢离子型态
hydrogen gas disperser 氢气分散器
hydrolysable collector 水解性捕收剂
hydrolysis 水解作用
hydrolysis equilibrium 水解平衡
hydrolysis tank 水解槽
hydrolytic model 水解模型
hydrolytic precipitation reaction 水解沉淀反应
hydrolyzable surfactant 水解表面活性剂
hydrolyzed metal cation 水解金属阳离子
hydrolyzed metal colloid 水解金属胶体
hydromechanic data 流体力学数据
hydrometallurgical equipment 水冶[湿法冶金]设备
hydrometallurgical method 湿法冶金(方)法
hydrometallurgical operation 水冶[湿法冶金]生产
hydrometallurgical oxidation 水冶氧化(作用)
hydrometallurgical plant 湿法冶金厂
hydrometallurgical process 湿法冶金工艺,水冶法
hydrometallurgical processing 湿法(冶炼)处理
hydrometallurgical production 水冶生产
hydrometallurgical separation 水冶选矿
hydrometallurgical technique 湿法冶金技术
hydrometallurgical test 水冶试验
hydrometallurgical upgrading 水冶提高品位
hydrophilic copper containing species 亲水含铜物质
hydrophilic entity 亲水实体
hydrophilic material 亲水物料
hydrophilic particle 亲水颗粒
hydrophilic polymer 亲水聚合物
hydrophilic property 亲水性(质)
hydrophilic solid 亲水固体
hydrophilic surface 亲水表面
hydrophilic zinc sulfite 亲水亚硫酸锌
hydrophilic zinc sulfite film 亲水亚硫酸锌薄膜
hydrophilic zinc-bearing layer 亲水含锌层
hydrophobic association 疏水缔合
hydrophobic bonding 疏水键合
hydrophobic character 疏水性
hydrophobic effect 疏水效应
hydrophobic entity 疏水体
hydrophobic gold 疏水金
hydrophobic material 疏水物料
hydrophobic state 疏水状态

hydrophobic surface 疏水表面
hydrophobic surface layer 疏水表面层
hydrophobicity 疏水性
hydropholic particle 疏水颗粒
hydroset 液控[压]装置
hydroset cylinder cover plate 液压缸盖板
hydroset mantle positioning system 锥体外壳液控定位系统(圆锥破碎机的)
hydroset mantle setting system 锥体外壳液控系统
hydroset pressure 液控压力
hydroset pumping system 液压泵[液控扬送]系统
hydrosizer 水力分级机
hydrostatic lubrication 流体静力[压]润滑
hydrostatic oil pressure 静油压
hydrostatic pressure 流体静压力
hydrosulfide anion 硫氢阴离子
hydrosulfide ion 硫氢离子
hydrothermal action 热液作用
hydrous manganese oxide 含水氧化锰
hydrous nickel oxide 水合氧化镍
hydroxamate 氧肟酸盐,异羟肟酸
hydroxamic 异羟肟的,羟肟
hydroxamic acid 异羟肟酸,氧肟酸 $RC(O)NHOH$
hydroxamic acid copper 氧肟酸铜
hydroxide complex 氢氧络合物
hydroxide eluant 氢氧化物洗提液
hydroximic acid 羟肟酸
$R \cdot C(OH):NOH$
hydroxy(=hydroxyl) 羟基,氢氧(基,化物)HO—
hydroxy benzoic acid 羟基苯甲酸(可用作捕收剂)$OH \cdot C_6H_4 \cdot COOH$
hydroxy complex 氢氧络合物
hydroxy ethyl cellulose 羟乙基纤维素
hydroxy quinoline 羟基喹啉
hydroxyapatite 羟基磷灰石
hydroxyethyl cellulose 羟乙基纤维素
hydroxyimino(=oximido) 肟基
hydroxyl 羟(基)HO—
hydroxyl complex 羟基络合物
hydroxyl depression 氢氧基抑制
hydroxyl group 羟基,羟族
hydroxyl ion concentration 氢氧根离子浓度
hydroxyl-apatite 羟基磷灰石
hydroxylated surface 羟基化表面
hydrozinc 氢锌(还原剂)
hypalon 氯磺酰化聚乙烯合成橡胶,海帕伦
hypermanganate 高锰酸盐
hypodermic syringe 皮下注射器
Hyponate L 石油磺酸钠(品名,相对分子质量415~430,不含矿物油,可用作捕收剂、起泡剂)
hypophosphorous acid 次磷酸
Hyposulfite 连二亚硫酸盐(可用作闪锌矿,黄铁矿的抑制剂)$M_2S_2O_4$
hypothetic case study 假设的案例研究
hypothetical example 假设的例子

hypothetical grinding circuit 假设的磨矿回路

hypothetical plane surface 假想平面
hysteresis effect 滞后效应

I

I. H. C.－Cleaveland jig 旋转耙液压圆形跳汰机(荷兰 I. H. C. 海洋采矿公司与美国诺曼·克利夫兰德(Normon Cleaveland)合作制成,采用了诺曼·克利夫兰德专利旋转耙结构)
ice cover 冰覆盖
ideal condition 理想条件
ideal mixer 理想混合器
ideal system 理想体系
ideal variogram 理想变异函数
idealistic manner 理想方式
identical condition 相同条件
identical order 同样顺序
identifiable ore type 可以区别的矿石类型
identification number 识别号码
identification of cleavage plane 解理面鉴定
identification of gangue mineral 脉石矿物鉴定
identification of material 物料标号
identification of mineral composition 矿物成分鉴定
identification of ore mineral 矿石矿物鉴定
identification of ore property 矿石性质的鉴定
identity matrix 单位矩阵
idle cash 闲置现金
idle mill equipment 闲置[空闲]选厂设备
idler alignment 托辊校直
idler seal correction 托辊密封修正
idler series 托辊系列[组]
idler series selection chart 托辊系列选择图表
idler space 托辊间距
idler spacing 托辊间距
idler troughing angle 托辊的输送槽角
idler wing roll 翼形[侧边]托辊
idling power 空转功率
idling power loss 空转动力损失[功耗]
Igepal 表面活性剂品牌名
Igepal A 油酸磺化乙酯钠盐(同 Igepo-nA,可用作湿润剂)
$C_{17}H_{33}COOHCH_2CH_2SO_3Na$
Igepal B 十二烷基/十四烷基苯酚聚乙

二醇硫酸盐(可用作湿润剂、浮选剂)

Igepal C　十二烷基苯酚聚乙二醇醚(可用作湿润剂、乳化剂)
$C_{12}H_{25}-C_6H_4(OCH_2CH_2)_nOH$

Igepal CA　环氧乙烷与烷苯酚的缩合物(烷基苯酚聚乙二醇醚,可用作浮选剂、捕收剂、湿润剂)
$R \cdot C_6H_4(OCH_2CH_2)_nOH$

Igepal CA-630　叔辛基苯酚聚乙二醇醚(可用作捕收剂、湿润剂)
$R \cdot C_6H_4(OCH_2CH_2)_nOH, R=叔辛基 C_8H_{17}, n=8$

Igepal CA-710　叔辛基苯酚聚乙二醇醚(可用作选矿药剂)
$R \cdot C_6H_4(OCH_2CH_2)_nOH, R=叔辛基 C_8H_{17}, n=11\sim12$

Igepal CA-Extra　C8/C9-烷基苯酚聚乙二醇醚(可用作选矿药剂)
$R \cdot C_6H_4(OCH_2CH_2)_nOH,$
$R=C_8H_{17}-/C_9H_{19}-$

Igepal CO-210　壬基苯酚基聚乙二醇醚(可用作选矿药剂)
$R \cdot C_6H_4(OCH_2CH_2)_nOH, R=C_9H_{19}, n=1\sim2$

Igepal CO-430　壬基苯酚基聚乙二醇醚(可用作选矿药剂)
$R \cdot C_6H_4(OCH_2CH_2)_nOH, R=C_9H_{19}, n=4$

Igepal CO-530　壬基苯酚基聚乙二醇醚(可用作选矿药剂)
$R \cdot C_6H_4(OCH_2CH_2)_nOH, R=C_9H_{19}, n=6$

Igepal CO-610　壬基苯酚基聚乙二醇醚(可用作选矿药剂)
$R \cdot C_6H_4(OCH_2CH_2)_nOH, R=C_9H_{19}, n=8$

Igepal CO-630　壬基苯酚基聚乙二醇醚(可用作选矿药剂)
$R \cdot C_6H_4(OCH_2CH_2)_nOH, R=C_9H_{19}, n=9\sim10$

Igepal NA　十四烷基苯磺酸盐(可用作湿润剂、乳化剂)
$C_{14}H_{29} \cdot C_6H_4 \cdot SO_3M$

Igepal W　十二烷基苯酚聚二醇醚(可用作湿润剂、乳化剂)
$C_{12}H_{25} \cdot C_6H_4(OCH_2CH_2)_nOH$

Igepon　表面活性剂品牌名(中文名称:胰加漂或依捷邦)

Igepon A　胰加漂 A(油酸磺化乙酯钠盐,可用作湿润剂)
$C_{17}H_{33}COOCH_2CH_2SO_3Na$

Igepon AC　胰加漂 AC(油酸磺化乙酯钠盐,可用作湿润剂)
$C_{17}H_{33}COOCH_2CH_2SO_3Na$

Igepon AP　胰加漂 AP(油酸磺化乙酯钠盐,可用作湿润剂)
$C_{17}H_{33}COOCH_2CH_2SO_3Na$

Igepon B　胰加漂 B(月桂酸羟丙基磺酸钠,可用作选矿药剂)

Igepon C　胰加漂 C(油酰氨基乙基磺酸钠,可用作湿润剂、絮凝剂)
$C_{17}H_{33}CONHC_2H_4SO_4Na$

Igepon CN　胰加漂 CN(油酰环己氨基乙基磺酸钠,可用作选矿药剂)

RCON(R') C$_2$H$_4$SO$_3$Na, R = C$_{16}$H$_{33}$, R' = C$_{16}$H$_{11}$—

Igepon 702K 胰加漂702K(可用作捕收剂)RCON(R')C$_2$H$_4$SO$_3$Na,R来自50%豆蔻酸、50%硬脂酸,R'=CH$_3$

Igepon KT 胰加漂KT(椰子油酰甲氨基乙基磺酸钠,可用作捕收剂) RCON(R')C$_2$H$_4$SO$_3$Na,R来自椰子油及棕榈核油酸,R'=CH$_3$

Igepon T 胰加漂T(油酰甲氨基乙基磺酸钠,可用作捕收剂) RCON(R')C$_2$H$_4$SO$_3$Na,R来自油酸,R'=CH$_3$

Igepon TC 胰加漂TC(椰子油酰甲氨基乙基磺酸钠,可用作捕收剂) RCON(R')C$_2$H$_4$SO$_3$Na,R来自椰子油酸,R'=CH$_3$

Igepon TE 胰加漂TE(牛脂酰甲氨基乙基磺酸钠,可用作捕收剂) RCON(R')C$_2$H$_4$SO$_3$Na,R来自牛脂酸,R'=CH$_3$

Igepon TK 胰加漂TK(妥尔油酰甲氨基乙基磺酸钠,可用作捕收剂) RCON(R')C$_2$H$_4$SO$_3$Na,R来自妥尔油酸,R'=CH$_3$

Igepon TN 胰加漂TN(软脂酰甲氨基乙基磺酸钠,可用作捕收剂) RCON(R')C$_2$H$_4$SO$_3$Na,R来自软脂酸,R'=CH$_3$

igneous complex 火成杂岩
igneous dike 火成岩脉[墙]
igneous ore 火成岩矿石
igneous type 火成岩类型
ignition point 燃点
IHPA(=isohexyl phosphonic acid) 异己基磷酸(可用作锡石的捕收剂)
illumination condition 照明条件
illumination equipment 照明设备
ilmenorutile 钛铁金红石[黑金红石]
IM50 俄制烷基羟肟酸(品名,可用作捕收剂)
image analyzer 图像分析仪
image replication 图像[影像]复制[现]
image[image] analyzer 图像[影像]分析仪
imine 亚胺
imine derivative 亚胺衍生物
imine group 亚胺基
immediate comparative data 立即[直接]比对数据
immersed mechanism 沉没机构
immersion 浸入,浸渍,浸泡,油浸,沉浸,沉没
immersion element 浸入式元件
immiscible organic solvent 不溶混的有机溶剂
immovable property 不动产
impact belt conveyer 冲击胶带运输机
impact blow 冲击
impact crusher 反击式破碎机
impact desk 冲击面板,防冲挡板
impact dryer 冲击干燥机
impact energy 碰撞能

impact hammer crusher 反击锤式破碎机
impact panel 冲击面板
impact piston 冲击活塞
impact point 冲击点
impact rotor 冲击转子
impact separator 冲击分离器
impact strength index 冲击强度指数
impact unit 冲击设备[装置,元件,零件,构件]
impact value 冲击值
impact-cyclone separator 冲击旋流分离机
impact-type crusher 冲击式破碎机
IMPC-X-56 N-烷基磺化琥珀酰胺盐(品名,可用作锡石的捕收剂)
impeller and stator baffle configuration 叶轮和定子折流板结构
impeller assembly 叶轮组
impeller breaker 反击式破碎机
impeller centrifugal pump 叶轮离心泵
impeller depth 叶轮高度
impeller diameter 叶轮直径
impeller peripheral velocity 叶轮圆周速度
impeller power coefficient 叶轮功率系数
impeller region 叶轮区
impeller shrouding 叶轮罩
impeller spacing 叶轮间距
impeller speed 叶轮速度[转速]
impeller speed relationship 叶轮速度关系
impeller submergence 叶轮浸没深度
impeller tip speed 叶轮外缘速度
impeller-disperser clearance 叶轮扩散器间隙(浮选机的)
impeller-stator assemble 叶轮-定子装配[组合]
impeller-stator clearance 叶轮定子间间隙
impeller-tank aspect ratio 叶轮与槽的纵横比
imperial measuring system 英国度量制,英制测量系统
impervious upstream membrane 上游防渗薄膜护面
impingement and entrainment scrubber 冲击挟带式洗涤器
impingement baffle 防冲[冲击]挡板
implementation 实施,履行,实现,执行,工具,器具
implementation cost 实施成本,执行过程费用
implementation language 工具语言
implicit 隐式
implicit constraint 隐式约束,默认约束条件
implicit relationship 隐含关系
important decision 重要决策[定]
important feature 重要特征
important step 重要步骤
important unit operation 重要设备操作

imposed guideline 硬性规定的指标[准则]

improper flotation condition 不适当的浮选条件

improvement 改善[良,进],改进措施,经改进的东西

impulse feeder 脉冲给矿机

impulse sequence 脉冲序列

impure copper 杂铜

impure cresol 杂酚

impurity 杂质,混杂物

impurity level 杂质含量

in situ reserve 原地储量

in situ tonnage estimate 原地吨位估计量

inaccessible mineralized zone 未揭露矿化带

inactive zone 不活动带

inapplicable 不适用的

in-bin blending 仓内混料[合]

inboard bearing 内侧轴承

inboard position 内侧位置

incapsulation 封闭

inching device 寸动装置

inching drive 爬行[低速]传动

inching drive mechanism 低速转[驱,拖]动机构,微拖发动机构

inching motor 寸动[渐进]电机

inch-millimeter equivalents 英寸毫米等值[换算](表)

incident electron beam 入射电子束

incident light intensity 入射光强度

in-circuit inventory 回路中储存量

incline belt loading 倾斜胶带运输机负荷

incline hoist 斜井提升机

incline portion 倾斜[上斜]部分

incline variable 可变倾角[倾斜度]

inclined belt 倾斜胶带(运输机的)

inclined circular-motion machine 倾斜圆周运动筛[设备,机器]

inclined hoist 斜井提升[卷扬]机

inclined load station 倾斜装载站

inclined settling plate 倾斜沉淀板(浓缩机的)

inclined surface 倾斜面

inclined tank 倾斜槽

inclusion 包裹体

Incoloy alloy 825 因科洛伊825 [Incoloy 825]合金(一种添加了钼、铜和钛的镍—铁—铬固溶强化耐蚀合金)

income flow 收益流动[量]

income generating portion (可)产生的收益部分

income statement 收益表

income tax 所得税

Income Tax Act of Canada 加拿大所得税法

incoming breaker 进线断路器

incompetent soil 软土壤[土地,地面]

incorporated company 股份有限公司

incorporation 公司,团体,结[联,掺,混]合

increased sensitivity 增大的敏感度
increasing attractive energy 递增吸引能
increasing function 递增函数
independent consulting firm 独立咨询公司
independent mass balance equation 独立质量平衡方程
independent transporter 单独的搬运机
independent variable 自变量[数]
in-depth experience 深刻经验
index mineral 标准[指标]矿物
indexing operating lever 换位操纵杆，换位操作手柄
indicated integration 指定积分
indicated net value 指示净价值
indicated ore reserve 控制矿石储量
indicator of floatability 可浮性指示参数
indifferent electrolyte 协助电解物[质],惰性电解质
indifferent ion 协助(电解质)离子
indirect annual operating expense 间接年生产费用
indirect evidence 间接证据[论据]
indirect fired reactivation kiln 间接燃烧式再生窑
indirect method 间接法(用于可用微积分学计算所有导数的场合)
indirect process 间接法
indistinguishability 不可分辨性

individual 个人,个体
individual application 单独使用
individual control box 单独控制箱
individual control structure 单个控制结构
individual cost 单项[单一]成本
individual electrochemical process 单个电化学过程
individual productivity 个人(劳动)生产率
individual sample 单个试样
individual specific gravity 各个比重
indoor type 室内式
induced air flow 感生气流
induced calcite surface 感应[生]方解石表面
induced magnetic field 感应磁场
induction disk 感应(圆)盘
induction furnace 感应(加热)炉
induction period 感应阶段
inductotherm 感应电热
inductotherm furnace 感应电热炉
Indusoil 精制妥尔油脂肪酸(品名,可用作捕收剂)
Indusoil L-5 精制妥尔油脂肪酸(品名,含脂肪酸90%,松香酸5%,可用作捕收剂)
Indusoil sulfonate 蒸馏妥尔油磺酸盐
industrial accident cost 工业事故费用
industrial age 工业时代
industrial and mining establishments 工矿[厂]企业

industrial application 工业应用
industrial computer 工业(控制)计算机
industrial concentrator 工业(性)选矿
industrial control 工业控制(过程)
industrial credit project 工业贷款项目
industrial data 工业数据
industrial design 工业设计
industrial engineering 工业工程
industrial evaluation 工业鉴定[评价]
industrial experience 工业经验
industrial flotation circuit 工业浮选流程
industrial flotation system 工业浮选系统
industrial flotation unit 工业浮选单元[机]
industrial grade 工业品级[位]
industrial health monitoring agency 工业健康监督机构
industrial hydrocyclone 工业水力旋流器
industrial hygiene standard 工业卫生[保健]标准
industrial installation 工业型设备
industrial laboratory 工业实验室
industrial mineral 工业矿物
industrial mineral consultant 工业矿物顾问
industrial mineral product 工业矿产品
industrial mineral project 工业矿物项目
industrial operation 工业生产,工业运行
industrial point source 工业点源
industrial practice 工业实践
industrial process control equipment 工业过程控制设备
industrial revolution 工业革命
industrial revolution record 工业革命纪录
industrial rougher circuit 工业粗选回路
industrial scale experiment 工业规模试验
industrial scale grinding circuit 工业规模磨矿回路
industrial scale test 工业(规模)试验
industrial scrap recovery 工业废料回收
industrial size mill 工业规格磨机
industrial standard allowance 工业标准误差
industrial use 工业用途
industrial value 工业价值
industrial ventilation 工业通风
industrial waste 工业废料
industrial waste water 工业废水
industry division 工业处[部]
industry experience 工业经验
inefficiency multiplier 低效率系数
in-elastic body 非弹性体
inert dust 岩粉

inert film 惰性薄膜
inert material 惰性材料
inert mineral matters 无用[惰性]矿物质
inert substrate 惰性吸附体[基质,金属层]
inertia potential 惰性电位
inertial force 惯性力
inertial impaction 惯性冲击
inertial term 惯性项
infeed flange 给入(漏斗,管)口法兰
infeed section 给料工段
inferred ore reserve 推断矿石储量
infinite dilution 无限稀释
infinite lattice 无限晶格
infinite particle size 无穷大的粒度
infinite slope 无限斜率
infinite speed regulation 无级调速
inflation 通货膨胀
inflation and risk consideration 通货膨胀和风险问题
inflation pattern 膨胀方式
inflation rate 通货膨胀率
inflection point 拐[转折,曲折,反曲]点
inflow rate 流入流量[速率]
inflow valley 内流谷[凹部]
inflowing valley 内流谷
information circular 资料[信息]通报
information gathering 资料收集
information officer 情报员,高级情报职员

information on typical application 典型应用资料
information processing 信息[情报]处理,情报整理
information theory 信息理论
infrared absorption analysis 红外线吸收光谱分析
infrared absorption spectroscopy 红外线吸收光谱
infrared aerial photograph 红外线航测[空中摄影]照片
infrared band position 红外线谱带位置
infrared detection unit 红外线探测器,热探头
infrared dryer 红外线干燥机[器]
infrared group frequency shift 基团红外线频率偏移
infrared pattern 红外线图像
infrared radiation 红外线辐射
infrared spectra 红外线光谱
infrared spectroscopic study 红外线光谱研究
infrared spectroscopic technology 红外线光谱技术
infrared specular reflection technique 红外线单向反射技术
infrared study 红外线光谱研究
infrared transmission 红外线透射
infrasizer 空气粒析器,微粒空气分级器
infrastructure 基础[公共]设施,下部

构造
infrastructure cost 基础设施造价
infrastructure in mining area 矿区基础设施
inherent action 特有[本质,固有,内在]作用
inherent cost 固有成本
inherent flotation rate 天然浮选速度
inherent flotation rate constant 天然浮游速度常数
inherent inefficiency 内在的无效因素[率]
inherent instability 固有不稳定性
inherent operational quality 固有作业[操作]性质
inherent remixer mechanism 特有的再[二次]混合机械结构[装置]
inherent sampling error 固有取样误差
inherent stability 固有[天然]稳定性
inherent variability 固有变量[率]
inhibition kinetics 抑制动力学
inhibition of flotation 浮选(的)抑制
in-house (机构)内部的,自身的,固有的
in-house development 内部研制
in-house estimate 内部估算
in-house knowledge 内部知识
in-house pilot plant autogenous grinding equipment 内部半工业试验自磨设备
in-house study 室内研究
initial amenability testing 初步适应性试验
initial background information 原始[初始,最初]基础资料
initial beneficiation 初选
initial calibration 最初的标定,最初定标
initial capital expenditure 初始投资费用[基建投资]
initial chemisorbed layer 初始化学吸附层
initial cleaning and upgrading cleaner circuit 最初精选和提高品位精选回路
initial concentrate grade 初始精矿品位
initial cost 开办费
initial cylinder-cone transition 由圆筒转变[过渡]圆锥法,起端圆筒末端圆锥转变法
initial decision 起始决定
initial economic assessment 初步经济评价
initial economic evaluation 初步经济评价
initial feasible study 初步可行研究
initial feed analysis 原始给料分析
initial financial appraisal 最初[初步]财务评估
initial flow scheme 初步流程
initial fracture 初步破裂
initial grinding test 初磨试验
initial guess 初始推测值

initial hydration 初始水合
initial information 原始[最初]资料
initial installation expense 最初装备费
initial investment 初始投资
initial mass 初始质量
initial mechanical engagement 起[初]始机械啮合
initial orebody development stage 矿体初期开拓阶段
initial peak 初始峰值
initial period 初始时期
initial plant proposal 初期选矿厂计划[建议]
initial price 初始价
initial production 初期[始]产量
initial production period 生产初期
initial programming cost 最初程序设计费
initial research program 最初研究规[计]划
initial sampling 最初取样
initial simplex 起始单纯形
initial stage 初始阶段
initial start-up period 最初起动期间
initial state 起始状态
initial surface 起始表面
initial sweep 初始[步]扫描
initial test 最初试验
initial transient 起始瞬变
initial work index 初步的功指数
injection pump 喷射泵
injection valve 喷射[出]阀

injector sampler 喷射器型管道取样器
inland storage yard 内陆贮[堆]料场
inlet area 入口面积
inlet cut-off valve 入口关闭阀
inlet feed box 进料[矿]箱
inlet feed chute 进料[矿]溜槽
inlet flange 入口法兰
inlet nozzle 入口喷嘴
inlet pressure 入口压力
in-line 一列(的),排成行的,排成直线的,联机的,轴向的
inline manifold 直列歧管
in-line manifold system 直线多管分配系统
inline spoon cutter 轴向匙状切取器
inline type manifold 直列式[型]歧管
inner circumference 内圆周
inner complex salt 内络盐
inner cone 内锥
inner countershaft bushing 传动轴[副轴]内轴瓦
inner diameter 内径
inner eccentric bushing 偏心轴内衬套
inner film 内膜
inner film of strongly bound water 强结合水内膜
inner periphery 内圆周
inner radius 内半径
inner side 内侧
inner vortex 内旋涡
innovative design feature 富有创新精神的设计特点

inordinate contamination	过度污染
inordinate wear	过度[异常]磨损
inorganic acid	无机酸
inorganic anion	无机阴离子
inorganic compound adsorption	无机化合物吸附
inorganic dispersant	无机分散剂
inorganic exchanger	无机交换剂
inorganic filler	无机填料
inorganic material	无机物料
inorganic metallic species	无机金属类(药剂)
inorganic modifier	无机调整剂
inorganic modifying agent	无机调整剂
inorganic non-metallic species	无机非金属类(药剂)
inorganic salt	无机盐
inorganic salts	无机盐类
in-pit crusher	露天矿场破碎机
in-pit movable crushing-conveying system	露天矿移动式破碎运输系统
in-plant control	近距控制
in-plant process water system	厂内生产[工艺]用水系统
in-plant pumping system	厂内泵送系统
in-plant system	近距系统
in-plant test	现场试验
input data	输入数据
input feed rate	给入的给矿量
input feed side	入料侧
input impedance of 10^{14} ohms	$10^{14}\Omega$ 输入阻抗
input output device	输入输出装置
input to control amplifier	输入控制器放大器
input variable	输入变量
input-output	输入-输出
input-output channel	输入输出通道
input-output model	输入-输出模型
inquiry sheet	查询表
insecticide	杀虫剂
insensitiveness	钝性[感],不灵敏(性),不敏感
insensitivity	低灵敏度,钝性[感],不灵敏(性),不敏感
inside operating function	厂内操作职能部门,厂内操作职责
in-site roadway	厂区公路[行车道]
in-situ spectroscopic examination	就地[现场]光谱检验
in-situ technology	就地[现场]技术
in-situ test	现场试验
insoluble basic copper sulphate	不溶性碱式硫酸铜
insoluble content	不溶物含量
insoluble hydrocarbon collector	不溶性烃类捕收剂
insoluble metal collector salt	不溶性捕收剂金属盐
insoluble oxidation product	不溶性氧化产物
insoluble phase	不溶相
insoluble residue	不溶残渣

insoluble species 不溶产物
insoluble zinc hydroxide 不溶性氢氧化锌
insolvency 无偿付能力
inspecting engineer 验收工程师
inspection activity 检查活动
inspection cover 检查孔盖,人孔盖
inspection door 观察门[孔]
inspection hole 检查[观察]孔
inspection pit 检修[检车,修车,检查]坑,检验井,探坑[井]
inspection port 检查口
installation 装置[备],设备[施],安装,装配
installation cost 安装费
installation cost of plant and equipment 选矿厂和设备安装费
installation drawing 安装图
installation information 安装资料
installed capacity 安装容量[能力]
installed capital cost 安装投资费用
installed equipment cost 安装设备费用
installed grinding power 安装磨矿功率
installed horse power 安装马力[功率]
installed power 安装功率
installed power requirement 安装功率需要量
installer 安装程序,安装单位,安装工
instantaneous acceleration 瞬时加速度
instantaneous air dryer 快速空气干燥机

instantaneous capacity 瞬时生产能力
instantaneous feed rate 瞬时给矿量[速度]
instantaneous feedback 瞬时反馈
instantaneous flotation 瞬时浮选
instantaneous flow function 瞬时流动函数
instantaneous photovoltage 瞬时[快速]光电压
instantaneous rate 瞬时速度
instantaneous result 瞬时(分析)结果
instantaneous strength 瞬时强度
institute of mining technology 矿业学院
in-stream probe analyzer 浸入式探头分析仪
in-stream probe system 浸入式探头系统
in-stream probe X-ray analyzer system X射线在线分析装置
instruction manual 操作[校准,维修]手册
instrument control switch 仪表控制开关
instrument control system 仪表控制系统
instrument engineer 仪表工程师
instrument marshalling 仪表配置[排列]
instrument output 仪表输出
instrument parts 仪表部件
instrument selection 仪表[器]选择

instrument technologist 仪表技师
instrumental colorimetry 比色计法
instrumentation 仪器仪表,测试设备
instrumentation device 仪器装置[设备]
instrumentation point 仪表设置点
instrumentation staff 仪表人员
instrumented plant 仪表化工厂
insulated copper wire 绝缘铜线
insulated water jacket 绝缘水套
insulating washer 绝缘垫圈
insurance company 保险公司
insurance policy 保险政策[方针]
insurance premium 保险费
integer valued constant 整值常数
integral action 积分作用
integral method 积分法
integral number 整数
integral part 组成部分
integral part of reagent testing 药剂试验组成部分
integral sensing device 完整传感装置
integral stage 积分段
integral value 整数值,积分值
integrated data processing 综合数据处理
integrated error [累积]误差
integrated method 综合(考虑)方法
integrated panel 联控盘
integrating unit 积分装置
integration mass flow 综合质量流量
intense agitation 强力搅拌

intense alteration 强蚀变
intense conditioning 强力调整[搅拌]
intense mixing 强力[烈]搅拌
intensity of agitation 搅拌强度
intensive cement 强(度)黏结剂
intensive research 深入[集中]研究
intentional aeration 有意的充气[曝气]
intentional chemical oxidation 有意识的化学氧化
intentional oxidation 有意的氧化
interaction effect 交互效应
interaction variable 相互作用变量
interatomic and molecular distance 原子与分子间距离
interatomic spacing 原子间距离[间隔,间距]
inter-bank rate 银行同业汇率
intercept length 截取长度
interchain cohesion 链间内聚力
interchain cohesive bonding 链间内聚力[结合]
interchangeability of parts 备件的可互换性
intercom 对讲装置,内部通话设备[系统]
intercom system 对讲电话系统
interconnect schematic 互相接线图
interconnecting chutes 相互连接的溜槽
interest 兴趣,利[益]息
interest charge 利息费用[开支]

interest cost 利息费用
interest during construction 建设期间利息
interest equalization tax 利息平均税
interest of outstanding debt 未偿债务的利息
interest on debt 借款利息
interest payable quarterly 按季付息
interest period 利息期
interest rate 利率
interested reader 有兴趣的读者
interest-free deposit 无息存款
interface design 接口设计
interface device 接口装置
interface hardware 接口硬件
interfacial and wetting behavior 界面湿润状态
interfacial behavior 界面状态
interfacial energy valley 界面能谷
interfacial migration 界面迁移
interfacial parameter 界面参数
interfacial process 界面过程
interfacial region 界面区
interfacial surface tension 界面表面张力
interference substance 干扰物质
interfering element 干扰元素
interfering ion 干扰离子
intergranular cement 颗粒胶结物
intergrown structure 共生结构
intergrowth 连生体
interim circuit 中间回路

interim report 阶段[中期,临时性]报告
interionic spacing 离子间距
interior optima 内部最优
interior pilling type 内部起球型
interlayer ionic charge 层间离子电荷
interlayer surface 间层表面
interlock of equipment 设备连锁
interlocked 连[互]锁的
interlocked system 连锁制
interlocking component 连锁组件
interlocking device 连锁装置
interlocking logic circuit 连锁逻辑线路
interlocking method 连锁连生方法
interlocking mineral 连生矿物
interlocking sequence 连锁顺[程]序
interlocking signal 连锁信号
interlocking system 连锁系统
intermediary autogenous grinding 中间自磨
intermediate beneficiation step 中间选矿阶段[步骤]
intermediate category 中间类
intermediate cell weir （浮选机）槽间闸门堰
intermediate concentrate 中间精矿
intermediate concentration step 中间（选矿）阶段[步骤,工序]
intermediate cone crusher 中型[碎]圆锥破碎机
intermediate flotation product 浮选中

间产品,浮选中矿
intermediate heat 中间热
intermediate level diaphragm discharge 中间水平隔板排料[矿]
intermediate level diaphragm discharge grind mill 中间水平隔[档]板排料[矿]磨机
intermediate material 中间产物
intermediate mill 中间磨机
intermediate ore 中间[等]矿石
intermediate ore conveyor 中间矿石运输机,中继矿石运输机
intermediate ore storage facility 中间贮矿设施
intermediate ore storage pile 中间贮矿场
intermediate pump 中间泵
intermediate shaft bearing 中间轴轴承
intermediate size 中等规格[粒度]
intermediate spacer 中间隔板
intermediate test 中间(产品)试验
intermediate tonnage 中等处理量[吨位]
intermediate tungsten compound 钨中间化合物
intermediate variable 中间变量
intermedium dust-holding tank 中间储尘槽
intermittent operation 间歇运转[作业]
intermittent production 间歇生产

intermixing 中间混合
internal allocation of overhead 内部管理费摊派
internal baffling 内部稳流板
internal circulating water 内部循环水
internal circulation pattern 内循环形式
internal flow control 内部流量控制
internal flow pattern 内流图形[象]
internal mass flow 内部物料流
internal pipe diameter 管子内径
internal protrusion 内部突[隆]起(部分)
internal pulp short circuit 内部矿浆短路
internal rate of return 内部收益率
internal rate of return method 内部收益率法
internal recycle 内部循环
internal recycle flow 内循环矿流
internal rotational flow equation 内(部)旋转流量方程
internal rotational mass flow 内旋转的质量流(量)(磨机内装料的)
internal slurry density 内部矿浆浓度
internal slurry medium 内部矿浆介质
internal slurry viscosity 内部矿浆黏度
internal source 内部来源
internal source of fund 资金内部来源
internal spring bolt 内弹簧螺栓
internal surface 内表面
internal variability 内部可变化性

internal variability of grade 品位内部可变化性
international agency 国际组织
international division 国际处[部]
international metal 国际金属
international operations 国际业务
international standard 国际标准
inter-particle comminution 粒间粉碎,中间粒子粉碎
interparticle interaction 颗粒间互相作用
interpolated value 内插值
interpolating grade 内插品位
interpolation of data 数据插值法
interpretation of data (试验)数据的解释
interpreter 讲解员,译员,解释[翻译]程序,翻译机[器],解释器
interpretive language 解释语言
interpretive program 解释程序
inter-stage vibrating screen 段间振动筛
inter-stage water addition 各段间[中间阶段]添加水量
interstitial fines 填隙细粒[粉矿]
interstitial trickling 孔[空]隙滴流,填隙涓流
intervening layer 中间层
intractable problem 难以解决的问题
introduction 介绍,引进,前言,简介
introduction of new technology 引进新技术

introductive remarks 引言,开场白
introductory remarks 引言,开场白
Invadine 烷基苯磺酸钠(品名,可用作湿润剂)$R \cdot C_6H_4SO_3Na$
inventory 库存,存货(清单)
inventory calculation of grinding steel 磨矿钢材存留量盘存计算
inventory expense 储藏费
inventory flow 存货流转[流量]
inventory level 库存料位
inverse matrix 逆矩阵
Invert soap 转化皂(品名,烷基苄基二甲基氯化铵,可用作乳化剂、湿润剂)$(C_6H_5CH_2)(R)(CH_3)_2N^+, Cl^-$
invertebrate 无脊椎动物
invertebrate survey 无脊椎动物调查
inverted siphon 反[倒]虹吸
inverting equation 转换方程
investigation 调查[研究],探[审]查,勘测,试验,调查报告,研究论文
investigation of processes to treat oxide ores 处理氧化矿方法的研究
investigation process 研究过程
investment bank 投资银行
investment cost 投资费用
investment decision criteria 投资决策准则
investment for equipment 设备投资
investment fund 投资资金
investment group 投资集团
investment project 投资项目
investment proposal 投资建议[计划]

investor 投资人[方]	ionic field 离子场
investor's return 投资者收益	ionic group 离子团
invoice 发票,发货单	ionic halo 离子晕
involuted feed 渐开线给矿(旋流器)	ionic head 离子端
involuted type 渐开式(旋流器)	ionic head size 离子端部尺寸
involuted type cyclone design 渐开线型旋流器设计	ionic lattice 离子晶格
	ionic pair 离子偶
inward-upward direction 向内向上方向	ionic polar group 离子极性基
	ionic solid 离子型固体
Ioct sulfonate 油溶性石油磺酸盐(品名,可用作氧化铁矿的捕收剂)	ionic species 离子种类
	ionic surface hydration energy 离子表面水合能
iodine titration 碘滴定法	
iodometric titration 碘量滴定法	ionic surfactant 离子表面活化剂
iodometry 碘量滴定法	ionizable collector 电离捕收剂
ion exchange column 离子交换柱	ionizing electrode 离电极
ion exchange material 离子交换物料	ionomolecular complex 离子分子络合物
ion exchange mechanism 离子交换机理	ionomolecular composition 离子分子组成
ion exchange process 离子交换过程	
ion exchange system 离子交换系统	ion-pair formation 离子偶形成
ion exchange technique 离子交换技术	Iporit 烷基萘磺酸钠(品名,可用作乳化剂)
ion exchanger 离子交换剂	
ion pair complex 离子偶络合物	iron and steel industry 钢铁工业
ion-exchange membrane 离子交换膜	iron bearing material 含铁物料
ionic activity 离子活度	iron carboxylate 羧酸铁
ionic atmosphere 离子雾	iron ethyl xanthate 乙基黄原酸铁
ionic bombardment 离子轰击	iron hexyl xanthate 己基黄原酸铁
ionic bond 离子键	iron linoleate 亚油酸铁
ionic charge 离子电荷	iron linolenate 亚麻酸铁
ionic composition 离子组成	iron magnesium silicate 铁镁硅酸盐
ionic crystal 离子型结晶	iron mill 钢磨
ionic deformation 离子变形	iron oleate 油酸铁

iron ore beneficiation 铁矿石选矿

iron ore crushing and sizing plant 铁矿破碎筛分厂

iron ore industry 铁矿石工业

iron oxide mineral 氧化铁矿物

iron oxidizing bacteria 铁氧化菌

iron salt 铁盐

iron-bearing pulp 含铁矿浆

iron-cyanide complex 铁-氰络合物

iron-rich transitional zone 富铁过渡带

irregular clot 不规则集聚体[凝块]

irregular sampling plan 不规则取样方案[平面图]

irregular shaped particle 不规则形状的颗粒

irrigation water 灌溉用水

isdecanal 异癸醛

isinglass 鱼胶,鳔胶,(白)明胶,白云母薄片,云母(可用作抑制剂)

island of non-contact 无接触区

isoamyl(=isopentyl) 异戊基 $(CH_3)_2CHCH_2CH_2—$

isoamyl xanthate 异戊基黄药

isobutyl 异丁基 $(CH_3)_2CHCH_2—$

isobutyl acetone 异丁基酮

isobutyl methanol 异丁基甲醇

isochoric heat of adsorption 等容吸附热

isododecane 异癸烷

isoelectric point 等电点

isohexyl 异己基 $(CH_3)_2CH(CH_2)_2CH_2—$

isolated area 偏僻地区

isolated site 隔离[偏僻]的现场

isolation frame 隔振框[构,车,机]架

isomorphic iron 类质同象铁

isomorphously substituted impurity 类质同象置换杂质

iso-octyl acid phosphate 异辛基酸性磷酸酯(可用作捕收剂)

isopentyl 异戊基 $(CH_3)_2CHCH_2CH_2—$

isopentyl xanthate 异戊基黄药,异戊(基)黄(原)酸盐

isopropyl 异丙基 $(CH_3)_2CH—$

isopropyl ethylenethiocarbamate (Z-200) 异丙基乙烯硫代氨基甲酸盐[酯](即Z-200)

isopropyl thionocarbamate(i.e Z-200) 异丙基硫逐氨基甲酸盐(即Z-200)

isopropyl xanthate 异丙基黄药,异丙基黄原酸盐 $(CH_3)_2CHOCSSM$

iso-thiourea salt 异硫脲盐(可用作铜的捕收剂)

isotope decay 同位素衰变

isotopes level controller 同位素料位控制器

issued share 已发行的股份

item 条目,条款,项目,项次,物[产]品,东西,零[元]件

itemized estimation method 分项估算方法

iteration 迭代

iteration number 迭代数

iterative calculation 迭代计算
iterative method 迭代法
iterative process 重复过程
iterative solution 迭代解

J

jack shaft 中间轴,变速箱传动轴,溜煤眼,暗井盲下水井
jacking cradle 千斤顶支架
Jaguar 387 一种古耳胶(刺槐豆胶)制品(可用作絮凝剂)
Jaguar 503 合成絮凝剂(可用于浮选尾矿水的净化(絮凝))
Jaguar MD 合成絮凝剂(可用于浮选尾矿水的净化(絮凝))
Jaguar −PK3 合成絮凝剂(可用于浮选尾矿水的净化(絮凝))
Janney flotation cell 詹尼浮选槽[机]
jaw crusher operating procedure 颚式破碎机操作程序
Jeffrey vibrating feeder 杰弗里振动给矿机
Jerome Mercury Detector 杰罗姆水银探测器[检波器]
jet momentum 射流动量
jet pump 喷射泵
jet stream 喷射气[水]流
jewelers roller 宝石对辊机
jig circuit 跳汰回路[流程]
jig compartment 跳汰机室
jig concentrate 跳汰精矿
job description sheet 工作[职务]说明表
job duty and responsibility 工作任务和职责
job engineer 职业工程师
job number 任务[工作,工件]编号
job requirement 工作需要
Johnson drum 约翰逊鼓[筒]式选矿机
joint investment 联合投资
joint state−private enterprise 公私合营企业
joint venture 合资企业,联合经营
joint working committee 联合工作委员会
Joy rotary−percussion drill 乔伊型旋转冲击式钻机
judgment estimate 判断性(的)估算
judgment factor 判断因素
jumbo 凿岩台车
junction box 接线盒

K

2-kg batch sample　2公斤批样
K-101　硫氰酸铵(品名,可用作活化剂)NH_4CNS
Kaiser Engineers' standard practice　凯撒工程公司标准实践
karaya gum　刺梧桐树胶(可用作湿润剂)
Kawasaki portable crushing plant　川崎移动式破碎厂
Kaydol(=petrolatum)　矿脂,凡士林(品名,可用作捕收剂、絮凝剂)
Kaye disc centrifuge　凯耶盘式离心机
KBr cell　溴化钾电池
KBr pettet technique　溴化钾压片法
K-di-n-amyl dithiocarbamate　二-正-戊基二硫代氨基甲酸钾
Keithley Model 604 differential amplifier　吉时利604型差动放大器(美国)
Kelcosol　藻朊(品名,可用作絮凝剂)
Kelzanx C　高分子聚合物(品名,美国Kelco公司产品,可用作絮凝剂、分散剂)
Kelgin W　藻朊(品名,可用作絮凝剂)
kerosene-pine oil system　煤油-松油药剂制度
K-ethyl xanthate　乙基黄原酸钾(乙基钾黄药)
key cost factor　关键成本因素
key diagram　概略原理图,说明[纲要]图
key drawing　纲要[索引]图
key efficiency factor　主要[关键]有效因素
key engineering project　关键工程
key location　关键部位
key metallurgical factor　关键选矿[冶金]因素
key personnel　主要人员
key plan　索引图,主要平面图,总平面图
key unit　关键单位,关键设备
KFS model cone crusher (for fine crushing)　KFS圆锥破碎机(用于细碎)
KH-3　一种絮凝剂(品名)
$KHCO_3$　碳酸氢钾(浮选Nb(铌)矿时可以改善胺或二胺捕收剂的作用(活化剂))
$KHSO_4$　硫酸氢钾(可用作调整剂)
kidney shaped stockpile　腰形料堆
kilojoule　千焦耳
kilovar　千乏,无功千伏安
kilovar-hour　千乏小时,无功千伏安

小时
kilovolt 千伏(特)
kilovolt ampere 千伏(特)安(培)
kilovoltage 千伏电压
kilovolt-ampere-hour 千伏安小时
kilovoltmeter 千伏计
kilowatt 千瓦(特)
kilowatt meter 电力[千瓦]表
kilowatt unit 千瓦单位
kilowatt-hour 千瓦小时,度
kilowatt-hour meter 电度表
kinematic similarity 运动学相似性
kinetic analogy 动态模拟
kinetic model 动力模型
kinetic parameter 动态参数
kinetic treatment 动力处理方法
kinetics of activation 活化动力学
kinetics of capture rate 捕捉[集]速度动力学
kinetics of depression 抑制动力学
kinetics of flotation 浮选动力学
K-iso-amyl xanthate 异戊基黄原酸钾
kit 工具箱,成套工具[用品]
Kleinbentink machine Kleinbentink型煤矿浮选机(荷兰产倒圆锥形截面、大直径浮选机)
Knapp and Bates centrifugal diamon pan 克纳普·贝茨型离心(分选)金刚石盘
know-how 专门技能[技术,知识],实践知识,(生产)经验,技术诀窍
knowledge of accounting terminology 会计术语知识
known ore 已知[标准]矿石(作比对试验用)
known power consumption 已知功(率消)耗
known value 已知值
known weight 一定[已知]重量
KOD 絮凝剂(二氯乙烷与乌洛托品的缩合物)
KODT 絮凝剂(二氯乙烷与乌洛托品的缩合物)
Koepe(mine)hoist 戈培式(矿井)提升机,矿用摩擦提升机(以德国人戈培(Koepe)的名字命名)
Kogasin 科尬油(品名,一种合成煤[汽]油的德文名称,合成洗涤原料油)
Kontakt 烷基磺酸盐(品名,可用作捕收剂)
Korenyl 30 烷基苯磺酸盐(品名,可用作湿润剂)
Korenyl 56 烷基苯磺酸盐(品名,可用作湿润剂)
Korenyl 80 烷基苯磺酸盐(品名,可用作湿润剂)
Krebs cyclone 克雷布斯旋流器,多管旋流器(美国克雷布斯工程公司制造)
Kreelon AD 选矿用药剂(烷基苯磺酸盐,可用作湿润剂)
Kronitex 选矿用药剂(三甲酚基磷酸盐,可用作捕收剂、调整剂、活化剂)$(CH_3 \cdot C_6H_4O)_3PO$

Krupp flotation machine 克雷伯型浮选机(德国 Krupp 公司制造)
krypton 氪(Kr)
Kue-Ken jaw crusher 库肯颚式破碎机(美国 Kue-Ke 破碎机制造厂生产)
Kylo 27 矿泥真空过滤絮凝剂
Kynch method 肯奇法(确定有效分离浓缩机面积的实验室试验方法)

L

La retarder 杜邦公司出品的一种抑制剂
lab mill 实验室磨机
labor cost 人工成本
labor crew 劳动人员
labor fringe benefit cost 劳动附加福利费
labor protection 劳动保护
labor rates 人工费率
labor relation 劳务关系
labor requirement 劳动力需要量
laboratory 实验室
laboratory analysis technique 实验室分析方法[技术]
laboratory and pilot-scale flotation 实验室和半工业规模浮选
laboratory approach 实验室方法
laboratory batch grinding test 实验室分批[批量]磨矿试验
laboratory calcining equipment 实验室煅烧设备
laboratory circuit 实验室回路
laboratory circuit arrangement 实验室流程[回路]配置
laboratory classifier 实验室分级机
laboratory column leaching unit 实验室浸出柱装置
laboratory consumables 实验室消耗品
laboratory crushing facilities 实验室破碎设施
laboratory director 实验室主任
laboratory experiment 实验室试验
laboratory flotation cell 实验室浮选槽
laboratory flotation pulp density 实验室浮选矿浆浓度
laboratory flotation response experiment 实验室浮选反应试验
laboratory flotation work 实验室浮选(试验)工作
laboratory for titration 滴定实验室
laboratory investigation 实验室试验
laboratory leaching procedure 实验室浸出过程
laboratory level 实验室试验水平

laboratory locked-cycle test 实验室闭路试验

laboratory machine 实验室设备[浮选机]

laboratory material 实验室材料

laboratory metallurgical result 实验室选矿效果

laboratory mill 实验室磨机

laboratory model X-ray unit 实验室模型X射线分析装置

laboratory personnel 实验室人员

laboratory pressure filter 实验室压滤机

laboratory procedure 实验室方法

laboratory reactivation test 实验室再生试验

laboratory research 实验室研究

laboratory rod mill 实验室棒磨机

laboratory rotating tube furnace 实验室管式回转炉

laboratory scale 实验室规模的,小型的

laboratory scale batch data 实验室小型试验数据

laboratory scale batch hydraulic classification 实验室规模批量水力分级

laboratory scale bench data 试[实]验室小型试验数据

laboratory scale jig 实验室规模跳汰机

laboratory scale separator 实验室规模分选机

laboratory scale testing 实验室规模试验

laboratory scale vanner test work 实验室规模带式流槽试验工作

laboratory separation test 实验室分选试验

laboratory separation time 实验室选别[浮选](停留)时间

laboratory staff 实验室人员

laboratory technique 实验室操作技术

laboratory test 实验室[小]规模试验

laboratory test data 实验室试验数据

laboratory test jig 实验室试验跳汰机

laboratory test program 实验室试验方案[计划,程序]

laboratory test work 实验室试验工作

laboratory testing procedure 小型[实验室]试验方法[步骤]

laboratory vacuum filter 实验室真空过滤机

laboratory wet table test work 实验室湿式摇床试验工作

laborer 工人

lack of maintenance 缺乏维修,欠修理

lacquer 漆,漆器

lactic acid(2-Hydroxy propionic acid) 乳酸(2-羟基丙酸,丙醇酸,可用作浮选抑制剂)$CH_3CHOHCOOH$

Ladal X-ray centrifuge Ladal X射线(离心)分离器[机]

ladder safety guard 梯子安全罩[栅]

lag effect 滞后效应

lagged pulley drive 带有外套的胶带轮驱动

lagging current 滞后电流

Lagrange multiplier method 拉格朗日乘数法(以法国数学家约瑟夫·路易斯·拉格朗日命名,是一种寻找变量受一个或多个条件所限制的多元函数的极值的方法)

Lagrange's method 拉格朗日法(又称随体法,是描述流体运动的方法:跟随流体质点运动,记录该质点在运动过程中物理量随时间变化的规律)

lamella clarifier 薄片[层]澄清器(俗称倾斜浓密箱)

lamella thickener 倾斜板浓缩机

lamella type unit 薄片[层]型设备

lamellar structure 层状结构

Lamflo sluice Lamflo 型溜槽

laminated plastic 层压塑料(板)

land formation 土地形成[复原]

land formation and revegetation 土地恢复和再植被

land owner 土地所[拥]有人[者]

landholder 土地所[拥]有人[者]

land-pebble phosphate rock deposit 陆地磷灰岩砾矿床

landscape bureau 园林局

landscape engineering 绿化工程

landscaping (环境)绿化,美化

Lang launder Lang 型洗矿槽

Langmuir(Irving Langmuir) 兰格缪尔[朗缪尔,兰米尔](美国物理化学家,1881~1957)

Langmuir adsorption 兰格缪尔吸附特性

Langmuir adsorption isotherm 兰格缪尔吸附等温线[式]

Langmuir equation 兰格缪尔方程

Langmuir isotherm 兰格缪尔(吸附)等温线

Langmuir isotherm type plot 兰格缪尔等温线类型图

Langmuir relationship 兰格缪尔关系式

large and complex open pit copper mine 大型复杂露天铜矿

large commercial range 大工业型[类](按使用规格分类)

large commercial size plant 大型工业选矿厂

large dosage 大剂量

large dose 大剂量,大药量

large high-capacity equipment 大型高生产力设备

large medium pressure venture scrubber 大型中压文丘里洗涤器

large organization 大机构

large porphyry copper deposit 大型斑岩铜矿矿床

large porphyry-type deposit 大型斑岩型矿床

large radius elbow 大弯曲半径弯管[头]

large scale 大规模

large scale flotation test 大规模[大型]浮选试验

large scale milling operation 大规模选矿作业[生产]

large scale mining operation 大规模采矿作业[生产]

large tonnage 大吨位

large unit operation 大型设备运行

large-diameter hollow trunnion 大直径中空[空心]耳轴

large-diameter mill end cover 大直径磨机端盖

large-diameter ring gear 大直径环形齿轮

large-diameter shallow tank 大直径浅槽

large-scale operation 大规模作业[生产]

LAROX 芬兰 LAROX[拉罗克斯]公司品牌名(主要产品有用于冶金、矿山、化工、食品、医药行业的各种型号的拉罗克斯压力过滤机)

LAROX automatic press filter LAROX[拉罗克斯]自动压滤机

LAROX press filter LAROX[拉罗克斯]压滤机

LAROX screw pump LAROX[拉罗克斯]螺旋泵

last-in first-out cost 后进先出成本

latch pin 插销

latch spring 锁紧弹簧,碰锁弹簧

Latecoll AS 絮凝剂(西德产)

lateral binding energy 侧向结合能

lateral boundary 水平边界

latest availability 最新的可利用性

latitudinal tilt 横向坡度

latter mineral 后生成[后来]的矿物

lattice 晶格,格子[架、栅],支撑桁架

lattice atom 晶格原子

lattice defect 晶格缺陷

lattice dimension 晶格尺寸

lattice distortion 晶格畸变

lattice energy 晶格能

lattice imperfection 晶格缺陷

lattice ion 晶格离子

lattice point 格点

lattice rearrangment 晶格重新排列

lattice replacement reaction 晶格置换反应

lattice square 正方晶格

lattice substitution 晶格置换

lattice sulphur 晶格硫

launder sampler 槽式取样器

laundry 洗衣房,送去洗的衣服

lauric 月桂,桂树脂酸

lauric aminoacid 月桂胺酸[氨基酸] $R_{12}CONHCH_2COOH$

lauric acid 月桂酸,十二烷酸 $CH_3(CH_2)_{10}COOH$

lauryl(=dodecyl) 月桂基,十二(烷)基 $CH_3(CH_2)_{10}CH_2-$

laurylamine 月桂胺,十二(烷)胺 $CH_3(CH_2)_{10}CH_2NH_2$

laurylamine hydrochloride 月桂基胺

氢氯化物,十二烷胺盐酸盐 $C_{12}H_{25}NH_2 \cdot HCl$

laurylsulfoacetate 十二烷基磺化乙酸酯 $C_{12}H_{25}OOC \cdot CH_2SO_3M$

laxative effect 通便作用

laydown and repair area （设备）存放和检修场地

laydown area 堆放区

layered pile 分层料[矿]堆

layout 布局,布置,安排

Lazrin reference electrode 拉兹仑参比电极

leach box 浸出箱

leach configuration 浸出（设备）配置

leach feed box 浸出给矿箱

leach on-off cycle 浸出开停周期

leach pad 堆浸场基

leach pad area 堆浸场地[面积,区域]

leach precipitation tank 溶浸沉淀槽

leach residence time 浸出停留时间

leach residue 浸出渣

leachant 浸出剂

leachant concentration 浸出[溶浸]剂浓度

leached and oxidized capping 溶滤[淋溶]氧化覆盖层

leached ore disposal 浸过矿石的处理,浸渣堆存

leached pulp 浸出矿浆

leachfield drainpipe 堆浸场排液管

leaching and precipitation unit 浸出置换装置

leaching characteristic 浸出特性

leaching kinetics 浸出动力学

leaching pad 堆浸场基(层)

leaching process 沥滤法,浸出工艺

leaching section 浸滤段

leaching step 浸出阶段[步骤]

leaching test 浸出试验

leaching time 浸出时间

leaching treatment method 浸出处理法

leaching-solvent extraction-electrowinning copper plant 浸出溶剂萃取电积铜厂[车间]

leach-ion exchange-flotation process 溶浸离子交换浮选联合法

leach-wash cycle 浸滤洗涤循环

lead acetate 醋酸铅

lead area 铅区(选铅工段)

lead butyl dithiophosphate 丁基二硫代磷酸铅

lead carbonate 碳酸铅

lead column 居首位的柱,最前面的柱,前导柱

lead dithiocarbamate phase 二硫代氨基甲酸铅相

lead dross furnace 铅渣炉

lead elimination 去[除]铅

lead ethyl dithiophosphate 乙基二硫代磷酸铅

lead hexyl xanthate 己基黄原酸铅

lead ion 铅离子

lead isoamyl dithiophosphate 异戊基

二硫代磷酸铅
lead lining 铅衬里,铅内衬
lead m-cresyl dithiophosphate m-甲苯基二硫代磷酸铅
lead methyl dithiophosphate 甲基二硫代磷酸铅
lead mold 铅模
lead o-cresyl dithiophosphate o-甲苯基二硫代磷酸铅
lead oxide 氧化铅
lead p-cresyl dithiophosphate p-甲苯基二硫代磷酸铅
lead phase 铅相
lead phenyl dithiophosphate 苯基二硫代磷酸铅
lead producer 铅生产厂
lead propyl dithiophosphate 丙基二硫代磷酸铅
lead sulfate 硫酸铅
lead sulfite film 亚硫酸铅薄膜
lead sulphide 硫化铅
lead thiosulphate 硫代硫酸铅
lead vacancy 铅空位
lead xanthate phase 黄原酸铅相
lead yield 铅生产率
leading article 社论
leading current 超前电流
lead-out facility 外运装载设施
lead-out station 装车站
lead-out tipple 装车用自动倾卸装置
lead-rich electrode 富铅电极
lead-zinc zone 铅锌矿带

leaf clarifier 叶片澄清机
lean middlings 低品位中矿
learning curve 学习[认知]的曲折过程[曲线]
learning period 学习时间[期]
lease agreement 租赁协议
lease financing 租赁融[筹]资
least square 最小平方[二乘方]
least square estimator 最小二乘法估算量
least square fitting procedure 最小二乘方求律程序
least square method 最小二乘法
least squares analysis 最小二乘分析法
least squares criterion 最小二乘法则
least squares technique 最小平方[二乘方]技术[法]
lecithin 卵磷脂(可用作氧化矿的捕收剂)$C_{42}H_{84}O_9PN$
ledge 突出部分,凸耳[缘],横档,壁架
Leed & Northrup pH meter Leed & Northrop 型 pH 值计(配有超声波清扫器)
Leeds automatic laboratory flotation cell 利兹实验室自动浮选槽(英国利兹大学研制)
Leeds cell 利兹型浮选槽(英国利兹大学研制的实验室设备)
left-handed screw axes 左旋轴
legal commitment 法律承诺
legal cost 法定费用
legal detail 法律细节

legal guarantee 法律保证

legal person regiment[company] board of directors 法人团[公司]董事会

legal repercussions 法律的影响

legal service cost 法律服务费

legend 图例,符号一览表(插图,照片,地图,图表等上的),说明

legislation 法律,立法,制定法律

legislative requirement 法律规定

legman 到处奔波的人,(现场)采访记者

Leiocom(=Leiogomme) 土豆淀粉(品名,可用作絮凝剂、抑制剂、调整剂)

Leiogomme 土豆淀粉(品名,可用作絮凝剂、抑制剂、调整剂)

lender 贷方

length conversion constant 长度换算常数

length inside liners 磨机衬板内长度

length of sample 取样[样品]长度

lengthwise slope 沿长边方向的坡度

Lensen-American filter press 莱塞-美利坚压滤机

Leonil 表面活性剂品牌名

Leonil C 月桂醇聚乙二醇醚(可用作湿润剂、乳化剂)
$C_{12}H_{25}(OCH_2CH_2)_nOH$

Leonil FFO 乙基/庚基-β-萘酚聚乙二醇醚(可用作湿润剂、乳化剂)
$R—C_{10}H_6(OCH_2CH_2)_nOH, R=C_6H_{13}—$ 及 $C_7H_{15}—$

Leonil O 油醇,十六烷醇聚乙二醇醚(可用作湿润剂、乳化剂)
$R(OCH_2CH_2)_nOH, R=C_{18}H_{33}—$ 及 $C_{16}H_{33}—$

lepidine 二甲基喹啉,勒皮啶(可用作捕收剂、湿润剂)

less abrasive rock 不易磨剥的岩石

less common metal 稀有金属

less common ore 稀有矿石

less-common technique 非一般性技术

lessee 承租人

lessor 出租人

lethal concentration 致死浓度

lethal level 致死含量

letter code 字母代号[符号,编码,代码]

leucine 亮氨酸,白氨酸
$(CH_3)_2CHCH_2CH(NH_2)COOH$

leucocratic quartz diorite 淡色英石闪长岩

leucoxene 白楣石

leukol 喹啉(可用作捕收剂、湿润剂)

level 水平,标准,水平面,级别,等级

level control 料位[水平]控制

level control loop 料位控制环路

level controller 料位控制器

level detection 料位[液面,水平,标高]探测

level detector 料位探测器

level detector probe 料位探测器探头[针]

level limit mode 料位极限控制方式

level limit over-ride 料位极限补偿

level of accuracy 精度水平,准确程度
level of ball addition 加球量
level of confidence 置信水平
level of ionization 电离电平
level of required profitability 所需利润水平
level of significance 显著性水平
level override signal 料位超(越)限(度)信号
level probe 料位探针
level prober 水平面[水位,液面]探测器
level sensor 液[料]位传感器
LI dry magnet (LI = low-intensity) 弱磁场[低强度]干式磁选机
LI wet magnet (LI = low-intensity) 弱磁场[低强度]湿式磁选机
liability 负债,债务
liaison group 设计联络组
liberated particle 解离颗粒
liberation characteristic 解离特性
liberation factor 解离系数[因数]
liberation requirement 解离度要求
liberation size 解离粒度
liberation size analysis 单体分离粒度分析
lien 扣押权,留置权
life expectancy 预期寿命
life hour 寿命[使用]小时
life repair cost 使用期间修理费
lift liner 提升衬板
lifting component 提升分力

lifting device 提升装置
lifting mechanism 提升机构
lift-out floor panel 可提起的地板
light diffraction 光衍[绕]射
light fraction 轻粒级部分
light frother 轻(油)起泡剂
light gravity particle 比重[密度]轻的颗粒,小比重[密度]颗粒
light hydrocarbon oil 轻烃油
light intensity 光强度
light material 轻质材料
light microscope 光学显微镜
light particle 轻颗粒
light scattering method 光线散射法
light source 光源
light steel construction 轻型钢结构
lighting condition 照明条件
lighting equipment 照明设备
lighting in plant 厂房照明
Lightnin mixer 莱宁混料[搅拌]机(美国莱宁搅拌设备公司产品)
lightweight aggregate 轻集[骨]料
light-weight secondary crusher 轻型二次破碎机
lignin 木质,木素,木质素
lignin sulfonate 木(质)素磺酸盐(可用作捕收剂、湿润剂、抑制剂、分散剂)
Lignite (=brown coal) 褐煤
Lignosulfonate 木素磺酸盐(可用作捕收剂、湿润剂、抑制剂、分散剂)
lime conveyor elevator 石灰输送提升机[升降输送带]

lime dump dust control system 石灰卸料(斗)粉尘控制系统
lime hopper 石灰漏斗
lime process 石灰法
lime pump 石灰泵
lime slaker 石灰消化器
lime slurry 石灰浆
lime treatment 石灰处理
lime unloading 石灰卸载
lime-flocculated slime 石灰絮凝的矿泥
lime-flotation process 石灰浮选法
lime-silicate rock 钙硅岩,钙矽岩
lime-silicate system 石灰-硅酸盐制度
lime-soda treatment 石灰苏打处理
limestone bed 石灰石矿床
limestone calcining plant 石灰煅烧厂,烧石灰厂
limestone crushing and screening plant 石灰岩破碎筛分厂
limestone masonry 石灰石砌筑
limitation 限制,限度,界[极]限,局限性,缺陷
limitation inherent 内在局限性
limitation of activation 活化限度[界限]
limited amount 有限数量
limited availability 有限利用度
limited carrying capacity 有限的运送[载]能力
limited liability company 股份有限公司

limited mesh 限制网目
limiting condition 极限状态
limiting factor 限制性因素
limiting stress 极限应力
limonene 苎[芋]烯,柠檬烯,萜二烯-[1,8] $C_{10}H_{16}$
Linatex 耐纳特(知名的抗损橡胶产品制造公司的名称及注册品牌名)
linatex 防锈胶乳
Linatex rubber 耐纳特橡胶(优质抗损橡胶)
line blockage 管线堵塞
line construction 线路架设[施工]
line erection 线路架设
line loop system 管线回路系统
line of credit 信用额度,授信额度
line unit 线路单元,线路装置
line velocity 管线速度(矿浆管线内速度)
linear 直线的,由线构成的,线状,细长的,成直线排列的,以线条为主的,直链的
linear absorption coefficient 线性吸收系数
linear breakage kinetics 线性破碎动力学
linear breakage program 线性破碎过程
linear compound 直链化合物
linear cut 直线切[截]取
linear dependence 线性关系
linear dimension 线性维数

linear expression 直线(方程)式	linearized loading signal 线性化荷载信号
linear form 线性形式	
linear increase 线性增加[量]	lined bottom chute section 衬底溜槽段
linear interpolation 线性内插法	liner breakage 衬里击坏,击坏衬里
linear interpolation method 线性内插法	liner configuration 内衬[衬板]图形[形状,构造形式]
linear kinetic grinding model 线性动态磨矿模型	liner consumption 衬板消耗(量)
	liner material 衬里材料
linear lattice 线性晶格	liner material wear rate 衬里材料磨损率
linear lumped parameter ball mill model 线性集中参数球磨机模型	
	liner profile 衬板形状
linear lumped parameter grinding model 线性集总参数磨矿模型	liner structure 衬板结构
	liner vessel 班轮,定期船只
linear lumped parameter model 线性集总[中]参数模型	liner wear rate 衬里[板]磨损率
	lining hypalon two partitions storage tank 衬氯磺酰化聚乙烯合成橡胶的两格贮池[罐]
linear lumped parameter system 线性总参数系统	
linear model 线性模型	linkage bearing 连杆轴承
linear motion 直线性运动,直线运动	link-belt vane feeder 链带叶片给料机
linear motion screen 直线运动筛	linoleate 亚油酸盐 $C_{18}H_{35}OOM$
linear programming 线性规划	linolenate 亚麻(油)酸盐
linear receiver 线性接收器	linseed oil 亚麻仁油
linear regression 线性回归	Lipal 40 九聚乙醇单油酸酯(品名,可用作非离子型脉石分散剂) $C_{17}H_{33}CO(OCH_2CH_2)_9OH$
linear regression analysis 线性回归分析	
linear relationship 线性关系	Liqro 粗塔尔油 (品名,Liqro GA 公司生产)
linear response 线性反应(曲线)	
linear transformation 线性变换	liquefied blend amine 液化混合胺
linear velocity gradient 线性速度梯度	liquefied flow 液化流
linear vibration 直线振动	liquid bridging 液体桥连[接]
linear way 直线形式	liquid chlorine 液态氯
linearization 线性化	liquid chlorine cylinder 液态氯罐

L

[钢瓶]

liquid circulation 矿浆循环

liquid circulation intensity 矿浆[液体]循环强度(矿浆循环量/槽容积)

liquid circulation intensity(liquid circulation/cell volume) 矿浆循环强度(矿浆循环量,槽容积)

liquid circulation path 矿浆[液体]循环通道

liquid circulation rate 矿浆[液体]循环量

liquid composition 液体成分

liquid concentration 液体浓度

liquid density 液体浓度

liquid droplet 液体微滴

liquid dynamics 液体力学

liquid film 液体薄膜,薄液膜

liquid flow condition 液体[矿浆]流动条件[状态]

liquid hydrocarbon 液态烃

liquid ion exchange 液体离子交换

liquid ion exchange process 液体离子交换法

liquid layer 液层

liquid matte 液态冰铜[锍]

liquid measure 液量单位制,液体测量

liquid medium 液体介质

liquid particle 液体颗粒

liquid phase 液相

liquid reagent 液体药剂

liquid recirculation flow 矿浆[液体]循环量

liquid rise velocity 矿浆上升速度

liquid rising speed 液体上升速度

liquid rosin 塔尔油

liquid sample 液态矿[试]样

liquid slag phase 液态渣相

liquid split 液体分流

liquid sulfur dioxide 液态二氧化硫

liquid sulphide 液态硫化物

liquid viscosity 液体黏度

liquid vortex 液体漩涡

liquid water 液态水

liquid-advancing contact angle 液体-前展接触角

liquid-receding contact angle 液体-后移接触角

liquid-solid ratio 液-固比

liquid-solid separating technique 液固分离技术

liquid-solid separation 液固分离

liquid-solid separation process step 液固分离工序

liquid-solid separation technique 液-固分离工艺[技术]

liquid-solid system 液-固系统

Lissapol 表面活性剂品牌名(英国ICI公司产品)

Lissapol A and C 十二烷基硫酸钠(可用作分散剂、湿润剂、乳化剂)

$C_{12}H_{25}OSO_3Na$

Lissapol N 烷基苯酚聚氧乙烷醚(可用作起泡剂、湿润剂、絮凝剂)

Lissapol N.D.B 辛基甲酚与环氧乙烷

的非离子缩合物(可用作起泡剂、湿润剂、絮凝剂)
Lissolamin(e)V 十六烷基三甲基溴化铵(品名,可用作捕收剂)
$C_{16}H_{33}(CH_3)_3N \cdot Br$
list of drawing 图纸目录表
list of drawings 图纸目录表
list of new publications 新书[刊]目录
literature 文学,文献,著作
literature review 文献评论[回顾,综述]
literature search 文献搜索[寻找,调查]
live load 活动负载[荷载]
live load energy 有效载重[荷载]能力(矿仓的)
live ore storage 有效矿石储存量
live storage 有效储(藏)量
live top cover 活动式顶盖
livestock 家畜
livestock watering 家畜饮用水
living conditions 生活条件[环境]
living quarters 生活区,宿舍区
LiX reagent LiX试[药]剂
load box car 装料箱车
load carrying capacity test 负荷承载力比对试验
load center 负荷中心
load density 充[装]填密度(磨机的)
load fraction 充填[负载]率
load pattern 负载图形[模式,特性曲线]

load point 负荷点
load recorder 负荷记录器
load size 载量
load station 装载站
load transducer 负荷传感器
load transducer output 负荷传感器输出信号
load variation curve 负荷变化曲线
loaded carbon 载金炭
loaded carbon assay 碳载金量分析,载金炭分析
loaded resin 载金树脂
loaded rubber-tired equipment 荷载的橡胶轮胎设备
loader 载入程序,装货设备,装载机
loader bucket 装载机铲斗
loading 载入,加载
loading dock 装车码头
loading gantry 装载台架(龙门吊车的)
loading operation 装载作业
loading point 装矿点
loading program 输入[加载]程序
loading zone 载荷[荷重]区[段]
load-out (外运)装车,码头装船
load-out conveyer 装车胶带机
loan arrangement 贷款安排
loan fund 贷款资金
loan limit 贷款限额
local anodic site 局部阳极区
local condition 当地条件
local control 局部控制

local control unit	局部控制装置
local currency	地方[当地]货币
local data	局部数据
local feature	局部特征,地方特色
local labor	当地劳动力
local land administration authorities	地方土地管理当局
local loop control	局部回路控制
local mechanical personnel	当地机械[机修]人员
local pulp environment	局部矿浆环境
local regulations	地方[当地]规章制度[条例]
local reserve	局部储量
local sampling variability	局部取样变率
local scale	局部规模
local stop-starting station	局部停开台[站]
local stream	当地河流
local tap-water	当地自来水
local taxation	地方税收
local variable	局部变量
local variation	局部变化
local water	当地水
locally fractured nature	局部破碎性质
location	位置,地点,定位
location of concentrator	选矿厂位置
location of orebody	矿体位置
location of plant site	厂址位置
lock link	防松连杆
lock link sleeve	防松连杆套
lock post	锁闭立柱(圆锥破碎机的)
lock post seat	防松杆底座
locked container	加锁的容器
locked cycle batch laboratory test	闭路循环单元实验室试验
locked cycle continuous test	闭路循环连续试验
locked cycle flotation test	闭路循环浮选试验
locked steel cover	锁[封]闭式钢盖
locked-cycle	闭路循环
locked-cycle data	闭路试验资料[数据]
locker room	衣帽间,更衣室
Lockhorst jig	洛克赫斯特圆形跳汰机
locking collar	防松[锁紧]圈[箍,环]
locking collar capscrew	防松圈带帽螺钉
locking nut cover	防松[锁紧]螺母盖
locknut	锁紧螺母
lode	矿脉,矿藏
lode gold ores	脉金矿石类
lode tin ore	脉锡矿石
log energy	记录能量
log mill energy input	记录磨机能量输入
log molar concentration	克分子浓度对数
log normal form	对数标准式
log product size	记录产品程度
log washer	斜槽式洗矿机
logarithmic concentration diagram	

对数浓度图
logarithmic equilibrium oleate concentration 油酸盐对数平衡浓度
logarithmic function 对数函数
logarithmic relation 对数关系
logging-control system 记录-控制系统
logic circuit 逻辑电路
logic control philosophy 逻辑控制原理
logic hardware 逻辑硬件
logic manner 逻辑方式
logic test 逻辑试验
logical method 合乎逻辑的方法
logical place 理所当然[合理]的地方
logical step 逻辑步骤
logic-controlled sequential computer 逻辑控制时序计算机
log-log plot 双对数坐标图
log-polynomial functional form 记录多项函数式
Lokomo G610 crusher 罗克玛G610破碎机(芬兰,Lokomo公司生产)
Lomar PW 萘磺酸钠(品名),可用作湿润剂
London inter-bank 伦敦银行同业
London inter-bank market 伦敦银行同业市场
London inter-bank rate 伦敦银行同业利率
long chain alcohol 长链(乙)醇
long chain alkyl group 长链烷基
long chain amine 长链胺

long chain electrolyte 长链电解质
long chained alcohol 长链醇
long distance pipeline 长距离管线
long measure 长度单位制
long primary mill 长筒式第一段磨机
long radius bend 大半径弯管
long scavenger section 长扫选段
long slot type woven wire cloth 长孔型金属丝编织筛布
long sweep elbow 大曲率半径弯头
long term 长期
long term forecast 长期预测
long vane feeder 长叶片式给矿机
long-chain alcohol 长链醇
long-chain carboxylate 长链羧酸盐
long-chain ionic surfactant 长链离子表面活化剂
long-chained carboxylic acid 长链羧酸
long-chained neutral molecule 长链中性分子
longer chained homologue of collector 长链捕收剂同系物
longer-term trend 较长时间(变化)趋势
long-hole drill sample 深孔凿岩机[钻]试样
longitudinal belt conveyor 纵[轴]向胶带运输机
longitudinal tilt 纵向坡度
long-range force of attraction 长程吸引力

long-run funds 长周期资金
long-term contract 长期合同
long-term debt 长期借款
long-term direct capital loan 长期直接资本贷款
long-term financing 长期集资
long-term profits 长期[远]利益[利润]
long-term rehabilitation planning 长期复原计划
long-term sales contract 长期销售合同
long-term stability 长期稳定性
long-term support 长期支持
long-term trend 长期趋势(市场)
long-term usage 长期使用
loop drawing 回路图
loop test 回路试验
loose cluster 松散晶簇
loose material 松散[疏松]物料
loose-leaf operation and control manual 活页操作控制手册
loosened ore 疏松[松散]矿石
Lorol 混合脂肪醇(含15%辛醇,40%癸醇,30%月桂醇,15% C_{14} 醇,15% $C_{16~18}$ 醇)(品名,可用作捕收剂、起泡剂)
Lorolamine 辛胺,癸胺与月桂胺的混合物(品名,可用作捕收剂)
Los Angeles abrasion test 洛杉矶磨耗[损]试验
loss payable clause 损失支付条款

loudness 响度,音量
louver 天窗,百叶窗
low acid pregnant liquor 低酸富液
low cost labor 低价劳动力
low density mineral matter 低密度矿物物质[体]
low density soil 低密度土壤
low efficiency collector 低效率集[收]尘器
low energy primary radiation 低能一次辐射
low energy source 低能源
low energy surface 低能表面
low frequency electric field 低频率电场
low grade core 低品位地核
low grade deposit 低品位[贫]矿床
low grade dump ore 低品位废石堆矿石
low grade high tonnage mill 低品位高吨位磨矿车间
low grade product 低品位产品
low head type 低型,矮型
low homologue of xanthate 低级黄药类
low intensity magnetic separator 弱磁选机
low level assembly language 低级汇编语言
low level detector 低料位[液面]探测器
low level diaphragm discharge 低水平

隔板排料[矿]

low level diaphragm discharge grind mill 低水平隔[档]板排料[矿]磨机

low marketability 滞销

low molecular weight fatty acid 低分子量脂肪酸

low output rate 低生产率,低输出量

low power binocular microscope 低倍双目显微镜,低放大率双目显微镜

low power number 低功率数

low pressure air equipment 低压[空气]设备

low pressure air jet 低压喷气嘴,低压喷射器

low pressure plant 低压压缩空气厂

low probability screening 低概率筛分

low profile liner 低型衬板[里]

low ratio of reduction 低破碎比

low ratio of reduction ball mill 低破碎比球磨机

low shear pump 低剪力泵

low silver-containing base metal sulfide ore 低银的贱金属硫化矿

low specific gravity ore 低比重[密度]矿石

low speed gyratory crusher 低速旋回破碎机

low speed shaft bearing 低速轴轴承

low sulphur medium volatile recoverable washed coal 低硫中挥发分可回收的洗煤

low surfactant concentration 低表面活性剂浓度

low temperature flexibility 低温柔[韧]性

low turbidity region 低混浊度区

low valued metal 低价值金属

low velocity and self-cleaning duct system 低速自动清扫管道系统

low volatile bituminous coal 低挥发分烟煤

low voltage equipment 低压设备

low weir classifier 低堰式分级机

low-energy operation 低能量作业

low-energy source 低能源

lower constraint 约束域下限

lower curve 下部曲线

lower deck 下层筛网

lower flotation edge 下部浮选边界

lower holder 下部夹子,下托架

lower level 下层,下能级

lower limit 下限

lower mantle 下部锥衬套

lower relay 低限继电器

lower screen 下层筛

lower spring segment 下部弹簧垫块

low-flow water-lubricated gland 低流水润滑密封盖

low-grade concentrate 低品位精矿

low-grade material 低品位物料

low-grade ore 低品位矿石

low-grade porphyry 低品位斑岩矿

low-intensity magnetic concentration 弱磁选

low-level probe 低料位探测器(用于矿仓)
low-pressure steam 低压蒸汽
low-temperature 低温
low-velocity duct system layout 低速管道系统布置
low-volatile group 低挥发分组
lsolox 变性酒精
lube ball bearing 润滑球轴承
lubricant grade molybdenite 润滑剂级辉钼矿
lubrication industry 润滑工业
lubrication truck 注油[润滑油]车
Lubrol 表面活性剂品牌名
Lubrol MOA 脂肪醇与环氧乙烷的缩合物(可用作絮凝剂)
Luggin capillary 鲁金毛细管(测量有电流通过电极时电极电势变化的一种工具)
luminescence spectroscopy 发光光谱

luminous intensity 光强度
Lumorol 表面活性剂品名(烷基苯磺酸盐与烷基硫酸盐的混合物,可用作湿润剂、起泡剂)
lump mill 块磨机
lump milling 块磨,[粗]砾磨(用粗块矿石磨同类矿石)
lump sampler 块矿取样机
lump size 块度,矿块粒度
lump sum average 总平均数[值]
lumped model 集总模型
lump-free slurry 无块泥浆
lumpy hydrocarbon sheet 块形烃片
lumpy material 块状物料
lunch room 午餐室
Lyofix 一种季铵油(品名,季铵盐)
Lytron 886 乙烯基醋酸盐与顺丁烯二酐共聚物(品名,可用作絮凝剂)
Lytron 887 乙烯基醋酸盐与顺丁烯二酐共聚物(品名,可用作絮凝剂)

M

18 megohm water 18 兆欧水
20 mesh crush 20 目破碎
2-mercapto-4,4,6-trimethyl-1,3,4-dihydropyrimidine 2-巯基-4,4,6-三甲基-1,3,4-二氢嘧啶(捕收剂,可用于硫化矿浮选)

2-mercaptobenzothiazole 2-巯基苯骈噻唑(可用作捕收剂)
4-methyl-2-pentanol 4-甲基-2戊醇
M & S cost index for mining and milling 采选工程 M & S 成本指数
M & S index M & S 指数(用于采矿和

选矿)

M. S. machine　M. S. 型底吹式浮选机

M. S. Owen machine　M. S. 欧文浮选机(无封闭叶轮底吹式)

M. S. subaeration flotation cell　M. S. 型底吹式浮选槽[机]

M. S. subaeration machine　M. S. 型底吹式浮选机

M. T. E. radial jig　M. T. E. 型径向跳汰机(取消了水平自调装置与旋转耙),荷兰采矿及运输工程公司(MTE)制造

MAAPO　50%甲基戊醇与松油混合物(品名,起泡剂,可用于从海矿浮选金红石)

maceral　煤素质,煤(岩)显微组分,煤中矿物组分的细微组织

maceral group　煤素质组,煤(岩)显微组分组

maceral separation　煤素质分选,煤(岩)显微组分分离

machine arrangement　机械构造[配置]

machine assembly language　机器汇编语言

machine characteristic　机械特性

machine design　(浮选)机械设计

machine designer　(浮选)机械[器]设计师[者]

machine family　(浮选)机械种类

machine hydraulics　机械水力学

machine language　机器语言

machine performance　(浮选机)机械性能

machine performance parameter　机器性能参数

machine productivity　设备生产率[能力,量]

machine sizing　设备规格确定,确定设备规格[尺寸,大小]

machine-oriented analytical procedure　与设备有关的分析过程

machinery component　机器零件

machinery repair workshop　机修车间

machinery room　机房

machining　机械加工

mackinawite　马基诺矿,四方硫铁矿 $Fe_{n+1}S_n$

macrocrystalline　粗粒结晶

macro-kinetics　宏观动力学

made-to-order　定制的

Madlung constant　马德伦常数(晶体结构的一个重要特征参数,是一个数值常数)

Madlung sum　马德伦和数

Magchar(=magnetic activated carbon)　磁性碳,磁性活性碳

Magnafloc　絮凝剂品牌名(英国布拉福德 Allied Colloids 公司产品)

Magnafloc 139[155,156,E24]　阴离子型絮凝剂

Magnafloc 140[292,352,455]　阳离子型絮凝剂

Magnafloc 351　非离子型絮凝剂

Magnafloc LT20　非离子型聚丙烯酰胺

絮凝剂

Magnafloc LT22[LT24,LT225] 阳离子型聚丙烯酰胺絮凝剂

Magnafloc LT25[LT26,LT29] 阴离子型聚丙烯酰胺絮凝剂

magnesium chloride 氯化镁(可用作絮凝剂)$MgCl_2$

magnesium sulfate 硫酸镁(可用作絮凝剂)$MgSO_4$

magnet hydrodynamic separator 磁流体动力分选机

magnet hydrostatic separator 磁流体静力分选机

magnetic coil 磁性线圈

magnetic core memory 磁心存储器

magnetic dewaterer 磁力脱水槽

magnetic drum 鼓形磁选机

magnetic feeder 电磁振动给料[矿]机

magnetic flow transmitter 电磁流量变送[传感]器

magnetic head pulley 磁首轮,磁主滑轮(胶带运输机的)

magnetic nose piece 磁头

magnetic pulley 磁滑轮

magnetic state 磁性状态

magnetic stirrer-heater 磁力搅拌加热器

magnetic stream 磁流

magnetic tramp-iron protection 磁性夹杂(碎)铁保护(装置)

magnetic vibrating feeder 电磁振动给料[矿]机

magnetic-induction flowmeter 磁感应流量计

magnetics 磁(力)学,磁性元件[材料]

magnetism dewaterer 磁力脱水槽

magnitude of circulating load 循环负荷量

magnitude of error 误差值

mahogany acid 红酸,油溶性石油磺酸(可用作捕收剂)

mahogany soap 红酸钠皂,油溶性石油磺酸钠[皂](可用作捕收剂)

Mahogany soap F445 红酸钠皂(品名,油溶性石油磺酸钠[皂],可用作捕收剂)

mahogany sulfonate 油溶性石油磺酸钠(含30%~36%芳烃的石油馏分经硫酸磺化的产物,可用作重晶石的捕收剂)

main building of concentrator 选矿厂主厂房

main compressed air line 压缩空气总管线

main control unit 主控单元

main effect 主效应

main floor 主要工作间[场地,楼面,楼层]

main flow 主料流,主矿流

main frame 主机架[座]

main frame cap 主机座帽(圆锥破碎机的)

main frame computer 主体计算机

main frame dowel 主机座连接销

main frame flange 主机座法兰(圆锥破碎机的)
main frame liner 主机座衬里
main frame memory 主体存储器
main frame seat liner 主轴座衬板
main frame socket bushing 主机座球窝座套
main haulage level 主要运输水平
main line railway 铁路干线
main memory 主存储器
main mineral processing method 主要选矿方法
main pipe rack 干线管架
main processor 主处理机[单元]
main program 主程序
main routine 主程序
main shaft assembly 主轴装置[组件]
main shaft nut 主轴螺母
main shaft sleeve 主轴套筒
main stage program 主要阶段计划
main stream 主物料流
main substation 主变电站
maintainability 可保养[维护,维修]性
maintenance access 检修(出入)口
maintenance activity 维修活动
maintenance and repair worker 维修工人
maintenance condition 维修条件
maintenance cost 维修[维护,修理,保养]费
maintenance crew 维修班组[人员]
maintenance department 维修部门

maintenance flexibility 维修灵活[机动]性
maintenance foreman 维修工长
maintenance function 维修职能部门
maintenance labor 维修[修理]劳动力[工人]
maintenance labor cost 维修用工费用
maintenance office 修理办公室
maintenance overhead 维修管理费
maintenance parts 维修部件
maintenance personnel 维修人员
maintenance power 维修动力[功率]
maintenance priority list [优先]维修顺序表
maintenance program 检修程序[计划]
maintenance requirement 维修要求
maintenance stores 检[维]修备品
maintenance strategy 维修[保养,维护]对策[策略,方法]
maintenance supervisor 维修监督[检查]人员
maintenance supplies 维修材料[用品],补给
maize oil soap 玉米油皂(可用作捕收剂)
maize starch 玉米淀粉(可用作锰矿抑制剂、絮凝剂)
major addition 大量添加
major category 主要类别
major consideration 应考虑的主要事项[问题]

major continuous flow 主要连续流
major equipment 主要设备
major expense item 主要费用项目
major intermittent flow(bypass line) 主要间歇流(旁通线)
major plant equipment cost 选矿厂主要设备费用
major political change 重大政治变化
major project 重点工程[项目]
major repair job 大修任务
make of equipment 设备制[构]造
make-up ball diameter 补充[给]球形料(或钢球)的直径
make-up ball size 补充[给]球形料(或钢球)的尺寸[粒度,规格]
make-up balls feed size 补充[给]球形料(或钢球)的给入料尺寸[规格](给料粒度)
make-up rods feed size 补充[给]棒形料的给入料尺寸[规格](给料粒度)
make-up water 补加水
malachite 孔雀石,石绿 $CuCO_3 \cdot Cu(OH)_2$
malachite green 孔雀绿(染料),碱性孔雀绿(可用作捕收剂、抑制剂、活性剂)$C_{23}H_{25}N_2$
malfunction 失灵,发生故障
malic acid 苹果酸,羟基丁二酸 $(COOH)_2CH_2CHOH$
malononitrile 丙二腈 $CH_2(CN)_2$
Mammut crusher 玛莫特破碎机(德国)

man power requirement 劳动定员
manageable froth 易控泡沫
manageable size 易处理[控制]的粒度
management attention 管理注意力
management fee 管理费
management flexibility 管理灵活性
management method 管理法
management office 管理办公室
management organization 管理机构
manager 经理,管理人员
manager of process control 过程控制经理
manager of publication 出版社经理
manager operational audit 管理人员业务审计[检查]
managing bank 承办银行
mandatory prepayment 强制预支
mandatory repayment 强制还款
maneuvering 操纵
manganese 锰 Mn
manganese dioxide(manganese ore) 二氧化锰,软锰矿
manganese steel 锰钢
manganese sulfate 硫酸锰(矿浮选 MnO_2 的活化剂)
manipulable process variable 操作过程变量
manipulable variable 操作[可控]变量
manipulated variable 操作控制变量
manipulation 操作[纵],控制,处理
manlift 电梯
manner of investigation 调查[研究]

方式
manning 配备人员,人员配备
manning chart 人员[定员]表
manning schedule 定员计划表
manning table 定员表
mannogalactan 甘露半乳聚糖(可用作调整剂、絮凝剂、分散剂、抑制剂)
manoeuvring 调遣,部署,操纵
manpower quality 人力素质
manpower requirement 劳动力[人力]需要量
manpower supply 人力来源
mantel hydraulic support pressure 锥体外壳液压支点[撑]压力
mantle cone 破碎机可动圆锥
mantle diameter 动锥直径(旋回破碎机的)
mantle head 可动破碎圆锥颈
mantle life 可动锥体寿命
mantle liner 动锥衬板
mantle position 锥体位置
mantle skirt 定锥罩边缘,锥体外壳边缘
mantle support pressure (圆锥破碎机)锥体外壳[定锥罩]支点[撑]压力
mantle-position detector 锥位探测器(破碎机的)
manual calculation method 手工计算法
manual computation 笔算,手算,用手摇计算机计算
manual control 手控

manual for evaluation of industrial projects 工业项目评价手册
manual plant 手工操作车间
manual pumping unit 手动抽水[油]机
manual solution 人工求解
manual take-up 人工拉紧装置
manually controlled needle valve 手控针型阀
manufacture (机械)制造,生产,加工,制造业,(pl.)制成品,产品,工厂
manufacture of fertilizer 肥料制造
manufactured item 工业产品
manufactured sand 人造砂,人工砂
manufacturer's brochure 制造厂家宣传册
manufacturer's bulletin 制品报告[通报]
manufacturer's equipment 厂家设备
manufacturer's literature 制造厂家的资料[文献]
manufacturer's performance specification 制造厂家的技术性能说明[性能标准]
manufacturer's recommendation 制造厂家推荐[建议]值
manufacturer's catalogue 制造厂商品目录
manufacturer's data 制造厂家数据
manufacturer's family 制造厂的产品系列
manufacturing cost 制造成本
manufacturing plant 制造厂家

map of mine area 矿区地图
Maracell 表面活性剂品牌名
Maracell E 木素磺酸钠(部分脱磺酸的木素磺酸钠,可用作分散剂(用于矿泥))
Marasperse 表面活性剂品牌名
Marasperse C 木素磺酸钙(可用作抑制剂)
Marasperse CB 木素磺酸钠(可用作抑制剂)
Marasperse N 木素磺酸钠(含磺酸钠基团14.3%,可用作抑制剂)
Marathon extract 木素磺酸盐(品牌名,可用作捕收剂、抑制剂、调整剂)
Marcy ball mill 马西型球磨机(美国)
Marcy density gauge 马西型浓度计(美国)
Marcy density meter 马西型浓度计(美国)
Marcy grate discharge ball mill 马西型格子排矿球磨机(美国)
Marcy grate discharge pebble mill 马西型格子排矿砾磨机(美国)
Marcy regrind mill 马西型再磨机(美国)
Marcy rod mill 马西型棒磨机(美国)
marginal equipment 边缘设备,勉强及格设备
marginal facies 边缘相
marginal particle 边缘[界]粒度
marginal specification concentrate 边缘技术规格的精矿(即技术条件勉强合格的精矿)
marine environment 海洋环境
marine insurance 海运保险
marine plywood 适于海水用的胶合板
marine type sediment 海相沉积(物)
marked test ball 有标记[记号]试验钢球
market condition 市场情况
market demand 市场需求
market economy country 市场经济知识领域
market information 市场情报[信息]
market input 市场投入(量)
market planning manager 市场策划经理
market potential 市场潜力
market price 市场价(格)
market requirement 市场要求[需求]
market specification 市场规范
market study 市场[销路]研究
market supply 市场供应(量)
marketability 市场性,适销性
marketable (适合)市场(销售)的,销路好的
marketable coarse product 畅销的粗粒产品
marketable collateral 可销售的抵押品
marketable concentrate 商品精矿
marketable grade concentrate 商品品位精矿,畅销品位精矿,适合市场销售品位的精矿
marketable security 可销售的证券

marketable value 市场价值
marketing 销售,销路
marketing and capital investment requirement 营销与基建投资需求
marketing channel 销路
marketing cost 销售[推销]费用
marketing information 商业[市场,销路]信息[情报]行情
marketing of product 产品销售
marketing-sales personnel 推销出售人员
market-place 市场
Marshall & Swift cost index (formerly Marshall & Stevens cost index) 马歇尔和斯维夫特[M & S]成本指数(原名马歇尔和史蒂文斯成本指数,是一种以时标估算工程项目的投资成本的工具)
martensite 马氏体(黑色金属材料的一种组织名称,最先由德国冶金学家Adolf Martens于19世纪90年代在一种硬矿物中发现)
martensitic 马氏体的
martensitic structure 马氏体结构
martensitic white iron 马氏体白口铸铁
Martin's diameter (Martin diameter) 马丁直径(沿一定方向将颗粒投影面积等分的直径)
Marxwell cell 马克斯韦尔浮选槽
Marxwell flotation cell 马克斯韦尔型浮选机[槽]

Marxwell flotation machine 马克斯韦尔浮选机
mass 块(状物),团,堆,片,群,体,质量,大量,成批
mass absorption coefficient 质量吸收系数
mass analyzer 质谱仪
mass balance 平衡重量,配重,质量[物料]平衡
mass balance calculation 质量平衡计算
mass balance closure 质量平衡闭合
mass balance expression 质量平衡表达式
mass balance smoothing 质量平衡调[修]匀
mass classification 按质量分级
mass data 大量数据
mass density function 质量密度函数
mass diagram 土(方累)积图,质量曲线
mass distribution curve 质量分布曲线
mass feed rate of ore through mill 通过磨机的总给矿量[率]
mass flow 质量流,质量流量
mass flow bin 大质量[整体]流动仓
mass flow calculator 质量流量计算器
mass flow cone 大质量[整体]流动锥
mass flow design 质量流(量)设计
mass flow hopper 大质量[整体]流动漏斗
mass flow insert 整体[大质量]流动

(小仓)嵌入板
mass flow meter 质量流计,质量流量计
mass flowmeter 质量流量计
mass flow metering system 质量[物料]流量计量系统
mass flow rate 质量流(量)比[率]
mass flowrate 质量流速[率]
mass-flow silo 整体[大质量]流动筒仓
mass fraction 质量分量[数]
mass fraction of material 物料质量百分数
mass fraction of particles 颗粒的质量百分数
mass hardness 全部过硬
mass hold-up of material 物料充填量
mass memory 大容量存储器
mass migration 大质量[规模]迁移
mass movement 大质量[块体]移动
mass of calcite (mg) 方解石质量(毫克)
mass production 大量[批量]生产
mass rate 质量流[比]率,质量流量
mass rotational flow rate 整体[总]旋转流量
mass run 大量生产
mass separation (按)质量分级
mass storage 大容量存储器
mass transfer 质量交换[传递],物质传递
mass transport 质量[大量]运输

mass weight 质量加权
massive copper-lead-zinc ore 块状铜铅锌矿石
massive orebody 大型[块状]矿体
massive sampler 块矿取样机
massive solar distill 大型太阳能蒸馏器
massive sulfide deposit 块状硫化物矿床
massive sulfide ore 块状硫化矿石
master composite sample 主组合试样
master electrician 电力主任,电力主任技师,电力工长
master totalizer 主累加器
match line 对称线
material 材料,原料,物料,物质[资],(技术)资料
material balance equation 物料平衡方程式
material breaking and reduction 物料破碎
material build-up 物料堆积(如传送胶带上的)
material certificate 材料合格证
material characteristics 材料特性
material consumption 材料消耗
material density 物料密度
material distribution system 物料分配系统
material fall distance 物料落下距离,物料落差
material flow control 物料流量控制,

进料控制
material flow-transport problem 物流运输[转运]问题
material fracture characteristic 物料破裂[断裂]特性
material friability 物料脆性
material gradation 物料粒度分级
material handling 材料处理[搬运]
material handling component 物料装运[卸]组成部分(矿仓,转运站的)
material handling crane 运料吊车
material handling operation 物料搬运[输送]作业
material handling plant 物料搬运设备
material handling system 物料搬运系统
material hangup 物料悬挂,挂料
material moisture content 物料水分含量
material of construction 建筑材料
material of liner construction 衬板构造[制作]材料
material overload angle 物料超载角
material parameter 物料参数
material process system 物料加工系统
material process variance 物料加工[生产]过程变化
material repose angle 物料安息[休止]角
material residence time 物料停留时间
material residence time distribution 物料停留时间分布

material service section 材料服务[供应,管理]部门
material specification 材料规格
material surcharge angle 物料超载角(胶带运输机的)
material transport 物料运输
material transport path 物料流通路线
material travel 物料移动
material travel rate 物料[矿石]移动速度
material-in-process inventory 工艺过程中的材料库存[存储]
mathematic analysis 数学分析
mathematic expression 数学表达式
mathematic model 数学模型
mathematic model of process system 过程[工艺]系统数学模型
mathematic modeling 数学模型化
mathematic processing 数学处理
mathematic programming 数学规划
mathematic relation 数学关系
mathematic structure 数学结构
mathematic treatment 数学处理
mathematical constant 数学常数
mathematical crystal 数学晶体
mathematical crystal model 数学晶体模型
mathematical description 数学描述
mathematical expression 数学表达式
mathematical formula 数学公式
mathematical framework 数学结构
mathematical manipulation 数学处理

mathematical model 数学模型
mathematical modeling of experiment 实验的数学模型
mathematical procedure 数学方法
mathematical process model 数学过程模型
mathematical programming 数学规划
mathematical relationship 数学关系
mathematical representation 数学表达式
mathematical simulation 数学模拟
mathematical strategy 数学策略
mathematical structure 数学结构
mathematical surface 数学面
mathematical symbolism 数学符号体系
Matheron model (spherical model) 马瑟龙[马特隆]模型(又称球状模型,法国学者 G. Matheron 创立)
Matheron's approach 马瑟龙[马特隆]近似法
Matheson flowmeter Matheson 流量计(美国 Matheson 公司产品)
matrix (岩,矿)基质,填质,脉石,模型,矩阵
matrix constant 矩阵常数
matrix effect 脉石[基体,基质]效应
matrix magnet 钢毛介质磁选机
matrix model 矩阵模型
matrix multiplication 矩阵乘法
matrix notation 矩阵记号,矩阵符号(表示法)
matrix of enhancement factor 增强因子矩阵
matte 锍,铜锍(Cu_2S 和 FeS 的混合物)
matte separation process 冰铜分离过程[工艺]
matte smelting 冰铜[锍]冶炼,造锍熔炼
mature fish 成年[熟]鱼
maturity date 归还日期,到期日
maxiouter dimension 最大外形尺寸
maximal flotation 最佳浮选
maximum abrasion resistance 最大抗磨力
maximum angle of conveying (胶带机)最大运输角
maximum constraint 最大限制
maximum conveyor grade 运输机最大坡[角]度
maximum efficiency 最大效率
maximum flotation 最佳浮选
maximum grain size 最大粒度
maximum grinding efficiency 最大磨矿效率
maximum information 最大信息量
maximum lump size 最大块尺寸,最大块粒度
maximum material height 物料最大高度
maximum participation 最大分享额

maximum particle size 最大(颗粒)粒度
maximum plant production 最大[高]的[工厂]车间产量
maximum rated power 最大额定功率
maximum recommended horse power 推荐的最大马力
maximum size media 最大尺寸介质
maximum spacing 最大间隔
maximum throat opening 最大给矿[料]孔[口]
maximum throughput 最大处理量
maximum tight side tension 最大的紧边张力
maximum value of protection 防护最大值
maybe dcision 两可决策
MBT(=Mercaptobenzothiazole) 巯基苯骈噻唑
mean cost 平均费用
mean diameter 平均直径
mean particle size 平均粒度
mean square 均方
mean square method 均方法
mean thickness 平均厚度
meaningful basic data 有意义的基本数据
meaningful data 有意义的数据
meaningful error 有意义的误差
means of adjustment 调整方法[手段]
measure of classification efficiency 分级效率的计量法
measure of length 长度单位[测量]
measure of pressure 压力单位[测量]
measure of reliability 可靠性度量[尺度]
measure of weight 重量单位[测量]
measured assay 被测试样
measured concentration profile 实测浓度分布[曲线]图
measured data 被测数据
measured data point 被测数据点
measured ore reserve 探明[测定]矿石储量
measured power 实测功率(除磨矿功率外,包括磨机本身及球或棒带动和驱动摩擦等所消耗的功率)
measured reserve 探明储量
measured variable 测量变量
measurement circuit 测量电路
measurement procedure 测量方法
measurement vector 测量矢量
measurement zone 测量带
measuring box 计量槽,量斗,检测试剂盒
measuring channel 测量道
measuring circuitry 测量电路图
measuring cylinder 量筒
measuring device 测量装置
measuring element 测定元件
measuring system 测量系统
measuring tape 测[卷,皮]尺
mechanical agitation 机械搅拌

mechanical agitator 机械搅拌器
mechanical air induction type flotation cell 自吸式机械搅拌浮选槽[机]
mechanical air induction type flotation mechanism 自吸(空气)机械搅拌式浮选机构[槽]
mechanical and electrical shop 机电修理间
mechanical brake caliper engagement 机械制动卡尺测量啮合
mechanical chain drive 机械链传动(装置)
mechanical classifier 机械分级机
mechanical compactor 机械压实机
mechanical damage 机械损坏
mechanical design 机械设计
mechanical designer 机械设计师[者]
mechanical discharge 机械排矿
mechanical distributor 机械分配器
mechanical drive component 机械驱动部[元,组]件
mechanical effect 力学[机械]效应
mechanical engineer 机械工程师
mechanical factor 机械因素
mechanical failure 机械故障
mechanical feeding device 机械给料[矿]装置
mechanical froth skimmer 机械泡沫刮板
mechanical gravity separator 机械重力分选机
mechanical handling and casting equipment 机械搬运铸造设备
mechanical interlocking 机械连锁
mechanical knocker 机械敲击[振动]器
mechanical loss 机械损失
mechanical mixer 机械搅拌器[机]
mechanical problem 机械问题
mechanical rake classifier 机械耙式分级机
mechanical self-aspiration type flotation cell 自吸式机械搅拌浮选槽
mechanical services 机械维修[保养]
mechanical shock 机械冲击
mechanical spiral classifier 机械螺旋分级机
mechanical stirring 机械搅拌
mechanical stirring device 机械搅拌装置
mechanical system drawing 机械系统图纸
mechanical trade 机械贸易
mechanical utilization 机械利用率
mechanical vibration 机械振动
mechanically agitated beaker 带机械搅拌的烧杯
mechanically agitated cell 机械搅拌浮选槽
mechanically aided scrubber 机械辅助式洗涤器
mechanically cleaned coal 机械(选出的)精煤,机选精煤
mechanically operated reagent splitter 机械驱动的药剂分配器

mechanical-pneumatic cell 充气-机械搅拌式浮选槽[机]

mechanical-pneumatic flotation machine 充气机械搅拌浮选机

mechanics of deactivity 去活动[活化]力学

mechanism 机制[理],手法,技巧

mechanism design 机械[机械结构]设计

mechanism draw power 机构牵引功率

mechanism for success 成功机制

mechanism geometry 机构几何形状

mechanism hydraulics characteristic 机构水力学特性(空气流量,矿浆循环量,矿浆上升速度和功率强度)

mechanism hydrodynamic relationship 机构的流体动力学关系

mechanism of apatite collection 磷灰石捕收机理

mechanism of attachment 附着机理

mechanism of charge development 电荷演变机理

mechanism of charge generation 电荷产生机理

mechanism of deactivation 去活动[活化]机理

mechanism of fabric filtration 织物过滤机理

mechanism of mineral collection 矿物捕收机理

mechanism operating condition 机构工作条件(浮选机的)

mechanism operation 机构运转

mechanism-cell combination 机构与槽子组合

mechanism-cell interaction parameter 机构与槽子相互作用参数

mechanistic approach 机械型工作设计法,机械化途径

mechanistic method 机械(学)方法

mechanistic model 机械(学)模型

mechanized beneficiation plant 机械化选矿厂,机选厂

mechanized mining 机械化开采

media carry-over 带出介质

media competency test 介质能力试验

media configuration 介质组态[排列,组合]

media flow path 介质流动路线[道]

media number 滤布号

media rotational flow 介质旋[环]流(流量)

media size 介质规格

media sized material 介质类[级,粒度]物料

media sized particle 定为介质粒度的颗粒,属介质粒度的颗粒(指自磨机中的砾石)

media voidage percent 介质孔隙百分数[率]

Medialan 表面活性剂品牌名(瑞士Claricant公司产品)

Medialan A 十八烯酰-N-甲氨基乙酸钠(可用作捕收剂) $C_{17}H_{23}CON(CH_3)CH_2COONa$

Medialan KA 椰子油酸基-N-甲氨基乙酸钠(可用作捕收剂)

Medialan LL-99 月桂酸基-N-甲氨基乙酸钠(可用作湿润剂)
$C_{11}H_{23}CON(CH_3)CH_2COONa$

Medialan LP-41 油酸基-N-甲氨基乙酸钠(可用作湿润剂)
$C_{17}H_{33}CON(CH_3)CH_2COONa$

Medialan LT-52 硬脂酸-N-甲氨基乙酸钠(可用作湿润剂)
$C_{17}H_{35}CON(CH_3)CH_2COONa$

median value 中间值,中值

medium crushing cavity 介质[中间物料]破碎腔

medium drainage 介质脱除,脱介

medium duty cutter 中型切[截]取器

medium duty service 中等负荷运转[行]

medium grade ore 中等品位矿石

medium hard ore 中(等)硬度矿石

medium recovery 介质回收

medium shape factor 介质形状因素

medium sized 中等大小的,中型[号]的

medium term borrowing 中期借款

medium-range sizing separation 中等粒级筛分分离

mega 兆,百万

mega electron volt 兆电子伏特

mega ohm 兆欧,百万欧姆

megohm 兆[百万]欧姆

megawatt 兆瓦(特)

Mekhanobr machine 米哈诺布尔浮选机(苏联)

melaniline 蜜苯胺,均二苯胍
$(C_6H_5NH)_2C:NH$

Melavin B 仲烷基硫酸盐(品名,波兰,可用作捕收剂)

melnikovite 胶黄铁矿

melting furnace 熔化[炼]炉,热熔炉

melting process 熔化过程

membrane process 薄膜处理方法[过程]

memory space 存储空间

memory utilization 存储器利用

mepasin 加氢合成煤油

mercaptan 硫醇 RSH

mercapto 巯基,氢硫基

mercaptobenzothiazole 巯基苯并噻唑(可用作捕收剂)

mercaptobenzothiazole (i. e. Cyanamid R404, R425) 巯基苯并噻唑(即氰胺404号,425号浮选剂)

merchant copper 商品铜

merchant house 商号[户],商品房

mercury fume 汞烟

mercury hygiene 汞保健[卫生]

mercury removal 除[脱]汞

mercury vapor 汞蒸汽

mercury-gold amalgam particle 汞-金合金[金汞齐]颗粒

merger 合并,并购,兼并

Merpisap AB 烷基苯磺酸盐(品名,可用作湿润剂)

Merpol 表面活性剂品牌名

Merpol C 烷基硫酸盐(可用作选矿药剂(湿润剂))

Merrick feed weight ore feeder 梅里克型给料自动称量给矿机(美国Merrick公司生产)

Merrill-Crowe deaeration tower 麦瑞尔·克洛韦脱[除]氧气塔(金银回收)

Merrill-Crowe gold and silver system 麦瑞尔·克洛韦金银回收法[系统](锌粉置换法)

Merrill-Crowe precipitation unit 麦瑞尔·克洛韦(锌粉)置换装置

Merrill-Crowe process 麦瑞尔·克洛韦(金银回收)法[工艺](用金属锌或铝置换沉淀法从氰化溶液中回收金银)

Merrill-Crowe system 麦瑞尔·克洛韦(金银回收)法[系统](氰化溶液经真空脱氧及锌粉置换沉淀法回收金银)

Merrill-Crowe zinc dust precipitation method 麦瑞尔·克洛韦(金银回收)锌粉置换沉淀法(在含金银的贵液中加入锌粉,通过锌与金银的置换反应使金银沉淀的方法)

Merrill-Crowe zinc dust precipitation plant 麦瑞尔·克洛韦(金银回收)锌粉置换厂

Merrill-Crowe zinc dust precipitation circuit 麦瑞尔·克洛韦(金银回收)锌粉置换沉淀回路

Mersol 30 选矿药剂(烷基磺酰氯,含量为30%)$R-SO_2Cl, R=C_{14}\sim C_{18}$烷基

Mersol D 选矿药剂(烷基磺酰氯,含量为82%)$R-SO_2Cl, R=C_{14}\sim C_{18}$烷基

Mersol-H 选矿药剂(烷基磺酰氯,含量为45%)$R-SO_2Cl, R=C_{14}\sim C_{18}$烷基

Mersolate 阴离子型鲸蜡基磺酸盐(品名,可用作捕收剂、起泡剂,如:浮选铁矿的捕收剂)$C_{16}H_{33}SO_3Na$

Mersolate 30 选矿药剂(烷基磺酸钠,由Mersol 30水解得来)$R-SO_3Na, R=C_{14}\sim C_{18}$烷基

Mersolate D 选矿药剂(烷基磺酸钠,由Mersol D水解得来)$R-SO_3Na, R=C_{14}\sim C_{18}$烷基

Mersolate-H 选矿药剂(烷基磺酸钠,由Mersol-H水解得来)$R-SO_3Na, R=C_{14}\sim C_{18}$烷基

Mertax 选矿药剂(巯基苯骈噻唑,可用作捕收剂)$C_6H_3CHNS\cdot SH$

mesh data 网格数据

mesoiden 中生代山脉

Mesozoic clastic rock 中生代碎屑岩

metabolic cycle 新陈代谢循环

metal accumulation 金属累积[加,计]

metal assay 金属含量(分析),金属品位(分析),金属化验

metal atom 金属原子

metal bisulfite molecule 金属亚硫酸氢盐分子

metal carboxylate 羧酸金属盐

metal chelates 金属螯合物

metal collector 金属捕集器

metal collector compound 金属捕收剂化合物	metal plating 金属电镀
metal collector salt 金属捕收剂盐	metal price 金属价格
metal complex 金属络合物	metal production 金属生产[产量]
metal consumption 金属消耗	metal reclamation 金属回收
metal corrosion 金属腐蚀	metal recovery 金属回收率
metal cyanide complex 金属氰化络合物	metal recovery installation 金属回收装置
metal deposit 金属沉积物,电积金属	metal salt 金属盐
metal detector 金属探测器	metal sulfide 金属硫化矿
metal dissolution 金属溶解	metal sulfonate 磺酸金属盐
metal distribution rate 金属分配率	metal technology 金属工艺学
metal dithio carbamate 二硫代氨基钾酸金属盐	metal wear 金属磨耗[损]
metal extraction 金属萃取	metal xanthate 金属黄原酸盐
metal grade 金属品位	metal xanthate solubility data 金属黄原酸盐溶解度数据
metal housing 金属外壳	metal-accumulation 金属聚集[积聚](作用)
metal impeller 金属叶轮	metal-collector complex 金属捕收剂络合物
metal industry 金属工业	metal-collector compound 金属捕收剂化合物
metal inventory 金属库存(量)	metal-cyano complex 金属-氰络合物
metal ion 金属离子	metallic aluminum 金属铝
metal ion activator 金属离子活化剂	metallic copper 金属铜
metal lined centrifugal pump 金属衬里离心泵	metallic ion concentration 金属离子浓度
metal loading capacity 金属荷载能力	metallic ions 金属离子
metal of interest 目的金属	metallic iron 金属铁
metal oleate 油酸金属盐	metallic iron electrode 金属铁电极
metal oxide 金属氧化矿	metallic manganese 金属锰
metal oxy-sulfur species 金属的氧-硫产物	metallic ore 金属矿石
metal particle 金属颗粒	metallic oxide mineral 金属氧化矿物
metal plate work piece 金属板工件	

metallic oxide ore 金属氧化矿石
metallic state 金属状态
metallic zinc 金属锌
metalliferous ore 金属矿石
metallurgical accounting 冶金[选矿,工艺指标]计算
metallurgical and grinding power efficiency 选磨功率效应[率]
metallurgical application 冶金[工艺]应用
metallurgical axiom 选矿原则
metallurgical balance 金属[冶金]平衡
metallurgical calculation 冶金计算
metallurgical circuit downstream 选矿流程下游
metallurgical composite 冶金合成物成分
metallurgical composite sample 冶金[选矿]综[组]合试样
metallurgical compromise 选矿工艺折中方案[办法],选矿工艺综合平衡[考虑]
metallurgical consideration 选矿问题
metallurgical consultant 选矿冶金顾问
metallurgical control 选矿控制
metallurgical control testing 选矿检验试验
metallurgical data 选矿[冶金]数据
metallurgical data base 冶金[选矿]数据库
metallurgical detail 选矿[选冶]细节
metallurgical division 冶金室[科]
metallurgical efficiency 选矿[冶金]效率
metallurgical efficient flowsheet 有效的选矿流程
metallurgical engineer 冶金[选矿]工程师
metallurgical evaluation 工艺[冶金]评价
metallurgical factor 冶金[工艺,选矿]因素
metallurgical flowsheet 选矿流程
metallurgical flowsheet development 选矿流程编制
metallurgical grade 冶金品位[级]
metallurgical grade concentrate 冶金级[品位]精矿
metallurgical grade fluorspar 冶炼级萤石
metallurgical grade performance 选矿品位指标
metallurgical index 选矿指标
metallurgical industry 冶金[选矿]工业
metallurgical laboratory 冶金[选矿]实验室
metallurgical measurement 选别指标测量
metallurgical method 选矿方法
metallurgical operation 选矿[冶金]作业

metallurgical ore 入选[冶金]矿石
metallurgical ore dressing research 选矿研究
metallurgical performance 工艺性能
metallurgical performance similarity 工艺性能的相似性
metallurgical personnel 冶金[选矿]专业人员
metallurgical plant 选矿厂[车间]
metallurgical plant flowsheet 选矿[冶]厂流程图
metallurgical point of view 冶金[选矿]观点
metallurgical practice 冶金[选矿]实践
metallurgical process 选矿方法[工艺]
metallurgical processing plant 选冶加工厂
metallurgical profession 冶金[选矿]专业
metallurgical program 冶金规划
metallurgical recovery performance 选矿回收率指标
metallurgical research 冶金[选矿]研究
metallurgical research team 选矿研究小组
metallurgical response 冶金[选矿]效应[反应]
metallurgical result 选冶试验结果
metallurgical staff 冶金[选矿]工作人员
metallurgical study 选矿[冶金]研究
metallurgical summary 选别指标总结
metallurgical task 冶金[选矿]任务[工作]
metallurgical technology 金属工艺学
metallurgical test 冶金[选矿]试验
metallurgical test data 选矿[冶金]试验数据
metallurgical test function 选矿[冶金]试验职能部门
metallurgical test work 选矿[冶金]试验工作
metallurgical testing 冶金[选矿]试验
metallurgical testing procedure 选矿[冶金]试验过程[方法]
metallurgist 选矿[冶金]工程师
metal-specific exchanger 特效金属离子交换剂
metal-sulphur bond 金属-硫键
metal-thiolate 金属硫醇盐
Metalyn 塔尔油酸甲酯(品名)
Metalyn sulfonate 塔尔油酸甲酯磺酸盐(品名,可用作捕收剂)
metamorphic sedimentary 变质沉积
metanil yellow (酸性)间胺黄(可用作捕收剂) $C_6H_5NHC_6H_4N:NC_6H_4SO_3Na$
metaphosphate 偏磷酸盐(可用作捕收剂、抑制剂、分散剂) MPO_3
metathetical displacement 复分解置换
metathetical reaction 复分解反应
Metaux flotation cell (实验室用)机械浮选槽

meteorological condition 气象条件

meter of slurry 矿浆扬程

meter reading 仪表读数

metering and dosing pump 计量配量泵

metering equipment 计量[测量,测定,调节]设备

metering pump 计量泵

metering system 测量系统

metering tube 计量管

methanol(= methyl alcohol) 甲醇 CH_3OH

method of application 使用方法

method of feeding 给矿方法

method of grinding 磨矿方法

method of heating 加热方法

method of successive approximation 逐步逼近法

methodology 方法,方法学,方法论

methyl 甲基 CH_3-

methyl alcohol 甲醇 CH_3OH

methyl amyl acetate 乙酸甲基戊酯,乙酸甲代戊基酯,甲基戊基乙酸脂(可用作起泡剂) $CH_3COO \cdot CH \cdot (CH_3) \cdot CH_2 \cdot CH(CH_3)_2$

methyl amyl alcohol 甲基戊醇(即甲基异丁基卡必醇,同 MIBC,可用作起泡剂、分散剂、调整剂) $(CH_3)_2CHCH_2CH(OH)CH_3$

methylated quartz 甲基化石英

methylated silica 甲基化硅石

methylation 甲基化,甲基化作用(加甲基作用和加甲醇使酒精变性的作用)

methylcellulose 甲基纤维素(可用作絮凝剂、抑制剂、湿润剂、捕收剂、调整剂)

methyl cotton blue(= methyl blue) 甲基蓝,棉蓝(生物染料,可用作脉石矿物浮选的抑制剂、捕收剂、活化剂)

methyl cyclohexanol 甲基环己醇,六氢化甲酚 $CH_3C_6H_{10}OH$

methyl dithiocarbamate 甲基二硫代氨基甲酸酯(可用作捕收剂)

methylene 甲叉,甲撑,亚甲(基) CH_2-

methylene blue(MB) (碱性)亚甲蓝(简称 MB,可用作染料、捕收剂及抑制剂,浮选铜钼矿时作为辉钼矿的抑制剂) $C_{16}H_{18}ClN_3S \cdot 3H_2O$

methylene blue active substance 亚甲蓝活性物质

methylene iodide 二碘甲烷 $CH_2 : I_2$, CH_2I_2

methyl eosin 甲基曙红,浮选铜钼矿时作为辉钼矿的抑制剂

methyl group 甲基

methyl isobutyl carbinol 甲基戊醇(即甲基异丁基卡必醇,简称 MIBC,可用作起泡剂、分散剂、调整剂) $(CH_3)_2CHCH_2CH(OH)CH_3$

methyl violet 甲基紫(可用作捕收剂、抑制剂、活化剂) $C_{24}H_{29}ON_3$

metric measuring system 公[米]制测量系统

metric system 公制,米制

metric system of measurement 公制度量法
metric thread 公制螺纹
metric tons of copper 铜公吨数
metric work index 公吨功指数(由邦德功(可磨碎性)指数换算出来的,邦德功指数以短吨计)
Metso 抑制剂(偏硅酸钠,可用于脉石矿物浮选)$Na_2SiO_3 \cdot 5H_2O$
Metso granular 粒状硅酸钠盐
MG-98A MG-98A 阳离子捕收剂(是8和10碳链烷基氧化丙胺混合物)
MGP 750 MGP 750 药剂(甲氧基聚乙二醇,聚二醇单甲醚,可用作调整剂及矿泥的分散剂)$CH_3(OCH_2CH_2)_nOH$
Miazol 咪唑(品名),2,3-二氮杂茂(可用作捕收剂、絮凝剂)$C_3H_4N_2$
MIBC(methyl isobutyl carbinol) 甲基戊醇,甲基异丁基卡必醇(可用作起泡剂、分散剂、调整剂) $(CH_3)_2CHCH_2CH(OH)CH_3$
mica-sericite 云母绢云母
Micate 塔尔油皂(品名,可用作捕收剂)
micellization 胶粒[束]化
Micol 捕收剂(十六烷基三甲基溴化铵)$C_{16}H_{33}N^+(CH_3)_3Br^-$
micrinite 碎片体(煤岩的)
micro pilot plant 微型试验厂
micro-autoradiographic study 显微自动射线照相研究
microbial life form 微生物生命形态

microbial treatment 微生物处理
microbiological treatment 微生物处理
microbubble activation 微泡沫活化作用
microbus 小面包车
microcell potential 微型电池电位
microcomputer 微型计算机
microcomputer system 微型计算机系统
microcrystalline 微粒结晶的,微晶的,微晶质
microcrystalline molybdenite 微晶辉钼矿
microcrystalline nature 微结晶性质
microelectrophoretic technique 微电泳技术
microflotation technique 微量[型]浮选法[技术]
micrograph 显微照片
microline 微斜长石
micromeritics automatic surface area analyzer 微晶[尘]学自动表面积分析器
micromesh 微网目
micromesh sieve 微网目筛
micrometer 微米
micrometer eyepiece 测微计[器]目镜
micrometer spinder 测微计[器]测量杆
micromole 微克分子量
micron size 微米大小,微米粒度(即以微米计的粒度)

micronmeter 微米
micropore 微孔隙
microprobe 微探针
microprobe technique 微探针技术
microprocess 微观过程
microprocesser 微处理器
micro-processor 微处理机
microprocessor system 微处理机系统
micro-processor system 微处理机系统
microscopic count 显微镜计数
microscopic examination 显微镜观测[检查]
microscopic model 显微模型
microscopic observation 显微镜观察
microscopist 显微镜工作者[专家]
microscreening module 微筛组件
mid size 中间粒度
middle curve 中间曲线
middle position 中间位置
Middle Tertiary 中第三纪
middling association 中矿共[连]生体
middling circuit 中矿回路
middling flotation 中矿浮选
middling flotation circuit 中矿浮选回路
middling product 中矿(浮选的)
middling recirculation 中矿(再)循环
middling regrinding ball mill 中矿再磨球磨机
middling service 中矿作业
middling thickener 中矿浓缩机
middling thickener underflow percent solids 中矿浓缩机底流固体百分数
middlings 中矿
middlings extinction type test 消灭中矿(形式)试验
middlings reconcentration 中矿再选[复选]
middlings regrinding circuit 中矿再磨流程[回路]
mids 中矿
midstream 中流
mil 密耳(用作测量金属线直径的单位,等于0.001英寸);密位,角密耳,角密度[角位](等于1/6400周角),千分角;毫升,立方厘米
mild activator 弱活化剂
mild carbon steel 低碳钢(软钢)
mild depressant 弱抑制剂
mild iron 软铁
mild leach 常温常压浸出
mild steel 低碳[软]钢
mild steel backing plate 低碳钢垫板
mild steel ball 软钢钢球
mild steel mill 软钢磨机
mild steel rod 软钢钢棒
mild steel test mill 软钢试验磨机
mill bearing loss 磨机轴承损失
mill building 选矿厂房
mill configuration 磨机构造(形式,形状)
mill construction 厂房(轧机)结构,耐火构造
mill content 磨机内物料

mill control office 选矿厂控制室
mill density 磨矿浓度
mill design 选矿厂设计
mill design engineering 选矿厂设计工程
mill designer 选矿设计人员
mill diameter effect 磨机直径效应[影响]
mill diameter inside liners 磨机衬板内侧直径
mill diameter inside shell 磨机筒体内侧直径
mill downtime 磨机停车时间
mill equipment 选矿厂设备
mill equipment list 选矿厂设备表
mill expansion 选矿厂扩建
mill feed box 磨机给料箱
mill head assay 选矿厂原矿分析
mill head grade 选矿厂原矿品位
mill head water 选矿厂给矿(加)水,磨机给矿(加)水
mill hydrostatic lubrication system 磨机流体静力[压]润滑系统
mill layout 选矿厂配置
mill liner shape factor 磨机衬板形状因素
mill liner wear rate 磨机衬板磨损率
mill lining configuration 磨机衬里[内衬]构造[型]
mill load density 磨机装料浓度
mill load sizing 磨机装料分级
mill loading 磨机充填[负载]率

mill loading percent 磨矿充填百分数[率]
mill lube oil pressure 磨机润滑油压
mill maintenance crew 选矿厂维修人员
mill model calculation 磨机模型计算
mill operator 选矿厂操作[经营]人员
mill power consumption 磨机功率[动力]消耗
mill power correction 磨机功率校正
mill power draft 磨机汲取功率
mill power draw 磨机拖动功率
mill power relationship 磨机功率关系式
mill pressure drop 磨机压力降
mill process 选矿厂工艺
mill process rate 选矿厂[选矿]加工量
mill productivity 磨机生产率[量,能力]
mill quality control 选矿厂质量控制
mill recovery 选矿[厂]回收率
mill rotational speed 磨机旋转[转动]速度
mill scoop box 磨矿机勺箱
mill shift foreman 选矿厂值班工长
mill site 厂址
mill solution 磨矿溶液(金氰化法中逆流倾析所得的贫溶液,送磨矿机或搅拌中使用)
mill solution tank 磨矿机溶液槽,选液贮槽(氰化)
mill stream 磨矿矿流

mill superintendent	选矿厂厂长
mill support facilities	选矿厂辅助设施
mill testing	选矿厂试验
mill throughput	选矿厂[磨机]处理量
mill throughput rate	选矿[厂]生产率
mill transformation calculation	磨机转换计算
mill transformation matrix	磨机转换矩阵
mill transport	(选矿)厂内运输
mill utilization	磨机利用率
mill water quality	选矿水质量
mill water system	选矿厂用水系统
milled ferrosilicon	磨碎硅铁
millicurie	毫居(里)
milliequivalent	毫(克)当量
millimole	毫(克)分子(量),毫摩尔
milling capacity	选[磨]矿能力
milling circuit	选[磨]矿流程[回路]
milling cost	选[磨]矿费用
milling division	选矿车间[段]
milling engineer	选矿工程师
milling facilities	选矿(厂)设施
milling power	磨矿功率
milling practice	选矿实践
milling procedure	选矿工艺过程
milling process	磨碎[碎矿]过程
milling rate	选矿量
milling rate fluctuation	磨矿速度波动
milling reserve	可选[选矿]储量
milling reserve tonnage	可选储量吨位
milling technique	磨矿技术
milliohm($m\Omega$)	毫欧(姆)
millipore	微孔隙
Millipore super-Q system	微孔隙超-Q系统
millivolt	毫伏,10^{-3}伏
Miltopan D503	起泡剂(一种脂肪烷基硫酸酯)
mimic panel	模拟盘
minable area	可采矿区
mine and mill complex	采选联合企业
mine and smelter corporation	矿冶公司
mine area	矿区
mine compressor	采矿空压机
mine development	矿山开拓
mine development process	矿山开发过程
mine dilution grate	可采混入废石品位,开采贫化品位
mine equipment size	矿山[采矿]设备规格
mine exploration	矿山勘探
mine financing	矿山筹款
mine ore inventory	矿山的矿石存货
mine planner	矿山计划人员
mine planning	矿山平面布置,矿山工作平面布置
mine planning group	矿山计划组
mine preproduction cost	矿山预生产费用
mine preproduction schedule	矿山生产前准备工作进度计划

mine process rate 矿山生产量
mine product 矿山产品
mine production rate 矿山生产量
mine reserve tonnage 可采储量吨位
mine run crushed coal 破碎原煤,已破碎原煤
mine run ore 原矿
mine run stockpile 原矿料堆
mine site 矿区,采矿场
mine site pilot plant 矿区半工业试验厂,矿山现场半工业试验厂
mine stripping 矿山剥离
mine survey 矿山测量
mine's technical staff 矿山技术人员
mined out cut 采空堑沟
Minemet flotation machine 米涅梅特浮选机(锥形叶轮,中空轴底吹式,不用定子的浮选机)
mine-mill complex 采选综合企业
miner 矿工
minerographic microscopy 矿相显微技术
mineral 矿物,矿产;无机物;矿泉水
mineral acid 无机酸,矿物酸
mineral aggregate 矿料,矿质集料,骨料,石料
mineral assembly 矿物总体
mineral association 矿物共生(体)
mineral beneficiation 选矿
mineral beneficiation area 选矿领域
mineral beneficiation department 选矿部门

mineral beneficiation field 选矿领域
mineral beneficiation process 选矿过程[方法,工艺]
mineral beneficiation technique 选矿方法[工艺]
mineral black 石墨
mineral blend 混矿,配矿
mineral bloom 晶簇石英
mineral claim 矿产要求权
mineral coke 天然焦
mineral commodity 矿产品
mineral composition 矿物成分
mineral compound 无机化合物
mineral content 矿物含量
mineral cotton 矿(渣)棉
mineral crystal 矿物晶体
mineral crystal shape 矿物晶体[结晶]形状
mineral crystal size 矿物结晶粒度
mineral description 矿物描述
mineral dispersion 矿物分散
mineral distribution 矿物分布
mineral division 矿物(分)部,矿业(分)部,矿产(分)处[部]
mineral dust 矿尘
mineral engineer 选矿工程师
mineral engineering 矿物工程
mineral examination 矿物鉴定[检定]
mineral exploration 矿石[产,床]勘探
mineral floatability 矿物可浮性
mineral form 矿物形态
mineral fraction vector 矿物组分向量

mineral identification 岩矿鉴定
mineral industry 矿物工业(石油,煤炭,水泥,黑色和有色矿物)
mineral lake 铬酸锡玻璃
mineral lattice 矿物晶格
mineral liberation 矿物解离
mineral locking 矿物连生
mineral makeup 矿物组成
mineral occurrence (state) 矿物赋存状态
mineral oil collector 矿物油捕收剂
mineral particle 矿物颗粒
mineral preparation (入送前)矿物准备
mineral process engineer 选矿工程师
mineral process laboratory 矿物加工实验室
mineral processing and research laboratory 选矿研究实验室
mineral processing circuit 选矿系统[流程,回路]
mineral processing cost 选矿费
mineral processing division 选矿部分
mineral processing engineer 选矿工程师
mineral processing flowsheet 选矿流程
Mineral Processing Handbook 选矿手册
mineral processing industry 选矿工业
mineral processing method 选矿方法
mineral processing operation 选矿作业

mineral processing phase 选矿阶段
mineral processing plant 选矿厂
mineral processing plant design 选矿厂设计
mineral processing plant design course 选矿厂设计过程
mineral processing plant design symposium 选矿厂设计会议
mineral processing plant practice 选矿厂实践
mineral processing plant variable 选矿厂变量
mineral processing technology 选矿技术
mineral processing textbook 选矿教科书
mineral processing unit 选矿设备
mineral processor 矿加工设备,选矿人员
mineral product 矿产品
mineral production 矿物生产
mineral property 矿物性质,矿权
mineral research and test 矿物研究与试验
mineral research center 矿物研究中心
mineral right 矿权
mineral sand 矿砂
mineral sands concentrator (海边)砂选矿厂
mineral sands deposit 矿砂沉积矿床,砂矿床
mineral scientist 矿物科学家

mineral seal oil 重质灯油	mineralogical analysis 矿物分析
mineral selectivity 矿物选择性	mineralogical approach 矿物学方法
mineral separation 选矿	mineralogical association 矿物共[连]生体
mineral separation plant 选矿厂	
mineral separation process 选矿工艺,选矿工艺流程,选矿法	mineralogical change 矿物特性变化
	mineralogical characteristic 矿物学特性
mineral separation size range 矿物分选粒级范围	
	mineralogical component 矿物成分
Mineral Separations machine MS 浮选机	mineralogical composition 矿物组成
	mineralogical content 矿物含量,物质组成
mineral solubility 矿物溶解度	
mineral surface 矿物表面	mineralogical data 矿物学资料[数据]
mineral surface modification 矿物表面改变(处理)	mineralogical definition 矿物定义[鉴定]
mineral surface reaction 矿物表面反应	mineralogical examination 矿物检测
	mineralogical factor 矿物因素
mineral tar 矿柏油,矿质焦油,软沥青	mineralogical parameter 矿物学参数
mineral texture 矿物结构	mineralogical sampling 矿物采样
mineral varnish 石漆	mineralogical structure 矿物结构
mineral-collector combination 矿物捕收剂结合作用	mineralogical study 矿物学研究
	mineralogical study technique 矿物学研究技术
mineralized emulsion 矿化乳浊液	
mineralized face 矿化面	mineralogical technique 矿物(学)技术
mineralized froth 矿化泡沫	
mineralized material 矿化物料	mineralogical viewpoint 矿物学观点
mineralized particle 矿化颗粒	mineralogical zone 含矿带
mineralized sample 矿样	mineralogy 矿物学,矿物性质
mineralized species 矿物化矿种	mineralogy of mill head 入选原矿的矿物性质
mineralized surface 矿化面	
mineralized trend 矿化(带)走向	minerals yearbook 矿物年鉴
mineral-matter-free 无矿物质	mineral-water interface 矿-水界面
mineralogical 矿物(学)的	Mineral-water interface tension

矿物-水界面张力

mineral-water-air system 矿物-水-空气体系

Minerec 捕收剂(黄原酸甲酸酯) ROC(S)S·COOR′

Minerec 1331 collector Minerec 1331 捕收剂(双烷基硫)

Minerec 27 捕收剂(复黄原酸,双黄药,选择性浮选铜)

Minerec A 捕收剂(复黄原酸,双黄药,在酸性或碱性介质中浮选铜)

Minerec B 捕收剂(复黄原酸,双黄药,在石灰介质中浮选铜)

mine-run lump material 原矿块状物料

mini plant 实验室规模试生产用小型设备[装置]

MINI$_3$ 锡矿捕收剂(品名,澳大利亚雷尼森选矿厂1978年使用,回收率80%~90%)

miniature conduit 小[微]型管道

minimum allowance 最小富余量[余量]

minimum capital expenditure 最小投资[基本建设]费用

minimum dust-conveying velocity 最小粉尘运输速度

minimum edge allowance 最小允许误差量(如运输胶带的)

minimum number of increment 最少(小批)取样次数

minimum recovery level of solids 最小固体回收量

minimum saleable concentrate grade 最低出售的精矿品位

minimum weight of increment 最少(小批)取样重量

minimum working capital ratio 最低周转金比率

mining 开[采]矿,开采[挖],矿业

mining analysis 矿山分析

mining and beneficiation plant arrangement 磨矿及选别车间配置图

mining and metallurgical industry 矿冶工业

mining and metallurgy complex 采冶联合企业

mining and milling complex 采选联合公司

mining and milling industry 采选工业

mining and processing operation 采选生产

Mining Chemicals Handbook 矿用化学药剂手册

mining chemicals research department 矿用药剂研究室

mining company 采矿[矿业]公司

mining core tube 岩心管,采矿机身管

mining cost 采矿费用,开采成本

mining cut 开采堑沟

mining dilution 开采混入废石量,采矿贫化

mining district 采矿区[地带]

mining duty 矿山工作

mining equipment sales manual 采矿设备销售手册
mining fact sheet 采矿记录表
mining firm 矿山企业
mining heading 采矿平巷
mining man 矿工
mining method 采矿方法
mining operation 矿山生产
mining phase 采矿阶段
mining plan development 开采计划编制
mining planner 采矿计划人员
mining process 采矿[开采]过程
mining property 矿山资产
mining rate 开采速度
mining reserve grate 可采储量品位
mining screen 矿用筛分设备
mining sequence 采矿程[顺]序
mining survey 矿山测量
mining system 采矿系统
mining taxation 矿山税收
mining unit 采矿设备
mining-metallurgical complex 采选(冶)复杂矿石
mining-metallurgical industry 采选[矿冶]工业
mining-milling-refining complex 采选冶联合企业
minor addition 少量添加
minor constituent 次要成[组]分
minor constituent mineral 次要矿物组分
minor continuous flow 次要连续流
minor intermittent flow (bypass line) 次要间歇流(旁通线)
minor modification 小修改,小改动
Minsk oil Minsk 起泡剂(氧化煤油,浮选闪锌矿时作为起泡剂)
mint 造币厂,薄荷
Mintek/RSM size analyzer Mintek/RSM 粒度分析仪(南非 Mintek 公司研制)
minute gold 微(小)粒金
minutes of meeting 会议记录
Miocene 中新世,中新统,第三纪中新世
Mipolam 麦波郎(耐磨塑料)
Mirapon F30 一种起泡剂
Mirapon RK 湿润剂(磺化脂肪酰胺衍生物)
mirror finish 镜面光洁度,镜面抛光
mirror image 镜像
mirror image vertex 镜像顶点
miscellaneous buildings 其他建筑物
miscellaneous business 杂务
miscellaneous division 总务科
miscellaneous equipment 其他设备
miscellaneous expenses 杂费
miscellaneous information 其他资料[情报]
miscellaneous items 其他事项
miscellaneous machinery 杂项机械
miscellaneous security fund 其他证券资金

miscellaneous service 杂务

miscellaneous storage room 杂品贮藏室

mischmetal 含铈的稀土,混合稀土,稀土金属混合物

misplaced particle 错配[逸失]颗粒,混粒

misplaced undersize particle 错配的筛下颗粒(即筛上产品所含的细粒)

misrepresentation 虚报

miss-tracking detector 跑偏探测器

mist eliminator 烟雾消除器

mist precipitator 烟雾集尘器[沉淀器]

mix alkyl hydroxamate 混合烷基羟肟酸

mixed amyl amine 混合戊胺(25~100克/吨,阳离子辅助捕收剂)

mixed ionic forces 混合离子键力

mixed layer 混合层

mixed oxide and sulfide ore 氧化与硫化混合矿石

mixed oxide-sulfide ore 氧化硫化混合矿石

mixed potential 混合电位[势]

mixed sulfide-oxide system 硫化矿-氧化矿混合组织

mixed sulphide ore 混合硫化矿

mixer-in series-model 混入系列模型

mixer-settler solvent extraction system 混合沉淀器熔剂萃取系统

mixers-in-series model 串联混合机组模型

mixers-in-series residence time distribution 串联混合机组停留时间分布

mixing agitator 混合搅拌槽

mixing behavior 混合行为

mixing cell 搅拌[混合]槽

mixing chamber 搅拌室(浮选机的)

mixing characteristics of machine 浮选机混合特性

mixing constant 混合常数

mixing stage 混合[搅拌]阶段

mixing zone 混合区

mixture 混合(物,体,气,料,剂,比,状态),配料,炉(混合)料成分

MK-Ⅱ Cleaveland jig 马克Ⅱ型克里夫兰德圆形跳汰机(取消了旋转耙,美国克里夫兰德公司制造)

mobile crane 移动吊车[式起重机]

mobile home 移动[活动]式住宅

mobile mechanical device 移动式机械装置

mobile scaling platform 移动式登高台(用于挑顶等)

mobile test laboratory 移动实验室

mobile unit 移动式装置[设备]

mobile workshop 移动修理车间

mode of breakage 破碎方式

mode of cell operation 浮选槽操作方式

mode of occurrence 赋存[产出]状态

mode of operation 操作[运行]方式

mode of power generation 发电方式

model accuracy 模型精确度
model building 模型建立
model building plan 模型建立计划
model building stage 模型建立阶段
model constant 模型常数
model development 模型研制
model equation 模型方程
model form 模型形式
model machine 样机
model number 型号
model of ball mill 球磨机模型
model of bin 矿仓模型
model of capture rate 捕捉[集]速度模型
model of crusher 破碎机模型
model of filter 过滤机模型
model of flotation cell 浮选机[槽]模型
model of pump 泵的模型
model of pump box 泵池模型
model of rod mill 棒磨机模型
model of screen 筛子模型
model of thickener 浓密机模型
model parameter 模型参数
model prediction 模型预测
model set 模型
model structure 模型结构
model verification 模型验证
moderate alteration 中等蚀变
moderate service line 中等(负荷)作业线
modern concentrator facilities 现代选矿厂设施
modern grinding era 现代磨矿时代[阶段]
modern industrialized countries 现代工业化国家
modern method 现代方法
modern methodology 现代方法
modern milling practice 现代选矿实践
modern mineral processing engineer 现代选矿工程师
modern practice 现代生产实践
modern single roll crusher 现代单辊破碎机
modern vibrating screening 现[近]代振动筛分
modernization trend 现代化趋势
Modicon 584 module 莫迪康584模块(施耐德电气的品牌产品)
modification 改进[良,善,变,建,装],修正,调整
modified equation 修正方程式
modified hallimond microflotation cell 改进型哈里蒙特微型浮选槽
modified Proctor test 改进的普氏试验
modified simplex direct search strategy 修正的单项直接检索对策
modified simplex search 修正的单项检索
modifier 调整[改性]剂,调节器
modifying agent 改变(矿物表面)药剂
modifying factor 修正[调整]系数[因素]

modifying reagent 调整剂
modular design method 模块化设计法
modular form 模块方式
modular type bolt-in tray 标准型螺栓固定底盘
modulation of flotation 浮选的控制[调节]
modulator 调整剂,调幅[制]器
module 模块,组件,模数
modulus of elasticity in compression 压缩弹性模数
Mogensen divergator 莫根森型分选机(德国)
Mogensen sizer 莫根森型筛分机(德国)
moil 十字镐,鹤嘴锄
moisture condensation problem 水分冷凝问题
moisture content 水分含量
moisture migration 水分移[流]动
moisture test 水分试验
moisture trap 气水分离器
molar 摩尔的,容模的
molar extinction coefficient 摩尔消光系数
molar ratio 摩尔比(率)
molar solution 容模溶液,(体积)摩尔溶液
mole 摩尔,防波堤
molecular 分子的,由分子组成的
molecular formula 分子式
molecular orientation 分子定向,分子取向
molecular oxygen 分子氧
molecular proportion 分子比,分子比例
molecular ratio 分子比率
mole fraction 摩尔分数
mole number 摩尔数
molten slag 熔渣
molten sulphide phase 熔融硫化物相
molyb-alloy grease 钼合金润滑剂
molybdenite insoluble locking 辉钼矿不溶物包裹体
molybdenite recovery plant 辉钼矿回收厂
molybdenite recovery procedure 辉钼矿回收方法[过程]
molybdenite vein 辉钼矿矿脉
molybdenite-hydrocarbon oil flotation system 辉钼矿-烃油浮选系统
Molybdenum-ochre 钼华
molybdenum trioxide 三氧化钼
molybdenum-bearing mineral 含钼矿物
molybdenum-copper 钼铜合金
molybdic oxide 三氧化钼
molybdite 钼华
moly-chrome castings 钼铬铸件
moly-cop rod 钼铜棒
molysite 铁盐
momentary velocity 瞬时速度
momentum balance 动量平衡
momentum transfer 动量传递

momentum vector 动量矢量
Monamid 表面活性剂品牌名
Monamid 150CE 椰油脂肪的二乙醇酰胺(可用作硅酸盐捕收剂)
Monawet 表面活性剂品牌名
Monawet MO 双-2-乙基己基磺化琥珀酸钠(同 Aerosol OT,可用作湿润剂)
monetary contraction 银根紧缩
monetary system 货币制度
monetary unit 货币单位
monionic 单离子型(的)(由相同电荷离子组成)
monitor well 监[检]测井
Monitor's error 监控器误差
monoabietylphthalate 单松脂基邻苯二甲酸酯钠盐(可用作起泡剂)
monoacetate 一醋酸酯,单醋酸酯
monoammonium phosphate 磷酸二氢铵,磷酸一铵 $(NH_4)H_2PO_4$
monoaryl dithiocarbamate 单芳基二硫代氨基甲酸盐(可用作捕收剂) R-NHCSSM
monodisperse material 单一[分散]物料
monodispersed system 单分散体系
monoethanolamine 一[单]乙醇胺(可用作捕收剂及脉石的抑制剂) $NH_2CH_2CH_2OH$
monograph 专题论文,专著论文单行本
monograph text 专题版[文]本
monolayer 单分子层,单层

monolayer absorption 单分子层吸附
monolayer capacity 单分子层容量
monolayer coverage 单分子层覆盖,单分子覆盖层
monolayer level 单分子层量
monomeric sulphur 单体硫
monomineral sample 单一矿物样品
mono-phosphonic acid ester 单膦酸酯
monopole soap 硫酸化蓖麻油酸钠皂,单极皂可用作捕收剂
monothiocarbonate 一[单]硫代碳酸盐
monovalent 一价的,单价的
monovalent ion 一价离子
monovalent metal 一价金属
monovalent metal n-butyl xanthate 一价金属的正丁基黄原酸盐
Montanol 300 一种起泡剂(可用于选煤,德国赫斯特化工厂生产)
Monte Carlo simulation method 蒙特卡罗模拟法(统计模拟方法,随机模拟法)
monthly cost 月费用
monthly progress report 月进度报告
monthly rate 每月费率
monzonit 二长岩
monzonite-dacite stocks 二长英安岩株
monzonitic rock 二长岩
monzonitic-porphyry 二长斑岩
monzonitic type rock 二长岩类型的岩石

Morcowet 表面活性剂品牌名

Morcowet 469 烷基萘磺酸盐(可用作湿润剂)

more complex mixture 多元混合矿,多金属混合矿

more even length-diameter ratio type of mill 更长的长径比类型磨矿机

morning shift 早班

mortar lining 砂浆衬里

mortgage bond 抵押证券

mortgage loan payable 应付抵押贷款

mortgage-backed securities 抵押证券

Mossbauer Effect 穆斯堡尔效应(一种原子核无反冲的γ射线共振散射或吸收现象。这个效应是由德国物理学家穆斯堡尔首次在实验中实现的)

Mossbauer spectroscopy 穆斯堡尔(光)谱(学)(是应用穆斯堡尔效应研究物质的微观结构的学科)

motor central control room 电动机中心控制室

motor contactor 电动机接触器

motor control 电动机控制(器)

motor control center 电动机控制中心

motor current draw 电机牵引电流

motor cut-out 电动机保险装置

motor failure warming 马达事故警报

motor grader 机动平路机,自动平地机

motor horse power 电动机[马达]马力

motor load 电动机负荷

motor mounting style 电动机安装方式

motor nameplate 电机铭牌

motor output shaft 电动机输出轴

motor patrol 电动养路车

motor power draw 电动机牵引功率

motor pulley 电动机皮带轮

motor rating 电动机额定值[功率]

motor reducer 电机减速器

motor running 电机运行

motor running horse power 电机运行马力

motor speed reducer 马达减速机

motor trip 电动机[马达]跳闸

motorized traveling tripper 电动移动式卸矿器

motor-mechanical room 马达机械室

motor-V-belt drive 马达三角皮带驱动

mount 固定伞板(圆锥破碎机的)

mounting dimension 安装[装配]尺寸

movable jaw depth 可动颚板高度

movable tripper 移动卸料[矿]小车

move 移动,步骤,迁居,感动

movement 运动,移动

moving equipment 运[活]动设备

moving interface 移动界面

moving point 移[活]动针,动点

Mozley table 莫兹利摇床(英国)

MP-189 捕收剂(烷基萘磺酸盐) RSO_3Na

mucin(=**mucoprotein**) 粘朊,可用作矿泥的抑制剂

multi rope Koepe hoist 多绳戈培式提升机

multibladed molded rubber or neoprene impeller 多叶片模压橡胶或氯丁(合成)橡胶叶轮

multi-can system 多罐式装置[设备,系统]

multi-cell machine 多槽浮选机

multi-channel X-ray analyzer 多路X射线分析仪

multi-compartment Denver wet reagent feeder 多室丹佛型湿式给药机

multi-cyclone 多管式旋流器

multi-dimensional distribution function 多维分布函数

multi-facetted control philosophy 多方面的控制原理

multifeed 多点给料[矿],多点供油

multifilament 多纤维,复丝

multi-function simulation program 多功能模拟程序

multilayer adsorption 多层吸附

multi-layer adsorption 多层吸附

multilayer coverage 多层覆盖

multilayer film 多层薄膜

multi-layers of copper hydroxamate 氧肟酸铜多分子层

multiphase-inhomogeneous system 多相不均质体系

multiple analog controller 多功能模拟控制器

multiple cell circuit 多槽浮选回路

multiple compartment ball mill 多室球磨机

multiple cut system 多产品提取系统

multiple cycloning 多级旋流

multiple data sets 多重数据组

multiple double bond 多双链

multiple element programmable controller 多路元件可编程序控制器

multiple filter 多节[多重]滤波器,复式过滤器

multiple hearth dryer 多膛[床]底火式干燥机

multiple hopper 复合式漏斗

multiple impeller 多级叶轮

multiple impeller and stator mechanism 复合的叶轮和定子机构

multiple mechanism 复合机构

multiple metal flotation 多金属浮选

multiple particle-multiple impact breakage 多颗粒多冲击破碎

multiple ply 多层

multiple product test 多产品试验

multiple pyramid type hopper 多角锥型漏斗

multiple regression 多次[重]回归

multiple regression analysis software 多元回归分析软件

multiple sluice concentrator 复式溜槽选矿机

multiple strategy capability 复合策略功能

multiple stream analyzer 多路物料流

分析仪
multiple treatment technique 多种方法处理[选别]技术
multiple-hood branch 多通风罩支管
multiple-outlet silo 多(排)口筒仓
multiple-reflection absorption 多次反射吸收
multiple-stage elutriator 多段淘[水]析器
multiple-stage open circuit ball mill 多段开路球磨(均开路)
multiplying factor 放大(倍加)系数,乘数,淬透系数
multi-stage air dispersion 多段空气分散
multi-stage cleaner floats 多次精选泡沫
multi-stage cleaning 多段精选
multi-stage crushing circuit 多段破碎流程
multistage crushing plant 多段破碎车间
multi-stage fashion 多段形[方]式
multi-stage grinding 多段磨矿
multi-stage problem 多级[段]问题
multistage process 多段工艺过程[工序]
multi-stage process 多段处理法
multi-stage regrinding and cleaning operation 多段磨矿和精选作业
multi-stage sampling system 多段取样系统
multi-stage system 多段[级]系统

multistage washing 多段洗矿
multi-station linear indexing model 多工位直线运动[变换]模型
multi-stream process control X-ray quantometer(PCXQ) 多流工艺控制X射线光谱分析仪(光子计算机,辐射强度测量计)
multi-surface screen 多层(筛面)筛
multi-tasking real-time control program 多任务实时控制程序
multi-unit commercial plant 多系列工业性选矿厂
multi-unit single-stage primary mill 多台[系列]单段粗磨机
multi-use interactive operating system 多用途交叉操作系统
multi-use process control system 多用途过程控制系统
multi-user interpretative operational environment 多用户解释操作环境
multi-user time-sharing system 多用户分时系统
multivariable control 多变量控制
multivariable control theory 多变量控制理论
multivariable generalization 多变量通式[则]
multivariable optimization technique 多变量最佳化技术
multivariable search technique 多变量搜索技术
municipal waste water 城市污水

municipal water 自来水
muriatic leach 氯化[盐酸]浸出
mutation 变种[化]
mutual intergrowth 相互共生关系

mylar 迈拉(聚酯薄膜)
mylar window 迈拉(聚酯薄膜)窗孔
myristate 十四[豆蔻]酸盐

N

N-(n-dodecyl)-propyl-1,3-diamine N-(正十二烷基)-丙基-1,3 二胺

N-(tridecyl ether-n-propyl)-propyl-1,3 diamine-monoacetate N-(十三烷基醚-正丙基)丙基-1,3 二胺一醋酸盐

N,N,N-trimethyl-n-dodecyl ammonium bromide N,N,N-三甲基-正十二烷基溴化铵

N,N-dimethyl-n-dodecylamine N,N-甲基-正十二烷基胺

Na ethyl xanthate 乙基黄原酸钠

$Na_2Cr_2O_7$ 重铬酸钠(分子式,可用作 pH 值调整剂)

Na_2CS_3 三硫代碳酸钠(分子式,可用作氧化铅、铜矿的活化剂)

Na_2O_2 过氧化钠(分子式,可用作活化剂)

Na_2S 硫化钠(分子式,硫化剂,可用作辉钴矿、闪锌矿及金矿的抑制剂,与硫酸铜一起作为闪锌矿活化剂)

$Na_2S_2O_5$ 焦亚硫酸钠(分子式,可用作硫化铁、硫化锌矿的抑制剂)

Na_2SiF_6 氟硅酸钠(分子式,可用作石英、长石的抑制剂及含镍的磁黄铁矿、氧化铁矿的活化剂)

Na_2SiO_3 偏硅酸钠(分子式,水玻璃,可用作脉石、方解石、辉钴矿的抑制剂及矿泥的分散剂,浮选锰矿时作铁的抑制剂)

Na_2SO_3 亚硫酸钠(分子式,可用作黄铁矿、闪锌矿的抑制剂及用作 pH 值调整剂)

Na_2SO_4 硫酸钠(分子式,可用作调整剂、活化剂)

Na_3PO_4 磷酸三钠(分子式,可用作抑制剂)

$Na_4Fe(CN)_6$ 亚铁氰化钠(分子式,可用作硫化矿的抑制剂)

$Na_4P_2O_7$ 焦磷酸钠(分子式,浮选菱镁矿时作方解石的抑制剂)

$Na_5P_3O_{10}$ 三聚磷酸钠(分子式,可用作分散剂)

Nacconol 表面活性剂品牌名(美国

Stepan 公司)

Nacconol HG 烷基芳基磺酸钠(可用作捕收剂)R·ArSO$_3$M

Nacconol LAL 十二烷基磺酸醋酸盐(可用作捕收剂)MO$_3$SCH$_2$COOC$_{12}$H$_{25}$

Nacconol NR 煤油烷基苯磺酸钠(相当于 C$_{14}$烷基,可用作捕收剂、起泡剂、湿润剂)C$_{14}$H$_{29}$—C$_6$H$_4$—SO$_3$Na

Nacconol NRSF 烷基苯磺酸盐(可用作湿润剂)

Nacconol SSO 烷基苯磺酸盐(可用作湿润剂)

NaCl 氯化钠(分子式,可用作调整剂)

NaClO 次氯酸钠(分子式,可用作抑制剂、活化剂)

NaCN 氰化钠(分子式,可用作硫化矿、辉钴矿的抑制剂)

NaCO$_3$ 碳酸钠(分子式,可用作抑制剂、分散剂及调整剂,浮选菱镁矿时作为白云石的抑制剂,矿泥分散剂)

NaF 氟化钠(分子式,可用作黄铁矿、闪锌矿的抑制剂,石英的致钝剂)

Nagahm flotation machine 纳嘎姆浮选机(曲面叶轮,定子叶片底吹式)

NaH$_2$PO$_4$ 磷酸二氢钠(分子式,浮选锰矿时作为 Fe 的抑制剂)

NaHCO$_3$ 碳酸氢钠(分子式,可用作抑制剂、调整剂、活化剂)

NaHSO$_3$ 亚硫酸氢钠(分子式,可用作黄铁矿、闪锌矿的抑制剂)

naked eye 肉眼

Nal-153 抑制剂(木素磺酸盐)

Nalcamine 表面活性剂品牌名

Nalcamine G-11 月桂基羟乙基咪唑啉(结构式同 Amine 220)

R—C$_3$H$_4$N$_2$—C$_2$H$_4$OH,R 来自椰子油脂肪酸

Nalcite 离子交换树脂品牌名

Nalcite HCR 磺化聚苯乙烯阳离子交换树脂

Nalcite HDR 磺化聚苯乙烯阳离子交换树脂

Nalcite HGR 磺化聚苯乙烯阴离子交换树脂

Nalcite SAR 强碱性阴离子交换树脂

Nalcite WBR 弱碱性阴离子交换树脂

Nalco 表面活性剂品牌名(美国 NALCO 公司品牌名,中文名称:纳尔科,)

Nalco 600 阳离子聚合物(可用作絮凝剂)

Nalco 614 铝酸盐(可用作絮凝剂)Na$_2$Al$_2$O$_4$·3H$_2$O

Nalco 650 加工的蒙脱石(可用作絮凝剂)

Nalco 680 铝酸钠(可用作絮凝剂)Na$_2$Al$_2$O$_4$·3H$_2$O

Nalco 1801 纳尔科 1801 絮凝剂

Nalcolyte 110 絮凝剂(非离子型高分子聚合物)

Nalcolyte 960 絮凝剂(非离子型极高分子聚合物)

n-aliphatic alcohol n-脂族醇

n-alkylamine 正烷基胺

nano 纳,毫微

NaNO$_2$ 亚硝酸钠(分子式,pH值调整剂)

nanometer 毫微米,纳米(10^{-9}米)

Nansa 表面活性剂品名(烷基苯磺酸盐,可用作湿润剂)

nantokite 铜盐,氯化亚铜矿 Cu_2Cl_2

NaOH 氢氧化钠(分子式,浮选菱镁矿时作为白云石的抑制剂)

naphtha 石脑油,(粗)挥发油,粗汽油

naphthalene 萘 $C_{10}H_8$

naphthalene wash oil 萘洗油(可作为重金石的捕收剂)

naphtha solvent 液体石油烃(溶剂,与脂肪酸一起使用,浮选钛铁矿或金红石时作为辅助捕收剂)

naphthenic acid 环烷酸,环酸,环脂己酸(可用作捕收剂、起泡剂)

naphthol 萘酚 $C_{10}H_7OH$

naphthyl 萘基 $C_{10}H_7-$

naphthylamine 萘胺(可用作捕收剂) $C_{10}H_7NH_2$

narrative report 叙事[陈述式]报告

narrow band 狭窄范围

narrow end 狭窄端

narrow size interval 狭窄粒度间隔

narrow size range 窄粒级范围

National pan conveyor National 平板[盘式]运输机

national security 国家安全

national taxation 国家税收

nationalization 国有化

native gum rubber 天然橡胶

native metal 天然金属

native plant 自然植物

native rubber 天然橡胶

natural activating mechanism 自然活化机理,自然活化作用过程

natural aeration capacity 自然充气能力

natural air flow rate 自然空气流量

natural apatite 天然磷灰石

natural blue-green 天然蓝绿色

natural breakage pattern characteristic of material 物料自然破碎模式[典型]特性

natural condition 自然条件

natural crystal 自然晶体

natural differentiation 性质差别,特性差别

natural direction of airflow 气流的自然流向

natural dissolved solid 天然溶解固体

natural environment 自然环境

natural floatability 天然可浮性

natural flocculation 自然絮凝

natural gas fired Nichols-Herreshoff multi-hearth dryer 燃气尼科尔斯-赫雷肖夫型多膛干燥机

natural grain size 天然粒度

natural grain size condition 天然颗粒[粒度]状态

natural hydrophobicity 天然疏水性

natural inherent moisture 天然内在水

(分)
natural isotope decay 天然同位素衰变
natural logarithm base 自然对数底
natural moisture content 自然水分
natural pH 天然 pH 值
natural physical parameter 自然物理参数
natural radioisotope decay 天然放射性同位素衰变
natural refrigeration 天然冷凝
natural science 自然科学
natural specimen 天然样品
natural sulfide 天然[原生]硫化矿
natural tannin agents 天然丹宁类药剂
natural variation 自然变化
natural water 天然水
natural weathering 天然风化
naturally floatable sulfide 天然可浮硫化矿物
naturally hydrophobic 天然疏水的
nature constraint 制约性
nature of decision criteria 决策准则性
nature of gangue-ore boundary 脉石与矿石接触[边缘]性质
nature of material transport through the mill 物料通过磨机的传输性能
nature of sample 试样[样品]性质
navigable water 航行水域
N-brand liquid 标准商品[标]液体
n-butyl xanthate 正丁基黄原酸盐
N-dichlorothiocarbanilide N-二氯代二苯基硫脲(可用作捕收剂)

n-dodecyl amine 正十二烷胺
n-dodecyl chloride 正十二烷基氯化物
near aperture material (粒度)接近筛孔(尺寸)的物料(即难筛物料)
near gravity material 接近分选比重[密度](在±0.05 倍分选比重之间的物料)
near size 接近筛孔(尺寸)的粒度(0.5~1.5 倍筛孔尺寸之间)
near size fraction (粒度)接近筛孔(尺寸)的部分
near-screen-size material (大小)接近筛孔尺寸[筛分粒度]的物料
neat device 精巧[致]装置
necessary and sufficient criteria 必要与充分准则[判据,标准]
necessary plant 必需设备
negative area 负区
negative cash flow 负现金流量
negative counter ion 负配衡离子
negative disjoining pressure 负分离压
negative logarithm 负对数
negative photovoltage 负光电压
negative site 负电表面区
negative space charge 负空间电荷
negative surface charge 负[阴性]表面电荷
negative surface potential 负表面电位
negatively charged bivalent sulfate anion 负电荷二价硫酸盐离子
negatively charged potential 负充电势,负充电位

negatively charged site 负电表面区

negatively charged surface 带负电荷(的)表面

Nekal 表面活性剂品牌名

Nekal A 双异丙基萘磺酸盐(可用作捕收剂、湿润剂、乳化剂)
$[(CH_3)_2CH]_2C_8H_5SO_3M$

Nekal B 异丁基萘磺酸盐(可用作捕收剂、湿润剂、乳化剂)
$(CH_3)_2CHCH_2C_8H_6SO_3M$

Nekal BX 双异丁基萘磺酸盐(可用作湿润剂、乳化剂)

Nekal NF 二烷基萘磺酸盐(可用作湿润剂、乳化剂)

Nelson refinery construction cost index 纳尔逊炼油厂建设费用指数

Nelson sprinkler head 尼尔森喷洒头(美国尼尔森有限公司(L. R. Nelson Corp.)产品)

NEMA electrical enclosure 符合 NEMA 标准的电气外壳

Neo-fat 表面活性剂(脂肪酸类)品牌名

Neo-fat 10 癸酸(可用作捕收剂)
$C_9H_{19}COOH$

Neo-fat 12 月桂酸,十二碳(烷)酸(可用作捕收剂)$C_{11}H_{23}COOH$

Neo-fat 139 精制棉籽油(可用作捕收剂)

Neo-fat 14 十四碳(烷)酸(可用作捕收剂)$C_{13}H_{27}COOH$

Neo-fat 140 油酸(44%)、亚油酸(55%)、亚麻油酸(1%)的混合物

Neo-fat 16 十六碳(烷)酸(软脂酸)(可用作捕收剂)$C_{15}H_{31}COOH$

Neo-fat 16-54 软脂酸与硬脂酸的低共熔混合物(可用作捕收剂)

Neo-fat 18 十八酸(硬脂酸)(可用作捕收剂)$C_{17}H_{35}COOH$

Neo-fat 255 蒸馏的椰油脂肪酸(可用作捕收剂)

Neo-fat 265 脂肪酸混合物(其中含月桂酸25%,可用作锡石的捕收剂)

Neo-fat 3 精馏的油酸及亚油酸

Neo-fat 3R 精馏的油酸及亚油酸(可用作捕收剂)

Neo-fat 42-07 分馏的塔尔油(含树脂酸7%,可用作捕收剂)

Neo-fat 42-12 分馏的塔尔油(含树脂酸12%,可用作捕收剂)

Neo-fat 42-06 分馏的塔尔油(含树脂酸6%,可用作捕收剂)

Neo-fat 55 蒸馏的椰子油酸(可用作捕收剂)

Neo-fat 65 蒸馏的动物油脂肪酸(可用作捕收剂)

Neo-fat 8 辛酸(可用作捕收剂)
$C_7H_{15}COOH$

Neo-fat 88-18 油酸(76%)、亚油酸(10%)、硬脂酸(7%)、软脂酸(6%)等的混合物(可用作捕收剂)

Neo-fat 92-04 精制油酸(可用作捕收剂)$CH_3(CH_2)_7CH:CH(CH_2)_7COOH$

Neo-fat D-142 由塔尔油中分出的油

酸与亚油酸的混合物(可用作捕收剂)

Neo-fat D-342 塔尔油的蒸馏残渣(可用作捕收剂)

Neo-fat D-343 塔尔油的蒸馏残渣(可用作捕收剂)

Neo-fat DD-Animal 动物油脂肪酸(可用作选矿药剂)

Neo-fat DD-Corn oil 玉蜀黍油脂肪酸(可用作选矿药剂)

Neo-fat DD-Cottonseed 棉籽油脂肪酸(可用作选矿药剂)

Neo-fat DD-Linseed 亚麻仁油脂肪酸(可用作选矿药剂)

Neo-fat DD-Palmoil 椰油脂肪酸(可用作选矿药剂)

Neo-fat DD-Soybean 大豆油脂肪酸(可用作选矿药剂)

Neo-fat S-142 由塔尔油中分出的油酸与亚油酸的混合物(可用作捕收剂)

Neo-lene 300 十二烷基甲苯磺酸盐(可用作湿润剂)
$C_{12}H_{25} \cdot C_6H_3(CH_3) \cdot SO_3M$

Neo-lene 400 十二烷基苯磺酸盐(可用作湿润剂) $C_{12}H_{25} \cdot C_6H_4 \cdot SO_3M$

Neomerpin N 湿润剂、捕收剂(烷基萘磺酸盐,可用作氧化铁矿辅助捕收剂)
$RC_{10}H_6 \cdot SO_3M$

Neopen SS 药剂(磺化松香酸,可用作氧化铁矿捕收剂)

Neopol 洗涤剂(同 Medialan A,十八烯酰-N-甲氨基乙酸钠,可用作捕收剂)
$C_{17}H_{23}CON(CH_3)CH_2COONa$

Neopol T 洗涤剂(油酰基甲基牛磺酸钠,可用作湿润剂)
$R \cdot CON(CH_3)CH_2CH_2SO_3Na$,
$R \cdot CO =$油酰基$(C_{17}H_{33}CO—)$

neoprene 氯丁(二烯)橡胶

neoprene rubber 氯丁橡胶

Neothane 热塑性聚氨酯橡胶[树脂]

nerve center 神经中枢

net cash flow 净现金流量

net cash movement 净现金运转状态

net cash outflows 净现金支出(额)

net earning 净收入

net effective screening area 净有效筛分面积

net finished product 净最终产品(闭路破碎筛分的筛下产品)

net fixed assets 固定资产净值

net foreign exchange revenue 净外汇收入

net hydrophobic effect 疏水净效应

net income 纯收益,净收入

net mill power 净磨机功率

net mill power draft 磨极汲取的净功率

net mine site realization 矿山兑现净值(付清与出售精矿有关的一切缴税和扣款后所得总金额)

net operating profit 净生产利润

net power 净功率(扣除机械和电力损失后的磨机实耗功率)

net power consumption 净功率消耗

net power rate 净单位功率消耗量

[率],净功率消耗率
net present value 净现值
net profit 净利润
net revenue 纯收益,净收入
net revenue vs. particle size curve 净收入与粒度关系[对比]曲线
net surface charge 净表面电荷
net system availability 系统有效[净,实际]利用率
net taxable income 净应税收入
net width 净宽
net work index value 净功指数值
net working capital 净周转资本
network analysis 网络分析
network calculation 网络计算
network chart 网[络]路图
network form 网络形式
network of tetrahedrally bonded zinc and xanthate groups 四面键合的锌和黄酸基网络
network operation (calculation) 网络操作[运算]
network prediction 网络预报
neutral atom 中性原子
neutral collector molecule 中性捕收剂分子
neutral collector molecule theory 中性捕收剂分子理论
neutral dimer 中性二聚物
neutral disulfide 中性二硫化物
neutral electrolyte 中性电解质
neutral fine textured soil 中性细粒结构土壤
neutral group 中性基团
neutral hydrocarbon oil 中性碳氢[烃]油
neutral long-chained substance 中性长链物质
neutral medium condition 中性介质条件
neutral molecular hydrocarbon fuel oil 中性分子碳氢燃料油
neutral molecular oils 中性分子油类
neutral molecule 中性[不带电]分子
neutral molecule theory 中性分子理论
neutral organic molecule 中性有机分子
neutral pH region 中性pH范围
neutral pH solution 中性pH溶液
neutral region 中性区
neutral solution 中性溶液
neutral surfactant molecule 中性表面活性剂分子
neutral synthetic flocculent 中性的合成絮凝剂
neutral system 中性制
neutralization number 中和值
neutron activation 中子活化作用
neutron activation analysis 中子活化分析(法)
neutron activation source 中子活化源
neutron pile 中子堆
Neutronix 醚、酯类非离子表面活性剂品牌名

Neutronix 331 脂肪酸聚乙二醇醚(可用作分散剂、湿润剂)

Neutronix 333 脂肪酸聚乙二醇醚(可用作分散剂、湿润剂)

Neutronix 600 烷基苯酚聚乙二醇醚(可用作分散剂、湿润剂) $R \cdot C_6H_4(OCH_2CH_2)_nOH$

new approach to technology 新技术途径

new crack length 新裂缝长度

new crack tip length 新破裂的长度

new era 新时代[纪元]

new feed rate 新给矿量

new phase 新相,新阶段

new processing facilities 新工艺设施[备]

new stage response function 新级响应函数

new technology 新技术[工艺]

new water line 新水管线

newly developed digital controller 新研制的数字控制器

newly discovered mineral deposit 新发现的矿床

Newton–Raphson method (= Newton's method) 牛顿-拉福森法(又称牛顿法、牛顿迭代法,即用牛顿迭代法求解非线性方程,是把非线性方程线性化的一种近似方法)

Newton's formula 牛顿公式

Newton–Rittinger relation 牛顿-雷廷格关系式(颗粒在介质中运动阻力与阻力系数的关系公式)

next decade 下一个十年,未来十年

next processing step 下一步加工阶段,下一工序

NGP 药剂(2-硝基-4-甲苯胂酸) $O_2N \cdot C_6H_3(CH_3) \cdot AsO_3H_2$

NGT 捕收剂(对位甲苯胂酸)(可用作锡石的捕收剂)$CH_3 \cdot C_6H_4 \cdot AsO_3H_2$

N–hexanol N-乙醇

Nichols – Herreshoff multi – hearth roaster 尼科尔斯-赫雷斯霍夫多膛焙烧炉

nickel 镍(Ni)

nickel electrowinning 电积法提镍,镍电解

nickel ore handling plant 镍矿石处理厂

nickel-bearing laterite 含镍红土矿

nigrosine(aniline black) 尼格(洛辛),苯胺黑(可用作染料及锡石的抑制剂、分散剂)

nigrosine sulfonate 磺化苯胺黑(可用作分散剂)

Ni-hard 硬镍合金,含镍白口铸铁(3C, 4Ni, 2Cr, 0.5Mn, 0.4Si, 余量 Fe)

Ni-hard case 硬镍合金壳(泵的)

Ni-hard cast iron 镍硬铸铁,硬镍铸铁(4Ni~4.5Ni, 1.5Cr)

Ni-hard impeller 硬镍合金叶轮

Nilin 磺酸盐类表面活性剂(烷基芳基磺酸盐,可用作湿润剂)

NIM simulator NIM 模拟程序

Ninate 表面活性剂品牌名
Ninate 402 十二烷基苯磺酸钙(含油50%,可用作选矿药剂)
Ninol 表面活性剂品牌名
Ninol 128 脂肪酸与二乙醇胺缩合物(可用作湿润剂、乳化剂)
Ninol 200 脂肪酸与二乙醇胺缩合物(可用作捕收剂、湿润剂、乳化剂)
Ninol 400 脂肪酸与二乙醇胺缩合物(可用作捕收剂、湿润剂、乳化剂)
Ninol 517 脂肪酸与二乙醇胺缩合物(可用作捕收剂、湿润剂、乳化剂)
Ninol 521 脂肪酸与二乙醇胺缩合物(可用作捕收剂、湿润剂、乳化剂)
Ninol 57A 脂肪酸与二乙醇胺缩合物(可用作捕收剂、湿润剂、乳化剂)
nitrate 硝酸盐,硝酸酯,硝酸(根)
nitrate anion 硝酸盐阴离子,硝酸根阴离子
nitrogen atom 氮原子
nitrogen gas 氮气
nitrogen saturated solution 氮饱和溶液
nitrogen-purged borate solution 吹氮硼酸盐溶液
nitroso 亚硝基(的)
N-methyldodecylamine N-甲基十二烷胺,十二烷基甲胺
N-methyl-n-dodecylamine N-甲基-正十二烷基胺
NMR investigation 核磁共振研究
N-nitroso-amine N-亚硝基胺

no quantitative method 不定量法
noble mineral 惰性矿物,珍贵矿物
noble potential 贵电势,(金属的)高电位,惰性电位
noble sulfide mineral 惰性硫化矿物
n-octadecane 正十八烷
n-octyl amine 正辛基胺
noise control 噪声控制
noise hazard 噪声危害
noise level 噪声[干扰,杂音]电平,噪声级,噪声水平
noise magnitude 噪声值,噪声等级
Nokes reagent 浮选药剂(P_2S_5与NaOH及As_2O_3反应的产物,可用作硫化物抑制剂)
nomenclature 专门术语,命名法
nominal aliphatic xanthate 标准脂肪族黄药
nominal capacity 标称能力
nominal friction case 标称[公称,标定,额定]摩擦情况
nominal monolayer 标准单分子层
nominal residence time 公称的停留时间
nominal separation size 名义[标称]分离粒度
nominal size 公称[标称]尺寸[规格]
nominal split 标称分流
nominal value 公称[标称]值
nominal width 公称[标称]宽度
nomogram 列线[线解,线示,尺解,诺模,示]图,计算图表,图解

nomograph 列线[线解,线示,尺解,诺模,示]图,计算图表,图解

nomographic chart 列线图,算图

nomography 列线图解法[术],图算法,计算图表学

non-agglomerating 不烧结的

nonane 壬烷 C_9H_{20}

nonanoic acid 壬酸

nonanol 壬醇 $C_9H_{19}OH$

nonanone 壬酮 $C_9H_{18}O$

nonanoyl 壬酰(基)$(CH_3CH_2)_7CO—$

non-availability 不能利用率,不可利用性,非有效量

non-available 不适用的

non-cash charge 非现金费用

non-cash deduction of income 收益非现金扣除额

non-circulating surface 非环流表面

non-collecting tendency 无捕收倾向[可能性]

noncombustible mineral substance 不可燃矿物质

non-competent ore 不合格矿石(不适于作为磨矿介质的矿石)

non-conductive tailings 非导电尾矿

non-conductor re-treat configuration 非导体物料再[精]选形式

non-consolidated associated company 临时联合公司

non-contact instrument 无触点仪表

non-contact radiation type detector 非接触[无触点]放射类型探测器

non-contact type 非接触式,无触点型

noncontact type level control 无触点式水平[液面,料位]控制

non-contact weighing system 无触点称量系统

non-convergent iterative calculation 不[非]收敛的迭代计算

non-corrosive material 非腐蚀性物料

non-corrosive material of construction 非腐蚀性的建筑材料

non-degrading solids 不碎裂的物料

non-economic 不经济的

non-electrochemical process 非电化学过程

non-electrochemical reaction 非电化学反应

non-electroneutral plane 非电中性平面

non-electroneutral surface plane 非电中性表面的平面

non-equilibrium adsorption 非平衡吸附

non-equilibrium condition 不平衡条件

nonferrous metal ore concentrating facilities 有色金属矿石选矿设施

nonferrous mineral division 有色金属矿物部门

nonferrous minerals 有色金属矿物(类)

non-first order rate expression 非一阶速度表达式

non-float concentrate 不浮产品精矿
non-float product 不浮产品
non-floatable component 不浮游成分
non-floatable condition 不可浮状态
non-floatable constituent 不可浮成分
non-floatable mineral 不浮矿物
non-floating product 不浮产品,底流（浮选机的）
non-flowing solids 不流动固体物料［颗粒］
non-freeflowing solids 非自由流动固体物料
non-gas flotation 不充气浮选
non-gradient method 非梯度法
non-hazardous 无害
nonhomogeneous mine 非均质矿山
non-homogeneous surface 非均匀表面
non-interest bearing collateral trust bond 带抵押品的无息信用券
Nonion 表面活性剂品牌名
Nonion TM-50 聚乙二醇醚类(可用作脉石的活化剂)
nonionic compound 非离子化合物
non-ionic dextrin molecule 非离子型糊精分子
non-ionic extra high molecular weight polymer 非离子型超高分子量聚合物
non-ionic flocculent 非离子絮凝剂
nonionic process 非离子过程
non-ionizing electrode 非电离电极
non-linear equation 非线性方程
non-linear form 非线性式

nonlinear model 非线性模型
non-linear model 非线性模型
non-linear proportion 非线性比例
non-magnetic material 非磁性物料
non-magnetic stream 非磁流
non-magnetic tramp metal 非磁性夹杂金属
non-magnetics 非磁性部分［物料］
non-mass flow bin 非整体流动［变量流］矿仓
non-mechanical sample cutter 非机械化样品截取器
non-mechanical sampler 非机械化取样器
non-metallic and metallic oxide promoter 非金属和金属氧化矿促进［活化］剂
non-metallic beneficiation plant 非金属选矿厂
non-metallic flotation system 非金属矿物浮选体系
non-metallic mineral 非金属矿物
non-metallic mineral flotation 非金属矿物浮选
non-metallic ore 非金属矿
non-metallic roughing circuit 非金属矿粗选回路
non-normalizable component 不可标准化分量
non-operating cost 非生产费用
non-ore 废石,无矿
non-oxidative dissolution 非氧化溶解
non-oxidizing condition 非氧化条件

[状态]
nonpolar 非极性[化]的
non-polar behavior 非极性[特性,性能,状态]
non-polar character 非极性特性
non-polar edge 非极性棱
non-polar face 非极性面
non-polar group 非极性基
non-polar organic solvent 非极性有机溶剂
non-polar portion 非极性部分
non-polar quality 非极性性质
non-polar reaction product 非极性反应产物
non-polar solvent 非极性溶剂
non-polar specie 非极性产品[物]
non-polar surface 非极性(表)面
non-pollution 无污染
nonporous product layer 无孔隙产物层
non-random error source 非随机误差源
non-rational tool 非有理法,非比率方法
non-reactive ionic solid 非反应性离子型固体
non-reactive tracer 非[无]反应性示踪物[剂]
non-selective collector 非[无]选择性捕收剂
non-selective dispersant 无选择性分散剂

non-selective flotation 非选择性浮选
non-slip coupler 非滑动[移]耦合器
non-standard component 非标准部件
non-stoichiometric nature 不能用化学法计量的性质
non-stoichiometric semiconductor 非当量化学半导体
non-stoichiometric surface region 非当量化学表面区
non-stretching 不延伸
non-sulfide ore 非硫化矿石
non-sulfide system 非硫化矿体系
non-technical personnel 非技术人员
non-time-oriented decision rule 不考虑时间的决策规则
non-toxic additive 无毒添加剂
non-uniform discharge 不均匀排矿
non-uniform distribution 不均匀分布
non-uniform orientation 不均匀定向[位]
non-valuable mineral 无用矿物(即废石)
nonyl 壬基 $CH_3(CH_2)_7CH_2-$
nonyl alcohol 壬醇 $C_9H_{19}OH$
nonyl phenyl tetraglycol ether 壬基苯基四甘醇醚
Norbak 品牌名(美国 Norbak 公司产品)
Norbak high impact backing Norbak 耐高冲击填料
Norbak locking compound Norbak 连接[防松]剂

Norbak trowel mix　Norbak 泥刀混合刮抹剂

Norbak wearing compound　Norbak 耐磨[抗磨损]剂

Nordberg　诺德伯格跨国集团公司(专业生产破碎筛分设备)

Nordberg conveyer screen　诺德伯格运输筛分机

Nordberg crusher data sheet　诺德伯格破碎机数据表

Nordberg feeder grizzly　诺德伯格给矿格筛

Nordberg GP screen　诺德伯格 GP 筛

Nordberg gyratory　诺德伯格旋回破碎机

Nordberg gyratory crusher　诺德伯格旋回破碎机

Nordberg heavy duty jaw crusher　诺德伯格重型颚式破碎机

Nordberg heavy-duty EHD screen　诺德伯格重型 EHD 筛

Nordberg heavy-duty GP screen　诺德伯格重型 GP 筛

Nordberg jaw crusher　诺德伯格颚式破碎机

Nordberg overflow discharge ball-mill　诺德伯格溢流型球磨机

Nordberg primary gyratory crusher　诺德伯格型粗碎旋回破碎机

Nordberg Process Machinery Reference Manual　诺德伯格工艺设备参考手册

Nordberg rod deck screen　诺德伯格棒条筛

Nordberg rod mill　诺德伯格棒磨机

Nordberg screen line　诺德伯格筛系列[筛组]

normal　正常的,规定的,标准的,当量(浓度)的,正(链)的

normal alkyl pyridinium salt　正烷基吡啶盐

normal binding energy　正常结合能

normal commercial contract　普通商业合同

normal commercial rate　正常商业率

normal complement　标准套[组](仪表)

normal diffusion rate　正常扩散速度

normal dose　正常药[剂]量

normal equation　正规方程,法方程

normal equipment operation　设备正常操作[运转]

normal gold ore　正常(一般)金矿石

normal industrial practice　正常工业生产实践

normal instrumentation standard　正常仪表标准

normal operating supply　正常工作[生产]消耗品

normal operation　正常生产

normal orbit　正常轨道

normal orbital　正常轨道

normal orbital position　正常轨道位置

normal ores　正常矿石类

normal pebble milling　标准砾(石)磨

(矿)
normal range 正常范围
normal starting torque 正常起动转矩
normalizable component 可标准化分量
Norman Cleveland jig 诺曼·克里夫兰(圆形)跳汰机(美国)
northern climate 北方气候
notation 符[记]号,标志,注释,符号表示法,标志法,记[计]数法
notification 通知(单,书)
notorious ore type 众所周知的难选矿石类型
novel new instrument 新型仪表
novel process 新型[颖]工艺
Novonacco 湿润剂(烷基萘磺酸钠)
nozzle air aeration system 喷嘴空气充气系统
nozzle opening 喷嘴孔
nozzle orifice 喷嘴孔
n-propyl alcohol 正丙醇
N-sodium silicate N-硅酸钠
N-Sol(A,B,C,D) 絮凝剂(活性硅溶胶)$SiO_2 \cdot xH_2O$
N-substituted-n-dodecylamine N-取代-正十二烷基胺
n-type semiconducting character n型半导体特性
n-type semi-conduction n型半导电性
n-type semiconductor n型半导体
n-type specimen n型样品
Nuchar 纽恰尔牌活性炭(美国 Mead-Westvaco 公司产品)
Nuclear Chicago pulp density meter 芝加哥核矿浆密度计
nuclear density gauge 核子浓度计
nuclear density meter 核密[浓]度计
nuclear force 核子力
nuclear level switch 核料位开关
nuclear magnetic resonance study 核磁共振研究
nuclear physical 原子核物理
nuclear reactor cladding [原子]核反应堆金属包层
nuclear regulatory agency 核管理机构
nuclear sediment density meter 核子沉积密度计[表]
nuclear sensing system 核子检[探]测装置
nuclear source 原子能源
nuclear weigh scale 核子秤
nucleonic density meter 核子浓度计
nugget effect 块金效应
nugget trap 金块捕集器
nuisance to adjacent property 邻近财产的公害
Nujol 浮选药剂(石油软蜡,精制白油,可用作捕收剂、絮凝剂、调整剂)
Nullapon(=EDTA) 表面活性剂(乙二胺四乙酸,可用作脉石的抑制剂)
number of cleaner stage 精选段数
number of coefficients 系数个数
number of coordination 配位数
number of counter-current stages

逆流段数
number of determination 测定次数
number of man-shift 人班数
number of mill revolutions 磨机转数
number of moles 克分子数,摩尔数
number of pieces 个[件]数
number of run 运算次数
number of samples 样品[试样]数
number of spacing 间隔数
number of step 段[步骤]数
number of trains volume 系列数量
numbering and coding system 编号编码系统
numeric code 数字代码
numerical approximation 近似数值,

数值近似
numerical model 数字模型
nutrient 养料
nutrient solution 营养液
Nyfapon 湿润剂(十八烯基/十六烷基硫酸盐)
nylon 尼龙,酰胺纤维
nylon ply 尼龙层
nylon reinforced chlorinated polyethylene plastic membrane 尼龙强化的聚氯乙烯塑料薄膜
nylon screw 尼龙螺钉
nylon sling 尼龙吊环[索,绳具],尼龙链钩

O

O ring seal O形环密封
O-alkyl dithiocarbonate (xanthate) 氧-烷基二硫代碳酸盐(黄药)
objectionable odor 令人不愉快的[不良]气味
objective element 目的元素
objective function 目标函数
objective of control 控制目标
objective of research 研究目标
oblique angle triangle 斜三角形
observation 观察,观[探,实]测,监

[注]视,遵守,观测值,短评,按语,意见
obsolete price 过期价格
occupancy cost 营运成本,运行成本
occupational radiation protection 职业性辐照防护
occupational safety and health regulation 职业安全和保健条例[规程]
occupied area 占地面积
ocean environment 海洋环境
ocean shipping company 海运公司
ochre(ocher) 赭石($Fe_2O_3 \cdot Al_2O_3(SiO_2)$),

含有大量的砂和黏土),赭石类(泛指成土状产出的金属氧化物矿物);赭[黄褐]色

ochre brown 赭褐(色)的

octadecane 十八(碳)烷 $C_{18}H_{38}$

octadecanoic acid 十八(烷)酸,硬脂酸 $CH_3(CH_2)_{16}COOH$

octadecyl 十八(烷)基 $C_{18}H_{37}-$

octadecylamine 十八(烷)胺,硬脂胺(可用作捕收剂、湿润剂、絮凝剂) $CH_3(CH_2)_{17}NH_2$

octadecylamine acetate 十八胺醋酸盐,硬脂胺醋酸盐(可用作捕收剂、湿润剂、絮凝剂) $C_{18}H_{37}NH_2 \cdot CH_3COOH$

octaldoxime 辛醛肟(可用作捕收剂)

octane 辛烷 C_8H_{18}

octanoic acid 辛酸

octyl 辛基 $CH_3(CH_2)_6CH-$

octyl amine 辛胺 $CH_3(CH_2)_7NH_2$

octyl decyl amine acetate (ODAA) 辛基癸基醋酸胺(缩写 ODAA)

octyl hydroxamate 辛基氧肟酸盐

octyl hydroxamic acid 辛基氧肟酸

octyl pyrophosphoric acid 辛基焦磷酸(可用作氧化铁矿的捕收剂) $C_8H_{17}HP_2O_7$

odor threshold 气味[嗅觉]阈值

O-Emulsifier 乳化剂(石油磺酸盐,可用作铝土矿的捕收剂)

off grade product 等外品,次品

off period 关闭[断开]期间

off-center load 偏心负荷

off-center shaft 偏心轴

off-center weight 偏心重

off-grade product 等外产品

office 办公室,办事处,营业[事务]所,政府机关,业务组织

office building 办公楼

office clerk 职员

office consumables 办公消耗品

office heating 办公室采暖

office of commercial counselor 商务参赞处

office supplies 办公用品

office-lunch room 办公兼午餐室

off-line analysis 离线分析

off-line batch test 离线分批试验

off-line optimization 离线最优化

off-line optimization program 离线调优程序

off-line optimization strategy 离线最优化对策

off-line optimization study 离线调优[最优化]研究

off-line optimization technique 离线调优技术

off-line program 离线程序

off-line strategy 离线策略

off-line technique 离线技术

off-line test 离线[间接]试验

off-line [on-line] data processing 脱机[联机]数据处理,间接[直接]数据处理

off-loading station 卸料[矿]站[台]

off-set inlet feed chute 分岔[旁通]进[给]料溜槽
offshore currency market 国外货币市场
offshore purchase 国[海]外采购
often-ignored aspect 常被忽视的方面
ohmic contact 欧姆接触,电阻性接触
ohmic loss 电阻损耗
oil absorption 油吸收
oil black-out 油封(在矿浆表面形成不溶性药剂薄膜,以阻止起泡)
oil burner 油燃烧器
oil droplet 油滴
oil droplet size 油滴大小
oil emulsifier 油乳化剂
oil flinger 抛油环[圈]
oil flinger housing 抛油环[圈]套
oil flow 油流(量,速度)
oil immersed clutch disk 油浸离合[连接]圆盘
oil lubricating pumping system 油润滑泵[扬送]系统
oil retaining ring 挡[护]油环
oil tempered steel 油回火的钢
oil trap 油阱,捕油器
oil well drilling 油井钻探
oily contaminant 油沾污物
oily liquid 油质液体
oily oxidation product 油类氧化产品
oily phase 油相
OK 16 machine OK 16 型浮选机(容积 $16m^3$,芬兰奥托昆普公司产品)

OK 3 cell OK 3 型浮选机[槽](容积 $3m^3$,芬兰奥托昆普公司产品)
OK floatation cell OK 型浮选机[槽](即芬兰的奥托昆普 OK 型浮选机)
OK floatation machine OK 型浮选机(即芬兰的奥托昆普 OK 型浮选机)
OK-mechanism OK 机械装置(OK 浮选机的)
old model 旧型号,旧模式
oleate 油酸盐[根,酯]
$CH_3(CH_2)_7CH:CH(CH_2)_7COOM$
oleate adsorption 油酸根吸附
oleic acid 十八烯酸,油酸
$C_{17}H_{33}COOH$
oleylalcohol 油醇(可用作捕收剂、发泡剂)$CH_3(CH_2)_7CH:CH(CH_2)_8OH$
oleyl sulfate 油醇硫酸脂,油烯基硫酸脂(可用作捕收剂、乳化剂、湿润剂)
$CH_3·(CH_2)_7·CH:CH·(CH_2)_7·CH_2OSO_3M$
Oliver continuous filter 奥利弗连续式过滤机(美国奥利弗过滤器公司生产)
Omo 表面活性剂(烷基芳基磺酸盐,可用作絮凝剂)
on line analytical instrument 在线分析仪表
on period 接通期间,工作期间
on site 在工作位置[工地],当[原,就]地
one line diagram 单线图
one piece 一[单]片的,单体的
one shift operation 一班生产[运行]

one stage crushing plant 单段破碎车间
one-component system 单组分体系
one-dimensional 一维的
one-product mill 单一产品的选矿厂
one-sided intake turbofan 一侧进气透平风机
one-time cost 一次性成本
one-time investment 一次性投资
one-unit operation 一个单元作业,单一作业
on-going changes 运行中的变化,正在进行的变化
on-going mine 正在开采的矿山[区]
on-line analyzer 在线分析器
on-line instrumentation 在线测试仪表
on-line optimization 在线最优化
on-line optimization program 在线最优化程序
on-line optimization strategy 在线最优化对策
on-line parameter 在线参数
on-line particle size analyzer 在线[流线,线上]粒度分析仪
on-line particle size measurement 在线粒度测量
on-line process measurement 在线过程测量
on-line program development 在线程序研发
on-line sensing 在线检测
on-line sensor 在线传感器
on-line single detector 在线单探头

on-line size assessment 线上粒度查定
on-line X-ray analysis equipment 在线X射线分析设备
on-line X-ray analyzer 在线X射线分析仪
on-schedule within-budget completion 按计划按预算完成
on-site 现场,就地
on-site cost 现场费用
on-site development 现场研发
on-site engineering personnel 现场[矿区]工程技术人员
on-site installation 现场安装
on-site office 现场办公室
on-site operation 现场作业[生产]
on-site programmer 现场程序设计员
on-site test 现场试验
on-stream analysis 载流[在线,流程,连续]分析
on-stream analysis unit 载流[在线]分析装置
on-stream analyzer 载流[在线]分析仪
on-stream composition analysis 载流[在线]成分分析
on-stream particle size analyzer 截载流[在线,联机]粒度分析器
on-stream system 载流[在线]系统
on-stream system X-ray system 载流[在线]系统的X射线系统
on-stream X-ray analyzer system 载流[在线]X射线分析仪系统
on-stream X-ray system 载流[在线]X

射线系统
on-the-job instruction 工作相关的指导
on-the-job training 在职[岗位]培训
OP-10 表面活性剂(辛基苯酚聚乙二醇醚,可用作捕收剂、湿润剂)
$C_8H_{17}\cdot C_6H_4\cdot(OCH_2CH_2)_{10}OH$
OP-100 表面活性剂(氧化煤油,一种沸点260~350℃的高馏分石油的液氧化物,可用作捕收剂)
opaque carbonaceous matter 不透明含碳物质
OPE-16 表面活性剂(叔辛基苯酚聚乙二醇醚,可用作湿润剂)
$C_8H_{17}\cdot C_6H_4\cdot(OCH_2CH_2)_{16}OH$
OPE-20 表面活性剂(叔辛基苯酚聚乙二醇醚,可用作湿润剂)
$C_8H_{17}\cdot C_6H_4\cdot(OCH_2CH_2)_{20}OH$
OPE-30 表面活性剂(叔辛基苯酚聚乙二醇醚,可用作湿润剂)
$C_8H_{17}\cdot C_6H_4\cdot(OCH_2CH_2)_{30}OH$
open account 未(结)清账目
open air stockpile 露天堆场[矿堆]
open air storage 露天堆场
open area 筛孔面积,露天场地,过流面积
open area factor 筛孔面积系数
open beaker 开口[无盖]烧杯
open building 露天建筑物
open cascade design 开路串级[梯流]布置的设计
open circle 开圆
open circuit 开路
open circuit ball mill 开路球磨
open circuit cleaning 开路精选
open circuit condition 开路状态
open circuit continuous(at steady state) behavior 开路连续流程(稳态)特性
open circuit cycloning 开路旋流器分级
open circuit flotation bank 开路浮选机
open circuit inefficiency multiplier 开路低效率系数
open circuit mill design calculation 开路磨机设计计算
open circuit operation 开路作业[操作]
open circuit oxidation 开路氧化
open circuit product curve 开路产品曲线
open circuit rod mill 开路棒磨
open circuit rod-ball mill 开路棒球磨
open flow 无阻流量,直流式
open line of credit 开口信用额度
open loop control 开环[式]控制
open market 公开市场
open navigation season 开航季节
open pan vibrating conveyer 敞式[无罩]槽形振动输送机
open pit deposit 露天采矿床
open pit operation 露天生产[开采作业]
open side discharge setting 宽边排矿口开度
open side setting 开边[最大]排口宽度
open spiral construction 开式螺旋结构
open squirrel cage type rotor 敞开式鼠

笼型转子

open stockpile 敞开式[无盖,露天]矿堆

open type machine 敞流式浮选机

open water bath 敞口水浴

open-air construction 露天建筑[施工]

open-circuit condition 开路条件

open-circuit rod mill 开路棒磨

open-circuited rod milling 开路棒磨磨矿(方法)

open-ended discharge trunnion 开端排料式耳轴

open-ended rod mill 开端式棒磨(排矿口直径大于给矿口)

open-flow flotation machine 敞流式浮选机

opening column 开孔[筛孔]栏

opening mobile 开口移动

opening size 开孔[筛孔]尺寸

open-space filling 孔隙充填

operate 操作,驾驶,运转,经营

operating and maintenance personnel 操作维修人员

operating area 生产区域

operating availability 运转率

operating cabin 操纵[作]室

operating characteristic 操作[运行]特性

operating concentrator 生产选矿厂

operating condition 工作[操作]条件

operating constraint 操作限制

operating cost 使用费,营运[经营]成本

operating cost comparison 生产费用比较

operating cost estimate 生产成本估算

operating cost estimation 经营费用估算

operating cost from feasibility study 可行性研究得出的生产[运行]费用

operating cost item 生产成本项目

operating cost maxim 生产成本准则

operating cost study 生产费用分析[研究]

operating cycle 生产周期

operating data 生产[操作]数据

operating day 工作[生产,运转]日

operating decision 生产[运营]决策

operating efficiency 操作[作业,运转]效率

operating efficiency factor 操作[工作,运转,作业]效率因素

operating equipment 生产设备

operating experience 操作[工作,生产]经验

operating flowsheet 生产[操作]流程图

operating frequency 操作频率

operating grade 作业品位

operating grade-recovery curve 作业品位-回收率曲线

operating guide 生产[运行]指南[指导原则]

operating horse power 生产功率[马力]

operating hour 运转时间

operating humidity 工作[操作]湿度

operating income 生产收益

operating instruction 操作规程[说明

书,指令]
operating instruction manual 操作规程[手册,说明书]
operating labor 生产劳动力[工人]
operating labor cost 生产工人费用[工资]
operating level 操作水平
operating license 营业执照
operating life 生产[使用]寿命[年限]
operating light 操作指示灯
operating limitation 使用[工作,操作,有效]限度
operating manpower 生产人员
operating manual 操作手册
operating material 生产材料
operating objective 生产[运转,运行]目标
operating overhead 生产管理费
operating parameter 操作参数
operating people 操作人[者]
operating performance 经营业绩,操作[运行]性能
operating permit 生产许可证
operating personnel 操作[生产]人员
operating philosophy 操作原理
operating plant 生产[选矿]厂
operating point 操作点
operating policy 操作方针
operating power 运转功率
operating prediction 生产预测
operating principle 操作原理
operating problem 生产[操作]问题

operating question 操作问题
operating rate 工作速率
operating recommendation 操作建议
operating record 操作记录
operating recovery 作业回收率
operating requirement 作业[工序]要求
operating return 作业[生产,营业]收益[利润]
operating routine 日常操作[生产]工作
operating schedule 生产制度
operating shift 生产班
operating spectrometer 运行中的光谱仪
operating speed 运行速度
operating station 操作站
operating statistics 生产统计
operating store 生产备件
operating strategy decision 生产策略决定[判定]
operating strategy option 生产策略选择方案
operating system 操作系统
operating technique 操作技术
operating temperature 操作[作业,生产]温度
operating typewriter 工作打字机
operating value 工作数据[值]
operating variable 运行[操作]变量
operating water expense 生产用水费用
operating work index 作业[操作]功指数
operation 运转,运行,操作,作业,操作

法;经营,营业;手术

operation and control philosophy 操作和控制原理[准则]

operation and maintenance training 操作和维修[护]培训

operation area 工作[操作]区域

operation cycle 操作[工作]循环

operation function 操作职能部门

operation hour 运行[作业]时间

operation manual 操作手册,操作说明书

operation mode 操作方式

operation of concentrator 选矿厂的运行[生产]

operation parameter 操作参数

operation rate 生产[开采]速度

operation specification 操作规程

operation stage 生产期[阶段]

operational change 操作变化

operational complexity 操作复杂性

operational considerations 操作(需要考虑的)事项[问题,条件]

operational data 操作数据

operational factor 操作因素,运算因子

operational final control actuator 运行终端控制执行机构[传动装置]

operational flexibility 生产灵活[机动]性

operational performance 使用特性,作业绩效

operational requirement 操作要求

operational setting 操作调节

operator 经营[操作,生产]人员;运算符

operator efficiency 操作人员效率

operator keyboard input 操作员键盘输入

operator training 操作人员培训

operator's console 操作人员控制台

operator's manual 操作(人员)手册

Opoil 塔尔油与脂肪酸的混合物[脂肪酸](品名,可用作黄铁矿的捕收剂)

opportunity cost of investment 投资机会费用

opportunity identification technique 机会[遇]识别技术

opposing dipole 相斥偶极

opposite direction 相反方向

oppositely charged ion 电性相反的离子

optic filter 滤光器(片)

optic instrument 光学仪器

optic microscope 光学显微镜

optic microscopy 光学显微术

optic signal 光信号

optical detector 光探测器

optical method 光学方法

optical reflection coefficient 光(线)反射系数

optimal beneficiation procedure 最佳选矿方法[工艺]

optimal circuit arrangement 最佳的回路布置

optimal condition 最佳状态

optimal controller settings 最佳控制器给定值

optimal flowsheet design 最佳流程设计
optimal strategy 最优策略
optimal value 最佳值
optimistic assumption 最有利的假设
optimistic side 最有利的方面
optimization method 最优化方法
optimization model 最优[佳]化模型
optimization procedure 最优化方法[程序]
optimization program 最佳化程序
optimization purpose 调优目的
optimization routine 最优化程序
optimization strategy 最优化[调优]对策
optimization study 最佳化研究
optimization technique 最佳化技术
optimizer 调优化程序,优化控制器
optimizing control strategy 最优化控制对策
optimizing plant design 最优[佳]工[选]厂设计
optimizing reagent use mode 调佳药剂使用方式
optimum alkalinity or acidity 最佳酸碱度
optimum choke-feed level 最佳挤满给矿料位
optimum circulating load 最佳循环负荷
optimum condition 最佳条件[状态]
optimum economy 最佳经济
optimum efficiency 最佳生产效率
optimum equipment utilization 最佳设备利用率
optimum feed size 最佳给矿粒度
optimum fineness 最佳细度
optimum moisture 最佳水分[湿度]
optimum operating condition 最佳操作条件
optimum performance 最佳操作[性能,成绩]
optimum plant performance 最佳车间运行[设备运行]
optimum process mechanics 最佳工艺力学
optimum profitability 最佳利润率[赢利性]
optimum rougher metallurgy 最佳粗选指标[效果]
optimum strategy 最优策略
optimum value 最佳值
option agreement 选择权协议
optional step 可选步骤
Ora 含 C_{15} 伯胺(36.2%)及50.8%的碳氢化物(品名,可用作烧绿石的捕收剂)
Oranap 表面活性剂(烷基萘磺酸钠,可用作氧化铁矿的捕收剂)
$RC_{10}H_6SO_3Na$
Oranit B 表面活性剂(多烷基萘磺酸盐,可用作湿润剂)
order 次序,顺序,秩序,指[命]令,指示,规则,等级,阶,次,数量级,种类,目,族,整理[顿],安排,订货[购](单)
order of ability 能力(大小)次序
order of decreasing floatability 可浮性

递降次序
order of magnitude 数量级,绝对值的阶
ordering 订购
order-of-magnitude capital cost estimate 数量级基建投资估算
order-of-magnitude cost 数量级费用
order-of-magnitude estimate 数量级估算
order-of-magnitude operating cost 数量级生产费用
order-of-magnitude study 数量级研究
ordinary pyrite 一般[普通]黄铁矿
ore assaying 矿石分析
ore axial flow 矿石轴向流(量)
ore bin grate 矿仓格栅
ore blend 配矿,混矿
ore blending 矿石混合,配[混]矿
ore block 矿块[段,体]
ore car 矿车
ore carrying idler 承载矿石的托辊
ore change 矿石(性质)变化
ore column 矿[料]柱
ore composition 矿石组成[成分]
ore description 矿石描述
ore dressing industry 选矿工业
ore dressing laboratory 选矿实验室
ore dressing metallurgist 选矿专家
ore dressing method 选矿法
ore dressing practice 选矿实践
ore dressing research staff 选矿研究人员
ore feed complexity 给矿复杂性
ore feed size 给矿粒度
ore flow 矿石流量
ore grindability 矿石可磨性
ore grindability difference 矿石可磨性差距
ore hardness variation 矿石硬度变化
ore head 入选原矿
ore lot 矿石堆
ore lump 矿块
ore mass 矿体
ore material 矿石料
ore matrix 原矿
ore milling 选矿
ore mineralogy 矿石矿物学[研究]
ore nest 矿巢
ore processing facilities 选矿设施
ore processing unit 选矿设备
ore producing area 矿石开采区
ore production 矿石生产,出矿
ore properties 矿石性质
ore quality 矿石质量
ore reclaim 矿石回收装矿点
ore reserve 矿石资源
ore reserve determination 矿石埋藏量确定
ore reserve estimation 矿石储量估算
ore reserve method 矿石储量法
ore resources 矿石资源
ore sample 矿样
ore slurry 矿浆
ore storage and reclaiming system 矿石堆取系统

ore storage area 储矿场[区]
ore storage facility 矿石存储设施
ore storage system 矿石储存系统
ore test 矿石试验
ore testing laboratory 矿物[选矿]实验室
ore transfer pile 矿石转运堆
ore treatment 选矿
ore treatment cycle 选矿操作周期[循环]
ore treatment rate 矿石处理量
ore variation 矿石(性质)变化
ore-bearing rock 含矿岩石,矿岩
orebody 矿体
orebody continuity 矿体连续性
orebody definition 矿体界限确定
orebody mineralogy 矿体的矿物学性质
orebody sampling 矿体采[取]样
ore-clay mixture 矿石黏土混合物
ore-waste unit sampling cell 矿石废石组合取样单元
Orfom 美国菲利普石油公司生产的选矿药剂品牌名
Orfom CO300 氢硫基类(Sulphydryl)药剂(为多种硫化矿捕收剂,用于辉钼矿浮选时,与其他捕收剂有协合作用,能增加铜和钼的回收率)
Orfom CO400 辉钼矿的优良捕收剂(对辉钼矿浮选很有效,与Orfom MCO类似)
Orfom CO800 氢硫基药剂(可用作钼、铅、锌硫化矿的捕收剂)
Orfom D8 水溶性有机物(成分为二羧甲基三硫代碳酸钠,辉钼矿和硫化矿浮选时用作铜钼分离及各种矿物硫化物的选择性抑制,可单独使用,代替诺克斯药剂、硫化钠和氰化钠,或与上述药剂结合使用,其毒性小)
Orfom MCO 由几种石油化工产品用专门方法合成的非极性复合烃类药剂(是辉钼矿的优良捕收剂,一般加入一段磨矿,能有效提高特别是斑岩型铜钼矿中钼的回收率)
organic agent 有机剂
organic constituent 有机成分
organic dispersant 有机分散剂
organic electrolyte 有机电解质
organic extractant 有机萃取剂
organic extraction technique 有机(溶剂)萃取法
organic flotation agent 有机浮选剂
organic flotation reagent 有机浮选剂
organic group 有机基
organic material 有机物质
organic modifier 有机调整剂
organic modifying agent 有机调整剂
organic phase 有机相
organic phase composition 有机相组成
organic phosphate 有机磷酸盐
organic polyelectrolyte 有机聚[高分子]电解质
organic solvent 有机溶剂

organic species 有机类(药剂)

organic sulfur 有机硫

organization chart 组织表

organometallic species 有机金属化合物类(药剂)

organophosphate 有机磷酸盐

organo-poly-siloxanes 有机聚硅烷,有机多分子硅醚(可用作捕收剂)

orientation test 定向试验

orientation test for dip 测斜试验,测定倾角方位试验

origin of potential 电位起因[源]

origin of sample 试样来源

original bid 原始报价

original data 原始数据

original equipment 原设备

original feed 原给矿[料]

original publication 原文[版]

original purchase agreement 原采购协议

original simplex 原始单项

original specification 原技术规范

original value 原值

original weight 原重

Ornite 表面活性剂(烷基芳基磺酸盐,可用作湿润剂)

ornithin 鸟氨酸,二氨基戊酸(可用作抑制剂)

$H_2N(CH_2)_3CH:(NH_2)COOH$

Oronite 一种低沸点汽油(品名,可用作起泡剂)

Oronite D-40 表面活性剂(烷基苯磺酸盐,可用作湿润剂)

Oronite D-60 表面活性剂(烷基苯磺酸盐,可用作湿润剂)

Oronite purified sulfonate L 水溶性石油磺酸盐(品名,可用作重晶石的捕收剂)

Oronite S 表面活性剂(四聚丙烯—苯基磺酸钠盐,同十二苯基磺酸钠盐,可用作洗涤剂及浮选锰矿的湿润剂)

Oronite wetting agent 湿润剂(油溶性石油副产品,也可用作捕收剂)

Oronite wetting agent S 水溶性阴离子型湿润剂(51%烷基磺酸钠,7%硫酸钠Na_2SO_4,42%水,也可用作乳化剂、湿润剂)

Orso 植物油中性皂(品名,可用作浮选白钨矿捕收剂)

orthogonal direction 正交方向

orthogonal experiment design 正交试验设计

orthophosphate 双辛基磷酸酯(正磷酸盐,可用作捕收剂)M_3PO_4

orthophosphoric acid 正磷酸 H_3PO_4

orthophosphoric acid bath 正磷酸液池

orthorhombic crystal structure 斜方晶体结构

orthorhombic structure 斜方晶结构

Orvus Es-paste 表面活性剂(伯烷基硫酸盐,烷基大于C_{12},可用作湿润剂)

OrvusWA 表面活性剂(十二烷基硫酸盐,可用作湿润剂)

Orzan A 表面活性剂(木素磺酸铵,可用作抑制剂及稀土矿捕收剂)

OSA stream sequence(OSA=on-stream analysis) OSA 采样程序(OSA=在线分析)

other depressants 其他抑制剂

other fringe labor cost 其他附加劳工成本

other interested parties 其他关注方

other(various)data 其他(各种)数据

out flight diameter 螺旋外直径

out of way position 边远位置,离开(输送)路线的位置

outage 停机(时期,状态),运输中断,停[断]电,供电中断,运输[贮压]中损失量,储运损耗

outboard bearing 外侧轴承

outboard position 外侧位置

outdoor step-down transformer 室外降压器

outdoor storage pile 露天储堆

outdoor type 露天式

outdoor-indoor type 露天-室内式

outer countershaft bushing 传动轴[副轴]外轴瓦

outer eccentric bushing 偏心轴外衬套

outer radius 外半径

outer side 外侧

outer spring bolt 外弹簧螺栓

outfit 全套设备[工具]

outflow rate 流出流量[速率]

outflowing valley 外流[排料]谷槽

outgassing temperature 排[逸]气温度

outgoing stream 溢[逸出]流量

outlet diameter 排口直径

outline and mounting dimension 外形与安装尺寸

outline dimension 外形[轮廓]尺寸

outline of content 内容提纲

out-of-pocket cost 付现成本(即用现金支付的成本),现金成本(指企业在生产经营过程中以货币形态存在的资产)

out-of-pocket expense 现金支付,实际支出,付现费用

out-of-way 边远的

Outokumpu flotation machine (芬兰)奥托昆普浮选机

Outokumpu rotor type impeller 奥托昆普转子式叶轮

output device 输出装置

output pulse signal 输出脉冲信号

output rate signal 输出速度信号

output RPM 输出转/分

output signal 输出信号

output value 产值

output variable 输出变量

outside 外部,外侧,外界,外表

outside engineering company 外包工程公司

outside firm 企业外部
outside line 外部线路
outside operating function 厂外操作职能部门,厂外操作职责
outside party 外界团体
outside storage 露天堆场
outstanding amount 待结金额,未偿还借款额[数额]
outstanding debt 未偿债务
outstanding loan 未偿还贷款
outstanding not payable 未偿付的应付票据
outstanding share 未清偿的股份
outward radial direction 向外径向
over simplification 过于简单化
overall analysis 全分析
overall automation scheme 总体自动化方案
overall capital cost 总投资费用
overall criterion 总指标(如回收率或每吨成本)
overall design 总体设计
overall economic condition 总体经济情况
overall economic consideration 整个经济问题
overall effect 总效应[效果,作用]
overall expansion program 全面扩建计划
overall fractional recovery 总分布回收率
overall macro-kinetics 一般宏观动力学
overall objects of simulation 所有模拟对象
overall operating system 整个操作系统
overall optimum profitability 全面最佳利润率
overall planning 全面规划
overall plot plan 总平面图,全厂总体规划
overall polarity-non-polar character 总极性-非极性特征
overall power cost 总电费
overall process efficiency 整个生产[工艺]过程的效率
overall project cost 整个工程项目投资费用
overall project plant 综合的工程装置
overall result 总[综合]结果
overall size 外形[轮廓]尺寸,外部最大尺寸
overall system design 整个系统设计
overall utility 全面应用
overall utilization 总利用率
over-blasting 超[过度]爆破
overbreak material 过碎物料
over-crushing 过破碎
overcurrent protection 过电流保护
overdesign 保守[过于安全,过余量]的设计
over-design parameter 保守[过于安全,过余量]的设计参数
overdrain classifier 顶流分级机

over-fed 加[给]料过多[量]
overfeed surge 给矿过多[超量给矿]的波动
overfilling 超充填
overflow % solids 溢流固体的百分量
overflow box 溢流箱[槽]
overflow capacity consideration 溢流量应考虑的问题
overflow clarity 溢流液澄清度
overflow condition 溢流条件
overflow density 溢流浓度
overflow discharge 溢流排料[矿]
overflow particle size 溢流粒度
overflow percent solids 溢流固体百分率[数]
overflow pressure 溢流压力
overflow product 溢流产品
overflow rod mill 溢流棒磨机
overflow slurry 溢流矿浆
overflow upcomer 粗砂溢流管
overflow velocity 溢流速度
overflow volume 溢流容量[体积]
overflow water 溢流水
overflow weir crest 溢流堰顶脊
overflowing 溢出
overflowing froth 溢流泡沫
overfrothing condition 起[发]泡过多情况
over-frothing condition 过量[分]起泡情况
overhang 伸出(物),延伸量,悬垂(物)
overhead 高架[架空]的,管理费
overhead charge 管理费,基本费用,日常开支
overhead clamshell crane (蛤壳式)抓斗吊车[起重机]
overhead conveyer 架空运输机
overhead cost 间接成本,杂项开支
overhead eccentric jaw crusher 高架偏心颚式破碎机
overhead expense 管理费
overhead line 架空线路
overhead screen 高架[架空]筛
overland conveyer system 地面运输机系统
overland flow technique 地面径流处理法
overland transport 陆地运输
overload 超[过]载
overload factor 过载系数
overload indicator 过负荷指示器
overload ratio 超载率
overload tonnage 过载吨位数
overloading 过(装)载
overpowering effect 难以抗拒[强烈]的作用
overpressure wave 剩余压力波
overrun 超支,超速
overs 筛渣,筛除物
overs product 筛上产品
over-sampling 超额取[采]样
overseas 国[海]外
overseas investment 海外投资
oversize correction factor 过大颗粒修正系数
oversize feed factor 过大粒度给矿

系数
oversize scissor conveyor 筛上(产品)返回运输机(剪刀形)
over-sized 超规格的
overstable froth 过稳定泡沫
overtime allowance 加班费
overxanthating 黄药过量
owner 所有者,产权人
owner preference 产权人的偏爱
ownership 所有权
ownership percentage 所有权百分数[率,比]
ownership retention agreement 所有权保有协议
oxalate 草酸盐
oxalic acid 草酸乙二酸 HOOC·COOH
oxidation condition 氧化条件
oxidation control 氧化(作用的)控制
oxidation cycle 氧化周期
oxidation depth 氧化(作用)深度
oxidation potential 氧化电位
oxidation process 氧化工艺[过程]
oxidation product 氧化产物
oxidation segment 氧化段
oxidation state 氧化状态
oxidation tank 氧化槽
oxidation zone 氧化区
oxidation-reduction treatment 氧化-还原处理法
oxidation-titration 氧化-滴定法
oxidative dissolution 氧化溶解
oxide 氧化物
oxide copper analysis 氧化铜分析
oxide leaching plant 氧化矿浸出厂
oxide molybdenum hydrometallurgical plant 氧化钼水冶厂
oxide pilot-plant campaign 氧化矿半工业试验厂试验期间
oxide surface 氧化表面
oxide-sulfide copper ore 氧化-硫化铜矿石,混合铜矿石
oxide-sulfide transition zone 氧化-硫化(物)过渡带
oxidization depth 氧化深度
oxidized capping 氧化覆盖
oxidized coal 氧化煤
oxidized form 氧化形态
oxidized free gold 氧化的自然金[砂金]
oxidized galena 氧化方铅矿
oxidized kerosine 氧化煤油,可用作捕收剂
oxidized layer 氧化层
oxidized lead ore 氧化铅矿石
oxidized paraffin 氧化石蜡
oxidized paraffin soap 氧化石蜡皂 RCOOH
oxidized petrolatum 氧化软蜡
oxidized recycle 氧化重整煤油(可用作捕收剂)
oxidized taconite 氧化铁燧岩
oxidizing condition 氧化条件
oxidizing treatment 氧化处理
oxime 肟 —CH(:NOH)
oximido(=hydroxyimino) 肟基
oxine 8-羟基喹啉(可用作捕收剂) HO·C_9H_6N

oxyethyl cellulose 羟乙基纤维素(可用作抑制剂、絮凝剂)
oxygen bridge 氧桥
oxygen deficiency 缺氧
oxygen demand 氧气需用量
oxygen gas 氧气
oxygen meter 氧量计
oxygen overpressure 剩余氧压
oxygen potential 氧含量[潜力]
oxygen reduction 氧还原
oxygen reduction activity 氧还原活性
oxygen reduction current 氧还原电流
oxygenated site 氧化区
oxygenated solution 充氧溶液,氧合液
oxygenated water 含氧水
oxygenation 用氧处理,(使)氧化,用氧饱和,充[供]氧作用
oxygen-bearing water 含氧水
oxygen-free 无[不含]氧的
oxygen-free solution 无氧溶液
oyster 牡蛎,蚝
oyster shell 牡蛎壳
ozone addition 充臭氧
ozonization 臭氧化

P

6 pocket hydrosizer 六室水力分级机
80% passing size in microns 通过量占80%的粒度(微米为单位)
P. E. flotation oil (= Petroleum-extract flotation oil) P. E. 浮选油(低沸点石油馏分和裂解产物,不溶于水,不含有极性基团,可用作石墨、硫黄及煤的捕收剂)
P. M. Colloid No. 4V 絮凝剂(氧化的动物蛋白质衍生物)
P. P. B. (pentadecyl pyridinum bromide) 十五烷基溴化吡啶(可用作捕收剂)
Pachuca 帕丘卡空气[气升,压气式]搅拌浸出槽(因为首先应用于墨西哥帕丘卡金矿而得名)
Pachuca agitator 帕丘卡(气升式)搅拌槽[器]
Pachuca tank 帕丘卡(空气搅拌浸取)槽[罐]
Pacific dryer Pacific 型干燥机
pack 包装,包裹,压紧,捆扎,塞满;组,组合(件);一堆,一伙
packaged operating system 成[配]套操作系统
packaging 包装
packed bed-downflow type column 封闭式底流型萃取塔[柱]
packed bed-downflow type of column

充填床层下流型交换柱

pad 场地垫层[地基]

pad construction 场基建筑[构筑,建造]

pad heap leaching 有垫层的堆浸

paddle abrasion machine 叶片磨损机

paddle abrasion test 叶片磨损试验

page number 页号(数,码)

paint industry 颜料工业

paint system 油漆方法

Palatinol 表面活性剂(邻苯二甲酸二丁酯,可用作起泡剂、活化剂、调整剂)

Palconate 五棓子酸钠(品名,浮选长石和白钨时抑制方解石)

Palcotan 抑制剂(来自红木皮的木素磺酸盐,也可用作分散剂)

palcoton 红木木质素制品(可用作浮选辰砂的细泥抑制剂)

Pale neutral oil vis 100 100号灰色中性油(可用作浮选辉钼矿的捕收剂)

Paleozoic 古生代的

Paleozoic era 古生代

Paleozoic limestone 古生代石灰岩

Paleozoic siltstone 古生代粉砂岩

pallet 托盘,货板,平板架,货架[盘],码垛盘

pallet truck 堆积起重车

palletizing machine 造球[制粒,渣块压制,团矿]机械

Pamak fatty acid 塔尔油脂肪酸(品名,可用作氧化铁矿、锰矿的捕收剂)

Pamform 絮凝剂(聚丙烯酰胺与甲醛的共聚物)

pan feeder 盘式给矿机

pan type 盘式

pan width 槽宽(给矿机的)

Pan-American jig 泛美跳汰机

panel indicator 仪表盘指示灯

panel meeting 专家小组会议

panel meter 面板仪表,嵌镶式仪表

panel mounting 屏式安装

Pang 絮凝剂(聚丙烯酰胺)

panning 淘金盘淘金

panning evaluation 淘(金)盘测定[评定]

pant leg chute 脉[振]动腿[支架]溜槽

pantograph mounting 受[导]电弓安装

pan-type(precipitate)dryer 盘式(沉淀物)干燥机

paper industry 造纸工业

paper mill by-product tall oil 造纸厂副产品塔尔油

paper money 纸币

papermaking 造纸

par value 面值(规定的货币票面价值)

para- 对(位);仲;聚(合);副

para-acid 仲酸(无机酸),对位酸(有机酸)

parabolic 抛物线的

parabolic minimum 抛物线最小值

parabolic rate equation 抛物线速度方程

parabolic rate law 抛物线速度定律
parabolic relation 抛物关系
parabolic way 抛物线形式
parallel flotation circuit 平行浮选回路
parallel mechanism 平行机理
parallel reaction 平行反应
parallel rod deck 平行棒条筛面
parallel sizing zone 平行分级[压碎]带
parallel tunnel 平行地道
parallel zone 平行带[区]
parameter calculation 参数计算
parameter estimation 参数估算
parameter estimator 参数计算机
parameter improvement 参数修正值
parameter of optimization 优化参数
parameter vector 参数矢量
parametric design data 参数设计数据
para-tolyl arsonic acid 对甲苯胂酸
parent company 母公司
parent population 母体
parking area 停车场
Parnol 洗涤剂(烷基甲苯磺酸盐)
part container ship 半集装载船
part winding 部分绕组,分线圈
partial derivative 偏导数,偏微商
partial ionization 部分电离
partial pressure 分压
partially riffled deck 局部(铺)格条床面
participation 参与,分享
particle aggregate 矿[颗]粒集合体

particle bubble aggregation 颗粒气泡聚合
particle concentration 颗粒浓度
particle count method 颗粒计数方法
particle density variation 颗粒浓度变化
particle dimension 颗粒尺寸
particle dispersion 颗粒扩散
particle kinetic energy 粒子动能
particle locking 颗[矿]粒连生
particle mass 颗粒质量
particle migration 颗粒迁移[移动]
particle mineralogy 粒度矿物学
particle morphology 颗粒形态[结构]
particle orientation 颗粒取[定]向
particle population statistics 颗粒数统计
particle settling rate 颗粒沉降速率
particle shape factor 颗粒形状[态]因素[系数]
particle size analysis 粒度分析
particle size analyzer 粒度分析仪
particle size based grinding control 以粒度为基础的磨矿控制
particle size constraint 粒度限制
particle size control 粒度控制
particle size correction 粒度校正
particle size distribution (颗粒)粒度分布
particle size distribution transmitter 粒度分布[组成]变送[传送]器
particle size monitor 粒度监测器

particle size of material 物料粒度
particle size output signal 粒度输出信号
particle size setting 粒度给定值
particle species 矿种[各种矿物]颗粒
particle structure 粒子结构
particle suspension 颗粒悬浮
particle type 颗粒类型
particle-bubble adhesion mechanism 颗粒[矿粒]-气泡附着机理
particle-bubble aggregation 颗粒气泡聚合
particle-bubble attachment mechanism 颗粒气泡附着机理
particle-bubble encounter 颗粒气泡相碰撞
particle-fluid entrainment mechanism 颗粒-流体夹带机理
particle-particle interaction 颗粒间互相作用
particular assembly 粒子组成
particular beneficiation problem 特殊选矿问题
particular block of ore 特定矿石块段
particular circumstances 特定环境[情况]
particular concentrator 特定选矿厂
particular condition 独特条件
particular flotation circuit 特定浮选回路
particular material 特定物质[料]
particular mineral deposit 特殊矿床
particular option 特殊方案
particular particle size 特定粒度
particular separation 特殊分离
particular situation 特殊[定]情况
particular strategy 特殊策略
particular terminology 特种术语
particular time 特定时间
particulate bed electrode 颗粒层电极
particulate concentration 粉尘[微粒]浓度
partition factor 区分[分配]因子
partitioned matrix 分块矩阵
partner 伙[同]伴,合作者,合伙人
partners' contribution 合伙人的出资
partnership 合伙[作]关系,合伙组织
parts availability 部件可用[有效,实用]性
parts distribution department 零件供应部
parts inventory 库存配[零,部]件
Parvalux shunt-wound dc motor 帕瓦鲁克斯并激直流马达(英国 PARVALUX 公司产品)
passive layer 钝化层
passive pyrite surface 钝态黄铁矿表面
passive zone 不活动带
patch 矿斑,小砂矿,补丁,碎片,地段
Patent blue A 专利蓝 A(三苯甲烷类染料,可用作选矿药剂(抑制剂、捕收剂、活化剂))
patent certificate 专利证书
patent infringement 专利侵权

patent pending	专利申请期间
pattern move	模型步骤
pattern of cash inflow	现金流入量方式
pattern of cash outflow	现金流出量方式
pattern of operation	生产[作业]模式
paved highway	铺面公路
pavement structure	路面结构
PAX 350	戊基黄药 350（可用作捕收剂）
PAX A350	戊基钾黄药 A350（可用作捕收剂）
pay off	偿还[清],付清
pay out	支付,付[偿]还,补偿
payback	偿还
payback method	偿还法
pay-back period	偿还期
payback time	偿还时间
paying bank	付款银行
payment of cash dividend	现金股息付款
payment of rent	租金支付
payment schedule	支付计划
payroll	薪资总额
PB-1	间硝基苯偶氮水扬酸（品名,可用作锡石的捕收剂）
PB-2	对硝基苯偶氮水扬酸（品名,可用作锡石的捕收剂）
pct(percent)	百分比[数]
peak efficiency	最大效率
peak load	高峰负荷
peak load control	高峰负荷控制
peak load control system	高峰负荷控制系统
peak magnitude	峰值大小
peak material load evaluation	高峰物料负荷计算
peak potential	峰值电位
pebble mill	砾磨机
pebble milling	砾(石)磨(矿)
pebble phosphate industry	砾状磷酸盐工业
pectin	果胶（浮选石英时作为石灰石的抑制剂）
pegmatite	伟晶岩,伟晶花岗岩
pegmatitic	伟晶岩的
pegmatitic mass	伟晶岩矿体
pegmatitic pod	伟晶岩扁豆体
Peirce-Smith converter	皮氏[P-S]卧式转炉（以法国人皮尔斯(W. H. Peirce)和史密斯(E. A. C. Smith)的姓氏字头命名）
pelargonic acid	壬酸（可用作捕收剂）$CH_3(CH_2)_7COOH$
pellet	球[条形,丸型]团矿,粒状产品,丸,小球,(团)球团,(压)模坯
pellet	球团矿冷却器
pellet fabrication	造球,制粒
pellet plant	团矿厂
pellet screen	球团筛
pelletizer	制粒[丸]机,造球[挤压]机
pelletizing plant	造球[制粒,渣块压制,团矿]设备,团矿[制粒]厂
pelletizing plant washdowns	制粒厂冲洗物

penalty function 补偿函数,罚函数
pen-chart recorder 钢笔图表记录仪
pendant push button control 悬吊开关按钮控制
pendulum conveyor 摆式运送机
pendulum impact test apparatus 摆锤冲击试验仪器[装置,设备]
pendulum set 摆锤式(试验)装置
pendulum tester 摆锤试验机[器]
pendulum type test device 摆锤式试验装置
Penford Gums 200 and 300 絮凝剂(玉米淀粉的环氧乙烷衍生物)
Pennsylvanian formation 宾夕法尼亚层系
penny mine business 小型矿山企业
pension fund 养老基金,退休基金
pension plan 养老金计划
pentanol 戊醇 $C_5H_{11}OH$
Pentasol frother 124 起泡剂(戊醇混合物),泡沫变态剂
Pentasol frother 26 起泡剂(合成戊醇),泡沫变态剂
Pentasol xanthate 戊基黄药,戊(基)黄(原)酸钾(可用作捕收剂) $C_5H_{11}OCSSK$
percent absorption 吸收率
percent capacity 生产能力百分数
percent critical 临界(速度)百分数
percent critical speed 临界速度百分数
percent reduction 减缩率
percent solids 固体百分数
percent solids by volume 固体体积百分数
percent solids by weight 固体质量分数
percentage completion 完成百分比
percentage critical speed 临界速度百分数
percentage drop 百分数下降
percentage ownership 百分比所有权
perchloric acid 高氯酸 $HClO_4$
perchloric acid solution 高氯酸溶液
Percol 182 一种具有较高分子量的阳离子聚丙烯酰胺类助留助滤剂,也可用作絮凝剂
Percol 351 一种分子量为 $(1300~1400)×10^4$ 的非离子型聚丙烯酰胺絮凝剂
Percol 47 一种具有中等偏高分子量的阳离子聚丙烯酰胺类助留助滤剂
percolate 渗滤,渗出,过滤,浸透,渗出液,
percolated solution 渗滤液(金)
percolation leaching 渗滤浸出(法)
percolation tank 渗滤(浸出)槽,(用于氰化法)
percussion drillhole 冲击式钻孔
perforated plate screen surface 冲孔板筛筛面
performance 性能,绩效,执行
performance and production data 技术性能和生产(能力)数据
performance and productivity of process machinery 工艺设备的技术性能和生产能力[率]
performance characteristic 工艺[运转,操作]性能

performance chart 性能[工作特性,操作性能]图
performance consideration(s) 操作[运行,作业]需要考虑的事项
performance data 生产[性能]数据
performance measurement 性能测定,工作状况的测定
performance of flotation circuit 浮选流程性能
performance of flotation network 浮选网络性能
performance of flotation plant 浮选设备[厂]的性能
performance period 执行周期
performance reduction 性能换算
performance requirement 性能要求
performance variability 指标[性能]变化
performance variable 生产成绩变量
performance variance (运转)性能变化
performing unit 执行元件[单元]
period of confinement 限制时间(对圆锥破碎机中的石料移动)
period of guarantee 保险[用]期
period of monetary contraction 银根紧缩时期
period of time 时间周期
periodic analysis 定期分析
periodic check 定期检验[校核,检查]
periodic cleanup 定期清理

periodic error-integration control function 周期误差-积分控制功能
periodic error-integration controller 周期性误差-积分控制器
periodic financial statement 定期财务报表
periodic interest payment 定期支付利息
periodic lease payment 定期租金付款
periodic pattern 同期性排列
periodic response 周期性响应
periodic sampling 定期取样
peripheral discharge rod mill 周边排料棒磨机
peripheral impeller velocity 叶轮圆周速度
peripheral overflow launder 周边溢流槽
peripheral overflow weir 周边溢流堰
peripheral screen 周边固定筛
peripheral speed 周边速度
peripheral stationary screen 周边固定筛
peripheral trash ring 周边废渣挡板(浓缩机的)
peripherals 外围设备
peripherals of various kinds 各种外围设备(计算机的)
Perkin-Elmer 577 double-beamIR spectrophotometer 珀金-埃尔默577型双束红外线分光光度计(美国Perkin-Elmer公司制造)

Perkin-Elmer frustrated multiple internal reflectance unit 珀金-埃尔默受抑多重内反射仪器（美国 Perkin-Elmer 公司制造）

Perkin-Elmer Model 337 infrared spectrophotometer 珀金-埃尔默 337 型红外线光谱分析仪（美国 Perkin-Elmer 公司制造）

permalloy 高磁导率铁镍合金，坡莫［透磁］合金

permanent change 永久变化［改变］

permanent crusher 永久［固定］式破碎机

permanent fund 永久性资金

permanent loan 永久贷款

permanent magnetic pulley 永磁滑轮

permanent magnet separator 永磁磁选机

permanent order 固定顺序

permanent precipitate 固定沉淀物

permanganate 高锰酸盐（可用作方铅矿、闪锌矿的抑制剂，浮选黄铁矿时作为砷黄铁的抑制剂）$MMnO_4$

permeametry 渗透测定［粒］法

permissible range of variable 变量允许变程［范围］

permit 执照，许可证，许可

permitting 允许，许可

Peronate K 石油磺酸钠（相对分子质量 440~450，34% 矿物油）

peroxide addition 添加过氧化物

Perrin filter press 佩林［Perrin］压滤机（德国）

persistent froth 稳定［固］泡沫

personal error 人为误差，操作人误差

personal factor 人为因素

personal judgment 人为判断

personality 个性

personnel 人员，员工

personnel administration 人事管理

personnel and public service department 人事与公共服务部门

personnel coordination 人员的协调

personnel enclosure 人员屏蔽室

personnel management 人事管理

personnel training 人员培训

perspective drawing 透视图

perspective view 透视图

pertinent data 相应的资料［数据］

pertinent information 有关资料

pertinent test work 相应的试验工作

Peru balsam 秘鲁香脂，浮选硫化矿时作为云母抑制剂

Peso 比索（拉丁美洲国家及菲律宾等国货币名）

pesticide 杀虫剂

pesticide industry 杀虫剂工业

Petepon 湿润剂（磺化油酸醇与十六醇混合物）

Petepon G 湿润剂（烷基硫酸盐，波兰产）

Peter cooper std. No. 1 絮凝剂（一种动物胶）

petrochemical industry 石油化工
petrographic study 岩相(学)研究
petrolatum soap oxidized 氧化石蜡皂(可用作浮选氧化钼矿的捕收剂)
petroleum derivative 石油衍生物
petroleum hydrocarbon 石油碳氢化合物
petroleum hydrocarbon oil 石油烃油
petroleum industry 石油工业
Petroleum N bases 烷基氮杂环化合物(可用作闪锌矿的捕收剂)
petroleum sulfonate 石油磺酸盐
petroleum sulfonate promoter 石油磺酸促进剂
Petronate 表面活性剂品牌名
Petronate H 石油磺酸钠(平均相对分子质量513,含33%矿物油,可用作捕收剂)
Petronate K 石油磺酸钠(平均相对分子质量440~450,含34%矿物油,可用作捕收剂)
Petronate L 石油磺酸钠(平均相对分子质量415~430,含33%矿物油,可用作捕收剂)
Petrosol 异丙醇(品名,是一种良好溶剂和化工原料)C_3H_8O
Petrosul 表面活性剂(石油磺酸钠类)品牌名
Petrosul 645 石油磺酸钠(相对分子质量455~465,35%矿物油,可用作捕收剂)
Petrosul 742 石油磺酸钠(相对分子质量415~430,35%矿物油,可用作捕收剂)
Petrosul 745 石油磺酸钠(相对分子质量440~450,35%矿物油,可用作捕收剂)
Petrosul 750 石油磺酸钠(相对分子质量505~525,35%矿物油,可用作捕收剂)
Petrov 750 接触剂(烷基苯磺酸,可用作絮凝剂)
Petrov contact 彼氏接触剂(其成分为烷基苯磺酸,可用作絮凝剂)
Petrov's method 彼得罗夫法(即浓浆高温法,是前苏联首先使用的白钨浮选方法,需要加温矿浆,选矿成本高)
pH control pH 值控制
pH measurement pH 值测量
pH modifier pH 调整剂
pH range pH 值[酸碱度]范围
pH region pH 范围
pH regulating agent pH 值调整剂
pH regulation pH 值调整
pH regulator pH 值调整[节]剂
pH value adjustment pH 值调整
pH variation pH 值变化
pharmaceutical white oil 制药白油,白色制药油
phase change 相变(化)
phase disengagement stage 相分离阶段
phase disengagement time 相分离[离

析]时间

Phase I and Phase II 第一阶段[期]与第二阶段[期]

phase partitioning treatment 相分配处理

phase rate 相率

phase ratio 相比

phase sensitive detector 相敏检测器

phased construction 分阶段[期]建设

phased development project 分段进行的研发项目

phased transfer of title 产权[开采权]分期转让

phenol （苯）酚 C_6H_5OH

phenol-aerofloat 双芳烃基二硫代磷酸盐(可用作捕收剂)

phenol hydroxyl 酚羟基

phenolic hydroxyl 酚式羟基

phenomenon of segregation 分凝[偏析,离析]现象

phenyl 苯基 C_6H_5-

philosophy of operation 操作原理

phoscorite body 磁铁橄磷岩矿体

phoscorite ore 磁铁橄磷岩矿石

Phosocresol 捕收剂(甲酚黑药)

phosphate agglomerate 磷酸盐团聚物

phosphate concentration plant 磷酸盐选矿厂

phosphate field 磷酸盐(矿)区

phosphate flotation plant 磷酸盐浮选厂

phosphate production 磷酸盐生产

phosphate rock beneficiation 磷酸岩选别

phosphate separating plant 磷酸盐分选厂

phosphatic sandstone 磷酸盐砂岩

phosphonic 膦酸的

phosphonic acid 膦酸 $RP(O)(OH)_2$

phosphorous （亚）磷的,含(三价)磷的

phosphorous acid 亚磷酸

phosphorus 磷(P);磷光体,发光物质

phosphorus pentasulfide 五硫化二磷

phosphorus pentoxide 五氧化二磷

photo 光,照相,照片

photo multiplier tube 光电倍增[放大]管

photoelectric absorption coefficient 光电吸收系数

photoelectric cell 光电管[池]

photograph 照片,拍照

photographic plate 照相[摄影]底片

photographic print 彩印

photometric principle 光度测定原理,光度原理

photometric sorter 光度分选[选矿]机

photomicrograph 显微照片

photomultiplier 光电倍增器

photomultiplier detector 光电倍增管探测器

photomultiplier pulse detector 光电倍增管脉冲探测器

photomultiplier tube 光电倍增器管

photon emission 光子发射

photon rate 光子率
photoreduction 光(致)还原(作用)
photoresponse 光响应
photosedimentation 测光沉淀法
photosedimentation technology 测光沉淀技术
photosensitive detector 感光探测器
photosensitive negative 感光阴极板
photo-sensitive negative plate 感光阴极板
photosensitivity 光敏性
photosignal 光信号
photovoltage 光电压
photovoltage-electrode potential curve 光电压,电极电位曲线
PH-PEP 一种苯乙烯膦酸(SPA)药剂(无色浅黄结晶粉末、无毒,是锡石优良捕收剂,也可用于黑钨、钽铁矿、锆英石和铌铁矿的浮选(Plaistere and Hanger 公司产品))
pH-value pH(酸碱度)值
phy-chemical property 物理化学性质
physical abnormality 身体异常
physical abrasive property 物理磨蚀性质
physical and chemical condition 理化条件
physical and chemical system 物理化学体系
physical arrangement 实际配置
physical aspect 物理方面[角度]
physical asset 有形资产

physical behavior 物理性质[特性]
physical beneficiation 物理选矿
physical characteristic 实际[自然]的特性
physical condition 物理条件
physical constant 物理常数
physical contact 物理接触
physical control action 实际[基本]控制作用
physical decay 机械损坏
physical description 物理[实物]描述
physical description of spectrometric process 能谱法的物理描述
physical effect 物理效应
physical environment 自然[物理]环境
physical examination 物理测定
physical factor 自然[物理]因素
physical handling 人工操作,物理学处理
physical implication 物理含义
physical input-output 实际输入输出
physical manipulation 实际[机械]处理[加工]
physical means 物理方法
physical model 物理模型
physical parameter 物理参数
physical phenomena 物理现象
physical plant layout 实际选矿厂布置
physical process 物理方法
physical reason 自然方面[界]的原因
physical science 自然科学
physical separation process 物理分

离法

physical size 实际尺寸[规格]
physical specification 物理特性
physical stability 物理稳定性
physical state 物理状态
physical structure of monomolecular layer 单分子层物理结构
physical system 物理体系
physical testing 物理试验
physical treatment 物理处理
physical variable 物理变量
physical wear 机械磨损
physical-chemical parameter 物理化学参数
physically absorbed water 物理吸附水
physically significant feature 物理主要特征[性]
physico-chemical basis of selectivity 选择性的物理-化学基础
physicochemical characteristic 物理化学特性
physico-chemical information 物理化学资料
physico-chemical mechanism 物理化学机理
physicochemical parameter 物理化学参数
physicochemical property 物理化学性质
physico-chemical property 物理化学性质
physisorbed film 物理吸附膜

physisorption 物理吸附
physisorption characteristic 物理吸附特性
picking belt idler 手选胶带托辊
pickling machine 酸洗机
pick-up 皮卡车,小吨位(货运)卡车,拾(起,取),拾波(器),拾音(器)
picogram 微微克(10^{-12}克)
picoline 皮考啉(甲基吡啶,可用作捕收剂)C_6H_7N
pictorial flow diagram 设备形象流程[系统]图
pictorial schematic diagram 形象联系图[示意图]
piece 块[片,件,只,个,根,支,篇,匹]
piece number 件号
piece rate 计件工资
piezometer 流体压力[强,缩]计,测压管[器],静压头计,微压表,流[水]压计,地下水位计
pig iron mold 铸铁模
pigment industry 颜料工业
pillow block bearing 架座
pilot closed circuit 半工业试验用的闭(合回)路
pilot concentration mill 实验选矿厂
pilot grinding circuit 半工业试验磨矿回路
pilot heap leaching program 半工业堆浸方案[计划]
pilot mill 半工业试验[试验性]磨机
pilot operation 半工业试验操作

[工作]

pilot plant 中间(试验性)工厂,小规模试验性工厂,(小规模)试验(性)设备,实验(性)装置,试验生产装置

pilot plant campaign 半工业性试验期间

pilot plant continuous test 半工业性工厂连续实验

pilot plant flotation cell 半工业性试验厂浮选槽

pilot plant investigation 半工业工厂研究

pilot plant metallurgical result 半工业试验选矿效果

pilot plant program 半工业试验规划[程序]

pilot plant research 半工业性工厂试验[中试]研究

pilot plant scale 试验[中间]厂规模

pilot plant stage 中试阶段

pilot plant study 半工业实验(研究)

pilot plant test 半工业性(工厂)试验

pilot plant test work 半工业性(工厂)试验

pilot plant testing 半工业性工厂试验

pilot plant unit 小型试验设备

pilot preheating plant 中试预热车间

pilot production 试生产

pilot scale data 半工业性试验数据

pilot scale program 半工业性试验规模程序

pilot test 半工业试验

pilot testing phase 半工业试验阶段

piloting 领港[航],(过程的)半工厂性检查

pilot-plant campaign 半工业试验厂运转期

pilot-plant comminution unit 半工业试验厂碎磨设备

pilot-plant concentrate product 半工业性试验厂精矿产品

pilot-plant data 半工业试验厂数据

pilot-plant flotation cell 半工业试验厂浮选槽(6~75立升)

pilot-plant metallurgy 半工业试验厂选矿[冶金]指标

pilot-plant range 半工业型[类]试验厂范围[规模]

pilot-plant scale 半工业试验厂规模

pilot-plant study 半工业试验厂研究

pilot-plant test 半工业试验厂试验

pilot-scale continuous data 半工业性连续试验数据

pilot-scale grinding circuit performance 半工业性磨矿回路性能

pilot-scale program 半工业规模试验计划

pilot-size metallurgical test 半工业规模冶金工艺试验

pinch valve 夹(紧,管)阀

pinched sluice 尖缩[扇形]溜槽

pinchout zone 尖灭区[带]

pine 松,松木

pine bark oil 松皮油(可用作起泡剂)

pinene 蒎烯(可用作起泡剂)$C_{10}H_{16}$
pine needle oil 松叶油(可用作起泡剂)
pine oil 松油
pine oil-Sytex-vapor oil rougher reagent system 松油-辛太克斯-汽化油粗选药剂制度[系统]
pine root oil 松根油(可用作起泡剂)
pinion shaft 小齿轮轴
pinion thrust washer 小齿轮止推垫圈
Pioche belt feeder 皮奥奇皮带给矿[料]机
pioneer jaw crusher 先锋型[先驱者]颚式破碎机
pioneering period 开拓期
pioneering work 先行[开垦]工程,首创性工作
pipe fitter 管道安装工,配管工
pipe fitting 管配件,管接头
pipe installation 管道安装[铺设]
pipe loop 管道环路,环形管道,环管
pipe rack 管道支架
pipe reactor 管式[管道]反应器
pipe tee 三通管
pipeline feed tank 管道给料槽
pipeline transport 管道输送
pipeline transport of concentrate 精矿管道输送
pipeline velocity 管道(内)速度
piping engineer 管道工程师
piping header 管路联箱
piping network 管网

piston membrane pump 活塞隔[薄]膜泵
piston metering pump 活塞计量泵
pit (地,凹,探,竖)坑,槽,洞,凹点[处,窝],(金属表面的)凹痕,砂[缩]孔,(汽车)检修坑,煤矿,矿井[坑],堑壕
pit pump sump 露天泵池
pit run screen analysis 原矿筛分分析(包括坑内)
pit workshop 露天采矿修理厂
pitman bearing 连杆轴承
pitot tube traverse 皮托管测定
pit-run low grade material 露天原矿低品位原[物]料
pit-run ore 露天矿的原矿
pitting corrosion 孔[坑]蚀
pivot point 枢轴点,旋转点,支点,轴心点
PK3 絮凝剂(一种合成的高分子化合物)
PL-164 絮凝剂(十二烷基甲基丙烯酸酯与二乙氨基乙烯—甲基丙烯酸酯共聚物)
placer gold 砂金
placer gold processing 砂金选矿(工艺、法)
placer miner 砂矿矿工
placer operation 砂矿操作[作业]
placer plant 砂矿选矿厂
plain drive pulley 平面驱动滚筒(胶带机的)
plain visual means 肉眼观察法

plane of shear 剪切面
plane of weakness 脆弱面
plane table 平面摇床
planner 计划[规划,策划]人员
planning matrix 计划矩阵
plant 工[选矿]厂,车间,设备,装置
plant access road 进厂道路
plant arrangement 工[选矿]厂配置[布置]
plant bottleneck 工[选矿]厂瓶颈
plant building 工[选矿]厂厂房
plant bulk 设备安装尺寸
plant capacity 工[选矿]厂生产能力,工厂能力,设备容量,电厂容量,设备能力
plant circuit 工[选矿]厂流程
plant commissioning 工[选矿]厂投产
plant component cost ratio method 工[选矿]厂各部分[分项]费用比推法
plant condition 工[选矿]厂条件(设备)
plant configuration 工[选矿]厂流程结构[设备配置]
plant configuration matrix 工[选矿]厂构形矩阵
plant construction 工[选矿]厂建设[施工]
plant cost 开厂费
plant cost ratio method 工[选矿]厂投资费用比推法
plant description 工[选矿]厂描述
plant design 设备平面布置,车间设计
plant design and engineering work 工[选矿]厂设计和工程技术工作
plant design problem 工[选矿]厂设计问题
plant designer 工[选矿]厂设计人员
plant economics 工[选矿]厂经济问题[情况,状态]
plant effluent 工[选矿]厂排出物[污水]
plant engineering 工[选矿]厂设备安装工程,选矿厂工程,设备工程
plant equipment size 工[选矿]厂设备规格
plant factor 工[选矿]厂设备利用率[利用系数]
plant feed 车间给矿
plant fencing 厂房围墙
plant flexibility 工[选矿]厂灵活性
plant flotation practice 选矿厂浮选实践
plant flowsheet 选矿厂流程图
plant growth 植物生长
plant hire cost 设备租金
plant housing 工[选矿]厂住宅,工厂住宅
plant layout 工[选矿]厂布置
plant layout stage 工[选矿]厂配置阶段
plant leaching practice 选矿厂浸出实践
plant manager 企业[工厂,选矿厂]经理

plant metallurgical result 生产[选矿]厂选矿效果

plant metallurgist 选矿厂冶金专业人员

plant operation procedure 选矿厂生产方法[程序,作业过程]

plant overhead 工厂管理费

plant performance 工[选矿]厂生产绩效[指标],设备性能

plant personnel 工[选矿]厂人员

plant planning 工[选矿]厂计划

plant power rate 工[选矿]厂单位功率消耗量

plant practice 工[选矿]厂生产实践

plant production 工[选矿]厂生产[产量]

plant record 工[选矿]厂[车间]记录

plant recovery 选矿厂[车间]回收率

plant reduction ratio 选矿厂[车间]破碎比

plant scale operation 工业规模生产

plant scale test 工业试验,选矿厂规模试验

plant services 工[选矿]厂服务设施

plant simulation 工[选矿]厂模拟

plant site 厂址

plant site development 工[选矿]厂址设计

plant size 工[选矿]厂规模

plant solution 选矿厂(生产)溶液

plant spill 选矿厂[车间]溢洒[溅出]物

plant test 工[选矿]厂试验

plant test work 工[选矿]厂试验工作

plant tonnage 选矿厂[车间]处理吨数

plant trial 工[选矿]厂试验

plant wash down 选矿厂[车间]冲洗[刷]

plant water 工厂用水,选矿厂用水

plant work index 选矿厂[车间]功指数

plant-wide control 全厂性控制

plant-wide control strategy 全厂性控制策略

plastic bag 塑料袋

plastic conduit 塑料管道

plastic feed line 塑料给药[料]管线

plastic film 塑料薄膜

plastic line 塑料管线

plastic lined barrel 内衬塑料圆桶

plastic sheeting 塑料薄膜

plastic tubing 塑料管道

plastic-covered cable 塑料包覆的铁索

plastic-lined channel 衬塑料的沟渠

plastic-lined ditch 衬塑料的明沟

plastic-lined settling pond 衬塑料的沉淀池

plate and frame filter press 板框压滤机

plate and frame press 板框式压滤机

plate electrostatic separator 板式静电分选机

plate filter 板框式过滤机

plate scrubber 板式洗涤器

plateau 高原,台(高)地,平顶,平稳[停滞]时期

plateau region 坪区,平坦区,高原地区

plateau response 平顶反应曲线

platework 板金加工

platform 台地,平台

platform scale 地磅

plating industry waste 电镀工业废水

platinum blade 铂片

platinum contact 铂接触

platinum electrode 铂[白金]电极

platinum foil counter-electrode 铂箔反电极

platinum wire voltage probe 铂丝电压探头

plenum fan 送气风扇

plenum system 压入式通风系统

plenum ventilation 压力通风

Plexiglass 普列克斯(耐热)玻璃(品名,成分为合成高分子有机化合物:聚甲基丙烯酸甲酯)

plexiglass 树脂[塑胶,有机,耐热有机]玻璃(源自普列克斯玻璃品牌名:Plexiglass)

plexiglass holder 有机玻璃夹[架]

plot plan 厂区[平面,位置,坝址平面]布置图

plot plan of surface plant 地表[面]厂房位置[平面]布置图

plow feeder 犁式给矿机

plow like turning device 犁状卸料[转动]装置

plug 塞子,栓(塞),堵塞物;插头;岩颈,沉积栓,井壁防水密封塞[圈];消防[给水]栓;火花塞

plug flow 单向[活塞式]流动,(活)塞[平推]流

plug flow reactor 单向[活塞流]连续反应器,(活)塞流反应器

plug flow model 塞流模型(所有物料在磨机中耗费相同时间)

plugged chute detector 溜槽堵塞探测器

plugging 插上插头,塞[堵]住

plumbing fixture 卫生器具[洁具]

plumbous (亚,二价)铅的

plumbous ion 铅离子

Pluramin S-100 湿润剂(脂肪酸与胺类的缩合物)

Pluronic 表面活性剂品牌名

Pluronic L62 聚氧化乙醇类醚醇(醇与环氧乙烷或环氧丙烷的缩合物,可用作起泡剂)

Pluronic L64 聚氧化乙醇类醚醇(醇与环氧乙烷或环氧丙烷的缩合物,可用作起泡剂)

Pluronic L68 聚氧化乙醇类醚醇(醇与环氧乙烷或环氧丙烷的缩合物,可用作起泡剂)

ply separation (胶带)层脱离[分裂]

pneumatic blaster 风力喷吹[爆破]器

pneumatic cell 充气式浮选槽

pneumatic conveying line 风力输送管线

pneumatic conveying system 风力输送系统,风动运输系统
pneumatic design 压气式设计(浮选机的)
pneumatic device 风力驱动装置
pneumatic drive 风力驱动(装置)
pneumatic dust handling system 风力烟[矿]尘输送系统
pneumatic gate 气动闸门
pneumatic pumping system 风力[空气]扬送系统
pneumatic suction hose 气压吸入软管
pneumatic tire 充气轮胎
pneumatic unloading system 气动卸载系统
pneumatically operated sampler 气动取样机
pneumo-mechanical flotation machine 充气机械搅拌浮选机
pneumo-mechanical machine 充气机械搅拌式浮选机
pocket 矿巢,口袋,衣袋
point of addition 加药点(浮选的)
point of concentrate sale 精矿销售点
point of delivery 交货地点
point of inflection 拐[转折,曲折,反曲]点
point of interest 有关点,兴趣点
point of loading 装矿点
point of precipitation 沉淀点
point of reagent addition 药剂添加点,加药地点

point-of-zero-charge 零电点
point-of-zero-charge value 零电点值
polar 极性[化]的
polar anion 极性阴离子
polar group 极性基
polar head 极性头[端]
polar interaction 极性相互作用
polar organic solvent 极性有机溶剂
polar reaction 极性反应
polar site 极性区间[地段,场地]
polarization curve 极化曲线
polarization experiment 极化试验
polarographic technique 极谱技术[法]
polish washing 抛光洗矿
polished disc 抛光(圆)面
polished electrode 磨光的电极
polished particle 磨光颗[矿]粒
polished sample 磨光试样
polished surface 磨光(表)面
polishing mill 磨[擦]光机
political event 政治事件
political force 政治势力
political interrelationship 政治关系
pollution control facilities 污染控制设施
pollution control revenue 污染控制收入
pollution control system 污染控制系统
pollution law 污染法,环境保护法
polyacrylamide 聚丙烯酰胺
polyacrylonitrile 聚丙烯腈

polycrystal 多晶体
polycrystalline material 多晶物质
polyester nylon 聚酯尼龙
polyethylene 聚乙烯
polyethylene glycol 聚乙二醇(可用作起泡剂)$HO(CH_2CH_2O)_nH$
Polyfon 木(质)素磺酸钠类表面活性剂品牌名(美国 Westvaco 公司产品)
Polyfon F 木(质)素磺酸钠(含磺酸钠基团32.8%,可用作抑制剂)
Polyfon H 木(质)素磺酸钠(含磺酸钠基团5.8%,可用作抑制剂)
Polyfon O 木(质)素磺酸钠(含磺酸钠基团10.9%,可用作抑制剂)
Polyfon R 木(质)素磺酸钠(含磺酸钠基团26.9%,可用作抑制剂)
Polyfon T 木(质)素磺酸钠(含磺酸钠基团16.7%,可用作抑制剂)
polyglycol (高分子)聚乙二醇
polyglycol ether 聚乙二醇醚
polyglycol frother 聚乙二醇起泡剂
polyglycol type 聚乙二醇类
polygon 多边形,多角形物体
"polygon of influence" technique 影响多边形技术[法]
polygonal block 多边形矿[地]块
polygonal method 多边形计算矿量法
polymer and surfactant interaction 高分子聚合物与表面活化剂互相作用
polymeric 聚合的,聚合体的
polymeric material 聚合物料
polymeric sulfonate 聚合磺酸盐

polymeric-crystalline structure 聚合结晶结构
polymorph 多晶型物,(同质)多形物[体],同质异象体
polymorphic 多晶[型]的,多形(性,态)的
polymorphic transformation 多晶转变,多形相变
polymorphic transition 多形转变,多形态变化,多晶转变,多晶型的转化
polymorphism (同质)多晶型(现象),多形性(现象),同素异构,同质异相
polynomial 多项式(的)
polynomial equation 多项式方程
Polyox 一种絮凝剂(品名)
polyoxyethylene 聚氧乙烯
polyphenol 多元酚
polyphenol group 多(聚)酚基团
polyphosphate 聚磷酸盐
polypropylene 聚丙烯
polypropylene glycol 聚丙二醇
polypropyleneglycol ether 聚丙二醇单醚(可用作起泡剂)
$R—(O—C_3H_6)_n—OH$
polypropylene sheet 聚丙烯板
polyurethane 聚氨基甲酸(乙)酯,聚氨酯,聚亚胺酯
polyvalent cation 多价阳离子
polyvalent metal ion 多价金属离子
polyvalent metal salt 多价金属盐
polyvinyl 聚乙烯,聚乙烯基
polyvinyl alcohol 聚乙烯醇,可用作钾

盐矿的捕收剂、湿润剂)
—[CH$_2$·CH(OH)]$_n$—

ponding 蓄液池[槽],人工池塘

pool end spiral flight 沉降区端部螺旋叶片

pool level 沉降区水位

pool region 沉降区

poor flotation practice 不良浮选实践

poor metallurgy 不良[低]选矿指标

poor road 劣质道路

population 人口(数),种群,居民,总体,全体

population balance grinding model 总体平衡磨矿模型

population balance method 总体平衡法

population balance model 总体平衡模型

population mean 总体平均值

population of organisms 有机生命群

population of parameters 参数群[组]

porcelain grind 瓷磨

pore gas pressure gradient 微孔气体压力梯度

pore wall 孔墙[壁]

porosity 孔隙[多孔]性

porosity mixture 多孔性混合状态

porous aluminum oxide tube 多孔隙的氧化铝管

porous false bottom 多孔假底

porous particle 多孔颗粒

porousness 气孔率[度],多孔性

porphyritic copper ore 斑岩铜矿石

porphyry copper ore 斑岩铜矿石

porphyry deposit 斑岩矿床

porphyry mineralization 斑岩矿化作用

porphyry ores 斑岩矿石类

porphyry type deposit 斑岩型矿床

porphyry-type copper ore 斑岩型铜矿石

port area 港口区

portable conveyer specification 移动式皮带运输机技术规格[说明书]

portable crusher 移动式破碎机,可移动[搬运]的破碎机

portable isotope X-ray fluorescence analyzer 手提[移动]式同位素X射线荧光分析仪

portable mill 移动式选矿厂

portable screening plant 轻便筛分成套设备

portable unit 移动装置

Portland cement 硅酸盐水泥[波特兰水泥](1824年英国人J.阿斯普丁用石灰石和黏土烧制成水泥,硬化后的颜色与英格兰岛上波特兰地方用于建筑的石头相似,被命名为波特兰水泥;从铁矿中浮选石英时作为石英活化剂,浮选菱镁矿、白云石时作为方解石抑制剂)

position flop gate 定位落下闸门

position of Fermi level 费米能级位置

positioning mechanism 定位机构

positive action [装置]

positive action 肯定[实际,确实,规定]的作用

positive and negative covenants 正,反面契约条款

positive area 正区

positive cash flow 正现金流量

positive circulating corrosion-proof pump 正循环防腐泵

positive circulating corrosion-proof pump and reservoir system 正循环耐腐蚀泵和储槽系统

positive constant 正常数

positive correlation 正相关[关联]

positive displacement metering pump 正排量计量泵

positive drive 正向传动

positive effect 积极[明显]效应

positive photovoltage 正光电压

positive seal 可靠[绝对]的密封

positive site 正电表面区

positive slope 正斜率

positive space charge 正空间电荷

positive stroke 强制冲程

positive surface potential 正表面电位

positive value 正值

positive volumetric type 规定容积式[类型]

positively charged functional group 带正电荷的官能团

positively charged oxide 带正电荷的氧化矿

positively charged surface 正电荷表面

possibility of improved recovery 改善回收率的可能性

possible alteration 可能的变更[改变]

possible alternatives 可能的选择方案

possible ore reserve 可能的矿石储量(勘探程度尚不充分)

possible reserve 可能的储量(即推断储量)

post 棒(棒形浮选机的叶轮棒),柱,杆;岗位;邮件[递]

post classification 后置分级(分级机置于磨机之后)

post-saturator 后置饱和器

postulated equation 假设方程

postulated reaction 假设[推测]的反应

postulation 假定[设]

postulation formula 假定公式

pot type furnace 坩埚炉

potable water 饮用[生活]水

potable water pump 饮用水泵

potable water system 饮用水系统

potash 钾碱,碳酸钾 K_2CO_3

potash feldspar 钾长石 $K[AlSi_3O_8]$ 或 $K_2O \cdot Al_2O_3 \cdot 6SiO_2$

potash industry 钾盐工业

potassium 钾 K

potassium bromide pellet technique 溴化钾压片法

potassium ethyl xanthate 乙基钾黄药,乙基黄盐酸钾 C_2H_5OCSSK

potassium hydroxamate 氧肟酸钾
potassium octyl hydroxamate 辛基氧肟酸钾
potassium stearate 硬脂酸钾
potassium sulphate 硫酸钾
potassium thiosulphate 硫代硫酸钾
potassium xanthate 黄(原)酸钾 $KS_2COC_2H_5$
potassium xanthate homologue 黄原酸钾同系物
potential benefit 潜在利益
potential coal depressant 煤潜在抑制剂
potential decay 电势衰变
potential determining ion 定位[决定电位]离子
potential distance curve 电位距离曲线
potential economic benefit 潜在的经济利益
potential flow 势流
potential gain 潜在性盈利
potential of thermodynamic stability 热力学稳定电位
potential problem 潜在问题
potential profit 潜在性盈利
potential project 潜在[可能]工程项目
potential range 电位范围
potential region 电位范围
potential saving 节约潜力
potential sweep 电位扫描
potential transformer 电压互感器
potential-determining anion 定位[确定电位]阴离子
potential-determining cation 定位[确定电位]阳离子
potential-determining electrolyte 定位[确定电位]电解质
potential-determining ion 定位[确定电位]离子
potential-distance dependency 电势与距离两者的依赖关系
potential-time characteristic 电位-时间特性
potential-time experiment 电位时间试验
potentiometric determination of surface charge 表面电荷的电位测定
potentiometric measurement 电位测定
potentiometric study 电位测定研究
potentiostat 稳压器,恒电势器,电势恒定器,电压恒定器
pottery grade 陶瓷品级
pour over the cell lip 跑槽(泡沫)
pouring froth 泡沫跑槽
pour-over 跑槽
powder metallurgy approach 粉末冶金法
powdered natural hematite 粉状天然赤铁矿
Powell direct search 鲍威尔直接搜索(共轭方向技术的变种,一种有效的共轭梯度方向法,只是导数不予计算)
Powell direct search method 鲍威尔直接搜索法

power and heat conversion constant 功热换算常数
power availability 功率可获量,动[电]力的可利用性
power band 功率带
power calculation 功率[动力]计算
power coefficient 功率系数
power consumption curve 功率消耗曲线
power control 功率控制
power control mode 功率控制方式
power controller 功率控制器
power correlation equation 功率关系方程[公]式
power correlation factor 功率关系数[因素]
power determination 功率测定
power device 功率测量装置
power diagram 功率图
power distribution and control 配电与控制
power distribution center 动力分配中心
power distribution system 配电系统
power draft 牵引[汲取]功率
power draft control 牵引[汲取]功率控制
power draught curve 牵引功率曲线
power draw 能耗,传动功率
power drawing capacity 功率抽取能力
power engineering 动力学[工程],力能学
power equation 功率方程式
power estimation 功率估计[估算,预测]
power failure 电[动]力故障
power formula 功率公式
power function 幂函数
power generation 发电
power generation and distribution 发电与配电
power gradient 功率梯度
power intensity 功率强度
power level 功率水平
power level control module 功率水平控制模数[式]
power limit over-ride 功率极限补偿,超越功率极限
power line tap 输电[电力]线分接头
power measurement 功率测量
power number 功率数
power outage 断电
power piping system 动力管道系统
power prediction 功率预测
power rate 功率比率,功率消耗率,单位功率(消耗量)
power rate energy parameter 单位功率消耗量能量参数
power rate unit 功率消耗量单位
power ratio 功率比
power ratio relationship 功率比关系
power recorder 功率记录仪
power relationship 功率关系

power requirement 动力[功率]需用量,电源要求
power select logic 功率选择逻辑
power substation 动力配电站[变电站]
power tester 功率测试器,功率计,瓦特表
power transducer 功率变送器
power transmission 动力传送
power transmission equipment 动力输送[输电]设备
power unit 能量[功率]单位,电源部分,电源组合,发动机,执行机构
power utilization 功率利用
powered rotary distributor 有动力驱动的旋转(矿浆)分配器
powerful effect 强烈效应
powerful frother 强起泡剂
powerful spring 强力弹簧
power-operated adjustment 电动调节
power-operated adjustment of close-side setting 最小[紧边]排口开[宽]度的电动调节
practical application 实际应用[用途]
practical consideration (生产)实际问题
practical contact angle 实际接触角
practical difficulty 实际困难
practical factor 实践因素
practical information 实用[践]资料
practical limit 实际使用范围
practical matters of economics 实际经济问题
practical operating range 实际操作范围
practical plant operating technology 选厂的实际生产[操作]技术
practical problem 生产实践问题
practical process control computer 实用的过程控制计算机
practical technique 实[应]用技术
practical textbook 实用教科书
practical time schedule 实际时间进度表
practical viewpoint 实用[践]观点
practice point of view 实践观点
pragmatism 实用主义,实验主义,实用观点与方法
preactivated 预活化的
pre-aeration tank 预充气槽
Pre-Cambrian 前寒武纪
Pre-Cambrian granite 前寒武纪花岗岩
Pre-Cambrian Grenville Series 前寒武纪格伦维尔统
precious metal beneficiation plant 贵金属选矿厂
precious metal cyanidation 贵金属氰化(法)
precipitate build-up 沉淀物积聚
precipitate filter 置换过滤机
precipitate flotation 沉淀浮选
precipitate press 置换压滤机
precipitate pressure filter 置换压滤机

precipitated copper 泥铜,沉淀铜
precipitating tank 沉淀池
precipitation 沉淀(作用),降水(量)
precipitation cone 沉淀[置换]圆锥
precipitation plant 沉淀车间
precipitation region 沉淀区
precipitation Stellar filter 置换管式过滤机
precipitation system 沉淀系统(氰化法)
precipitation tank 沉淀槽
pre-classification 前置分级(分级机置于磨机之前)
pre-coat filter 预涂助滤剂的过滤机
precoat tank 预涂层助滤剂搅拌槽
pre-collector 前级收尘器,预先收集器
preconcentrate 预富集物,预富集精矿
pre-concentration method 预选法,预先富集法
preconcentration process 预富集[预选]流程[工艺]
preconditioned 预调和[整]的,预处理的
preconstruction stage 建设[施工]前期
pre-cyanide 氰化前
predetermined fixed rate of speed 预定速度
predetermined limit 预定的限制
predetermined spacing 事先确定的间距[隔]
predetermined value 预定值

predicted grade 预测[计,期]的品位
predicted power consumption 预计的功率消耗
predicted product size distribution vector 预报的产品粒度分布向量
predicted recovery 预测[计,期]的回收率
pre-exploitation drilling program 开采前钻探程序[计划,方案]
prefabricated steel 预制钢构件[结构]
pre-feasibility study 预可行性研究
preferential adsorption 优先吸附
preferential comminution 择优磨[粉]碎
preferential dissolution 优先溶解
preferential grinding 优先粉[磨]碎
preferential price 优惠价格
preferred reduction ratio 优先选用的破碎比,优选[最佳]破碎比
preferred stock 优先股票
prefloat treatment 预浮处理
pregnant solution flow 贵液流量
pregnant solution storage basin 富液贮存池[槽]
pregnant solution tank 贵液[含金溶液]槽
pregnant strip solution 富萃取[洗提]液
preg-robbing characteristics 消耗含金溶液的特性,含金溶液消耗特性
preg-robbing species 含金溶液消耗物质,富液消耗物质

preheated air 预热空气
preliminary agreement 初步协议
preliminary analysis 初步分析
preliminary analysis on ore 矿石初步分析
preliminary bench scale flotation test 初步的实验室[小型]浮选试验
preliminary bidding 最初投标
preliminary capital cost estimate of processing plant 选矿厂初步投资费用估算
preliminary design 初步设计
preliminary design phase 初步设计阶段
preliminary design stage 初步设计阶段
preliminary drawing 初步设计图
preliminary economic viability study 初步经济生存能力研究
preliminary estimation 初步估算
preliminary evaluation 初步评价
preliminary feasibility investigation 初步可行性研究
preliminary feasibility study 初步可行性研究
preliminary flotation test 初步浮选试验
preliminary flotation test work 初步浮选试验工作
preliminary flowsheet 初步流程(图)
preliminary general arrangement drawing 初步总布置(蓝)图

preliminary geostatistical study 初步地质统计研究
preliminary investigation 初步调(查)研(究)
preliminary laboratory test 实验室初步试验
preliminary laboratory test work 初步实验室试验工作
preliminary layout 初步配置
preliminary market evaluation 市场[销路,行情]的初步评估
preliminary metallurgical evaluation 初步的选矿评价
preliminary metallurgical flowsheet 初步选矿流程
preliminary motor list 初步的电动机表
preliminary note 初步评论
preliminary prospectus 初步计划书
preliminary report 初步报告
preliminary sampling information 初步取[采]样资料
preliminary sketch 草图
preliminary structure design 初步的结构设计
preliminary study 初步研究
preliminary study procedure 初步研究过程
preliminary test 初步试验
preliminary test of alternatives 初步的方案试验
preliminary test run 最初试运转

[试车]
preliminary testing 初步试验
pre-loading surveyor 装货前检查员
pre-mineral movement 成矿前运[移]动
premium grade 高级[优质]的
premium quality product 优质产品
premium rate 保险费率
premix tank 预先混合槽
prepaid expense 预支费
preparation of documentation 文件的准备
preparation of minutes 记录的准备
preparation of ore 矿石准备
preparatory stage 准备阶段
preparatory technique 预先处理技术
prepayment 预支费
preponderance of fines 粉矿(数量)占优势
preprint 预印本,未定稿版
pre-process 预先处理
pre-production cost 预生产费
preproduction model 试制模型,样机
preproduction work 生产前准备工作
pre-qualification list 资格预审表
pre-qualified supplier 通过资格预审的供应商
prerequisite 先决条件,前提
pre-requisite 先决条件,前提
pre-run 预运转
pre-run procedure 预运行过程
pre-saturator 前置饱和器

prescaler 预脉冲计数[预定标,预分频,前置分频]器,预换算装置
pre-screened 预先筛分的
pre-screening 预先筛分
prescribed mill load level 规定的磨机料位
prescribed standard 规定的标准
pre-selected model 预选模型
present day flotation procedure 目前[现行,现代]的浮选方法
present value 现值
present value factor 现值因数
present worth 现值
presentation 展示,描述,介绍,赠送
presentation system 展示系统
present-day unit price 当前的单价
preset insert 预置嵌入件
preset limit 预先整定范围
pre-set tonnage 预定吨数
preset tonnage controller 预调吨位控制器
preset value 预先给定值
president 董事长,总裁,总经理,会长
prespecified percent solids 预定的百分固体含量
press feed pump 压滤机给矿[料]泵
press unit 压滤装置
pressure condition 压力条件
pressure correction factor 压力修正系数
pressure drop calculation 压降计算
pressure equation 压力方程式

pressure filtration 压滤,加压过滤(法)
pressure flow 压力流
pressure formula 压力公式
pressure number 压力数
pressure pot 压力槽,压力罐
pressure precoat filter 预涂层压滤机
pressure profile 压力分布图
pressure ratio relationship 压力比关系
pressure recorder 压力记录器
pressure strip 压力解吸
pressure stripping 加压解吸
pressure stripping equipment 压力[加压]解吸设备
pressure tap 压力旋塞
pressure transmitter 压力变送器
pressure tube sampler 压力管式取样器
Prestabit V 硫酸化或磺化脂肪酸钠盐(品名,法国产,可用作捕收剂)
pre-startup safety review(PSSR) 启动前安全检查(缩写 PSSR)
pre-thickening 预先浓缩
pretreatment procedure 预处理程序[方法]
pre-treatment procedure 预先处理法
preventive action 预防行动
preventive maintenance program 预防性维修计划
previous experiences 以往经历
price index 物价指数
primary alkylamine 伯烷基胺

primary amine 伯胺
primary amine salt 伯胺盐
primary area of study 主要研究范围[领域]
primary beam 一次束
primary bin 粗矿仓,原矿仓
primary breakage fragment 原始破碎屑
primary collector 粗选[主要]捕收剂
primary concentrate 最初[初级,第一阶段]精矿
primary concentration 初始富集
primary concentration route 主要选矿方法
primary copper sulfide mineral 原生硫化铜矿物
primary crushing facility 粗碎设施
primary crushing operation 粗碎作业,第一段破碎作业[操作]
primary crushing plant 粗碎车间
primary crushing station 粗碎站,第一破碎站
primary cut 第一段切取样
primary cutter 第一段切[截]取器
primary cyclone 第一段(水力)旋流器
primary cyclone overflow percent solids 第一段旋流器溢固体百分数
primary data 原始数据
primary difference 主要差异(区别)
primary dryer 一次干燥机
primary excitation 一次激发
primary fine 原生细泥

primary fines 原生粉矿
primary flotation cell 初浮选机
primary flotation circuit flowsheet 第一次[阶段]浮选流程
primary gold bearing sulfides 原生含金硫化矿物
primary grinding 粗磨,第一段磨矿
primary grinding equipment 第一段磨矿设备
primary highway 主要公路,干路
primary impact-type crusher 第一段冲击式破碎机
primary instrumentation 一次(测量,检测)仪表
primary magnesium 原生镁
primary matte flotation 粗冰铜浮选
primary metal 粗金属
primary mill loading 第一段磨矿机载荷
primary milling 第一段磨矿
primary milling stage 第一磨矿阶段,第一段磨矿
primary objective 首要目标
primary pebble milling 粗砾(石)磨矿
primary radiation 一次辐射
primary reactor 主反应器
primary resistor 主电阻器
primary road 干路
primary sample 原矿样
primary sampler 第一次取样器
primary sampler overhead drive 第一次取样机顶上驱动装置
primary sampling machine 第一次取样机
primary sand-slime separation 初步的泥砂分离
primary screening duty 第一段筛分任务
primary side 一次(线圈)侧
primary slime 原生矿泥
primary smelter 粗炼厂
primary solution 原液
primary sulfide 原生硫化矿
primary sulfide zone 原生硫化矿带
prime bank 主要银行
prime component 主要组成部分
prime cost 主要成本,首要成本,原价,进货价格
prime pump 起动注油泵
prime rate 优惠利率
Primene 81-R 捕收剂(脂肪胺,带支链的脂肪族第一胺)
Primene IM-T 捕收剂(脂肪胺,带支链的脂肪族第一胺)
primer 第一层;(防锈用)底层涂料,底漆,涂底剂
primer fluid 起动液
primer pump 起动泵
primer water 虹吸管引动水,准备水,引入水(入炉水)
primitive tool 原始工具
principal 本金,资金,基本财产,主题,首长
principal amount of debenture 债券本

身总额

principal design engineer 负责设计工程师

principal field engineer 现场主要工程师

principal sub-routine 主要子程序

principle financial statement 原则财务报告(表)

principle of gamma ray absorption 伽马射线吸收原理

principle of gravity classification 重力分级原理

principle of gyratory crusher work 旋回破碎机工作原理

principle of operation 工作[操作]原理

principle of optimality 最优化原理

principle of probability 概率原理

principle of reduction 破碎原理

principle of stratification 分层原理

print-out 打印(输出),印刷

prior conditioning 预先调浆

priori justification 事前验证

private communication 未公开的[内部,保密]通信

pro rata 按比例的,可按比例估价的

probability density function 概率密度函数

probability level 概率级别[水平]

probability of air bubble-floatable specie collision 气泡-可浮矿物[粒]碰撞几率

probability of collision 碰撞概率

probability of separation 分离概率

probability of unscheduled downtime 计划外停机[车,工]概率

probability relationship 概率关系

probability theory 概率理论

probability theory application 概率理论应用

probable accuracy 大概[概率]精确度

probable error 概差,或然误差,概率误差

probable reserve 设想[预计]矿石储量(勘探程度比较充分)

probe analyzer 探头[针]分析仪

probe measuring device 探头测量装置

proceedings 科研报告集,记录汇编,会刊,学报,会议记录,事项,项目,议程

proceedings at the meeting 会议事项

proceeds of concentrate shipment 精矿发货[出售]收入

process 加工,选矿,工艺,方法

process analysis 过程分析

process and instrument diagram 工艺仪表系统[形象]图

process area 工艺现场,生产区域

process arrangement 工艺过程布置[配置]

process background 工艺[流程]背景

process building 生产[工艺]厂房

process calculation 工艺计算

process center 生产中心

process condition 工艺条件

process configuration 生产过程结构,工艺配置[结构]
process consideration 过程研究
process control 过程控制
process control application 过程控制应用
process control computer 过程控制计算机
process control demonstration 过程控制示范
process control economic justification 过程控制经济合理性
process control input output subsystem 过程控制输入输出子系统
process control installation 过程控制装置
process control strategy (工艺)过程控制对策
process control system 工艺流[过]程控制系统
process controller 工艺过程控制器
process delay 过程滞后
process description 工艺流程[生产过程]说明
process design 工艺设计
process design criteria 工艺设计准则[标准]
process design data 工艺设计数据
process development 工艺[方法,过程]编制,工艺开展
process development and study 选矿[工艺]开发与研究
process development testing 工艺开发试验
process development work 工艺开发工作
process efficiency 工艺[生产]效率
process engineer 工艺[选矿]工程师
process equipment 工艺设备
process equipment capital cost 工艺设备投资费用
process equipment cost 工艺设备费
process equipment list 工艺设备表
process experience 过程控制经验
process feed requirement 生产[工艺,加工]过程给矿需要[需求]量
process flow 加工[工艺,生产,处理,操作]过程[流程,流线]图
process flowsheet 工艺流程图
process fluctuation 过程波动
process fundamentals 工艺基本原理
process index 工艺指标
process industry 加工工业
process instrument 工艺仪表
process instrumentation 工艺仪表
process interface 过程接口
process knowledge 工艺知识
process log printer 过程记录打印机
process machinery division 工艺机械部
process management factor 过程管理因子
process measurement 过程测量
process mechanism 工艺(作用)机理

process model 过程模型

process model parameter 工艺模型参数

process operation 过程操作

process output 过程输出

process piping 工艺管道

process plant 选矿厂

process plant specifications 选矿厂规程[范],选矿详细说明书

process pump 生产过程用泵

process rate 生产[加工,处理]量

process requirement 工艺要求

process residue 工艺残渣[废料]

process restriction 工艺过程限制

process result 生产结果[效果]

process scale 过程规模

process schedule 进度时间表

process scheme 流程[流向,作业,流程方框,工艺流程,操作程序]图

process selection 工艺选择

process simulation 过程模拟

process simulator 工艺模拟程序

process solution 工作[生产]溶液

process step 工序

process stream 作业[生产,加工]线,工艺流程线,工序流程

process structure 流程结构

process subjective 工艺[加工]目的

process tank 工艺[加工]用罐[槽]

process technology 生产工艺学,加工技术

process temperature 操作[加工,工艺过程]温度

process throughput 生产[处理,通过]量

process topology 过程拓扑

process unit 工艺[选矿,加工]设备,工作[过程]单元

process unit model 工艺设备模型

process unit response 工艺设备因变量

process variable 过程变量

process variance parameter 过程变化参量[数]

process waste water 生产废水

process water distribution flowsheet 工艺用水分配工艺流程

process water recycling trial 选矿[生产]水循环(使用)试验

process water sump 生产水池

processing 处理,加工,作业,选矿,工艺过程

processing alternative 选矿方案

processing equipment 选矿设备

processing facilities 选矿[工艺]设施

processing field 选矿工作现场

processing goal 工艺过程目标

processing industry 加工工业

processing line 生产系列

processing of data 资料整理,数据处理

processing plant 加工[选矿]厂

processing step 生产阶段[方法,步骤,工序]

processing technique 加工技术

processing time 处理时间

processing unit	运算器,处理部件,[化工]加工设备
processor	加工者,加工机械,加工[处理]程序,信息加工[处理]机
process-oriented staff	专攻过程的人员
Proctor C method	普罗克特C方法(确定货仓内矿物空隙最大含水量的方法)
Proctor compaction test	普氏[普罗克特]压实试验
Proctor cylinder	普氏[普罗克特]击实筒
Proctor method	普氏[普罗克特]压实法
procurement	采购
procurement activities	采购活动
procurement activity commitment	采购活动委托[承诺]
producer	生产厂家
product analysis	产品分析
product clarity	产品澄清度
product coarseness	产品粗度[粒度]
product collecting launder	产品收集流槽
product development	产品研制
product diagram	成品图形
product drying	产品干燥
product export	产品出口
product filtration characteristic	产品过滤特性
product fineness	产品细度
product gradation	产品粒级[粒度]
product gradation curve	产品分级[粒度]曲线
product grade	产品品位
product grade target	产品品位指标
product inventory	产品库存[存储]
product manager	产品经理
product output	产品产量
product process requirement	产品的工艺要求
product pump	产品(输送)泵
product quality	产品质量
product quantity and quality level	产品数质量水平
product sales	产品销售
product screen analysis	产品筛分分析
product settling rate	产品沉降速度
product size	产品粒度
product size control	产品粒度控制
product size distribution	产品粒度分布
product size distribution analysis	产品粒度分布分析
product size module	产品粒度模数
product size specification	产品粒度技术要求
product sizing	产品粒度分级
product specification	产品规格
product sump	产品(泵)池
product support literature	产品使用和维修说明书
product system operation	产品系统

操作
product value 产品价值
product vector 产品向量
product weight 产品重量
production area 开采[回采,生产]区
production bottleneck 卡生产脖子(的问题),生产难关,生产障碍,影响生产作业线的因素
production center 生产中心
production coordinator 生产协调员[主管]
production cost 生产成本
production curve 生产曲线
production data 生产数据
production disruption 生产中断
production environment 生产环境
production equipment 生产设备
production expansion program 扩大生产计划
production experience 生产经验
production goal 生产目标
production headframe 生产井架
production hoist 主井提升机
production level 生产水平
production limitation 产量限制
production loss 生产损失
production mill 生产选矿厂
production office 生产办公室
production payment 生产付款
production people 生产人员
production period 生产期间
production plan 生产计划

production planning 生产规划
production plant data 生产厂数据
production problem 生产问题
production program 生产计划[程序]
production requirement 生产要求
production revenue 生产收入
production scale 生产规模
production schedule 生产计划[进度表]
production size equipment 生产规模的设备
production specification 生产技术要求,产品技术规格
production staff 生产人员
production statistics 生产统计资料
production stoppage 停产
production target 生产目标
production trial 生产试验
production vacuum filter 生产用的真空过滤机
production worker 生产工人
productive consumables 生产消耗品
productivity 劳动生产率,生产能力
profession of mineral processing 选矿专业
professional 专业人员
professional engineer 专业[具有专门技术的]工程师
professional journal 专业(性)杂志
professional personnel 专业人员
professor 教授
professor of geological science 地质科

学教授

professor of mineral engineering 矿物工程教授
profile gate opening 纵剖面闸门开口
profile steel 型钢
profiled bar 异形棒材
profiled iron 型钢[铁]
profiled panel 成型面板[盘面]
profiled steel sheet 成型薄钢板
profit 利润,利益,盈利,获利
profit accrual 利润增加额
profit making 赢利
profit margin 利润率
profit maximization 利润最大化
profit potential 利润潜力
profit sharing 分红制
profit status 盈利[利润]情况
profit structure 收益结构
profitability 利润效率
profitability assessment technique 利润率评价法
profitability criterion 利润率标准
profitability impact 利润率影响
profitability potential 利润率潜力
profitability ranking 利润率等级
Profloc 絮凝剂(脱去纤维的可溶性干燥的血和血蛋白)
program design 编制程序,程序设计
program development 程序研[编]制
program development cost 程序研[编]制成本
program execution 程序执行
program flow 程序流程
program flow chart 程序流程图
program for scale-up prediction 按比例扩大预测程序
program input 程序输入
program input output device 程序输入输出装置
program installation 程序装置[机构]
program library 程序库
program listing 程序清单
program logic control system 程序逻辑控制系统
program module 程序模块
program structure 程序结构
program(me) 方案,程序,计划,大纲
programmable calculator 可编程计算器
programmable controller 可编程控制器
programmable logic control 可编程逻辑控制
programmable logic control system 可编程逻辑控制系统
programmer 程序[器,控制元件,设计人员]
programming 编程
programming aid 程序设计工具[助手]
programming capability 程序设计能力
programming console 程序设计控制台
programming format 程序设计格式

programming staff 程序设计人员
progress chart 进度(图)表
progress control 进度控制
progress estimate 进度估计
progress report 进度报告
project 项目,计[规]划,设计,方案
project administrator 项目管理人
project appraisal 工程评价
project approval 项目批准,立项
project contractor 工程承包商
project contractor study 工程承包商研究
project control estimate 工程控制估算
project control estimate of total cost 工程总费用控制估算
project control manager 项目[工程]控制经理
project coordinator 项目协调人员
project cost 工程项目成本
project criteria 工程标准[准则,规范,依据]
project description 工程描述
project design engineer 项目设计工程师
project development sequence 工程设计程序,工程项目设计进展程序
project engineer 项目工程师,(设计)主管工程师
project file index 项目档案索引
project financing 项目融资
project financing scheme 项目融资方案

project forecast 项目预测[报]
project group 工程项目组
project location 项目选址[定位,地点]
project management 工程(项目)管理
project management group 工程管理组
project management personnel 工程管理人员
project management structure 工程管理架构
project manager 项目经理
project material and equipment 工程项目物资和设备
project metallurgist 工程项目冶金[选矿]专家
project office facility requirement 项目办公设施要求
project organization 设计机构
project organization chart 项目组织图表
project plan 工程计划
project procedure manual 项目程序手册[细则,说明书]
project progress 工程[项目]进度
project purchasing agent 项目采购代理
project report 项目报告
project responsibility 工程责任
project schedule 工程[项目]计划
project secretary 项目秘书
project status review meeting 工程情

况检查会

project status summary 项目(设计)情况简报

project supplies (工程)项目物资供应

project team 工程项目组

project viability 工程服务年[期]限,工程寿命

project's capital cost 项目基建投资[费用]

project's desirability 工程项目的客观需要

projected cost 预期[计]成本

projected figure 投影图形,预测数字,设计数字[值]

projected load variation curve 设计[计划]负荷变化曲线

projected result 预期[计]的结果

projected-area-diameter 投影面积直径

projection 投射[影],规划,预[推]测

promoter(= promotor) 促进剂,助催化剂,助聚剂,助触媒;激发器

promoter-frother combination 捕收[促进]泡沫联合剂

proof testing 考核试验

propane 丙烷 C_3H_8

propane torch 丙烷喷灯

propanol 丙醇 $CH_3(CH_2)_2OH$

proper application 正确[合理]应用

proper cleaning 适当[正确]的清理

proper content 适当[正确]的内容

proper crusher selection 破碎机的正确选择

proper feed distribution method 适当的给矿分布方式[法]

proper load sharing balance 负荷分配适当平衡

proper process equipment 适当的工艺设备

property 性质,性能,财产,所有权

property acquisition 矿产[地产,不动产]的购置

property acquisition fee 产权费

property of orthogonality 正交性

proportion of fines 细粒比例

proportional effect 比例效应

proportional feed rate control 给矿量比例调节

proportional floating control action 比例浮动控制作用

proportional splitter gate 比例缩分闸门

proportional-integral algorithm 比例积分算法

proportionality constant 比例常数[恒量]

proportionality term 比例项

proportioning gate 配比闸门

proposal 推荐,申请,建议,提议

proposal for co-operation 协作建议

proposed flowsheet 推荐采用的流程

proposed loan 计划贷款

proposed method 推荐[建议]的方法

proposed mining and mineral processing

plant 拟建采选厂
proposed mining plan 建议的采矿方案
proposed mining project 拟建矿山工程
proposed mining venture 拟建采矿企业
proposed production flowsheet 推荐的生产流程
proposed project 所建议的工程项目
proprietary article 专利品,专卖品
proprietary information 专利[有]资料
proprietary surfactant 具有专利权的表面活化剂,专卖的表面活化剂
proprietary wet scrubber design 专利湿式洗涤器设计
propyl 丙基 $CH_3CH_2CH_2-$
propylene 丙烯 $CH_3CH:CH_2$,亚丙基 CH_3CHCH_2-
propylene diamine 丙烯[隣]二胺,1,2-二胺基丙烷 $CH_3CH(NH_2)CH_2NH_2$
propylitic alteration 青磐岩化蚀变,青磐变化
propylitized basalt 青[绿]磐岩化玄武岩
prospect for expansion 扩建远景
protection valve 安全阀
protective action 保护性作用
protective device 防护装置
protective light curtain 薄防护屏[帘]

(压滤机的)
protective oily coating 保护性油膜
protonization 质子化
prototype 原形[样机]设备
prototype automatic wet sieving device 原型自动湿筛装置
prototype calcining and smelting system 标准的煅烧熔炼系统
prototype machine 原形设备,样机
proven component 成熟[可靠]的部件
proven ore reserve 证实矿石储量
proven reserve 证实[探明]储量(即查明储量)
provisional cost 临时费用
provisional regulation 暂行规定
proviso 附带条件,限制性条款,附文
pseudo 假的,冒充的,赝,伪君子,假冒的人
pseudo depressant 假抑制剂
pseudo speed relation 假速度关系
PTAA(para-toluene arsonic acid) 对甲苯胂酸
p-type semiconducting character p 型半导体特性
p-type semiconductor p 型半导体
public bond 公债
public health 公众健康
public pension funds 公共养老基金
public water supplies 公用[共]水源
publication 发表[布,行],公布,出版物,刊物
publications committee 出版委员会

published capacity 公布[发表]的生产能力
published chart 发表的计算图表
published data 发表[公布]的数据[资料]
published literature 已发表[刊行]文献
published selling price 公布的销售价
publisher 出版者,发行人,发表者
publishing empire 大型出版企业
pulley 滑轮,皮带轮,滚筒
pulley center distance 皮带轮中心距
pulley cover 皮带轮罩
Pullin dc tachogenerator 普林型直流转速传感器
pulp ageing effect 矿浆老化效应
pulp circulation 矿浆循环
pulp circulation capability 矿浆循环能力
pulp conditioner 矿浆搅拌槽,调浆槽
pulp density relation 矿浆密度关系
pulp density table 矿浆浓度表
pulp density variation 矿浆浓度变化
pulp dilution 矿浆稀释度
pulp distributor feed launder 矿浆分配器给矿[料]槽
pulp flowrate 矿浆流速[量]
pulp height 矿浆高度
pulp level 矿浆液面
pulp level regulation 矿浆液面调节[整]
pulp phase 矿浆相

pulp sample 矿浆试样
pulp short-circuiting 矿浆短路(循环)
pulp specific gravity 矿浆比重
pulp transport characteristic 矿浆运输特性
pulp weight 矿浆重量
pulp X-ray unit 矿浆 X 射线装置
pulp-air injection 矿浆空气喷射
pulp-froth interface 矿浆泡沫界面
pulsating rotation 脉动旋转
pulsating water column 脉冲[动]水柱
pulse count measurement 脉冲计数测定
pulse duration and interval 脉冲持续和间歇时间
pulse energy 脉冲能量
pulse feeder 脉动式给矿机
pulse filter 脉冲过滤机
pulse height analyzer 脉冲高度分析仪
pulse of compressed air 压缩空气脉冲
pulse output 脉冲输出
pulse rate 脉冲速率
pulse scaler 脉冲定标器
pulse size 脉冲振幅
pulsejet bag house dust collector 脉冲[动]喷气布袋收尘器
pulse-jet fabric filter 脉冲喷射织物过滤器
pulse-jet fabric filter dust collector vender 脉冲-喷射织物过滤机收尘器供方
pulse-jet filter bag 脉冲喷射过滤袋

pulse-jet type	脉冲喷射型
pulsing valve	脉冲调节阀
pulsing valve size	脉冲阀规格[尺寸]
pulverized mineral	粉矿(粉碎的矿物)
pulverized sample	已研磨的[研磨好的]试样
pulverizing equipment	粉磨设备
pump box	泵池
pump box level	泵池液位
pump box level transmitter	泵池液位变送器
pump box model	泵池模型
pump centerline	泵中心线
pump characteristic curve	泵特性曲线
pump curve	泵曲线
pump design	泵设计
pump expert	泵业专家
pump gland water	泵密封[水封]水
pump head	泵扬量
pump horsepower indicator	泵功率指示器
pump loop system	泵回路系统
pump pad	泵垫
pump performance curve	泵特性曲线
pump suction line entrance	泵吸入管入口
pump sump	泵池
pump sump level	泵池液面[位]
pump surging	泵涌浪
pumpage	泵抽运量[工作能力],抽运能力
pump-box level	泵池液面[位]
pump-cyclone installation	泵和旋流器装置
pump-cyclone system	泵和旋流器系统
pumper cell	泵浮选槽(既可浮选也可用泵提升)
pumping test	泵输送试验
purchase cost	采购费用
purchase order	订购单,采购订单
purchase specification	采购明细表[说明书,技术规格]
purchase time	采购时间
purchaser of mineral product	矿产品买方
purchasing	采购
purchasing guide	采购指南
purchasing manager	采购经理
purchasing schedule	采购计划
pure cassiterite mineralogical sample	纯锡石矿样
pure galena	纯净方铅矿
pure gum rubber	纯天然橡胶
pure mineral	纯矿物
pure oxygen	纯氧
pure salicylaldehyde collector	纯水杨醛捕收剂
pure stress	纯应力
pure tension	纯张力
pure water	纯水
purely numerical factor	纯数字因数
purification circuit	纯[净]化回路
purified kerosene	精炼煤油

Purifloc C31 絮凝剂(聚乙烯胺)
Purifloc C32 絮凝剂(聚乙烯亚胺)
purpose of research 研究目的
purpose-built self-trimming single deck bulk carrier 特制的自行整平单甲板散装货船
purpurin 红紫素(1,2,4-三烃基蒽醌)(可用作脉石的抑制剂)$C_{14}H_8O_5$
push button 按钮
push button control panel 按钮控制盘
push button retract 按钮回程
push button start-stop control 按钮停开控制
push feeder 推式给矿机
push feeder-separator 推式分级给料机(分级给料一体机)
pushbutton control 按钮控制
push-button mineral processing 按钮控制选矿
push-loaded scraper 推装式铲运机
putrescine 腐胺,丁二胺-[1,4](可用作捕收剂)$NH_2(CH_2)_4NH_2$
PVC prefabricated mercury trap 聚氯乙烯装配式水银捕集器
PX-917 三甲酚磷酸酯(品名,可用作捕收剂、活性剂、调整剂)$(CH_3C_6H_4O)_3PO$
Pye combined glass and calomel electrode Pye 型玻璃甘汞混合电极
Pye-Unican pH meter Pye-Unican 型 pH 值计
pyramidal hopper 角锥漏斗

Pyrex 派热克斯[高耐火,高硬,高硅]玻璃
Pyrex beaker (派勒克斯)耐热[硬质]玻璃烧杯
Pyrex glass 派热克斯(耐热)玻璃,派热克斯硬玻璃
Pyrex glass cell 硬质玻璃槽
pyrex glass pipe 硼硅酸(耐热)玻璃管,(派热克斯)耐热[硬质]玻璃管
pyrite-arsenopyrite sulfide mineral 黄铁矿砷黄铁矿硫化矿物
pyritic calcine 硫[黄]铁矿烧渣
pyrogallic acid 焦性没食子酸,连苯三酚,焦棓酸,焦棓酚 $C_6H_3(HO)_3$
pyrogallol(=pyrogallic acid) 焦性没食子酸,连[邻]苯三酚,焦棓酚,焦棓酸(可用作抑制剂、起泡剂,浮选硫化锌矿时作为萤石的抑制剂及氧化铁矿的抑制剂)
pyrometallurgical extraction 火法冶金提取[提炼,精炼]
pyrometallurgical operation 火法冶炼操作[作业,生产]
pyrometallurgical process 火法冶金工艺,火治法
pyrometallurgical recovery method 火法冶金回收法
pyrometallurgical refining 火法精炼
pyrometallurgist 火法冶金工作者
Pyronate 捕收剂(石油磺酸钠,相对分子质量 350~370,15%矿物油)
pyrorefining 火法精炼

Q

8-quinolinol 8-羟基喹啉(可用作捕收剂)$C_9H_6N \cdot OH$
QNSN$_3$ 起泡剂(松油)
quadratic form 二次型[式],二次形式
quadrature equation 求积方程
quadrature procedure 积分[正交]法
quadrupole moment 四极矩
qualification 资质,资格证明书
qualified technician 胜任的技术人员
qualitative conceptual framework 定性概念框架
qualitative estimate 定性估计
qualitative information 定性资料
quality 质量,品质,高品质的,音色[质]
quality assurance program [plan] 质量保证程序[计划]
quality control 质量管理[控制]
quality control of sample method 取样法质量控制
quality control restriction 质量控制的限制
quality indicator 质量指标
quality inspection 质量检查
quality level 质量水平
quality of separation 分离[选]质量
quality of water 水质
quality specification 质量规范
quality standard 质量标准
quantitative approach 定量测定法
quantitative calculation 定量计算
quantitative correlation 定量相关关系
quantitative data 定量数据
quantitative description 定量描述
quantitative determination 定量测定
quantitative document 定量文件
quantitative estimate 定量估算
quantitative evaluation 量化[定量]评价[估]
quantitative expression 定量表达式
quantitative linear relation 定量线性关系
quantitative means 定量法
quantitative method 定量方法(分析)
quantitative mineragraphy 定量矿相学
quantitative mineralogical analysis 定量矿物分析
quantitative model 定量模型
quantitative output signal 定量输出信号
quantitative relationship 定量关系
quantitative spectrographic analysis 定量光谱分析

quantity of induced air flow 感生[诱发]空气流量
quantometer 光量计,辐射强度测量计
quarry run screen analysis 采石场石料筛分分析
quarry soft limestone 采石场软石灰石
quartering and coning 堆锥四分缩样法
quarterly interim unaudited statement 季度中期未经审计报表
quarterly production report 季度生产报告
quartz 石英 SiO_2
quartz diorite 石英闪长岩
quartz gangue 石英脉石
quartz monzonite laccolith 石英二长岩岩盖[盘]
quartzmonzonite-porphyry 石英二长斑岩
quartz pebble 石英砾石
quartz sand 石英砂
quartz veinlet 石英细脉
quartz-alkylammonium system 石英烷基铵体系
quartz-filled tabular structure 充填石英的板状构造
quartz-pyrite-sericite veinlet 石英-黄铁矿-绢云母细脉
quartz-siderite-muscovite facies 石英陨铁[菱铁矿]白云母相
quasi equilibrium reaction 准平衡反应
quasi-steady state 准稳(定状)态
quaternary amine 季铵
quaternary amine salt 季铵盐
quebracho 白雀树
quench tank 淬火槽,骤冷槽(碳浆法的)
quick analysis technique 快速分析技术[方法]
quick lime 生石灰
quick method 快速方法
quicksilver film 水银薄膜
quiescent period 平静期
quiescent region 静止区
quiescent solution 静止溶液
quiescent zone 静止区,稳定区[带](浮选机的)
Quix 絮凝剂(一种烷基硫酸盐,可用作选煤废水的絮凝剂)
quotation of price 报价(单)

R

5 roller high side mill 五辊高边磨矿机
R-10 双环已烷基二硫代氨基甲酸钠(品名,含3分子水)$(C_6H_{11})_2NC(S)SNa$
R-39(=**butoxy ethoxy propanol**) 起泡剂(丁氧乙氧基丙醇,可用作硫化铋

与砷化物的捕收剂(德国产))
Raconite 粗制丁基黄药(品名,可用作捕收剂)C_4H_9OCSSM
RADA(=Rosin Amine Denatured Acetate) 变性松酸胺醋酸盐(可用作捕收剂)
radial distance 径向距离
radial distributor 径向[放射状]分配器
radial feed distributor 辐射式给料分配器
radial horizontal arm 径向水平臂
radial impeller 径向叶轮
radial manifold 放射状歧[集]管
radial rotor blade 径向转子叶片
radial stacker 径向堆积机
radial stream 径向流
radial thru-flow hole 径向通流孔
radial tip impeller 径向尖头叶轮
radial type of manifold 放射型歧管
radiant energy 辐射能
radiation characteristic 辐射特性
radiation discrimination 辐射分辨[鉴别,区别,识别]
radiation effect 辐射作用[效应]
radiation flux 辐射流
radiation scattering method 辐射散射法
radiation source 辐射源
radio frequency probe 高[射]频探针
radio system 广播系统
radioactive decay detector 放射衰变探测器
radioactive principle 放射性原理
radioactive source 放射源
radioactive source radiation detector 放射源放射(矿物)探测器
radioisotope source 放射性同位素源
radioisotope technique 放射性同位素技术
radiometric sorting 放射选矿法
radiometric technique 放射性测量技术
radiotracer technique 放射性示踪技术
radius of particle 颗[矿]粒半径
raffinate oil 残油
rail grizzly 轨条筛
rail link 铁路连接(线)
rail tunnel dimension 铁路隧道[平硐]尺寸[大小]
railings 栏杆
railroad and truck scale 铁道和汽车秤(衡)
railroad scale 铁路磅秤
railroad spur line 铁路支线
railroad track scale 铁路轨道秤
railway hopper car 铁路漏斗车
railway private sidings 铁路专用线
railway system 铁路系统
railway trunk line 铁路干线
rainbird sprinkler head 雨鸟式喷洒头
rainbird-type sprinkler 鱼鸟式洒水车
rainbow trout 虹鳟鱼
rainfall 雨量

rainwater 雨水
rainy season 雨季
rake 返砂(螺旋分级机沉砂)
rake arm 耙臂
rake capacity 耙矿[返砂]能力(螺旋分级机的)
rake condition 返砂条件
rake lifting mechanism 耙齿提升机构
rake load 耙负荷,返砂量
rake material 扒出的物料,返砂(分级机的)
rake percent solids 返砂固体百分数
rake product 返[沉]砂产品(分级机的)
rake volumetric flowrate 返砂体积流量
raking capacity 耙料[返砂]能力
raking mechanism 耙料[集]机构
ramification roughing 分支粗选
ramp down 下坡(胶带机,道路等)
ramp grade 斜坡(路)
ramp on 上坡(胶带机,道路等)
Ramsey ball mill belt scale 拉姆齐球磨机皮带秤
Ramsey belt coil 拉姆齐胶带输送机探测线圈(用于连续测量胶带输送机上矿石的磁铁含量;测量并控制磁性矿石给入棒磨球磨机中)
Ramsey belt scale 拉姆齐皮带秤
Ramsey chain-weigh system 拉姆齐链压装置(测料位的)
Ramsey Engineering Vey – R – Weigh load transducer dual idler belt scale 拉姆齐工程公司 Vey-R-Weigh 负荷传感器型双托辊皮带秤(Vey-R-Weigh 为品牌名)
Ramsey Model 80-12B Automatic Crusher Feed Control System 拉姆齐 80-12B 型破碎机给矿自动控制系统
Ramsey nuclear vane feeder 拉姆齐核子叶片式给料器[机]
Ramsey Vey-R-Weigh(conveyor)scale 拉姆齐 Vey-R-Weigh(传送带)秤(Vey-R-Weigh 为品牌名)
Ramsey weigh belt feeder 拉姆齐称量带式给矿机
random bias 偶然偏差
random change 随机变化
random cut 偶然[随机]切取
random error 随机[偶然,不规律]误差
random fluctuation 随机波动
random fluctuation of sufficient magnitude 大幅度的随机波动
random independent variable 随机独立变量
random intercept method 随机截取法
random network 随机网络
random number 随机数
random process error 随机过程误差
random sample location 随机[不规则]取样位置
random spacing 任意[随机]间距,任意布置

randomly located drill hole 随机布置的钻孔
range 范围,幅度,变程
range of addition 添加范围
range of autocorrelation 自相关范围
range of fluctuation 波动范围
range of ore quality fluctuation 矿石质量波动范围
range of value 价值范围
range of values of operating parameters 操作参数值域
range of variable 变量范围
rapid analysis method 快速分析法
rapid chemical analysis 快速化学分析
Rapid disc magnetic separator 赖皮特盘式磁选机(高强度胶带电磁选矿机,英国矿物磁选专业公司,Boxmag-Rapid公司生产)
rapid dryer 快速干燥机
rapid estimation 迅速估算
rapid flotation 快速浮选
rapid on-site analysis 快速现场分析
rapid quantitative method 快速定量法
rapid response 迅速反应
rapid sand filter 快速沙滤器
rapid separation 快速分离
rapid-pass 快速通过(磨矿中矿物的快速通过)
rare earth geochemistry 稀土地球[质]化学
rare metal 稀有金属
rate (比)率,比(值),速率(度),等(级),费用
rate constant 速度(率)常数
rate control sub-loop 给矿量控制子控制环
rate controller 速率[速度]控制器
rate determining 速率[速度]确定
rate determining mechanism 确定浮选速率[速度]的机理
rate equation 速率[速度]方程式
rate limitation 速率[速度]限制
rate of application 使用比,适用率,用液比(氰化法)
rate of assay 分析[测试]速率
rate of capture 捕获比率
rate of charge 装载[给矿]量[率,速度]
rate of concentrate production 精矿生产率[量]
rate of consumption 消耗速度[率]
rate of curves 曲线斜率
rate of deactivation 去活速度[率]
rate of discount 贴现率
rate of escalation for production cost 生产费用上升[增长]率
rate of extraction 提取[萃取,分离]速度
rate of finished product 成品率
rate of flotation 浮选速度
rate of flow 流量[速,率]
rate of generation 生成率[速度]
rate of grinding 磨矿速度[量]
rate of interest 利率

rate of mining 开采速度
rate of reaction 反应速度
rate of return 投资偿还率
rate of return on investment 投资回报率
rate of rise 上升速度[速率]
rate of saponification 皂化速度[速率]
rate of travel 运行速度(筛上矿物的)
rate per shift 每班生产率
rate schedule 费用[价格]计划表
rate-controlling stage 速度控制阶段
rated capacity 额定处理能力
rated load 规定荷载,计算[设计,额定]荷载
rated production 额定生产量
rated volume 额定量
rathole diameter 管状腔(鼠洞)直径
ratholing problem 起管状腔问题
rating test 标定[评级]试验
ratio estimate 比率估算
ratio meter 比率计
ratio of adsorption density 吸附密度比
ratio of concentration 浓缩[密]比
ratio of density 密度比(值)
ratio of flow rate 流速比
ratio station 比值操作器
raw distillate 粗馏出物
raw material 原(材)料
raw material handling stage 原料处理阶段
raw material inventory 原料库存[存储]
raw survey data 原始测量数据
raw water 新[生]水,未净化水
Raymond mill 雷蒙德磨机
rayon production wastewater 人造纤维生产废水
Raytheon sonic level measuring device Raytheon 声波料位测量装置(美国 Raytheon 公司生产)
reacted pulp 反应矿浆
reaction heat 反应热
reaction plate weigher 感[反]应板式衡器,反馈板式衡器[秤]
reaction product 反应产物,反应后的生成物
reaction series 反应系列
reaction time 反应时间
reactivated resin 再生树脂
reactivation 再生,再活化
reactivation kiln (炭)再活化[性]窑
reactor design 反应器设计
reading 阅读,读数
readings high-intensity separator 带读数高强度磁选机(湿式)
read-out 读出
read-out unit 读出装置
Reagent "xx" 美国氰胺公司出品的浮选药剂品名
Reagent 40 (= oleic acid emulsion) R-40药剂(油酸乳剂,可用作萤石的捕收剂)
Reagent 60 R-60药剂(伯醇及仲醇类、

煤油等混合物,可用作起泡剂,萤石浮选时作为脉石抑制剂)

Reagent 107 R-107 药剂(石油磺酸盐,可用从萤石中捕收重晶石的药剂(捕收剂))

Reagent 301 R-301 药剂(仲丁基钠黄药,可用作捕收剂)

Reagent 303 R-303 药剂(乙基钾黄药,可用作捕收剂)

Reagent 322 R-322 药剂(异丙基钾黄药,可用作捕收剂)

Reagent 325 R-325 药剂(乙基钠黄药,可用作捕收剂)

Reagent 343 R-343 药剂(异丙基钠黄药,可用作捕收剂)

Reagent 350 R-350 药剂(戊基钾黄药,可用作捕收剂)

Reagent 404 R-404 药剂(巯基苯骈噻唑,可用作捕收剂)

Reagent 407 R-407 药剂,(石油磺酸盐,可用作铝土矿的捕收剂)

Reagent 425 R-425 药剂(巯基苯骈噻唑,可用作捕收剂)

Reagent 444 R-444 药剂(巯基苯骈噻唑,可用作捕收剂)

Reagent 505 R-505 药剂(绿黄色粉末,遇水反应放热,铜钼分离时作为硫化铜抑制剂)

Reagent 512 R-512 药剂(可用作白铅矿的捕收剂)

Reagent 610 R-610 药剂(有机胶体,成分可能是磺化木素类,可用作脉石分散剂)

Reagent 615 R-615 药剂(含有木素磺酸盐、糊精等成分,与 Reagent 620 相似,组成不同,可用作抑制剂)

Reagent 620 R-620 药剂(含有木素磺酸盐、糊精等成分,可用作抑制剂)

Reagent 633 R-633 药剂(含有木素磺酸盐、糊精等成分,与 Reagent 620 相似,组成不同,可用作抑制剂)

Reagent 645 R-645 药剂(为有机胶体,成分可能是磺化木素类,可用作碳质脉石、硫化砷、锑的抑制剂)

Reagent 710 R-710 药剂(植物油脂肪酸,可用作捕收剂)

Reagent 712 R-712 药剂(植物油脂肪酸,可用作捕收剂)

Reagent 723 R-723 药剂(植物油脂肪酸,不饱和度比药剂 R-710、712 增高,可用作捕收剂)

Reagent 765 R-765 药剂,(植物油脂肪酸,不饱和度比药剂 R-710、712 增高,可用作捕收剂)

Reagent 801 R-801 药剂(石油磺酸,可用作捕收剂)

Reagent 825 R-825 药剂(石油磺酸,可用作捕收剂)

Reagent 827 R-827 药剂(石油磺酸,可用作捕收剂)

reagent addition 药剂添加,加药

reagent addition level 药剂添加量[程度]

reagent addition rate 药剂添加量[率]

reagent balance 药剂平衡
reagent combination 药剂组合[配方]
reagent concentration 药剂浓度
reagent consumption 药剂消耗
reagent control 药剂控制
reagent cost 药剂成本
reagent dosage 药剂用量
reagent facilities 药剂设施
reagent for water treatment 水处理药剂
reagent fume control system 药剂烟气控制系统
reagent pattern 药[试]剂模式
reagent personnel 药剂人员
reagent policy 药剂方法[制度]
reagent preparation section 药剂制备段
reagent quantity 药剂(用)量
Reagent S-3019 S-3019 药剂,一种絮凝剂
Reagent S-3100 S-3100 药剂,一种絮凝剂
Reagent S-3257 S-3257 药剂(有机硅油类化合物,可用作捕收剂,用于硫化矿浮选)
Reagent S-3258 S-3258 药剂(有机硅油类化合物,可用作捕收剂,用于硫化矿浮选)
Reagent S-3259 S-3259 药剂(有机硅油类化合物,可用作捕收剂,用于硫化矿浮选)
Reagent S-3275 S-3275 药剂(有机硅油类化合物,可用作捕收剂,用于硫化矿浮选)
Reagent S-3292 S-3292 药剂(有机硅油类化合物,可用作捕收剂,用于硫化矿浮选)
Reagent S-3302 S-3302 药剂(戊基黄原酸丙烯酯,可用作捕收剂,用于浮选辉铜矿及硫化铜矿)
Reagent S-3315 S-3315 药剂(有机硅油类化合物,可用作捕收剂,用于硫化矿浮选)
Reagent S-3317 S-3317 药剂(有机硅油类化合物,可用作捕收剂,用于硫化矿浮选)
Reagent S-3346 S-3346 药剂(有机硅油类化合物,可用作捕收剂,用于硫化矿浮选)
reagent scheme 药剂计划[方案]
reagent splitter 药剂分配器
reagent system 药剂制度[系统]
reagent testing 药剂试验
reagent type 药剂类型
reagent utilization 药剂的使[应,利]用
reagent-grade borate 试剂(纯)级硼酸盐
reagent-grade HCl 试剂(纯)级 HCl (氯化氢,盐酸)
reagentized feed 用药剂调整的给矿
reagentized particle 受药剂作用的颗粒
real center point value 实际中(点)值

real estate 不动产,房地产
real estate mortgage 不动产抵押品
real experiment variogram 真实实验变异函数
real flotation system 实际浮选系统
real plant capacity 车间实际生产能力,选矿厂的实际生产能力
real surface area 实际表面积
real time application 实时应用
real unit 实际量
realistic simulation 实际模拟
real-size scale 实际尺寸比例
real-time bookkeeping 实时簿记
real-time clock 实时时钟
real-time disk-based operating system 实时磁盘操作系统
real-time event 实时事故
real-time extension 实时扩充
real-time operating system 实时操作系统
real-time particle size analyzer 实时粒度分析器
real-time particle-size measurement 实时粒度测定[量]
real-time sizing method 实时粒度[筛分]分析法
real-time work 实时(运行)工作[操作]
reappearing kinetic energy 再现动能
reappraisal 重新评价
rear end dumper 后卸车,尾部自动倾卸车

reasonable safety system 合理的安全系数
reassay 再分析(校对)
rebound pendulum 回弹摆锤
rebound pendulum maximum potential energy 回弹摆锤最大位能
rebound platen 回弹板
receding contact angle 后移接触角
receiver opening 受矿口
receiver set 接收机
receiving conveyor 受料运输机
receiving hopper 受矿漏斗
receiving opening 给矿口
receiving tub 受矿[料]桶
recessed chamber type center feed precipitate press 板框式中央给矿压滤机
reciprocal double layer 对应双电层
reciprocating feeder 槽式给矿机,往复式给料机
reciprocating plate feeder 往复板式给料机
recirculation belt conveyer 循环胶带运输机
recirculation fan 循环风机
recirculation of middlings 中矿循环
reclaim 回收,再生,重新使用,恢复,复原,收回,要求收回[恢复],开垦,填筑,翻造,革新
reclaim belt conveyer 回收胶带[装载胶带]运输机
reclaim conveyor 返矿运输机,(矿石

回收)装载运输机
reclaim hopper 回收[集矿]漏斗
reclaim tunnel 回收隧道
reclaim tunnel system 回收隧道系统
reclaim water pond 回水池
reclaim water sump 回水泵池
reclaim water system 回水系统
reclaimed land 复田,回填[开垦]土地,填海土地,
reclaimed mineral 回收的矿物
reclaimed rubber 再生橡胶
reclaimed water station 回水站,再生水站
reclaiming water 回收水
reclamation 复田,开垦,开拓,收复,回收利用,再生
reclamation work 复原[田]工作
recleaned copper concentrate 已再次精选的铜精矿
recleaner concentrate 再精选精矿
recleaning 再精选
Reco(primary collector) Reco粗选捕收剂(二甲酸基二硫代磷酸钠)
recommendation 推荐,建议
recommended acceptable level 建议[推荐]允许含量
recommended apex diameter 推荐的排矿口直径
recommended closed side settings 推荐的窄边开口调节(尺寸)
recommended equipment 推荐设备
recommended flowsheet 推荐流程

recommended installed power 推荐(的)安装功率
recommended method 推荐方法
recommended minimum product 推荐最小产品[量]
recommended power 推荐功率
recommended product size range 推荐产品粒度范围
recommended spare parts 推荐的备件
recommended underflow percent solids 建议的底流固体百分率
recommended vender 推荐的供应商
reconnaissance stage 踏勘阶段
recorder 记录器,记录装置
recoverable value 可回收价值
recovery 恢复,复原,回收(率)
recovery curve 回收率曲线
recovery depression 回收率下降
recovery improvement 回收率提高
recovery mechanism 回收机理
recovery operation 回收作业
recovery plant (金属)回收厂
recovery procedure 回收方法[作业过程,步骤]
recovery process 回收法
recovery set point 回收率给定点[值]
recovery system 回收系统
recovery time curve 回收率(浮选)时间曲线
recovery vs. particle size curve 回收率与粒度关系[对比]曲线
recovery-grade curve 回收率与品位

关系曲线
recreational area 娱乐区
recrystallized and deformed shale 重结晶变形页岩
rectangular block 矩形矿[地]块
rectangular channel 长方[矩]形槽
rectangular natural crystal 矩形自然晶体
rectangular opening 矩形孔
rectangular orifice 矩形进口
rectangular outlet 矩形排口
rectangular prism 矩形棱柱,直角棱镜
rectangular vibrating pan 矩形振动槽[盘]
recuperation 再生,回收,同流换[节]热(法),换热(作用);蓄热;继续吸热(法);余热利用,再生利用法
recurrent cost 经常费用[成本]
recycle 再生,再循环,重复利用;重整煤油,叠合油(石油不饱和烃加氢产品,沸点 130~350℃)
recycle conveyor 循环胶带运输机
recycle stream 再[重复]循环矿浆
recycle water 循环水
recycled classifier sand 分级机返砂
recycled water 循环水
red mud slurry 红[赤]泥泥浆(炼铝用)
red oil 红油(即工业油酸,可用作捕收剂)
redox couple 氧化还原电对[偶]
redox potential 氧化还原电位

reduced belt 减层胶带,减低强度胶带
reduced capacity 减低(输出)容量,减弱能力
reduced efficiency curve 折算的效率曲线
reduced efficiency curve equation 效率曲线折算方程,折算效率曲线方程
reduced particle recovery curve 折算的颗粒回收率曲线
reduced print 缩小的图纸[片],缩小的影印[打印]图纸
reduced recovery 折算[对比]回收率
reduced recovery curve 折算的回收率曲线
reduced size print 缩小图幅的图纸,缩小图幅的影印[打印]图纸
reduced yield 减产,降低产量
reducer drive 减速器驱动
reducer ply 减层
reducible species 还原物
reducing atmosphere 还原(性)气氛
reducing condition 还原条件
reducing energy 还原能量
reducing treatment 还原处理
reduction curve 还原曲线
reduction cycle 还原周期
reduction equipment 破碎设备
reduction factor 减缩[折减]系数
reduction gear ratio (传动)减速比
reduction peak 还原峰值
reduction rate 减速[传动]比,转速比
reduction ratio of speed reducer 减速

机减速比
reduction segment 还原段
reduction valve 减压阀
reduzate 还原[沉积]物
reef facies 礁相
reef material 石英脉矿石(含石英脉矿石)
reevaluation 重新评价
re-examination 再检查
reference article 参考文章[献]
reference drawing 参考图
reference information 参考资料
reference lattice 基准格点
reference library 资料室,图书参考室
reference manual 参考手册
reference material 参考材料
reference ore 标准矿石
reference performance 参考性能,工作正常情况
reference signal 参考信号
reference transmission spectrum 参比透射光谱
refined gold 纯金
refined kerosene 精炼煤油
refined lead 精(炼)铅
refined tall oil acid 精制塔尔油酸
refinement of T. B. Crowe T.B.克洛式提纯法
refinery area 精炼厂区
refinery loss 炼金损失
refinery plant 精炼厂,炼油厂,炼金室[车间]

refinery sweeping 精炼金银屑(精炼产生的)
refining cell 电解精炼槽
refining kettle 精炼锅
reflectance material 反射物质
reflectance measurement 反射测定
reflectance spectroscopy 反射光谱法
reflectance spectroscopy in visible, ultraviolet and infrared regions 可见光,紫外线和红外线反射光谱法
reflected light optical microscope 反光(光学)显微镜
reflecting material 反射物质
reflection grating 反射光栅
reformation 改革[造,良,善],革新
refractive index 折射率
refractory 不容易处理的(矿石),难选的(矿石)
refractory industry 耐火材料工业
refractory ore 难选矿石
refractory ores 难选矿石类
refractory oxide ore 难选氧化矿石
refractory species 难选[浮选]矿物[种]
refractory sulphide 难选硫化物[矿]
refrigerant cost 冷凝剂费用
refrigeration unit 冷冻设备
refuse storage 矸石堆场,废渣堆场
refuse waste 废弃矸石,废渣
regenerant 再生剂
regenerant chemical 化学再生剂
regenerant concentration 再生液浓度

regenerant inlet tube detail 再生液入口管详图
regenerant solution 再生剂溶液
regenerated carbon 再生碳
regeneration area 更新面积
regeneration cycle 再生周期
regeneration kiln (碳)再生回转窑
region of depression 抑制区
region of no flotation 不[非]浮选区
regional productivity 地区生产力
regionalized variable 区域化变量(品位,厚度,金属聚集等)
registered trademark 注册品牌名
regression analysis 回归分析
regression coefficient 回归系数
regression equation 回归方程
regression model of hydrocyclone 水力旋流器回归模型
regressive analysis 回归分析
regrind ball mill 再磨球磨机
regrind cyclone 再磨旋流器,与再磨配用的旋流器
regrind cyclone feed pump 再磨旋流器给矿泵
regrind cyclone feed pump box 再磨旋流器给矿泵池
regrind mill circulating load 再磨机循环负荷
regrind mill product size 再磨机产品粒度
regrind principle 再磨矿原则
regrind-feed splitter box 再磨给矿分配箱[分流槽]
regrinding plant 再磨车间
regrinding section 再磨工段
regrinding unit 再磨(磨矿)机
regular check 定期检查
regular disposition 规则分布[排列,布局]
regular disposition of sample 样品规则布置
regular drilling pattern 规则钻孔布置型式
regular feature 固定专栏[栏目],常规特征
regular flotation test 正规浮选试验
regular grade molybdenum concentrate 正规[正常,标准]品位钼精矿
regular interval 规则的间距[隔]
regular network 规则网络
regular pattern 规则布置
regular sample distribution 规则的试样分布
regular sample length 固定的样品长度
regular sample pattern 规则[常规]的样图
regular sampling 定期取样
regular sampling pattern 规则[常规]的取样网图
regular sampling plan 规则[常规]的取样平面图
regular square 正方形
regular square sample network 正方

形取样网络

regulated variable 被调变量

regulating-isolating transformer 调节隔音变压器

regulator （自动）调节器[闸,装置],调整器,控制器,节制闸,稳定器,稳[减]压器,校[调]准器,调节剂

regulatory agency 管理机构,监管机构

regulatory authorities 管理机关[当局]

regulatory restriction 政策的限制

rehabilitation area 复原区

rehabilitation cost 更新[修复,改建]费,复原费

rehabilitation of mined land 开采土地的恢复[复原]

rehabilitation program 复原计划

rehabilitation standard 复原标准

Reichert cone concentrator 赖克特圆锥选矿机

reimbursement of expense 费用偿还

reinforced concrete construction 钢筋混凝土结构

reinforced concrete drop box 钢筋混凝土跌水[缓流]箱

reinforced concrete stack 钢筋混凝土烟囱

reject bin 废渣仓,废石仓,矸石仓

reject bucket elevator 废渣[尾矿,废品,废弃料]斗式提升机

reject core 废岩芯

reject screen 轻产品[浮物,废石]筛

reject system 报废制度

related information 有关资料

relative accuracy 相对精确度

relative adsorption 相对吸附量

relative area 相对面积

relative depressing force 相对抑制力

relative efficiency 相对效率

relative energy 相对能量

relative error 相对误差

relative error range 相对误差范围

relative expense 相对费用

relative grindability factor 相对可磨性系数

relative humidity 相对湿度

relative intensity 相对强度

relative measure of slurry viscosity 矿浆黏度的相对度量

relative monetary unit 相对货币单位

relative potential 相对电位

relative power of reagent 药剂相对能力[量]

relative power requirement 相对（的）功率需要量

relative priority 相对的优先次序

relative solubility 相对溶解度

relative speed of response 相对的响应速度

relative standard deviation 相对标准偏差

relative standard error 相对标准误差

relative surface area(RSA) 相对表面积(缩写RSA)

relative work 相对功

relatively independent asynchronous task 相对独立的异步任务

relay unit 继电器单元

Releach 再浸出

reliability function 可靠性函数

reliability parameter 可靠性参数

reliability test 可靠性试验

relief deck 保险[护]筛网

relief shift foreman 安全值班工长

relocatable binary file 浮动二进位资料

remaining mine reserve 剩余[保留,保有,现有]矿山储量

remaining pebble conveyor 多余砾石胶带运输机

remaining property 剩余财产

remark 评语,意见,评论,备[附]注

remarks column 备注栏

remedial action 补救行动

remedial measure 补救措施

remedial method 补救办法

remelt furnace 再熔炉

remote adjustment 远距离调整

remote area 偏僻[远]地区

remote area [square field] 边[偏]远地区

remote control crane 遥控起重机

remote control tripper car 远距离控制卸矿[料]小车

remote geographical area 边远地理区域

remote indication 远距离[遥控]指示,遥测

remote process monitoring and automated control 远距离程序[过程]监视和自动控制

remote terminal 远距离终端设备

remoteness 遥[边]远,偏僻

removable access panel 活动[可拆卸]的入口面板

removable handrail 活动[可拆卸]的栏杆

removable panel 活动[可拆卸]的镶嵌壁板

Renex 非离子型表面活性剂品牌名(塔尔油脂,可用作非离子型矿泥分散剂)

Renex 648 壬基酚聚氧乙烯(5)醚化学结构式:

$$R-\langle\rangle-O(CH_2CH_2O)_n-H,$$
$$R=C_9烷基, n=5$$

Renex 688 壬基酚聚氧乙烯(8)醚化学结构式:

$$R-\langle\rangle-O(CH_2CH_2O)_n-H,$$
$$R=C_9烷基, n=8$$

Renex 690 壬基酚聚氧乙烯(10)醚化学结构式:

$$R-\langle\rangle-O(CH_2CH_2O)_n-H,$$
$$R=C_9烷基, n=10$$

rental income 租金收益

reorientation 再定向

reoxidation 再氧化(作用)
repair outfit 修[维]理工具
repair piece 备件
repair procedure 修理程序
repair tunnel 检修通道(磨机下)
repayment 偿还,还款
repayment of principal 还本,偿还本金
repayment schedule 偿付[还款]计划
repeated cleaning 再精选
repeated handling 反复操作
repeated sampling 重复取样
repeater totalizer 复示累加器
repeating unit 循环单元
repeating unit array 循环单元阵列
repellency toward water 疏水性
repetitive process 重复过程
replaceable liner of ceramics 可更换的陶瓷衬里
replaceable vortex finder 可置[更]换的涡流溢管
replacement of wear parts 磨损件的更换
replacement value of asset 资产更换价值
replating 再电镀,补镀
report drawing 报告书(图纸)
report proceedings 报告程序
report typewriter 报告打字机
reprecipitate 再沉淀物
representative slurry sample 有代表性的矿浆样品

representativeness 代表性
re-processing plant 再处理厂,再选厂
reproducible copy 可复制的副本
reproducible sample 重复[可再现]现试样
reproducible sheet 可复印[制]的表
reproduction 复制(品),复写,翻印[版]
reproduction center 复制中心
reproduction procedure and technique 复制程序和技术
repulp water circulator 再浆化水循环管[器]
repulsive energy barrier 推[排]斥能垒
repulsive energy carrier 推[排]斥能载体
repulsive force 排斥力
repulsive potential energy 排斥位[势]能
request for quotation 要求报价,询价
required daily capacity 需要的日处理能力
required delivery date 要求的交货日期
re-rating program 重新标定程序
Resanole 饱和 $C_{10} \sim C_{14}$ 脂肪酸混合物(品名,可用作捕收剂)
research agenda 研究议程
research and application engineering 研究与应用工程
research council 科学研究委员会

research department 研究部门[室]
research engineer 科研工程师
research establishment 科研机构
research expense 研究经费
research facilities 研究设施
research fee 研究费
research grant 研究拨款[补助金]
research group 研究(小)组
research metallurgist 选矿[冶金]研究人员
research method 研究方法
research paper 研究论文
research personnel 研究人员
research procedure 研究步骤
research process 研究过程
research program 研究程序[计划]
research project 研究项目
research route 研究路线
research target 研究目标
research team 研究组[团队]
research tool 研究工具
research worker 研究工作者
researcher 科研人员
reserve 储备(物),储量
reserve boundary 储量边界
reserve capacity 富裕能力
reserve estimation 储量估算
reserve fee 储备费
reserve fund 公积[准备]金
reserve requirement 准备金需要额
reserve storage 备用堆场[仓库]
reservoir system 水库[贮存]系统

residence time (磨机中物料)停留时间
residence time distribution 物料停留时间分布
residence time distribution calculation (物料)停留时间分布计算
residence time distribution information (物料)停留时间分布数据
resident office 常驻代表机构[办事处]
resident office of foreign enterprises 外国企业常驻代[办]表机构
resident representative office 常驻代[办]表处
residual charge density 剩余电荷密度
residual concentration 残余浓度
residual cyanide level 氰化尾矿含氰化物水平,氰化渣中氰化物含量
residual DPG (diphosphoglyceric acid) concentration DPG(二磷酸甘油酸)残余浓度
residual owner 剩余值所有者
residual sum of square 残差[剩余]平方和
residual sum of square function 残差[剩余]平方和函数
residual undepreciated value 剩余未贬低的价值
residual value 剩余值
residual variance 剩余方差
residue 残[滤]渣,尾矿
resilient shock absorbing suspension 弹性减震悬挂装置

resin batchwise 分批树脂
resin column 离子[树脂]交换柱
resin ion exchange process 树脂离子交换工艺[过程]
resin mesh size 树脂粒度
resin mold 树脂模
resin selectivity 树脂选择性
resin-in-pulp ion exchange Pachuca 矿浆树脂离子交换帕丘卡槽
resin-in-pulp ion exchange technique 矿浆树脂离子交换技术
resin-in-pulp procedure 矿浆树脂离子交换过程
resin-in-pulp technology 矿浆树脂离子交换技术
resinous type 树脂型
resinous type exchanger 树脂型交换剂
resistance 阻力,电阻,抵抗(力)
resistance loss 电阻损耗
resistance temperature detector 电阻温度探测仪[计]
resistance transducer 电阻传感器
resonance effect 共振效应
respiration 呼吸(作用)
response 响应,反应,感应,效应,回答
response contour 响应等高线
response function 响应函数
response matrix 反应矩阵
response radiation 响应辐射
response surface 响[反]应面
response test 响应试验
response time 反应时间

response value 响[反]应值
rest potential 残[剩]余电位[势]
resultant curve 综合曲线
resultant pulp 合成矿浆
resulting mill design 所完成的磨机设计
resulting sum 总和
resulting undepreciated value 最终不[未]折旧价值
retained earning 保有收入
retained income 保有收入
retained weight 筛上[保留]产品重量
retaining dam 拦水[土]坝
retaining ring 挡[定位]圈,扣环
retention time 停留时间
retirement plan 退休规划
retort plant 蒸馏车间
retractable belt feeder 可伸缩胶带给矿机
retractable fabric hood 伸缩式布罩
retreat 再选,撤退
retreatment operation 再选作业
retrofit project 翻新改进[造]项目
return 返回[矿]
return belt training idler 回空段导辊
return cathodic sweep 阴极回扫
return flow 返回矿流,返矿流
return idler 回空段托辊
return idler application factor 回空段托辊应用因素[系数]
return idler hanger 回空段托辊吊架
return idler serial number 回空段托辊

系列号

return product 返回产品,返矿

return pump 返回泵

return side 返回边[端]

return solution 回水,返回的溶液

return stream 返流,返回矿流(中矿)

return water 回(用)水

return water line 回水管线[道]

return water system 回水系统

reuse water 回用水,再生水

revegetation program (再)植被计划

revenue and expenditure 岁入和岁出

reverberatory furnace 反射炉

reverberatory furnace smelter 反射炉冶炼厂

reverse closed circuit 反向闭路(前置分级)

reverse flotation 反浮选

reverse leaching 逆浸出

reverse orientation 反向定向

reverse osmosis 反渗透

reverse process 逆过程

reverse side 反面

reverse spoon cutter 回动匙状切取器

reversibility of adsorption 吸附可逆性

reversible closed circuit 可逆闭路

reversible conveyor 可逆运输机

reversible flexible loader 可逆挠性装载器

reversible interchange 可逆交换

reversible interchange of ion 离子可逆交换

reversible potential 可逆电位[势]

reversible potential of xanthate-dixanthogen 黄原酸盐-双黄药电偶的可逆电位

reversible rubber belt conveyor 可逆胶带运输机

reversible transformation 可逆变换

reversible value 反值

reversible xanthate-dixanthate potential 可逆的黄药-二黄药电位

reversing starter 反向起动器

review 回顾,复习,检查,复审

review meeting of bid 投标评审会议

revolution in milling method 选矿方法革命[改革]

revolutionary excavator 旋转挖掘机

revolving credit 周转信贷,循环贷款

revolving drum conditioner 回转筒型调整槽

revolving helix 旋转螺旋

revolving period 周转期间

revolving trammel screen 回转圆筒筛

revolving-credit agreement 周转贷款协议

Rex rope frame conveyor idler Rex型绳索支架运输机托辊

Rex rope frame idler Rex型绳索支架运输机托辊

Rexnord gyradisc crusher 莱克斯诺型转盘破碎机

RF-54-38 德国产液体胺(可用作磷矿、云母及石英的捕收剂)

rheostat control 变阻器控制
rhodium 铑 Rh
rhombohedral shell 菱形壳
rhombohedral structure 菱面体结构
Riddick zeta meter 里迪克ζ[Z]电位计
Ridge structure 山脊构造
Riddick Type Ⅱ-UVA Plexiglass cell 里迪克Ⅱ型—UVA有机玻璃槽
riffled deck 铺格条床面
riffled launder gold trap 格条流槽捕金器,带格条的选金流槽
riffled sluice 来复条[格条]溜槽
riffling 沉砂,格槽缩样,用格条装置进行筛选(金矿中使用的)
rifle(lattice bar)launder 来复(格条)流槽
rifled launder 来复[格条]流槽
rifled launder gold trap 来复[格条]流槽金粒捕集器
right angle triangle 直角三角形
right machine 适用的设备
right to vote 投票权
right triangle 直角三角形
right-handed screw axes 右旋轴
rigid body 刚体
rigid conduit 刚性导线管
rigid specification 硬性[严格]技术要求
rigid systematic approach 严格的系统方法
rigid systematic handling approach (对试样的)严格系统加工法
ring gear 大齿轮,环形齿轮
rip detector 撕裂[裂口]探测器(用于胶带运输机)
rise velocity 上升速度
riser pipe 竖[立]管
rising belt conveyor 提升[上升]胶带运输机
rising speed 上升速度
rising velocity 上升速度
risk capital 风险投资
risk insurance premium 风险保险费
risk investment 风险投资
risk potential of investment project 投资项目的风险
Risor 经水蒸气蒸馏的松油(品名,可用作起泡剂)
river gravel crushing and screening plant 河砾破碎筛分厂
RKL Pinch Valve RKL夹管阀(美国Red Valve公司生产)
road bed loading 路基载荷[荷重]
road connection 道路连接
road elevation 路面标高
Roan Consolidated, Zambia 赞比亚罗昂联合企业
roaster calcine 焙烧炉焙砂
roasting 焙烧
roasting process 焙烧过程
roasting reaction 焙烧反应
robble duplex Edwards roaster 双搅拌式爱德华兹焙烧炉

rock box outlet 石仓出口
rock chip 石屑,岩石碎片
rock contour 岩石等高线
rock hook 岩石吊钩
rock meal 岩粉
rock-boxing 矿岩装箱作用[冲击]
rockering 摇动溜槽淘金
rod charger 装棒机
rod mill circulating load controller 棒磨机循环负荷控制器[调节器]
rod mill feed solids 棒磨给矿固体物料
rod mill feedwater proportioning valve 棒磨给水配量阀
rod mill operation 棒磨机操作
rod mill tonnage 棒磨机处理量
rod mill water 棒磨机给水量
rod mill-ball mill circuit 棒磨-球磨流程[系统]
rod mill-ball mill combination 棒磨球磨联合使用
rod mill-ball mill grinding section 棒磨-球磨磨矿工段
rod milling 棒磨
rod tangle 钢棒缠结,绞棒
rod tangling 缠棒
rod usage 棒使用量
role of oxygen 氧的作用
roll crusher setting 对辊破碎机(排口)调节
rolled bottle 滚动瓶
roller 滚轮,滚轴
roller chain drive 滚子链传动

rolling mill sludge of scale 轧钢机氧化[锈]皮残渣(轧钢产生的)
roll-out carriage 滚动拉出式托架
ROM ore 原矿矿石
roof member 屋顶构件
roof ventilator 屋顶风机
roofing industry 屋面材料工业
room temperature 室温
Roos feeder 鲁斯链型给矿机
rope condition 绳状排砂条件(水力旋流器的)
rope way 架空索道
ropeway loading station 索道装载站
roping 成束,绳结(旋流器底流)
roping condition 成束条件(旋流器底流)
roping constraint 成束约束(旋流器底流成绳索状约束)
Rosano surface tensiometer 罗萨诺表面张力计(美国产品)
rosin 松香,松脂
rosin amine 松香胺(可用作捕收剂)
rosin amine acetate 醋酸松香胺,松香胺醋酸盐
Rosin amine D 变性松香胺(品名)
rosin amine D acetate 变性松香胺醋酸盐(可用作捕收剂)
rosin amine denatured acetate 变性松酸胺醋酸盐
Ross chain feeder 罗斯链式给矿[料]机
Ro-Tap with timer 罗太普带定时器

的摇筛机(美国 Tyler 公司生产)
rotary belt concentrator 回转带式选矿机
rotary blast hole 旋转爆破孔
rotary blast hole drill 旋转钻机
rotary cut 旋转切[截]取
rotary drill 牙轮钻
rotary drum feeder 旋转筒式给矿机
rotary grinding mill circuit 旋转式磨机回路
rotary hole 旋转钻孔
rotary oil-fired reactivation furnace 回转燃油(炭)再活性炉
rotary optic filter 旋转滤光器(片)
rotary paddle feeder 转动桨叶给矿机
rotary plough feeder 回转刮板[犁式]给料机
rotary plow feeder 回转刮板[犁式]给料机
rotary precoat filter 预涂助滤剂旋转过滤机
rotary scrubber 回[旋]转式洗涤机
rotary strake 旋转刻槽皮带溜槽(如:澳大利亚花伯根造,刻槽矿泥摇床与皮带溜槽组合运动的细泥选别设备)
rotary table feeder 转[圆]盘给矿机
rotary type 旋转式
rotary valve 旋转阀
rotary valve feeder 旋转阀门给矿机
rotated disk electrode 旋转盘式电极
rotated electrode technique 旋转电极技术

rotating cutter blade 旋转切割刀片
rotating device 旋转装置
rotating heap leaching 轮流堆浸
rotating motor-driven paddle 马达驱动的旋转叶片
rotating paddle 旋转刮板
rotating tap device 旋转旋塞装置
rotating weight 旋转配重[重量]
rotational mass flow 旋转质量流量
ROTEX screen ROTEX 筛分机(美国 ROTEX 公司制造,是一种独特的水平回旋筛,进料端成圆周运动,而出料端为直线往复方式运动)
rotoclone wet type dust collector 旋风型湿式收尘器
rotor aspect ratio 转子纵横比(长度/直径)
rotor cavity vacuum 转子空腔真空度
rotor energy 转子能量
rotor mechanism 转子机构
rotor operating condition 转子操作条件
rotor rotational speed 转子转(动)速(度)
rotor speed 转子速度[转速]
rotor submergence 转子沉没度
rotor tip speed 转子叶尖速度
rotor top 转子上端
rotor's tip speed 转子末端速度(浮选机的)
rotor-cell bottom distance 转子与槽底距离

rotor-disperser region	转子-分散器区
rotor-standpipe cavity	转子竖管空腔
rough electrode	粗糙的电极
rough flowsheet	简陋[粗略]流程
rough plan	计算草图
rough service	笨重[艰巨]任务[工作]
rougher area pump sump	粗选区域泵池
rougher circuit	粗选回路
rougher concentrate	粗选精矿
rougher concentrate pump box	粗选精矿泵池[箱]
rougher concentrate regrinding unit	粗精矿再磨设备
rougher flotation circuit	粗选浮选回路
rougher flotation density	粗选浮选浓度
rougher flotation plant	粗选浮选车间
rougher grinding circuit	粗选磨矿回路
rougher operating	粗选作业
rougher performance	粗选指标
rougher regrind	粗选再磨
rougher tailings	粗选尾矿
roughers	粗选槽
rougher-scavenger banks	粗-扫选浮选机
rougher-scavenger behavior	粗-扫选特性[状态]
rougher-scavenger cell	粗-扫选槽
rougher-scavenger cell assembly	粗-扫选槽机组
rougher-scavenger flotation	粗-扫选浮选
rougher-scavenger metallurgical operation	粗-扫选选矿作业
rougher-scavenger system	粗-扫选系统
round port cock	圆口旋塞
rounding off	四舍五入,化成整数
round-table conference	圆桌会议
routine analysis	定期分析
routine basic test	常规的基本试验
routine basis	常规基础
routine engineering precaution	常规工程预防措施
routine error	常规误差
routine function	日常[程序]功能
routine information	（矿物分析）常规资料(即平均粒级,粒径分布,晶粒形状,脉石与矿石边界特性的共生体类型等)
routine inspection	例行[常规,定期]检查
routine library	程序库
routine maintenance	日常维修
routine mineral examination	常规矿物检测
routine plant activity	车间[工厂]常规运转
routine report	例行报告
routine sieving analysis	例行[定期,常规]筛分分析

routine storage 指令存储区
routine test 例行[定期,常规]试验
routine timing 常规定时
routine work 日常工作,常规作业
royalty 矿区使用费,版税,专利权税
RPM(revolutions per minute) 每分钟转数
RSA(relative surface area)feed 相对表面积给矿
RSA(relative surface area)product 相对表面积产品
rubber absorber 橡胶减震器
rubber ball mill linings 球磨机橡胶衬板
rubber belt conveyer 胶带运输机
rubber belt velocity transducer 胶带速度传感器
rubber cover 胶层(胶带面)
rubber covered steel liner 包胶钢衬
rubber disc return idler 回空段胶带托辊
rubber end holder 橡胶端座[架]
rubber hose 橡胶软管
rubber impeller 橡胶叶轮
rubber industry 橡胶工业
rubber lined Denver cyclone feed pump 橡胶衬里丹弗型旋流给矿泵
rubber lined pump 胶泵,衬胶砂泵
rubber material conducting hose 橡胶导流软管
rubber mill liner 磨机橡胶衬里
rubber molded wearing parts 成型橡胶磨损件
rubber mount 橡胶座
rubber pinch-valve 橡胶夹阀
rubber screen cloth 橡胶筛网
rubber shear mount 橡胶剪力座
rubber skirt 橡胶垫环[挡板]
rubber skirt board 橡胶拦矿板
rubber stopper 橡胶塞
rubber tired dozer 橡胶轮胎推土机
rubber wear element 橡胶耐磨元件
rubberized asphalt membrane 橡胶沥青膜
rubber-lined pipe 衬胶管道
rubber-lined pipeline 衬橡胶管道
rubber-lined tube 衬胶管
rubber-on-steel chute 衬胶钢溜槽
rubber-tired load-haul-dump unit 轮胎式装运卸设备
Ruggles Coles rotary dryer 拉格尔斯·科尔斯回转[转筒]干燥机
rule of thumb 经验法则[方法]
run 运转[动,算,行];驾驶,控制,操作;趋向,进[执]行;流量;熔化[铸]试车[转];趋势,倾[走,动]向;(斜,水)槽,管[坑]道;偏斜[不规则]矿体;联络巷道;流行,畅销;(连续)刊登;跑
run empty 空转
run of mine ore 原矿
run of pit waste 采场废石
run time 运转时间
running horse power 运转马力

running light 运转灯
running load 工作[运转,动]负荷
running without load 空运转
run-of-pit mine ore 露天矿原矿
run-of-underground mine ore 坑内[地下]原矿
run-up time 起动时间
rural area 农村地区
rusty metal 生锈金属
Ruth machine 鲁思浮选机(封闭叶轮,中空轴的底吹式)
Rutter's crusher 卢特破碎机

S

3-stage sampling system 三段取样系统
400 series promoter 400系列促进剂
700 series promoter 700系列促进剂(来自植物的脂肪酸型产品)
8 station rotary model 8流[工位]旋转形式
800 series promoter 800系列促进剂(为阴离子石油磺酸型)
S. H. S. 十六烷基硫酸钠(品名,可作为多金属硫化矿浮选的捕收剂)
S_3 醚醇起泡剂(庚醇与5克分子环氧乙烷的缩合物,罗马尼亚产)
S_4 醚醇起泡剂(甲醇与10克分子环氧乙烷的缩合物,罗马尼亚产)
S_5 醚醇起泡剂(杂醇油与4克分子环氧乙烷的缩合物,罗马尼亚产)
S-3019 絮凝剂(美国氰胺公司出品)
S-3100 絮凝剂(美国氰胺公司出品)
S-3302 选矿药剂(戊基黄原酸丙烯酯,可用作浮选辉钼矿及硫化合物的捕收剂) $C_5H_{11}OCSSCH_2CH:CH_2$
safe breath level 安全呼吸水平
safe operating tonnage 安全生产吨位数
safe procedure 安全措施
safe value 安全[保险]值
safety 安全,保险[安全]装置,安全设备
safety cable pull switch 缆式拉线安全[紧急]开关
safety consideration 安全需要考虑的事项
safety device 安全装置
safety insurance 安全保险措施
safety office 安全办公室
safety pull switch 拉线安全[紧急]开关
safety service department 安全业务部门

safety system 安全体系
safety trip cord 安全自动跳闸绳
SAG mill and ball mill configuration 半自磨和球磨结构
SAG mill sound 半自磨机声响[噪声]
Sala filter 萨拉型过滤机(瑞典 Sala 公司)
Sala flotation machine 萨拉浮选机(瑞典 Sala 公司)
Sala high gradient separator 萨拉高梯度磁选机(湿式)(瑞典 Sala 公司)
Sala vacuum filter 萨拉真空过滤器(瑞典 Sala 公司)
salable grade 可销售品位
salable product 可售[畅销]产品
salaried supervisor 领薪水的监督[检查]人员
salaries 工资,薪酬
sale and leaseback 出卖后反租,售后回租
saleable product 可销售[畅销]产品
sales agent 销售经纪人
sales contract 销售合同
sales director 营业主任
sales engineer 商业[推销]工程师
sales gross revenue 销售年总收入
sales letter 推销信函
sales letter of intent 推销意向书
sales literature 销售文件
sales net annual income 销售净年收入
sales office 销售部门
sales order 销售定单
sales price 售价
sales revenue 销售总[年]收入
sales staff 销售人员
sales tax 销售[营业]税
sales value of product 产品销售(价)值
sales volume 销售量
salicyl 水杨基,邻羟苄基 $HO \cdot C_6H_4 \cdot CH_2-$
salicylaldehyde 水杨醛,邻羟基苯醛
salicylaldoxime 水杨醛肟(可用作捕收剂)$HOC_6H_4CH-NOH$
salicylic acid 水杨酸(可用作石英抑制剂,浮选氧化铁时作为 pH 值调制剂)$HOC_6H_4CO_2H$
salicylic aldehyde solution 水杨醛溶液
saline water 盐水
salmon 橙红色,赭色,鲑鱼
salt compound 盐类化合物
salt ore 盐类矿石
salt type 盐类
salt type mineral 盐类矿物
same relative direction 相同方向
sample aging 试[矿]样时效
sample analysis 试[矿]样分析
sample balance 试[矿]样平衡
sample calculation 试[矿]样计算
sample can 试[矿]样罐
sample carrier 试[矿]样容器
sample chute 试[矿]样溜槽
sample collector 试[矿]样收集器
sample density 试[矿]样密度

sample deterioration 试[矿]样变质[坏]
sample dimension 试[矿]样尺寸
sample distribution 试[矿]样分布
sample drying oven 试[矿]样烘干箱
sample excitation 试[矿]样激发
sample extraction formula 试[矿]样(提取)量计算公式
sample extraction schedule 试[矿]样提取量计划表
sample identification 试[矿]样确[鉴]定,识别
sample information 试[矿]样资料
sample length 试[矿]样长度
sample lot 试[矿]样组[堆]
sample material 试[矿]样物料
sample method 取样法
sample mixing technique 试[矿]样混合技术
sample of ore 矿石试样
sample of waste 废石试样
sample pattern 取样点布置方式
sample piece 样件
sample plan design 取样方案[平面图,计划]设计
sample preparation room 试样制备室
sample preparation system 试[矿]样制备系统
sample preservation method 试[矿]样保存方法
sample pump 样品泵
sample pump sump 样品泵池
sample receipt 试[矿]样接受
sample roll 试样辊碎机,碎试样对辊机
sample size 试[矿]样尺寸[规格]
sample size requirement 试[矿]样规格要求
sample spacing 试[矿]样间隔
sample splitting 缩[分]样
sample splitting technique 试[矿]样缩分技术
sample storage bin 试[矿]样贮存仓
sample stream 试[矿]样流
sample tower 取样塔
sample transfer line 试[矿]样转运管线
sample weight 试[矿]样重量
sampler helper 取样工助手
samples of investigation 分析研究的试样
sampling arrangement 取样布置图
sampling aspect 取样方面
sampling belt 取样皮带
sampling calculation 取样计算
sampling equipment 取样设备
sampling equipment supplier 取样设备供应商[厂家]
sampling error 取样误差
sampling handbook 取样手册
sampling information 取样资料
sampling inspection 抽样检查
sampling machine 取样机
sampling machine drive 取样机驱[传]动装置

sampling method	取样方法
sampling methodology	取样方法[学]
sampling moil	取样十字镐[尖凿,鹤嘴锄]
sampling pattern	取样布置[网]图
sampling period	取样周期
sampling plan	取样平面[方案]
sampling plan design	采样平面设计
sampling plant	取样车间
sampling platform	取样平台
sampling population	抽样总体
sampling principle	取样原理
sampling procedure	取样工艺[过程]
sampling process	取样过程
sampling program	取样规划[计划,安排]
sampling rate	取样速度
sampling result	取样结果
sampling standard	取样标准
sampling survey	取样检查
sampling system	取样系统
sampling technique	取样技术[方法]
sampling theory	取[采]样理论
sampling variability	取样变率
sand bath	沙浴
sand bleed valve	粗砂排放阀,放砂阀
sand cleaner circuit	矿砂精选回路
sand fill system	水砂充填系统
sand filter feed pump	滤砂机给料泵
sand fraction	砂矿级别
sand hopper	砂斗[仓](充填用)
sand line	输砂管
sand manufacturing plant	砂制造厂
sand product	砂质产品
sand relief	放[泄]砂孔(浮矿机的)
sand screw classifier	砂螺旋分级机
sand table middlings	矿砂摇床中矿
sand tailings cyclone	砂矿尾液旋流器
sand tails	砂矿尾矿
sandbox	砂箱[仓]
Sandopan BL	选矿药剂(伯—烷基苯磺酸盐,可用作分散剂、湿润剂和乳化剂)
sand-slime separating technique	泥砂分选[离]方法
sand-slime separation	泥砂分选[离]
sand-slime separation technique	泥砂分选[离]技术
sand-slime technique	泥砂分选[离]方法
sand-slime treatment technique	泥砂处理技术
sandstone grain	砂岩颗粒
sandwich layer	夹心状层
sandwiched in	夹在两层之间的
sandwich-like	夹心状的
sandwich-like layer	夹心状层
sanitary	卫生[清洁]的
sanitary facilities	卫生设施
sanitary sewer	污水管[道],下水道
sanitary sewers and disposal system	污水管道和处理系统
sanitary standard	卫生标准
Santodex	选矿药剂(二烷基苯乙烯聚

合物,可用作矿泥抑制剂或黏土的捕收剂、絮凝剂)

Santomerse 表面活性剂品牌名

Santomerse 1 十二烷基苯磺酸钠(可用作氧化铁矿捕收剂或湿润剂)
$C_{12}H_{25}C_6H_4SO_3Na$

Santomerse 2 十二烷基苯磺酸钠(可用作氧化铁矿捕收剂或湿润剂)
$C_{12}H_{25}C_6H_4SO_3Na$

Santomerse 3 十二烷基苯磺酸钠(可用作氧化铁矿捕收剂或湿润剂)
$C_{12}H_{25}C_6H_4SO_3Na$

Santomerse D 癸基苯磺酸盐(德国产,可用作氧化铁矿捕收剂)
$C_{10}H_{21}C_6H_4SO_3Na$

Sapamine 表面活性剂品牌名

Sapamine CH, 二乙氨基乙基油酰胺(可用作捕收剂,浮选石膏时作为石英捕收剂及氧化铁矿捕收剂)
$C_{17}H_{33}CONHCH_2CH_2N(C_2H_5)_2$

Sapamine KW 二乙氨基乙基油酰胺(浮选石膏时作为石英,氧化铁矿捕收剂) $C_{17}H_{33}CONHCH_2CH_2N(C_2H_5)_2$

Sapamine MS 二乙氨基乙基油酰胺(可用作捕收剂,浮选石膏时作为石英捕收剂及氧化铁矿捕收剂)
$C_{17}H_{33}CONHCH_2CH_2N(C_2H_5)_2$

Sapogenat A 湿润剂(烷基苯酚聚二乙醇醚)$R \cdot C_6H_4(OCH_2CH_2)_nOH$

Sapogenat B 湿润剂(烷基苯酚聚二乙醇硫酸盐)
$R \cdot C_6H_4(OCH_2CH_2)_nOSO_3M$

Saponate 捕收剂(石油磺酸钠,相对分子质量 375~400,含 33%矿物油)

saponified fatty acid 皂化脂肪酸

saponified tall oil 皂化塔尔油

sapphire hard alumina ceramic bead 蓝宝石硬矾土瓷小球

saran 萨冉树脂,莎纶[赛伦](聚偏氯乙烯纤维或其共聚物纤维的统称)

saran screen 萨冉树脂筛

Sarcosyl(=**Sarkosyl**) 表面活性剂品牌名

Sarcosyl L 月桂酰-N-甲氨基乙酸(可用作捕收剂、乳化剂)
$C_{11}H_{23}CON(CH_3)CH_2COOH$

Sarcosyl LC 椰子油酰-N-甲氨基乙酸(可用作捕收剂、乳化剂)

Sarcosyl NL100 月桂酰-N-甲氨基乙酸钠(结晶性,可用作捕收剂、乳剂)
$C_{11}H_{23}C(O)N(CH_3)CH_2COONa$

Sarcosyl NL30 月桂酰-N-甲氨基乙酸钠(水溶性,可用作捕收剂、乳化剂)
$C_{11}H_{23} \cdot CON(CH_3)CH_2COONa$

Sarcosyl O 油酰-N-甲氨基乙酸(可用作捕收剂、乳化剂)
$C_{17}H_{33}CON(CH_3)CH_2COOH$

Sarcosyl S 硬脂酰-N-甲氨基丙酸(可用作捕收剂、乳化剂)
$C_{17}H_{35}CON(CH_3)(CH_2)_2COOH$

SAR value (**SAR = sodium adsorption ratio**) SAR 值(钠吸附比值)

satisfactory drying period 安全干燥期

satisfactory general flowsheet 满意的

整体流程
satisfactory performance 令人满意的指标
saturated homologue 饱和同系物
saturated hydrocarbon 饱和烃
saturated lime 饱和石灰
saturated lime solution 饱和石灰溶液
saturated oxygen concentration 饱和氧浓度
saturated water steam, saturated water steam 饱和水蒸气[水汽]
saturation adsorption 饱和吸附
saturation load 饱和负载
saturation pressure 饱和压力
saving 节约,保存;救助;(pl.)节省额,储蓄金
saw-tooth curve 锯齿形曲线
Saybolt universal viscosimeter 赛波特通用黏度计
Saybolt universal viscosity 赛波特通用黏度
scale formation 结垢[疤]
scale inhibition 防止结垢,防垢
scale inhibitor 防垢剂
scale of operation 操作规模
scale of turbulence 湍流度
scale platform 称量台
scale up 按比例放大
scaled parameter value 比例参数值
scale-down model 按比例缩小的模型
scale-forming constituent 结垢组分
scale-up assumption 按比例扩[增]大的假设
scale-up calculation 按比例扩[增]大计算
scale-up criterion 按比例放大的判据
scale-up design for tumbling-mill grinding circuit 滚磨机磨矿回路的按比例扩大设计
scale-up formula 按比例放大的(计算)公式
scale-up parameter 按比例放大参数
scale-up procedure 按比例放大法[步骤]
scale-up program 按比例放大程序
scale-up relationship 按比例放大关系
scale-up scheme 按比例放大方案
scale-up theory 按比例放大理论
scaling drawing 比例缩小[几何]图形
scaling-up model 按比例放大模型
scalped feed 筛除大块的给矿
scalping 筛除大块
scalping cyclone 脱粗旋流器
scalping duty 筛除大块任务
scalping grizzly 除大块格筛
scalping jig 筛粗跳汰
scalping screen 粗筛
scan rate 扫描率
scan value 扫描值
scandium oxide 氧化钪
scatter intensity 散射强度
scattered crystal 分散[散射]结晶
scattered light 散射光线
scattered radiation 散射辐射

scattered radiation channel 散射[扩散]辐射通道
scattering angle 散射角
scattering channel 散射道
scavenger 扫选浮选机
scavenger cell feed box 扫选槽给矿箱
scavenger circuit 扫选回路
scavenger cleaner concentrate 扫选精选精矿
scavenger cleaner tailings 扫选精选尾矿
scavenger collector 扫选捕收剂
scavenger concentrate 扫选精矿
scavenger concentrate regrinding 扫选精矿再磨
scavenger flotation 扫选(浮选机)
scavenger flotation circuit 扫选浮选回路
scavenger performance 扫选指标
scavenger regrind 扫选再磨
scavenger tailings 扫选尾矿
scavenger tails 扫选尾矿
scavenger variation 扫选变化
schedule 进度计划
schedule adherence 计划完成率[执行情况]
schedule update 计划修改[更新]
scheduled oil sampling 定期取油样
scheduled production 计划产量
scheelite flotation performance 白钨浮游性
schematic arrangement 示意[概略]配置图
schematic diagram 略[简,原理]图
schematic diagram of concentrator 选矿厂设备形象系统(示意)图
schematic diagram of process equipment 工艺设备形象系统(示意)图
schematic outline 概略[简,略]图
schematic presentation 示意图,简单描述
schematic representation 示意图
science-based technology 基于科学的技术
scientific basis 科学基础
scientific computing experience 科学运算经验
scientific method 科学方法
scientific research unit 科研机构
scientific researcher 科研人员
scientific term 科学术语
scientist 科学家
scintillation crystal 闪烁晶体
scissors belt 剪刀式胶带
scissors conveyor 剪刀型运输机
sclerotinite 菌核体(煤岩)
scoop box (磨机)给[挖]矿勺箱
scoop feeder box 斗式给矿机箱
scope of work 工作范围
scope test 范围检验
scoping test 范围检验
scrap iron 铁屑
scrap iron tank 铁屑(沉淀,置换)槽[罐]

scrap metal reclamation 废金属回收
scrap metal waste 金属废料
scraper 刮[铲]土机
scraper discharge 刮板排矿(过滤机)
scraping and shearing operation 刮光和剪切作业[工作]
scree ore 山麓碎石矿石
screen 筛子,筛网,筛分机
screen analysis report 筛析报告
screen area selection 筛网面积选择
screen cap 筛子盖
screen capacity formula 筛分机生产能力(计算)公式
screen cell 搅拌筛分槽
screen component 筛子部件
screen diameter 筛孔直径
screen electrostatic separator 筛分静电分选机
screen feed distributor 筛子给矿[料]分配器
screen fraction 筛分部分
screen launder 筛子流槽,流槽筛
screen mesh size 筛子网目尺寸,筛号
screen motion 筛子运动
screen opening size 筛孔大小[尺寸]
screen panel 筛面[网]板(俗称筛笼子)
screen performance 筛(生产)性能
screen requirement 筛分设备需要量
screen side plate 筛子侧[壁]板
screen size 筛分粒度,筛号,筛孔尺寸
screen size analysis 粒度筛分分析
screen specification 筛子规格
screen spray 筛分喷水
screen tower 塔式筛分厂房
screenability 筛分能力[效率]
screening anion 屏蔽阴离子
screening area 筛分面积
screening characteristic 筛分特性
screening division 筛分部[区]
screening duty 筛分任务[作用]
screening element 筛分元件,筛网
screening media 筛分介质,筛网
screening media specification 筛分介质[筛网]技术规格
screening medium 筛分介质
screening operation 筛分作业
screening process 筛分过程
screening test 筛分试验
screening-classification-gravity separation 筛分-分级-重力选矿
screenings 筛出[下]物,筛屑[渣]
screw auger 螺旋推运器,螺旋推进加料器
screw take-up 螺旋拉紧装置
scroll 涡形(管)
scrubber fume 洗涤器烟气
scrubber pump 擦洗泵
scrubbing section 擦洗段(圆筒筛)
scrubbing unit 洗涤器[机]
sea water 海水
seal drag 密封拉[曳]力

seal selection guide 密封选择指南[手册]
sealed and fixed head disk 密封固定磁头磁盘
sealed bid 密封投标
sealed container 密封容器
sealed drum 密封桶[筒]
sealed proposal 密封投标
sealing skirt 密封拦矿机
search 搜寻,调查,探求,检索,觅数
search routine 检索(例行)程序,常规检索
seasonal demand 季节性要求
seasonal peak 季节高峰
seasonal requirement 季节性要求
seasonal weather 季节性气候
second anodic peak 第二个阳极峰值
second aqueous phase 第二水相
second cleaner 第二段[次]精选
second derivative 二阶导数
second exploratory underground 第二次地下勘探
second metal plant 第二金属厂
second mineral flotation 第二种矿物浮选
second recleaner concentrate 第二次再精选精矿
second recleaner tailings 第二次再精选尾矿
second recleaning 第二次再精选
second screen cloth oversize 下[第二]层筛网筛上物

second shift 第二班
second stage regrinding mill 第二段再磨机
second steam heated heat exchanger 二次蒸汽加热的热交换器
secondary amine 仲胺,第二胺
secondary amine group 仲胺基
secondary amine salt 仲胺盐
secondary and tertiary crushing 第二,三段破碎
secondary and tertiary crushing circuit 中细碎回路
secondary attractive minimum 第二最小吸引能
secondary breakage 二次破碎[碎裂]
secondary breakage potential 二次破裂潜力
secondary chalcocite 次生辉铜矿
secondary chalcocite coating 次生辉铜矿覆盖层
secondary cleaner 二次精选槽
secondary cleaner bank 第二次精选[浮选]机
secondary cleaner tailings 第二次精选尾矿
secondary close side setting 二段破碎窄[紧]边排口(开口)宽度
secondary concave 二次破碎机固定锥体
secondary copper sulfide 次生硫化铜矿

secondary covellite coating 次生铜蓝覆盖层

secondary crusher discharge screen oversize product 二段破碎机排矿筛上产品

secondary crushing equipment 第二段破碎设备

secondary cyclone 第二段(水力)旋流器

secondary dryer 二次干燥机

secondary enrichment zone 二次[次生]富集带

secondary grinding equipment 第二次磨矿设备

secondary mantle 二次破碎机可动锥体

secondary mining 二次采矿(回采)

secondary pebble mill 第二段砾磨机

secondary processing plant 二次加工厂

secondary promoter 辅助[补充]促进剂

secondary reaction 副反应

secondary response function 次级响应函数

secondary sampler 二次取样装置[器]

secondary sampling machine 第二次联样机

secondary sampling system 二次取样系统

secondary vibration 二次振动作用

secondary-tertiary crusher building 二段三段破碎机厂房

second-order difference 二阶差分

second-order difference equation 二阶差分方程

second-order differential equation 二阶微分方程

second-order integral differential equation 二阶积分微分方程

second-order linear differential equation 二阶线性微分方程

second-stage flotation 第二段浮选

secretarial function 秘书的职能

section(al) construction 预制构件拼装结构,组合式结构

sections of different levels 不同含量区间

sections of various levels 各个含量区间

secured bond 有担保证券

securities 证券

security 安全,保安;证券,抵押物

security department 保卫部门

security hopper 安全(贮矿)漏斗

sedimentary apatite 沉积磷灰石

sedimentary ore 沉积岩矿石

sedimentary phosphate 沉积磷酸盐

sedimentary type 沉积岩类型

sedimentation operation 沉积作业

sedimentation pond 沉淀池

sedimentation process 沉积过程[法]

sedimentation promotion 促进沉积作用

sedimentation system 沉积系统

sedimentation vessel 沉淀[降]器

sedimentograph 沉降图像[解]
Sedomax F 絮凝剂(合成的有机聚电解质)
segregated solids 离析的固体颗粒
segregation effect 偏析[离析]作用
selection function 选择函数
selection function parameter 选择函数参数
selection of equipment sizes 设备规格选择
selection of model 模型选择
selection of technical process 流程选择
selection procedure 选择步骤
selective activator 选择性活化剂
selective anionic flotation 选择性阴离子浮选
selective collector 选择性捕收剂
selective depressing action 选择性抑制作用
selective flocculation flotation 选择性絮凝浮选
selective flocculation slime 选择性絮凝矿泥
selective flocculation-cationic flotation process 选择性絮凝-阳离子浮选法
selective flocculation-desliming flotation process 选择性絮凝-脱泥浮选法
selective flocculation flotation process 选择性絮凝浮选法
selective flotation machine 优先浮选机
selective flotation plant 优先浮选厂
selective froth 选择性泡沫

selective mining 选择性采矿
selective oiling 选择性加油处理
selective sequential flotation method 优先按序浮选法
selective solubilization 优先溶解(作用)
selective starch coating 选择性淀粉膜
selectivity function 选择性功能
selectivity of adhesion 粘附选择性
selectivity of depression 抑制选择性
self aeration machine 吸气式浮选机
self tuning control 自调控制
self-aerating flotation cell mechanism 自吸气浮选槽机械结构
self-aerating lab size flotation machine 自动充气实验室型浮选机
self-aerating type 自吸[充]气式
self-aeration 自吸[充]气
self-aeration mechanical machine 机械自动吸气浮选机
self-calculating chart 自动计算图表
self-cleaning iron magnet 自清式除铁器(电磁铁式)
self-cleaning magnet 自清式除铁器(电磁铁式)
self-contained rotary dual slurry sampler 自身配套的双截取器旋转矿浆取样机
self-contained rotary slurry sampler 自身配套的旋转矿浆取样器
self-contained slurry sampler 自身配套的矿浆取样器
self-contained unit 本身配套的设备

[装置]

self-propelled conveyer belt unit 自行[进]式胶带运输机组

self-regulating feedback control 自动调节反馈控制

seller 卖方

seller supply 供应商供货

selling agent 代销商[人,店]

selling and administrative expense 销售和行政管理费

selling price 售价

Selvyt 高级清洁抛光绒布品牌名（英国）

semi-annual cash requirement 半年现金需要量

semi-annual forecast 半年预测[报]

semiannual index 半年索引

semi-arid inland environment 半干旱内陆环境

semi-arid region 半干旱地区

semi-autogenous 半自磨的

semi-autogenous mill 半自磨机

semi-autogenous milling circuit 半自磨回路[系统,流程]

semi-autogenous secondary grinding 半自磨二段磨矿,第二段半自磨磨矿

semi-automatic control 半自动控制

semi-batch rate equation 半分批速度方程

semi-batch testing 半批量试验

semi-bulk flotation development 半混合浮选研究

semiconducting mineral 半导电[传导]矿物

semiconducting nature 半导体特性

semiconducting property 半导体性质

semiconductivity 半导体性能

semiconductor band theory 半导体能带理论

semiconductor property 半导体特性

semiconductor surface 半导体表面

semi-container ship 半集装箱船

semi-continuous batch analysis 半连续批量分析

semi-continuous nature 半连续性

semi-continuous operation 半连续操作[运转,生产]

semi-crystal position 半晶体位置

semi-empirical mathematic model 半经验性数学模型

semi-infinite lattice 半无限晶格

semi-infinite solid 半无限固体

semi-portable milling plant 半移动式选矿厂

semi-professional personnel 半专业人员

semi-quantitative calculation 半定量计算

semi-quantitative evaluation 半定量评价

semi-quantitative method 半定量法

semisoluble salt 半溶性盐

semi-variogram 半变异函数

semi-way unit 半工业装置

senior application engineer 高级应用工程师

senior mechanical engineer 高级机械工程师

senior operating staff 高级操作人员

senior researcher 高级研究员

senior vice president of engineering 高级工程副经理

sensible heat 显热

sensing device 传感装置

sensing element 感应元件

sensing instrument 感测仪器

sensing position 传感位置

sensitive analysis 敏感[灵敏]度分析

sensitivity analysis 敏感分析

sensitivity test 灵敏度试验

sensor probe 传感器探头

Separan 美国道氏公司聚丙烯酰胺类合成高分子絮凝剂品牌名(中文名称:赛帕隆[赛派栏])

Separan 2610 絮凝剂(聚丙烯酰胺,相对分子质量150万~170万,也可用作分散剂、抑制剂、捕收剂)

Separan 2910 絮凝剂(聚丙烯酰胺)

Separan AP30 絮凝剂(一种水解聚丙烯酰胺)

Separan MGL 絮凝剂(聚丙烯酰胺,相对分子质量$3×10^6 \sim 5×10^6$)

Separan NP10 絮凝剂(丙烯酰胺与丙烯酸钠共聚物,其中丙烯酸钠量10%)

Separan NP20 絮凝剂(丙烯酰胺与丙烯酸钠共聚物,其中丙烯酸钠量为20%)

Separan NP30 絮凝剂(丙烯酰胺与丙烯酸钠共聚物,其中丙烯酸钠量为30%)

separate anodic site 单独阳极区

separate bay 单独跨

separate cathodic site 单独阴极区

separate circuit 独立[单独]回路

separate computer program 单独的计算机程序

separate electrochemical step 独立[分段]电化学阶段[步骤]

separate experiment 单独试验

separate middling flotation step 单独[独立]中矿浮选段[步骤]

separate product 单独产品

separate regrind section 独立再磨段

separate sulphide and cement copper flotation 单独硫化矿与沉淀铜浮选

separating circuit 分选[分离]回路

separating cone (重介质)圆锥分选机,分选圆锥

separating medium 分选介质

separating principle 分选原理

separating technique 分选[选别]技术

separating vessel 分离[分选]器

separation 分离[裂,隔,开,散],释[析]出,分馏,选矿,间隔[距,隙],空隙

separation accuracy 分离准确性[度]

separation circuit 分离[选]回路

separation coal 精选煤

separation coupling	可拆联轴节
separation effect	分离效应
separation mesh size	分离粒度网目
separation of ions	离子距离[分离]
separation problem	分选[离]问题
separation scheme	分选[离]方案
separation size	分[级]离粒度
separation step	分选[离]步骤
separation technique	分选[离]技术
separation test	分选[离]试验
separator performance	分选[离]机性能
separatory funnel technique	分选[离]漏斗技术
septic tank	化粪池
sequence of control	控制程序
sequence of impulse	脉冲序列
sequence of operation	工序,操作[运行]程序[步骤]
sequence of reaction	反应系列
sequencing control	顺序控制
sequencing function	顺序功能
sequential control	顺序[连续]控制
sequential interlock	顺序连锁
sequential iterative solution	序贯迭代解
sequential method	序贯方法
sequential order	顺序
sequential sink-float analysis	连续[顺序]沉-浮分析
sequential stopping time control	顺序停车时间控制
sequestering	螯合,隔离
serial number	系列号
series arrangement	串联排列[配置]
series flotation	连续浮选
series of test	试验顺序
serious consequence	严重后果
serious economic consequence	严重的经济后果
serpentine – magnetite – apatite rock	蛇纹岩-磁铁矿-磷灰石岩
service	服务
service brake	常用闸,脚踏闸(汽车的)
service class	服务[工作]级别[等级]
service contractor	维修[操作,服务]承包方
service demand	维修要求,服务需求
service equipment	辅助设备
service factor	工作[操作,计算,使用]系数
service fee	服务费
service function	服务性[辅助]机构[任务,工作]
service lease	服务性出租
service letter	服务信函
service literature	维修说明书
service manual	维修手册
service platform	工作台
service power	服务[辅助]功率(电站内部所需能电量)
service report	维修报告
service routine	使用[服务,辅助]程序

service scope 服务[业务]范围
service staff 服[勤]务人员
service water 服务[辅助]用水(如泵水封水,生活用水)
serviceability 可[适]用性,使用可靠性,可服务性
servo-simulation 模拟伺服机构
session chairman 会议主席
session editor 会议编辑
set of configuration factors 流程排布因子集合
set of testing sieve 试验套筛
set point 设[给]定值,凝点,塑性变形点
set procedure 规定的步骤[程序]
set standard 规定的标准
set value 给定值
set-aside cash 闲置现金
set-point basis 给定值基础
setting alarm limit 整定报警极限
setting alteration 设定值变更(如排矿口开度,宽度变更)
setting indicator 设定值指示器(如排矿口开度,宽度指示器)
settleable oil emulsion 可沉淀的油类乳浊液
settled distance 沉降距离
settled product 沉淀产品(如沉砂)
settled underflow solids 沉降的底流固体颗粒[物料]
settling aid 助沉(降)剂
settling and dewatering bin 沉淀脱水仓
settling area dimension 沉降面积尺寸
settling characteristic 沉降特性[征]
settling pond 澄清池
settling pool 沉降槽[池](螺旋分级机)
settling test 沉积[淀,降]试验
settling time 沉降时间
set-up normal equation 建立标准方程
set-up X matrix 建立 X 矩阵
seven digit display direct-reading 七位数字显示直接读数
severe condition 严峻条件
severe problem 严重问题
sewage 污水,下水道
sewage disposal 污[废]水处理
sewage plant 污水厂
sewage purification 污水净化[处理]
sewage tank 污水(沉淀)池[罐]
sewage treatment plant 污水处理厂
SFT-15 一种波兰产起泡剂(α-萜烯甲醚)
shaft 竖井
shaft depth 竖井深度
shaft kiln 竖窑
shaft loading installation 竖井装载装置
shaker type 振动型
shale group 页岩族
shallow impeller design 浅(槽)叶轮设计
shallow impeller immersion 浅叶轮沉

[浸]没
shallow tank 浅槽[罐]
shape factor constant 形状因素[系数]常数
share operating system 共用操作系统
shared display control system 共同显示控制系统
shared display distributed control instrument package 共用显示分布控制仪表组件
shared display system 共用显示系统
share-holders' equity 股东权益
sharpness of classification 分级精度
shear force 剪力
shear plane 剪切面
shear rubber mount 剪力橡胶座
shear strength 切变(抗剪)强度
shearing deformation 剪切变形
shearing deformation of flow 流动剪切变形
shearing force 剪力
sheep's foot roller 羊足滚压机(压路用)
sheet silicate 层[片]状硅酸盐
shell and tube heat exchanger 管壳式热交换器
Shell aromatic solvent 54 壳牌芳香族溶剂54
shift basis 以班为基础
shift engineer 值班工程师
shift foreman 值班工长
ship loader 装船机,船用装载机

ship loading 装船
ship owner 船主
Ship's gear 船用仪表和设备,船舶装备
shiploader 装船[装货,装载]机
shiploader hopper 装载给料漏斗
shiploading scale 装船胶带秤
shipment 装货,运货,运载的货物
shipment rubber belt weigh 装船胶带秤
shipping agent 货运代理商
shipping company 海运公司
shipping cost 海运费用
shipping information 发货信息
shipping information sheet 装运数据表
shipping order 发货订单,装货(通知)单
shipping point 装运地点
ship-side 码头[船端]
ship-side loading 码头[船端]装载
ship-side storage 码头[船端]贮存
ship-type propeller agitated tank 螺旋桨式叶轮搅拌槽[罐]
shop and warehouse building 车间和仓库用房
shop assembly 车间装配
shop order 工厂订单
shop test 工厂试验
Shore hardness 肖氏[回跳]硬度
Shore hardness number 肖氏[回跳]硬度值

Shore-hardness 肖氏[回跳]硬度

Shore-hardness test 肖氏[回跳]硬度试验法(利用金刚尖锥的回弹做实验)

Shore-hardness tester 肖氏[回跳]硬度计

short chain alkali metal xanthate 短链碱金属黄原酸盐

short chain organic acid 短链有机酸

short chain salts 短链盐类

short chain thiol molecule 短链硫氢[醇]分子

short circuit of material 物料短路

short circuiting 短路(现象)

short introduction 简短介绍

short primary mill 短筒式第一段磨机

short rougher section 短粗选段

short term 短期

short-circuiting criterion 短路循环准则

short-cone hydrocyclone 短锥水力旋流器

short-cut method 简化法

short-head cone crusher cavity 短头圆锥破碎机腔室

short-ranged 短程的

short-term borrowing 短期借款

short-term debt 短期债务

short-term deposit 短期存款

short-term financing 短期集资

short-term government security 短期政府债券

short-term interference 短时间[期]干扰

short-term investment of funds 短期(资金)投资

short-term loan 短期贷款

short-term prepayment 短期预支[付]款

short-term source 短期(资金)来源

short-term toxicity test 短期毒性试验

short-term variability 短期易变性

short-wave limit 短波限

shot blast 喷丸处理

shoulder 线肩

shovel control 电铲控制

shovel size (电铲)铲斗尺寸[规格]

shovel tooth shank 电铲[铲斗]牙柄

shower room (淋)浴室

shrinkage 收缩(量),缩小[水],缩率[减]

Shriver filter press 史瑞沃型压滤机

Shriver press 史瑞沃压滤机

shrouded impeller 罩盖式叶轮

shutdown 停产

shutdown gold treatment plant 停产的金选厂

shutoff gate 封闭[断流]闸门

shutting off 停车,关闭,切断

shuttle belt conveyer 梭式胶带运输机

side adjacent 邻边

side bearing 端轴承

side draw-off 侧面排出(跳汰机)

side effect 副作用

side elevation 侧视[面]图

side entry dust collector 侧面进气收尘器
side loading 一侧给矿,给矿偏边
side opposite 对边
side travel detector 跑偏探测器
sidearm assembly 旁臂组件
sideway slope 侧向坡度
Siebtechnik mill 德国产西伯型磨机
sieve effect 筛分作用[效应]
sieve end chemical analysis 筛分末端化学分析(筛分的最后一级或最小的粒级分析)
sieve size 筛分粒度(一般指小型筛,实验室用筛的)
sight-flow meter 目[可]视流量计
sightseeing area 游览[观光]区
sign convention 符号规约[惯例]
signal analysis component 信号分析组件
signal cabinet 信号箱
signal converter 信号转换器
signal multiplexing 信号多路通道
signal processing 信号处理
significant cost 大量费用
significant decision 重大决定[策]
significant improvement 显著改善[进]
significant quantity of recycle water 大量循环水
silane 硅烷 Si_nH_{2n+2}
silanol group 硅烷醇基,硅羟基
silex lined (有)石英衬里的

silica flour 石英粉
silica flux 石英熔剂
silica-bearing fines 含硅细粒
silicagel-sol 可溶性二氧化硅胶(用脂肪酸作捕收剂浮选氧化铁矿时用作分散剂)
silicate 硅酸盐
silicate group 硅酸盐基
silicated limestone 硅酸盐灰岩
silicated, garnetized and chloritized limestone 硅酸盐化、石榴石化和绿泥石化石灰岩
siliceous barite 硅质重晶石
siliceous gangue 硅质脉石
siliceous gangue depressant 硅质脉石抑制剂
siliceous gangue mineral 含硅[硅质]脉石矿物
siliceous material 硅质物料[矿物]
siliceous opalite 硅化蛋白石
siliceous ore 硅质矿石
silicified rhyolitic ash-flow tuff 硅化流纹火山凝灰岩
silicon analyzer 硅分析仪
silicon carbide crucible 碳化硅坩埚
silicon carbide paper 碳化硅[金刚砂]纸
silicon-carbide-lined 碳化硅衬
silicon-dolomite 硅白云石
silicone 硅酮,聚硅酮类 R-Si(O)-R'(可用作捕收剂、分散剂、调整剂)
Silicone fluid F60 捕收剂(氯基甲基聚

硅酮)

Siliconemulsion Le40 硅酮乳剂(品名,含40%聚硅酮,1.2%乳化剂,58.8%水,可用作方铅矿、闪锌矿的捕收剂)

silicone oil 硅油、硅酮油、聚硅油,(可用作捕收剂)

sill 基石,底梁,平巷底(板),岩床,海底(山)脊

sill value 基台值

silo 筒仓,储料仓,地下仓库

silo failure 筒仓破坏

siloxane 硅氧烷(可用作捕收剂) $H_3Si(OSiH_2)_n OSiH_3$

silver butyl dithiophosphate 丁基二硫代磷酸银

silver chloride 氯化银

silver ethyl dithiophosphate 乙基二硫代磷酸银

silver ethyl xanthate 乙基黄原酸银

silver halide 银卤化物

silver hexyl xanthate 己基黄原酸银

silver iodide 碘化银

silver isoamyl dithiophosphate 异戊基二硫代磷酸银

silver material 银矿物

silver m-cresyl dithiophosphate m-甲苯基二硫代磷酸银

silver methyl dithiophosphate 甲基二硫代磷酸银

silver o-cresyl dithiophosphate o-甲苯基二硫代磷酸银

silver p-cresyl dithiophosphate p-甲苯基二硫代磷酸银

silver phenyl dithiophosphate 苯基二硫代磷酸银

silver plant 银[氰化]厂

silver prill 银粒

silver propyl dithiophosphate 丙基二硫代磷酸银

silver sulfide 硫化银

similar fineness of grind 类似磨矿细度

similar repetitive test 类似的重复试验

simple analytical solution 简单分析解法

simple automatic feed rate control system 简单的自动控制给矿量系统

simple grade estimate 简单的品位估算

simple line diagram flowsheet 简单线流程图

simple open circuit batch test 简单的开路单元试验

simple optical technique 简单光学技术[方法]

simple ore 简单矿石,单矿,单组分矿石

simple polynomial equation 简单多项式方程

simple salt 单盐(其阳离子和阴离子不水解)

simple salt mineral 单盐矿物

simple surface alteration 简单的表面变化

simplex 单形,单纯形
simplex crossflow classifier 单螺旋横流分级机
simplex Denver Vezin sampler 单截锥丹佛·维津型取样器
simplex machine 单螺旋设备(如分级机)
simplex screw classifier 单螺旋分级机
Simplex Self-Directing Evolutionary Operation(SSDEVOP) 单纯形自定向调优运算(缩写 SSDEVOP)
simplification 简化,精简
simplified band diagram 简化能带图
simplified diagram 简图
simplified line diagram flowsheet 简化[单线]流程图
simplified population balance kinetics 简化母体平衡动力学
simplified power equation 简化的功率方程
simplified practice recommendation 简化实践的建议
simulated behavior 模拟行为
simulated closed circuit grinding 模拟闭路磨矿
simulated data 模拟数据
simulated plant 模拟(选矿)厂
simulated response 模拟反应
simulation methodology 模拟[仿真]方法
simulation mode 模拟方式
simulation of entire circuit 整个系统[回路]的模拟
simulation program 模拟程序
simulation structure 模拟结构
simulation study 模拟研究
simulation technique 模拟方法[技术]
simulation test 模拟试验
simulator input 模拟器输入
simulator output 模拟器输出
simulator structure 模拟器结构
simultaneous construction 同时建设
simultaneous differential equation 联立微分方程
simultaneous equation 联立方程式
simultaneous linear equation 联立一次方程,联立线性方程
simultaneous solution 联立解法
sine 正弦
sine law 正弦定律
sine rule 正弦规则[定理]
single concentrating cone 单锥选矿机(赖克特圆锥选矿机,澳大利亚生产)
single diaphragm pump 单联[室]隔膜泵
single eccentric shaft 单偏心轴
single first order rate constant model 单个一阶速度常数模型
single flight belt conveyer 单索胶带运输机
single inlet 单侧进口
single input 单一输入
single line circuit 单系列流程[系统,回路]

single line diagram(SLD) 单线图(缩写SLD)
single matrix equation 单一矩阵方程
single output 单一输出
single overflow Fagergren cleaner 单槽溢流型法格古伦精选机
single overflow Fagergren flotation machine 单槽溢流型法格古伦浮选机
single particle-single impact breakage 单颗粒单冲击破碎
single pass regrinding 开路再磨矿
single performance index 单一性能指标
single pitch spiral 单节距螺旋
single point feed opening 单一给矿口
single power source 单一电源,单电源
single quadrant gate 单扇形闸门
single residence time distribution 单一存留时间分布
single rope Koepe hoist 单绳戈培式提升机
single run 一次作业
single sampling method 单一取样方法
single screw coarse material washer-classifier dehydrator 单螺旋粗料洗涤-分级装置
single screw fine material washer-classifier-dehydrator 单螺旋粉料洗涤-分级脱水装置
single shift system 单班制
single slurry sampler 简单矿浆取样器
single spiral 单螺旋

single spiral construction 单螺旋结构
single stage ball mill 单段球磨机
single stage ball-mill grinding 单级球磨机磨矿
single stage comminution circuit 单段破磨回路[流程]
single stage grinding 单段磨矿
single stage primary mill 单段粗磨机
single stope 单一采场(矿房)
single stream 单矿流
single variable elimination method 单变量消元方法
single variant loop 单变量回路
single vector 单矢量
single width 单面宽度
single-hood branch 单通风罩支管
single-outlet silo 单(排)口筒仓
single-phase 单相
single-phase liquid 单相液体
single-stage autogenous grinding 单段自磨
single-stage circuit 单段回路(浮选)
single-stage crusher control 单段破碎机控制
single-stage rod milling 单段棒磨矿
single-stage semi-autogenous milling 单段半自磨矿
single-stage wet vacuum pump 单段湿式真空泵
single-toggle jaw crusher 单肘板颚式破碎机
single-variable optimization 单变量最

优化

sink(tailing) product 槽底产品(浮选的)

sink and float plant 重介质选矿车间[厂]

sink of electron 电子汇(吸收剂)

sink product 沉品(重介质)

sink screen 重产品[沉物]筛

sink screen product 沉物筛分产品[重介质]

sinking fund 偿债基金,还债准备金

sinking fund provision 偿债基金条例

sinks-floats sieve bend (分离)沉浮产品弧形筛

sinter feed fine 烧结机给料粉矿

sintered metal Mott candle filter 烧结金属[金属陶瓷]Mott型过滤棒过滤机

sintering equipment 烧结设备

Sipex 表面活性剂品牌名

Sipex CS 十六烷基硫酸钠(可用作捕收剂、湿润剂)$C_{16}H_{33}OSO_3Na$

Sipex S 十二烷基硫酸钠(可用作捕收剂、湿润剂)$C_{12}H_{25}OSO_3Na$

siphon reagent feeder 虹吸式给药机

site characteristic 矿区特性

site characteristics 现场特征

site cleavage 就地解理

site cleavage experiment 就地解理实验

site facilities 现场设施

site improvement 现场[矿区]条件改善

site location 厂址位置

site location plan 厂区位置平面布置图,现场位置图

site personnel 现场人员

site preparation 场地平整[准备]

site preparation data 场地[现场]准备(工作)资料

site preparation phase 场地准备阶段

site preparatory work 现场准备工作

site service 工地[现场]服务

site survey 现场测量

Sitrex 水三钙柠檬酸的干馏产物(品名,可用作锡石捕收剂)

six compartment calcine furnace 六室焙烧电炉

six compartment Denver wet feeder 六室丹佛湿式给矿[药]机

six compartment wet reagent feeder 六室湿式给药机

six digit 六位数

six-tenths rule 十分之六准则

size band 粒度带[范围]

size change 粒度变化

size controlled grinding circuit (以)粒度(为基础的)控制磨矿回路

size criteria 规格尺寸(粒度大小)要求

size discreted selection function 粒度离散选择函数

size distribution curve 粒度[级]分配曲线

size distribution prediction 粒度分布预测

size factor	规模因素
size fraction	粒级,颗粒组
size fraction basis	颗粒级别基础,粒级基础
size independent	粒度无关
size interval	粒度间隔[区间]
size interval boundary	粒度间隔边界
size of bolt	螺栓尺寸
size of feed	给矿粒度
size of finished particle	最终(磨矿)粒度
size of ground product	磨矿产品粒度
size of loading equipment	装载设备规格
size of mesh	筛孔尺寸
size of operation	生产规模
size of the largest piece	最大件规格[尺寸]
size range applicability	粒度范围适用性
size range of particle	颗粒物粒径范围
size ratio	粒度比,尺寸比(例)
size reduction	粉碎,磨细
size reduction factor	破碎系数
size relationship	粒度关系
size screen	筛子规格
size segregation	粒度[级]析离[分聚]
size split	粒度裂解[劈裂]
size variation	粒度变化,尺寸变化
sized flotation	分级浮选
sized fraction	已分级粒级
sized ore	分级矿石
sizing	量[定,测]尺寸,定大小,(轧管)定径,定规格
sizing application	分级[筛分]应用
sizing calculation	筛分计算
sizing deck	筛面
sizing device	筛分装置
sizing methodology	分级方法
sizing specification	规格确定
sizing vibrating screen	分级振动筛
sketch	草[略,简]图
skill support	技术支持
skilled operating personnel	熟练的操作人员
skilled personnel	(技术)熟练人员
skilled programmer	熟练的程序员
skilled trades person	技术行业人员
skim region	刮泡区
skimming operation	撇渣作业[操作]
skirt board	拦矿板
skirt board containment	拦矿板封锁[遏制]
skirt board friction factor	导料[拦矿]板摩擦系数
skirt board zone	拦矿板段[区]
skirt plate	拦矿板
skirt seal	拦矿密封
skirt wall	挡矿[导料]板板面
slab-like domain	(电)板状磁畴
slag pot	渣罐
slag property	炉渣性质
slaker	消化器,消石灰器
slave unit	伺服装置,从属单元

slewing device 旋转装置
slide base 滑座
slide base support 滑轨座支架
sliding surface 滑动面
slight rectangular opening 略呈长方形（筛）孔
slime coating 矿泥覆盖膜
slime concentrating table 细泥摇床
slime contamination 矿泥污染
slime dispersant 细泥扩散剂
slime disposal 矿泥处理
slime flocculation 细[矿]泥絮凝
slime fraction 细泥部分[粒级]
slime generation 矿泥产生,产出矿泥
slime line 细泥管线[道]
slime loss 细泥损失(过磨造成)
slime overflow line 细泥溢流管线
slime Pachuca agitation 矿泥气升[压气,帕丘卡]搅拌
slime processing 细泥处理(选矿)
slime product 细泥[泥质]产品
slime settling area 矿[细]泥沉淀区
slime tails 泥矿尾矿
slime thickener 矿[煤]泥浓缩机
slime treatment plant 泥矿[选矿]车间
slime treatment plant circuit 细泥选矿回路
slimed material 泥化物料[矿石]
slime-free ore 脱泥矿石
slime-tailing thickener 矿泥尾矿浓缩机

sling 吊环[索,绳,链,具],链钩,吊重装置
slip form silo 滑模筒仓
slipping plane 滑动面
slop tank 污水箱,污油罐
slope 斜率
slope hoist 斜井提升机
slope stability 边坡稳定性
sloped side board 倾斜式侧板
sloping hopper 斜漏斗
sloping side 倾斜面
sloping static internal screen 倾斜固定式内装[槽内]筛(桥筛)
sloping surface 倾斜表面
slot configuration 长孔构造(形式)
slot end-slope 排料槽[口]末端坡度
slot factor 长条孔系数
slot feeder 槽式给矿机
slot outlet 长形排口
slot side-slope 排料槽[口]侧边坡度
slot type belt feeder 缝隙型胶带给矿机
slotted opening factor 长条筛孔系数
slotted panel 有缝隙的筛板
slow floatable component 慢速浮游成分
slow floating mineral 慢浮矿物
slow floating species 慢速浮游矿物[种],慢浮矿物
slow moving device 慢速移动[运动]装置
slow return valve 慢回流阀

slower assaying device 慢速分析装置
slow-floating 慢浮游的,浮游速度慢的
sludge discharge 泥[矿]浆排放
slug 棒,(嵌)条,金属块,金属片状毛坯,铁芯,芯子,弹丸
sluice box concentrate 流矿槽精矿
slurry 悬浮体[液],淤[矿]浆,矿泥
slurry composition 矿浆成分
slurry conveyance 矿浆输送
slurry density 矿浆[泥浆]浓度
slurry density measurement 矿浆浓度测量
slurry distributor 矿浆分配器
slurry electrode 矿浆电极
slurry equation 矿浆公式
slurry feed 矿浆给料
slurry flow characteristic 矿浆流动特性
slurry flow measurement 矿浆流量测量
slurry flow meter 矿浆流量计
slurry handling equipment 矿浆运载设备
slurry line elbow 泥浆管线弯头
slurry pulp density 矿[泥]浆浓度
slurry sample 矿浆样
slurry sampling 矿浆[泥]取样
slurry sampling system 矿浆取样系统
slurry test loop 矿浆试验回[环]路
slurry trap 矿浆捕集器,矿浆气水分离器
slurry velocity 矿浆速度

slurry viscosity 矿[泥]浆黏度
SM119 一种日本产有机亚砷药剂(可用作锡石捕收剂)
small and medium sized mill 中小型选矿厂
small batch sample 小批量样品
small business owner 小企业主
small commercial size plant 小型选矿厂
small diameter feed distributor 小直径给矿分配器
small mill test 小磨机试验
small process control computer 小型过程控制计算机
small scale continuous test 小型[规模]连续试验
small scale continuous testing 小规模连续试验
small scale laboratory test 小规模[小型]实验室试验
small-scale arrangement 小比例布[配]置图
small-scale batch test 小规模分批[单元]试验
small-scale frothing test 小型起泡试验
SMB-50 日本产疏基苯骈噻唑(钠盐)(药剂,可用作捕收剂)
smelter contract 冶炼(厂)合同
smelter cost 冶炼费
smelter flux 冶炼熔剂
smelter house 熔炼室

smelter operation 冶炼厂生产
smelter penalty 冶炼厂代价[损失量]
smelter slag waste 冶金熔炉废渣
smelter stack emission 冶炼厂烟囱排放物
smelting chamber 熔炼室
smelting cost 冶炼费用
smelting furnace 熔炼炉
smelting method 熔炼法
smelting procedure 熔炼步骤[方法,程序]
smelting rate 熔炼速度(率)
smelting reverberatory furnace 熔炼反射炉
smelting slag 冶炼渣
smelting technique 冶炼技术
smelting unit 熔炼设备
Smoluchowski equation 斯莫卢佐[霍]夫斯基公式(用于计算动电位)
Smoot medium pressure pneumatic conveying system 斯穆特型中等压力风力运输系统
Smoot rotary airlock feeder 斯穆特型旋[回]转风闸给矿[料]器
smooth curve 圆滑曲线
smooth surface 光[平]滑表面
smoothing 滤除[清,波],平滑,校[修]平,修匀
snap blow feature 快速[急速]鼓风[吹气]特性[特点]
snow content 雪含量
snowball 雪球
snowfall 下[降]雪
snub pulley 拉紧轮
SO_2 concentration 二氧化硫(SO_2)浓度
SO_2 cyanide destruction area 二氧化硫(SO_2)除氰区[工]段
soaking period 浸泡期
Sobragene 抑制剂(海生植物产品海藻粉末,可用作脉石抑制剂)
social cost 社会成本
social welfare 社会福利
society 社会,学会,团体
society transactions 学[协]会会[学]报,学会会刊
socket bushing 球窝座套
socket capscrew 碗形座带帽螺钉
socket connector 插座连接器,接插件
socket pin 碗形轴承销
socket switch 灯座开关,插座式开头
sodium 钠 Na
sodium adsorption ratio 钠吸附比值
Sodium aerofloat 钠黑药(干燥粉末状黑药,可用作锌矿捕收剂)
Sodium aerofloat B 钠黑药 B(其性质类似钠黑药)
sodium aluminate 铝酸钠
sodium aluminate liquor 铝酸钠液
sodium bicarbonate 碳酸氢钠(小苏打,可用作调整剂)$NaHCO_3$
sodium bicarbonate leaching procedure 碳酸氢钠浸出法
sodium bisulfite 重亚硫酸钠(亚硫酸

氢钠,可用作抑制剂)NaHSO₃
sodium bisulphate 硫酸氢钠,亚硫酸氢钠
sodium bisulphite 亚硫酸氢钠,重亚硫酸钠
sodium butanoate 丁酸钠(盐)
sodium butyrate 丁酸钠
sodium caprate 癸酸钠
sodium caprylate 辛酸钠
sodium carbonate 碳酸钠 Na_2CO_3
sodium carbonate-bicarbonate leach 碳酸钠-碳酸氢钠浸出法
sodium cyanide solution 氰化钠溶液
sodium decanoate 癸酸钠
sodium decyl sulfate 癸烷基硫酸钠
Sodium dibutyl dithiophosphate 丁钠黑药(二丁基二硫代磷酸钠,可用作捕收剂)
sodium dibutyl naphthalene sulfonate 二丁基萘磺酸钠
Sodium dicresyl dithiophosphate 25号钠黑药(二甲苯[酚]基二硫代磷酸钠,可用作捕收剂)
sodium diethyl dithiocarbamate 二乙基二硫代氨基甲酸钠 $(C_2H_5)_2NC(S)SNa$
sodium diethyl dithiophosphate 二乙基二硫代磷酸钠(钠黑药)
sodium diethyl thiophosphate 二乙基硫代磷酸钠
sodium diisoamyl dithiophosphate 二异戊基二硫代磷酸钠

sodium diisoamyl thiophosphate 二异戊基硫代磷酸钠
sodium diisobutyl dithiophosphate 二异丁基二硫代磷酸钠
sodium diisopropyl dithiophosphate 二异丙基二硫代磷酸钠
sodium di-sec butylthiophosphate 二仲丁基硫代磷酸钠
sodium dithiophosphate 二硫代磷酸钠
sodium dodecanoate 十二烷酸钠
sodium dodecyl sulfonate 十二烷基磺酸钠
sodium elaidate 反油酸钠
sodium fluoride(= **sodium fluorite**) 氟化钠(可用作硫化铁、硫化锌矿抑制剂)NaF
sodium fluorosilicate 氟硅酸钠 Na_2SiF_6
sodium hexadecanoate 十六烷酸钠
sodium hexadecyl sulfonate 十六烷基磺酸钠
sodium hexadecyl sulphate 十六烷基硫酸钠
sodium hexametaphosphate 六偏磷酸钠(可用作分散剂)$(NaPO_3)_6$
sodium hexanoate 己酸钠
sodium hydrosulphide(= **sodium hydrosulfide**) 硫氢[氢硫]化钠(可用作硫化铁、硫化锌矿抑制剂)NaHS
sodium hypochlorite 次氯酸钠(可用作抑制剂、活化剂)NaOCl
sodium isobutyl xanthate 异丁基钠黄

药(异丁(基)黄(原)酸钠)

sodium isopropyl xanthate 异丙基钠黄药(异丙(基)黄(原)酸钠) $(CH_3)_2CHOCSSNa$

sodium lattice point 钠格点

sodium lignin sulphate 木素[质素]硫酸钠

sodium metabisulfite 偏亚硫酸氢钠,焦亚硫酸钠

sodium metabisulfide (= sodium metabisulfite) 焦亚硫酸钠

sodium metasilicate 偏硅酸钠(水玻璃,可用作抑制剂) Na_2SiO_3

sodium naphthenate 环烷酸钠(可用作捕收剂)

sodium octadecyl sulfate 十八烷基(脂)硫酸钠(盐) $C_{18}H_{37}OSO_3Na$

sodium octadecyl sulfonate 十八烷基磺酸钠

sodium octanoate 辛酸钠

sodium octyl sulfate 辛基硫酸钠(盐) $C_8H_{17}OSO_3Na$

sodium octyl sulfonate 辛基磺酸钠

sodium palmitic aminoacetate 十六基氨基醋酸钠(棕榈酸甘氨酸钠) $C_{16}H_{33}NHCH_2COONa$

sodium peroxide 过氧化钠

sodium salt of fatty acid 脂肪酸钠盐

sodium secondary butyl xanthate 次[仲]丁基钠黄药

sodium silicofluoride 氟硅酸钠(可用作长石抑制剂,氧化铁矿活化剂)

Na_2SiF_6

Sodium site 钠格点

sodium stearate 硬脂酸钠

sodium sulfide (= sodium sulphide) 硫化钠 Na_2S

sodium sulfite 亚硫酸钠(可用作抑制剂) Na_2SO_3

sodium sulfite-sodium sulfide combination 亚硫酸钠硫化钠混合物[剂]

sodium sulphate 硫酸钠 Na_2SO_4

sodium tetraborate 四硼酸钠,硼砂

sodium tetraborate solution 四硼酸钠溶液

sodium tetradecanoate 十四烷酸钠

sodium tetradecyl sulfate 十四烷基硫酸钠

sodium tetradecyl sulfonate 十四烷基磺酸钠

sodium tetrametaphosphate 四偏磷酸钠

soft alkaline water 碱性软水

soft disk back-up 软盘备份

soft limestone 软石灰石

soft mineral 软矿物

soft money 钞票,软货币

soft ore 软矿石

soft ore mineral 软矿石矿物

soft polyurethane 软质聚氨基甲酸酯,软质聚氨酯

soft rubber lining 软胶内衬

soft start 软[缓]启动,挠性启动(凹形胶带机的启动)

soft start-up 柔性[软]起动
soft steel 低碳[软]钢
soft zone 软(矿)带
softening 软化
softening and settling furnace 软化沉淀炉
software 软件
software development 软件开发[研制,试制,设计]
software documentation 软件文件
software house 软件服务站
software organization 软件结构[机构]
software package 软件[程序]包
software system 软件系统
soil and rock box outlet 砂石仓出口
soil box outlet 砂仓出口
soil salinity 土壤盐[咸]度
Soil sealer SS-13 土壤密封剂 SS-13
soil sorption 土壤吸附(作用)
soja sludge 大豆渣(大豆加工副产物,可用作捕收剂)
solar distill 太阳能蒸馏器
solar evaporation 曝晒蒸发
solar evaporation system 暴晒蒸发系统
solar oil 太阳油(索拉油,可用作辅助捕收剂)
solar powered calculator 太阳能计算器
solenoid control valve 电磁控制阀
solenoid-controlled rubber pinch valve 电磁(控制)式橡胶夹管阀

solid arch 实体拱
solid concentration (溶液中)固体浓度
solid constituent 固体组分
solid curve 实线曲线
solid epoxy 固体环氧
solid epoxy resin 固体环氧树脂
solid feed rate 固体进[给]料速度
solid flow function 固体流动函数
solid fuel 固体燃料
solid fuel beneficiation plant 固体燃料选矿厂
solid glass rod 实心玻璃棒
solid ground 未开采的岩层,坚实的地面,直接接地
solid line 实线
solid magnetite 固体磁铁矿
solid material 固体物料
solid particle 固体粒子
solid polyurethane cyclone 硬聚氨基甲酸酯[聚氨酯]旋流器
solid process plant 固体处理设备
solid rock inclusion 原岩[硬岩石]包裹体[夹杂物]
solid sample 固体矿[试]样
solid section 坚固地段
solid solution 固溶体
solid sphere 固体球体
solid state change 固态变化
solid state diffusion process 固态扩散过程
solid state reaction 固态反应

solid suspension capability 固体悬浮能力
solid suspension characteristic 固体悬浮特性
solid third phase 固体第三相
solid throughput 固体处理量
solid weight 固体重量
solid woven belt 固体物编织的胶带
solid(s) 固体[态],立体,实心,固体燃料,(pl.)固体粒子[颗粒]
solid-fluid process plant 固体流体处理设备
solid-liquid ratio 固-液比
solid-liquid separating unit 固-液分离装置
solid-liquid separation 固-液分离
solid-liquid system 固-液系统
solids coagulation 固体凝聚[结]
solids concentration 固体浓度
solids contact reactor 固体接触反应器
solids flocculation 固体絮凝
solids production rate 固体生产速度
solids specific gravity 固体比重(矿物比重)
solid-solid separating equipment 固体与固体分离设备
solid-state detector 固态探测器
solid-to-solution ratio 固体-溶液比
solid-water interface 固-液[水]界面
Solox 起泡剂(变性酒精)
solubility 溶解度,溶解性,溶度,可溶性

solubility constant 溶度常数
solubility data 可溶度数据
solubility limit 溶解度极限范围
solubility product 溶度积
solubility product constant 溶度积常数
solubility product-criterion 溶度积判据
solubility property 溶解性能
soluble calcium 可溶钙
soluble complex ion 可溶性络离子
soluble constituent 可溶性组分
soluble copper 可溶[溶解]铜
soluble ethyl mono thiocarbamate species 可溶性乙基—硫代氨基甲酸盐(或脂)类
soluble ferric ion 可溶性高铁离子
soluble ferrous ion 可溶性亚铁离子
soluble fertilizer 水溶性肥料
soluble gold-cyanide compound 可溶氰化金化合物
soluble iron 可溶性铁
soluble manganese salt 可溶性锰盐
soluble metal species 可溶性金属产物
soluble mineral 可溶性矿物
soluble monothiocarbonate 可溶一硫代碳酸盐
soluble pollutant 可溶性污染物
soluble salt 可溶性盐
soluble sodium aluminate 可溶铝酸钠
soluble species 可溶产物
soluble starch 可溶性淀粉,可用作抑

制剂
soluble sulfide 可溶性硫化物
soluble zinc cyanide complex 可溶性氰化锌络合物
solute concentration 溶质浓度
solute recovery 溶质回收率,溶解物回收率
solution 溶液,溶解,解[法],解决
solution channeling 溶液沟流
solution chemistry 溶液化学
solution component 溶液组分
solution distribution system 溶液分配系统
solution lines key 溶液线例解
solution loss 溶解[液]损失
solution percolation 溶液渗滤[透](作用)
solution phase 溶液相
solution species 溶解物质
solution storage tank 溶液贮存槽
solution temperature 溶液温度
solution transfer 溶液输送
solvation 溶合作用,溶解[化],溶剂[化]作用
solvency 溶解状态[力],偿付能力
solvent extraction 溶剂萃取
solvent extraction process 溶剂萃取法
solvent extraction technique 溶剂萃取技术
Solvent L 制造酮基溶剂的副产物(主成分为二异丙基丙酮及二异丁基甲醇,可用作起泡剂)

sonic bin level control system 声波矿仓料位控制系统
sonic level detection system 声波料位探测系统
Sonneborn reagent 1 索诺邦1号药剂(石油磺酸盐,美国Sonneborn公司生产,可用作捕收剂)
Sonneborn reagent 2 索诺邦2号药剂(石油磺酸盐,美国Sonneborn公司生产,可用作捕收剂)
Sonneborn reagent 3 索诺邦3号药剂(石油磺酸盐,美国Sonneborn公司生产,可用作捕收剂)
sophisticated control algorithm 尖端控制计算法
sophisticated instrument 高级[尖端,精确]仪器
sophisticated plant 复杂的工厂[车间]
sophisticated programming technique 成熟[完善]的程序设计技术[方法]
sophisticated XRF and XRD analytical equipment 先进的射线荧光和射线衍射分析设备
sophistication 采用先进技术
sorption cell 吸收盒[管,室]
sorption phenomenon 吸附现象
sorption procedure 吸附法
sorptive capacity 吸附能力
sound analyzer 音响分析仪
sound attenuator 消音器
sound level intensity 音级强度
sour ore 含硫[酸性]矿石

source information 源信息
source language 源语言
source of dry waste heat 干燥废热源
source of fresh water 新水水源
source of fund 资金来源
source of internal financing 内部资金来源
source of power 电源
source of radiation 辐射源
source of software 软件资源
source of variability 变化根源
source of water 水源
source of work force 劳动力来源
source program 源程序
source unit (放射)源,源单元[装置]
Southwestern type pneumatic cell 西南型气升式浮选机
space available 可利用空间
space charge distribution 空间电荷分布
space charge effect 空间电荷效应
space charge layer 空间电荷层
space charge potential 空间电荷电位
space charge region 空间电荷区
space limitation 空间限制
space requirement 场地[空间]要求
spacing of samples 试样布置间距,取样间距
span length 跨距长度
spar salty mineral 晶石盐类矿物
spare parts 备品备件
spare parts list 备品备件表

spare screen 备用筛
spare stock 库存备品,备品存料[货]
sparger 喷雾[洒]器
spark source mass spectrometry 火花源质谱分析
sparse matrix inversion routine 稀疏矩阵逆求[反演]程序
spasmodic behavior (突然)阵发状态[情况]
spatial location 空间位置
spatially distributed assay value 空间分布化验值
spec. sand 特种[专用]砂
special adsorption effect 特殊吸附效应
special affinity 特别亲和力
special alloying system 特殊合金系统
special application 专门用途
special area of treatment 特定的处理范围[选矿领域]
special assay technique 特殊分析技术
special chemical action 特殊化学作用
special cloth 特殊滤布
special digital computer 专用数字计算机
special entity 特设实体
special flotation machine 专用浮选机
special in-house-designed proportional splitter gate 机构本身特[专]设的比例缩分闸门
special plexiglass holder 特制有机玻璃夹[架]

special problem 特殊问题
special problem of blending 匀矿专门问题
special steel 特种钢
special technique 专门技术
special type of controller 特种类型控制器
special vocabulary 特殊语汇
specialist tailing dam consultant 专业尾矿坝顾问,尾矿坝专家顾问
specialized application 特殊用途
specialized application knowledge 专用知识,学问
specialized circumstances 特定情况
specialized device 专用装置[设备]
specialized duty 特殊任务[功用]
specialized equipment 专业化设备,专用设备
specialized high level language 专用的高级语言
specialized high level process control language 专用的高级过程控制语言
specialized leasing company 专营出租[租赁]公司
specialized reduction machine 特殊型[化]破碎机
specialized sampling procedure 专业取样工艺[方法,过程]
specialized task 专门[专业]任务
specially adsorbed ion 特殊[性]吸附离子
specially woven cloth 特殊编织(的)滤布
special-purpose equipment 专用设备
specialty metal 特种金属
specie separation 矿物分离
species 物种,种类
species mass flow 矿种物料流
specific 专门,具体,准确,特定
specific adsorption 特殊吸附
specific adsorption potential 特殊吸附势
specific air flow 比空气流量(空气流量/槽容积)
specific application 具体[特定]应用
specific area 比面积,特定场合[区域]
specific asset 专门资产
specific capacity 比生产能力(每平方英尺小时吨数)
specific capacity curve 比生产能力曲线
specific case study 专题研究
specific category 专门类别
specific chelating agent 特定螯合剂
specific chemical interaction 特殊化学作用
specific collector 特定捕收剂
specific collector-mineral adsorption reaction 特殊捕收剂矿物吸附反应
specific computer 专用计算机
specific condition 特定条件
specific control program 专用控制程序
specific control response 特定控制

响应

specific criteria 具体准则

specific design condition 特定[具体]设计条件

specific digital computer control system 专用(的)数字计算机控制系统

specific energy consumption 电能[能量]比耗,单位电能[能量]消耗

specific energy input 比能量输入值(为产出需要的产品,每处理一吨原矿时磨机所消耗的能量)

specific energy requirement 单位能耗,单位能量需用量

specific field 特定范围[领域,方面]

specific field of interest 关注的具体领域

specific frequency 特定频率

specific gravity gauge 比重计

specific gravity of separation 分离比重

specific gravity range 比重范围

specific gravity system 比重系统

specific gravity test 比重试验

specific gravity tolerance 比重差容(许偏)差

specific gravity tolerance curve 比重容差曲线

specific gravity unit 比重单位[元]

specific hardware 专用硬件

specific idler 特殊托辊

specific information 专门资料

specific instruction 专门指令

specific investment cost 单位投资成本,比投资费用

specific ionic electrode method 特定离子电极法

specific item 特定项目

specific lip length 比泡沫溢流堰长度(浮选机用)

specific machine language coding 特殊的机器语言编码

specific measurement 具体测定(结果)

specific mineral deposit 特定矿床

specific opening 特定孔径

specific operation 特定生产(情况)

specific ore 特定的矿石

specific position 特定位置[处所]

specific power 比[单位]功率

specific power consumption 比[单位]功率消耗

specific process 专门方法

specific product size 规[指,特]定产品粒度

specific productivity 单位生产率

specific program 专用程序

specific project 特定工程

specific property 特定的性质(如矿石的)

specific pulp flow rate 比矿浆流速

specific quantity 比量,单位分量,特定数量

specific question 具体问题

specific rate constant 比速(率)常数

specific responsibility 特定[具体,明

确]的职责

specific rotational flow (flow per unit volume) 单位旋转流量(单位体积流量)

specific scattering coefficient 比散射系数

specific selection function 比选择函数

specific selection function relationship 特定选择函数关系式

specific situation 特定情况

specific size 特定粒度

specific surface free energy 比表面自由能

specific technique 专项技术

specific throughput 单位处理量[生产能力]

specific type of circuit 特定流程[回路]类型

specific volume capacity 单位体[容]积处理能力

specific volumetric flow rate 比体积流速

specific working capital level 特定的周转资金水平

specification 规格,规范[程],技术要求,(工序,设计)说明书

specification engineer 规范工程师

specification information 技术规格资料

specification sheet 细目[明细]表

specified base year 规定的基准年

specified coarse particle 特定粗(颗)粒

specified condition 特定条件

specified convergence limit 规定的收敛极限范围

specified criteria 明细规范

specified fine particle 特定细(颗)粒

specified gain 特定增益

specified horsepower 额定马力

specified load 规定荷载,计算[设计,额定,条件]荷载

specified micron size 特[规]定的微米粒度

specified number of samples 给[规]定的样品数

specified range 特定范围[界限]

specified rate 额定量

specified value 给定值

specimen of ore 矿石样[标]本

specimen of waste 废石样[标]本

spectro- [光,频,波,能]谱

spectro-chemical process 化学分光[光谱化学]法

spectro-grating 分光光栅

spectrometer 分光计

spectrometer-computer system 光谱仪-计算机系统

spectrometric process 能[光]谱法

spectrometric system 能[光]谱装置[系统]

spectrometry 光谱法

spectrophotometric examination 分光光度检验

spectrophotometric method 分光光度分析法

spectrophotometry 分光光度[学,测定法]

spectroscopic analysis 光谱分析

spectroscopic examination 光谱检验

spectroscopy 光谱法

specular reflection technique 镜面[单向]反射技术

speed changing gear 变速齿轮

speed distribution 速度分配

speed feedback mechanism 速度反馈机构

speed of suspension 悬浮速度

speedier assaying device 快速分析装置

speedup belt conveyer 高速胶带运输机

sphalerite 闪锌矿

sphalerite-pyrrhotite separating circuit 闪锌矿-磁黄铁矿分选回路

sphere of application 适用范围

spherical agglomeration process 球团聚法(选煤采用的)

spherical impermeable particle 球状不透水颗粒

spherical model 球形模型

spherical sand particle 球形砂粒

spherical socket bearing 球窝轴承

spider arm 星形臂

spider bushing lubricating system 十字支架套管润滑系统

spigot diameter 沉砂嘴直径

spigot pinch valve 塞栓[插口,套管]夹阀

spigotting tailings 粗粒尾矿(旋流器底流)

spike (高)峰值,最大值,(尖峰)脉冲

spinning air 旋转空气

spinning motion 旋转运动

spiral classifier operation 螺旋分级机作业[操作]

spiral classifier sizing 螺旋分级机分级

spiral counter-current washing classifier 逆流洗涤螺旋分级机

spiral drive 螺旋传动装置

spiral flotation 螺旋浮选

spiral path 螺旋形轨迹[途径]

spiral pattern 螺旋形[型]式

spiral pump 螺旋分级机用泵

spiral rake thickener 螺旋耙式浓缩机

spiral raking capacity 螺旋返砂能力

spiral return idler 回空段螺旋(形)托辊

spiral rotation speed 螺旋旋转速度

spiral screen 螺旋筛(在球磨机排口处)

spiral separation 螺旋选矿

spiral shaft 螺旋轴

spiral speed 螺旋速度

spiral start 螺旋头(螺旋选矿机的)

spiraling 螺旋选矿

splash skirt (防)喷溅喇叭口[裙板](用于控制旋流器底流)

splice length 交错接头长度(胶带运输机的)
spline function 样条[仿样]函数
splintery chalcopyrite 片状[裂片]黄铜矿
split 分离[样,裂,解],劈开
split drill core 劈开的钻孔岩心
split seal 分支[对开]密封
split seal arrangement 分支[对开]密封装置
splitter 缩分器
splitter arrangement 分流器配置
splitter blade 分料板
splitting crusher frame 组合[可拆]的破碎机机架[机座,固定锥]
spontaneous attachment 自发[然]附着
spontaneous oxidation 自然氧化
sporinite 孢粉体
spot welder 点焊机
spray chamber 喷雾室
spray outfit 喷雾设备
sprayer 喷雾[洒]器
spraying water 喷洒水
spring bolt 弹簧螺栓
spring roll crusher 弹簧辊式破碎机
sprinkler 洒水车
spudding-in 开(始)钻(孔)
spurious high reading 寄生[假的]高读数
square aperture size 方孔筛孔尺寸
square hole 方孔

square matrix 方阵
square measure 面积单位制
square opening 方孔
square opening woven wire deck 方形孔金属丝编织筛面
square outlet 方形排口
square prism 正方柱
square root formula 平方根公式
square screen opening 方形筛孔
squared difference 平方差
squared relative error 平方相对误差
squared variation 方差
squirrel cage 鼠笼式(电动机)
stability 稳定性
stability constant 稳定常数
stability field 稳定场
stability of suspension 悬浮液稳定性
stabilizing control 稳定控制
stabilizing effect 稳定效应
stabilizing property 稳定性
stabilizing spring 稳定弹簧
stable contact angle 稳定接触角
stable effect 稳定效应
stable empty rathole 稳定空心管状腔
stable rathole 稳定管状腔
stable solution species 稳定溶液产物
stable wettability 稳定湿润性
stack gas 烟囱废气
stacked arrangement 重叠配置
stacker belt conveyer 堆料机胶带运输机
stacker system 堆料机系统

stacker-reclaimer 堆取料机
stacking conveyor 堆积运输机[传送带]
staff assistants and clerks 科室辅助人员和办事员
staff cost 工资费用
staff salaries 职员薪资
staff training 人员培训
staffing 人员配备
stage additions of reagent 分段加药
stage construction 分期施工,多层面构造
stage flotation 阶段浮选
stage flotation type flowsheet 阶段浮选(式)流程
stage of evaluating a deposit 评价矿床阶段
stage of splitting 缩分段
stage of up-grading 提高品位阶段
stage process 阶段过程
stage recovery 阶段回收率
stagnant mass 停滞矿体[积矿堆]
stagnant region 停滞区
stagnant solids 停滞固体物料
stainless steel ball 不锈钢球
stainless steel beaker 不锈钢烧杯
stainless steel boat 不锈钢蒸发皿
stainless steel can 不锈钢罐
stainless steel capsule 不锈钢衬壳
stainless steel coarse grain spray purifier 不锈钢粗粒喷雾净化器
stainless steel coarse particle mist eliminator 不锈钢粗粒喷雾净化器
stainless steel Denver-type cell 不锈钢丹佛型浮选槽
stainless steel fine particle mist eliminator 不锈钢细粒喷雾净化器
stainless steel grinding rod 不锈钢磨矿棒
stainless steel pipe 不锈钢管
stainless steel rotary oil-fired reactivation furnace 不锈钢回转燃油(炭)再活性炉
stand pipe 竖[立]管
standard application information 标准使用资料
standard averaging technique 标准平均法[技术]
standard catalog format 标准样本格式
standard catalog information 标准样本资料
standard chemical procedure 标准化学分析法
standard classification 标准分级[类]
standard closed circuit 标准闭路(后置分级)
standard commercial range 标准工业型[类](按使用规格分类)
standard commercial size screen 标准工业规格筛
standard compiler 标准编译程序
standard concentrate grade 标准精矿品位

standard cone crusher 标准圆锥破碎机
standard cone crusher cavity 标准圆锥破碎机腔室
standard control system 标准控制系统
standard Crowe tower 标准克罗韦塔(即脱气[氧]塔)
standard cyanidation bottle roll test 标准摇动台烧瓶氰化试验,标准烧瓶滚动氰化试验
standard cyclone 标准旋流器
standard data handling function 标准数据处理功能
standard data sheet 标准数据表(格)
standard demand meter 标准需要量仪表
standard design self-trimming bulk carrier 标准设计自动整平散装船
standard deviation 标准偏差
standard distribution function 标准分布函数
standard drum type filter 标准筒[鼓]型过滤机
standard duct diameter 标准管径
standard fitting method 标准拟合法
standard float 标准浮选
standard flotation machine 标准浮选机
standard flotation technique 标准浮选技术
standard form 标准格式
standard formatted language 标准格式化语言
standard free energy 标准自由能
standard grindability 标准可磨性
standard grindability figure 标准可磨性数值
standard grindability table 标准可磨性表
standard grindability test 标准可磨性试验
standard grinding reference operation 标准磨矿的参考操作(法)
standard heavy-duty centrifugal industrial exhauster 标准重型离心式工业抽风机[排气通风机]
standard heavy-duty spiral rake thickener 标准重型螺旋耙式浓缩机
standard high level language 标准高级语言
standard installation 标准安装[装置,装配]
standard laboratory ball mill grindability test 标准实验室球磨机可磨性试验
standard laboratory equipment 标准实验室设备
standard laboratory grind and floatation flowsheet 标准试验磨浮流程
standard language compiler 标准语言编译程序
standard layout 标准配置
standard level 标准水平
standard material 标准物料
standard metal mining operation 标准

的金属矿山生产

standard motor speed 标准电动机速度

standard open area 标准筛孔面积

standard operating value 标准操作值

standard operation configuration 标准操作[生产]配置

standard power calculation 标准功率计算

standard practice 一般做法

standard pressure head 标准压头

standard process 标准工艺[方法]

standard rake capacity 标准返砂[耙矿]能力

standard range 标准量程[范围]

standard reagent 标准药剂

standard return idler 回空段标准托辊

standard RSA number 标准相对表面积数值

standard sample 标准样品[试样]

standard sample analysis 标准样分析

standard sea level condition 标准海平面条件

standard service line 标准(负荷)作业线

standard sink-float separation 标准沉浮[重介质]分离

standard size 标准尺寸

standard spectrophotometric method 标准分光光度[光谱]分析法

standard state 标准状态

standard tank length 标准槽长

standard thread 标准螺纹

standard three-roll idler set 标准的三辊式托辊组

standard titration procedure 标准滴定法

standard tray arrangement 底盘[板]标准布置[排列]

standard type 标准型

standard units 标准设备

standard wet chemical analysis 标准湿式化学分析

standard work index calculation 标准功指数计算

standardized laboratory procedure 标准(化的)实验室程序

standardized laboratory test procedure 标准化实验室试验程序

standardized procedure 标准化方法[步骤]

standardized settling and decantation procedure 标准沉淀倾析法

standardized shipping container 标准化船[海]运集装箱

standby application 备用

standby diesel boiler 备用柴油锅炉

standby Eurodollar 备用欧洲美元

standby fee 备用费

standby letter of credit 备用信贷许可证

standby line of credit 备用信用额度

stand-by power 备用功率

standby power source 备用电源

standby screen 备用筛
standing bench 永久[固定]台阶
standpipe-rotor-disperser combination 竖管-转子-分散器组合体
standpipe-rotor-disperser hydrodynamics 竖管-转子-分散器的流体动力学
stannic chloride 氯化锡
star rotor 星形转子
star shaped impeller 星形叶轮
starch depressor 淀粉抑制剂
starch derivative 淀粉衍生物
starch hydroxyl 淀粉羟基
starter dam 初始坝(尾矿坝的)
starting and stopping transient phase 开,停瞬变状态[阶段](胶带运输机的)
starting date 投产日期
starting design 起始设计
starting simplex 起始单纯形
starting torque 起动转矩
starting torque limit 起动[开车]转矩极限
starting value 起始值
start-stop system 启闭系统
start-up 开动,起动,开始工作[动转],投产
start-up ball charge 开车装球量
startup operation 投产
start-up service 投产[开车]服务
starvation addition 饥饿加药,缺量[限量]加药
starvation concentration 浓度不足

starvation switch 缺量加药开关,限量给料开关
state of aggregation 聚集状态
state of oxidation 氧化状态
state tax 国家税收
stated dividend 规定的股息
stated monetary face value 规定的货币面值
stated size 规定粒度
statement of earnings 收益表
statement of fact 事实说明,实际论据[点]
statement of monthly cost 月经费财务报表
static column test 静止萃取塔浸出试验
static condition 静止条件[状态]
static contact angle 静态接触角
static control 静态[定位]控制
static double layer 静电双电层
static electrical double layer 静电双电层
static leaching 静止[态]浸出
static mass balance 静态质量平衡
static peripheral screen 固定周边[圆周]筛
Statim constant temperature unit Statim 恒温装置
stationary air seal type 固定式空气密封型
stationary conveyer frame specification 固定运输机支架技术规格

stationary flat screen 固定平面筛
stationary grizzly 固定棒条筛
stationary jaw 固定颚板
stationary jaw depth 固定颚板高度
stationary load 静载荷
stationary peripheral and launder screen 固定式周边[圆周]溜槽筛
stationary region 稳定区
statistic (典型的)统计量,统计数值
statistic correlation 统计相关(性)
statistic data 统计资料[数据]
statistic data of random independent variable 随机独立变量的统计资料
statistical accuracy 统计精度
statistical analysis 统计分析
statistical analysis of data 数据统计分析
statistical approach 统计法
statistical assessment 统计评估
statistical assessor 统计鉴定计[器]
statistical data 统计资料[数据]
statistical design test 统计设计试验
statistical error 统计误差
statistical evaluation 统计评价
statistical figures 统计数字
statistical inference 统计推断
statistical inference 统计推论[理]
statistical mass balance smoothing 统计物料平衡修匀
statistical measure 统计测量[测定,估量,衡量]方法[措施]
statistical method 统计法

statistical model 统计模型
statistical optimization study 统计最佳研究
statistical sense 统计意义
statistical treatment 统计处理
statistical value 统计值
statistically based approach 基于统计的方法
statistically based method 以统计学为基础的方法
statistically designed study 统计设计研究
statistically oriented procedure 基于统计的方法[程序]
statistics 统计[学,资料]
statistics of independent variable 独立[自]变量统计
statistics of random variable 随机变量统计
stator blade 定子叶片
stator design 定子设计
stator-baffle 定子折流板
staveline 木管道
steady classifier operation 稳定的分级机操作,分级机稳定运行
steady head concrete tank 恒压混凝土(水)池
steady state 稳态
steady state batch test 稳态分批试验
steady state condition 稳定状态条件
steady state current 稳态电流
steady state data 稳[定]态数据

steady state input-output relationship 稳态输入-输出关系式

steady state measurement 稳[定]态测定

steady state mill discharge size distribution 静态磨机排矿粒度分布

steady state model 稳[静]态模型

steady state multiple product test 稳态多产品试验

steady state operation 稳态操作

steady state performance 稳[静]态特性

steady state rest potential 稳态残余电位

steady state rougher flotation performance 静[稳]态粗浮选特性

steady state vacuum 稳态真空

steady-state condition 稳态条件[情况]

steady-state flow expectation 稳定状态流量预期值

steady-state mill load 稳定状态的磨机负荷

steady-state optimization 稳(定)状态最优化

steady-state size distribution 稳态粒度分布

steam boiler 蒸汽锅炉

steam coil 蒸汽蛇形[盘]管

steam generation 蒸汽产生

steam heating process 蒸汽加热过程

steam hood 蒸汽罩

steam jacketed elution column 蒸汽夹套洗提塔

steam power plant 蒸汽动力厂[发电厂]

steam shattering 蒸汽破碎[破裂]

steamer vessel 蒸汽包

steaming 汽蒸,通入蒸汽

steaming process 蒸汽法,汽蒸工艺

stearic 硬脂的,硬脂酸的,十八酸的

stearic acid 硬脂酸(十八(碳)(烷)酸,可用作捕收剂)$CH_3(CH_2)_{16}CO_2H$

stearic aminopropyl acid 十八氨基丙酸 $C_{18}H_{37}NHCH_2COOH$

stearic effect 硬脂效应

Stearns flotation machine 斯特恩斯型浮选机

steel and alloy flue 钢合金烟道[管]

steel and fiberglass cladding 钢板和玻璃纤维(覆盖,保温)层

steel ball mill 钢制球磨(机)

steel building 钢结构建筑物

steel cable 钢丝绳,钢缆,钢索

steel coil spring 钢盘簧[螺旋弹簧]

steel density 钢密度

steel flushing pipe 冲水钢管

steel grind 钢磨

steel grinding 钢介质磨矿

steel hammer head 钢锤头

steel industry 钢铁工业

steel lined coupling 衬钢接头

steel liner 钢衬里

steel maker 制钢厂

steel media 钢介质
steel mill 钢磨
steel mixing tank 钢结构混合槽
steel plate screen 钢板筛
steel rod mill 钢制棒磨(机)
steel rod mill linings 棒磨机钢衬板
steel rolling mill 轧钢机
steel rolling mill scale 轧钢机轧钢的氧化皮
steel sash window 钢(框)窗
steel steamer feed chute 钢制蒸汽给料槽
steel toughener 钢增韧剂
steel wear[loss] 钢材磨耗[损]
steel wool cathode 钢丝棉[丝绒]阴极
steep ascent 陡坡
steepest ascent 最陡上升(法)
steepest ascent method 最陡上升法
steepest ascent technique 最陡升高技术
steepest direction 最陡方向
steepest path 最陡路径
Stein equation 斯坦方程式[斯坦公式](得到所需精度,需要附加钻孔数的计算公式)
SteinHall 斯坦霍尔法[工艺](美国 SteinHall 黏合剂制作工艺)
Stellar filter 恒星牌(Stellar)过滤器(美国恒星汽车集团(Stellar Automotive Group)产品)
step 台阶,方法,步骤,手段,措施,工序
step adjustment 阶跃调节

step bearing 端轴承,立式止推轴承,阶式止推轴承,踏板轴承,枢轴承
step change 阶跃变化
step length 步长
step response 阶跃响应
step response test 阶跃响应试验
step size 步长
step-down potential transformer 降压变压器
step-down transformer 降压变压器
Stephens-Adamson coarse ore reclaim pan feeder 史蒂芬斯-阿达姆逊型粗矿回收盘式给矿机
Stephens-Adamson pan feeder 史蒂芬斯-阿达姆逊型盘式给矿机
stepless speed variation 无级调速
step-to-step numerical method 逐步数值法
stepwise regression 逐步回归
stern layer 坚定[固定,腹]层
Stern layer 斯特恩层(电化学)
stern plane 坚定[固定]平面
Stern plane 斯特恩面(电化学)
Stern potential 斯特恩电位(电化学)
Stern-Graham double layer model 斯特恩-格雷厄姆双电层模型(电化学)
Stern-Grahame model of electrical double layer 斯特恩-格雷厄姆双电层模型(电化学)
Sterox 表面活性剂品牌名
Sterox CD 塔尔油与 12 克分子环氧乙烷的反应物,可用作分散剂,湿润剂

steward 乘务员,招待员
sticking probability 黏附概率
sticky material 胶黏[黏性]物料
sticky nature 黏性
sticky ore 胶黏矿石
stiffened steel plate 加强薄钢板
stipulated error limit 规定的误差范围
stipulated probability level 规定的概率级别
stir rate 搅动速度
stirrer speed 搅拌器速度
stirring action 搅拌作用
stirring condition 搅拌条件
stirring dependence 搅动相关性[依赖关系]
stirring intensity 搅拌强度
stirring rate 搅拌速度
stirrup 箍筋,钢箍,马镫,镫形物
stochastic behavior 随机行为[变化过程,工作情况,性能]
stock commission 证券委员会
stock exchange 股票交易[所]
stock farm 畜牧场
stock solution 储备溶液(指药剂),原液,储液
Stocke's formula 斯托克公式(用来确定球形固体颗粒在液体中下沉速度的公式)
Stocke's law 斯托克定律(是颗粒半径与颗粒在静水中自由沉降速率的关系式)
Stocke's law of sedimentation 斯托克沉降定律
stockholder 股票持有人,股东
stockpile 资源,蕴藏量,全国的原料物资贮存量
stockpile feed conveyor 堆场给矿胶带运输机
stockpile layout 堆场配置
stockpile staking and reclaiming operation 储矿堆堆取作业
stockpiling 堆放[存]
stoichiometric 化学计算[计量,当量,数量]的,理想配比的
stoichiometric composition 化学计量成分
stoichiometric compound 化学计量[定比]化合物
stoichiometric ratio 化学当量比
stoichiometric release 按化学比例析出
Stokes equation 斯托克斯方程式(微积分的基本定理之一,以英国数学家Stokes命名)
Stokes expression 斯托克斯公式(微积分的基本定理之一,以英国数学家Stokes命名)
Stokoflot PK 捕收剂(脂肪酸硫酸化皂,德国产)
Stockopol CN 湿润剂(烷基苯磺酸盐与烷基硫酸的混合物)
Stockopol FNF 湿润剂(烷基苯磺酸盐与脂肪酰胺缩合物)
stone sand plant 砂石厂
stop order 停机指令

stop watch 秒[停]表
Stopes-Heerlen system of classification 斯托普斯-海尔伦分类体系(煤岩成分的分类体系)
stop-go traffic signal 停止-通行交通信号
stoping cycle 回采循环
stoppered measuring cylinder 塞封[加塞,带塞]量筒
stop-start control 停-开自动控制
storage area 储存区
storage capacity 储矿能力,存储容量
storage cone 储矿漏斗
storage drum 储存(圆)桶
storage pit 储存仓[坑]
storage plant 储藏室,仓库
storage silo 储存筒仓
storage system 存储系统
storm water 暴雨水
storm water-sewers and control system (暴)雨水水沟和控制系统
stove oil 炉油
straight chain 直线链
straight chain sodium laurel sulfate 直链十二烷[月桂]基硫酸钠
straight cyanide circuit 纯[直接]氰化法
straight dashed line 直虚线
straight line basis 直线(计算)法
straight section of connecting duct 连接输送管直线部分
straight sizing operation 连续筛分作业
straight through transfer 全部直线传送
straight-forward application 直接应用
straightforward porphyry copper ore 简单[单纯]的斑岩铜矿石
strandline 勘探线,海岸[滨]线
strategic material 战略物资
strategic stage 战略阶段[步骤]
strategy 策略,对策
strategy selector matrix board 控制策略选择器矩阵板
stratified random network 分层[有层次的]随机网络
stratified random sampling 分层随机取样
stratified zone 层状地带
stratigraphic system 地层系统
stratum system 地层系统
stream scanning 电子流扫描
stream scanning technique 电子流扫描技术
streaming potential apparatus 流动电位仪
strength of addition 添加的强度(药剂)
stretching mode 伸展方式
striking blade 冲击叶片
strip chart recorder 带形图表记录仪
strip circuit 洗提[解吸]回路
stripped area 剥离区

stripped carbon 解吸[脱金]碳
stripping characteristic 解吸特性
stripping efficiency 解吸[萃取,洗提]效率
stripping operation 剥离[解吸,洗提,萃取]作业
stripping ratio 剥采比
stripping section 解吸工段[工区,部分]
stripping stage 洗提[解吸]段
stripping technique 解吸技术[工艺,方法]
stripping technology 解吸技术[工艺]
stripping test 解吸试验
stripping tower 反萃取[解吸]塔
stroboscopic illumination 频闪照明
stroke characteristic 冲程特性
strong affinity 强亲和力
strong bas resin 强碱性树脂
strong base resin 强碱树脂
strong bonded ore(s) 强结合[键合,黏着]矿石(类),硬矿石(类)
strong bubble contact 强(的)气泡接触
strong froth 稠泡沫
strong hydration of chromate ion 铬酸离子强烈水化作用
strong hydrogen bond 强氢键
strong market 畅销市场
strong metal container 坚固的金属容器
strong oxidant 强氧化剂
strong oxidizing agent 强氧化剂
strong promoter 强促进剂
strong reducing agent 强还原剂
strong room 保险库(贮存金锭用)
strongest bound water 最强结合水
strongly hydrated polar group 强水化(的)极性基团
strongly magnetic separator 强磁选机
structural alignment 构架平直性(胶带支架的)
structural analysis 结构分析
structural constitution 结构组成
structural description 构造描述
structural design 结构设计
structural design criteria 结构设计标准
structural element 结构元件
structural failure 结构破坏,破坏结构
structural formula 结构式
structural investigation 结构研究
structural steel frame 钢结构框架
structural steel with cement/sand blocks 钢筋混凝土/砂石结构
structural study 结构研究
structural supervisor 结构管理程序
structural support 结构支架
structure design 结构设计
structure detail 结构细节
structure difference 结构差异
structure of foam 泡沫结构
structure parameter 结构参数
structure part 结构件

style 式样,种类,类型[别],风格
styling variation 设计变量
styrene 苯乙烯 $C_6H_5CH:CH_2$
styrene phosphonic acid 苯乙烯膦酸
Styromel 絮凝剂(苯乙烯与失水苹果酸酐的共聚物铵盐,捷克产)
styryl 苯乙烯基
styryl phosphonic acid 苯乙烯膦酸
subaeration type 底吹式(浮选机)
subclassification 细[再]分类
sub-contract cost 分[转]包费用
subcontractor 分包商[方]
subfactor 分因子
subfreezing temperature 亚冻结温度
sub-freezing winter 冰点以下的冬季
sub-grade 低品位[品级,质量]
subjective application 主观应用
subjective weighted factor 主观加权系数
subjective weighting factor 主观加权因素
sublevel caving 分段崩落采矿法
sublevel interval 分段间距
submarginal local ore 亚边缘[极限]矿石(不值得开采的矿石)
submerged classifier 浸没式分级机
submerged plant 水下植物
submerged screen 浸没式筛
submerged spiral classifier 沉没式螺旋分级机
submergence (螺旋分级机)浸没区
submersible pump series 潜水泵系列

sub-micro particle 亚微粒
sub-micro particle size 亚微粒级
sub-microscopic film 亚微观薄膜
submonolayer quantity 亚单分子层量
sub-office 分局
sub-optimization 部分[局部]最优化
sub-phenomena 亚现象
sub-program of classification 分级子程序
sub-program of crushing 破碎子程序
sub-program of grinding 磨矿子程序
sub-program of material flow 料流子程序
subprogram(me) 子程序
sub-sampling stage 再取样阶段[步骤]
subscripted variance 标有下标的方差
subsequent crushing 后一段破碎机
subsequent free movement 随后的自由运动
subsequent test 随后[后继]的试验
subsidiary 子公司,附属机构
subsidiary company 子[附属]公司
subsoil investigation 地基下层土调查
substantial issue 具体问题
substantial program 实际[具体]计划
substation 变电站,分局
substation design 变电所设计
substation specification 变电所技术要求[详细说明]
substrata 下部地层,底层,基层
subsurface aquifer 地下蓄水层

subsurface calcium ion 表面以下钙离子
subsystem 子[分,次要]系统
sub-system 辅助[子]系统
subtotal 小计
successful bidder 成功的投标人
successive flotation 依次浮选
succinamate 琥珀酰胺酸盐
succinic acid 丁二酸,琥珀酸 $(COOH)_2C(CH_2)_2$
suction lift of pump 泵吸入高度[扬程]
suction main 吸入总管[干管]
suction manifold 吸入支管
suction piece 吸入连管
suction side 吸入侧(泵的)
suitable fashion 适宜[合适]的方式
Sulfanol 选矿药剂(二十烷基苯磺酸钠,可用作捕收剂) $C_{20}H_{41} \cdot C_6H_4-SO_3Na$
sulfate 硫酸盐 M_2SO_4
sulfate ion 硫酸盐离子
sulfate sulfur 硫酸盐硫
sulfated coconut oil 硫酸化椰子油
sulfated coconut oil emulsifier 硫酸化椰子油乳化剂
sulfated glyceride 硫酸甘油酯
sulfated long chain alcohol 硫酸化长链(乙)醇
sulfated red oil 硫酸化油酸(含4.5%的有机 SO_3),硫酸化红油
Sulfetal C 湿润剂(十二醇基硫酸盐)
Sulfetal O 湿润剂(油醇硫酸盐)
Sulfetal OC 湿润剂(十二醇基硫酸盐与油醇硫酸盐的混合物)
sulfhydryl(= suphydryl) 氢硫基,巯基 HS—
sulfhydryl collector 氢硫基捕收剂,巯基捕收剂
sulfhydryl ion 巯基离子
sulfide concentrator 硫化矿选矿厂
sulfide copper ores 硫化铜矿石类
sulfide gangue 硫化矿脉石
sulfide group 硫化矿族
sulfide material-aqueous solution interface 硫化矿物-水溶液界面
sulfide mill 硫化矿选矿厂
sulfide mill metallurgist 硫化矿选矿厂选冶工程师
sulfide mineral electrode 硫化矿(物)电极
sulfide operation 硫化矿生产
sulfide ore and metallic promoter 硫化矿和金属促进剂
sulfide plant facilities 硫化矿选矿厂设施
sulfide precipitant 硫化物沉淀剂
sulfide tailings 硫化物尾矿
sulfidized mineral 经硫化的矿物
sulfidizing agent 硫化剂
sulfite alcohol spent wash 由木屑制酒精时的废液(可用作稀土、六解石、重晶石的抑制剂)
sulfo-(=sulpho-) 硫代(词头),磺

基 —SO_3H

sulfonate 磺酸盐,磺化

Sulfonate 表面活性剂品牌名

Sulfonate AA 十二烷基苯磺酸钠(可用作捕收剂、湿润剂)
$C_{12}H_{25}$—C_6H_4—SO_3Na

sulfonated abietic acid 磺化松香酸(可用作分散剂、捕收剂)

sulfonated petroleum collector 磺化石油捕收剂

sulfonated red oil 磺化(或硫酸化)油酸(含45%有机SO_3)

sulfonated tall oil 磺化塔尔油(可用作捕收剂)

sulfonated tea-seed oil 磺化茶子油钠皂(可用作捕收剂)

sulfonated whale oil 磺化鲸油(可用作捕收剂)

sulfonate flotation 磺酸盐浮选

sulfonate group 磺酸基

Sulfonate N56 烷基苯磺酸钠(可用作湿润剂)R—C_6H_4—SO_3Na

Sulfonate OA-5 硫酸化油酸钠皂(可用作捕收剂)

sulfonium 锍基

sulfonium salt 锍盐

Sulfopar H 捕收剂(硫酸化的氧化石蜡铵皂)

Sulfopon OK 湿润剂(伯—烷基硫酸盐)

sulfo-salt (复)硫盐,硫酸盐,磺(酸)盐

sulfo-salt group 含硫盐类

α-sulfostearic acid α-磺化硬脂酸(可用作重晶石,绿柱石,氧化铁矿捕收剂)

sulfosuccinate (= **sulphosuccinate**) 磺基琥珀酸盐(酯),磺基丁二酸盐(酯)

sulfosuccinic acid 磺基琥珀酸

Sulframin 表面活性剂品牌名

Sulframin AB85,AS 烷基苯磺酸盐(可用作湿润剂)

Sulframin DR 羟烷酰胺醇硫酸盐(可用作湿润剂)

Sulframin KE 烷基苯磺酸盐(可用作湿润剂)

Sulframin N 二烷基萘磺酸钠(可用作湿润剂)

Sulfsipol 湿润剂(十二烷基硫酸钠)

sulfur dioxide diffuser 二氧化硫扩散器

sulfur dioxide gas 二氧化硫气体

sulfur removal 除硫

sulfuric acid leaching 硫酸浸出

sulfuric acid leaching procedure 硫酸浸出法

sulfuric acid process 硫酸生产工艺

sulfuric acid-ferric sulfate solvent 硫酸/硫酸铁溶剂

sulfydryl collector 巯基捕收剂

sulfydryl compound 巯基化合物

sulphated monoglyceride 硫酸(化)甘油一酸酯

sulphide combustion energy 硫化物燃烧能量

sulphide-rich surface 富硫表面
Sulpho-(=sulfo-) (拉丁字头)硫代,磺(酸)基 —SO_3H
sulpho-succinamate 磺化琥珀酰胺酸盐
sulphoxyl ion 硫氧基离子
sulphoxyl species 硫氧基产物
sulphur dioxide depressant 二氧化硫抑制剂
sulphur vacancy 硫空位
sulphuric acid circuit (制)硫酸流程[回路]
sulphur-oxygen band 硫氧谱带
sulphydric class 硫代化合物类
sum readout 和数读出
summarized statistic 累计统计数
summarized statistics 简要统计
summary comparison 总结对比[比较]
summary flowsheet 简明[概略]流程
summary generation program 总计生成程序
summary of loan 借款概要
summary of project 工程[项目]概述
summary sheet 摘要[概略]表
summation 求和(法)
summation operation 求和运算
summation series 求和级数
summator 加法器
summer run-off 夏季径流(量)
sump level indicator 泵池液面指示器
sump pump 地面污水泵

sump water 泵池给水量
sump water addition rate 泵池加水量
sundry service 杂务
sundry stores 各种备品
sunlight hours 日照时数
sunny days 晴天数
super charge machine 加压充气(浮选)机
super heavy element 超重元素
super power station 特大功率电厂
super station 特大功率电站
supercharged machine 充气式浮选机
super-critical machine 超临界机械
super-critical screen 超临界筛
superficial air velocity 表面[层]空气流速(浮选机)
Superfloc 美国氰胺公司产絮凝剂品牌名
Superfloc 84 絮凝剂(高分子量聚丙烯酰胺)
Superfloc 127 絮凝剂(高分子量聚丙烯酰胺)
Superfloc 16 絮凝剂(聚丙烯酰胺,颗粒状白色固体,相当于 Separan MGL)
Superfloc 20 絮凝剂(聚丙烯酰胺,相对分子质量高于 Superfloc 16,低于 Superfloc 84)
Superfloc 330 絮凝剂(阳离子聚合物,可用于絮凝和助滤)
Superfloc A95~A150 美制阴离子型絮凝剂(号码越大,其阴离子性越强)
Superfloc C521~C581 美制阳离子型

絮凝剂
Superfloc flocculent　Superfloc 絮凝剂
Superfloc N100　美制非离子型絮凝剂
Superfloc N100S　美制非离子型絮凝剂
superfluous model　过多的模型
supergene alteration　表生变化
supergene sulfide material　浅成硫化矿物
supergene zone　表生带
superimposed breakage data　叠加破裂数据
superimposed pendulum test　迭加摆锤试验
superintendent　管理[监督,指挥]人员,段长,车间主任,部门负责人
superintendent of maintenance　维修车间主任
supernatant clarity　上清[澄清]液澄清度
supernatant liquid　上清[澄清]液
supernatant solution　澄清溶液
Supernatine S　絮凝剂(阴离子型磺化甘油酯)
Supernatine T　絮凝剂(阴离子型磺化甘油酯)
superpanner　淘砂机,超型淘金机
superscript　上标
supervision　监督[理],管理
supervision cost　管理费
supervision of installation　安装监督[管理]
supervisor office　管理办公室

supervisory computer　监控计算机
supervisory computer application　监控计算机应用
supervisory control　监督控制
supervisory engineering staff　技术管理人员
supervisory function　监督作用
supervisory multivariable control tool　监控多变量控制工具
supervisory personnel　监督[理]人员,检查人员
supervisory program　监督控制程序
supervisory set point　监督给定值
supplier of operating material　生产材料供应商
supplier selection　供方的选择
supplier's representative　供方代表
supplier's staff　供方人员
supplier's programming team　供方的编程团队
supplier's project engineer　供货单位项目工程师
supplier's proposal　供方的建议
suppliers selection　厂商选择
supply　供给,补给,提供,供应品
supply firm　供应厂商
supply of spare parts　备件供应
supply unit　供电设备
supply voltage　供电电压
support pressure　支点压[反]力
supporting electrolyte　协助[支持]电解质

supporting frame 支架
supporting structure 支撑结构
supporting system 辅助系统
suppressant combination 抑制混合剂
suppressing nucleation 抑制成核[集结]作用
suppression equipment （粉尘）抑制设备
surcharge 附加费,超载
surcharge angle 超载角
surcharge on wages 工资附加费用
surface 表面,表层,外观
surface acid group 表面酸基
surface active nature 表面活性性质
surface active organics 表面活化有机物
surface active reagent 表面活性剂
surface active substance 表面活性物质
surface activity 表面活性
surface affinity 表面亲和性
surface area 表面积[区域]
surface area estimate 表面积估算
surface area measurement 表面积测定[量]
surface area ratio 表面积比
surface binding energy 表面结合能
surface carbonation 表面碳化作用
surface characteristic of solid 固体表面性质
surface charge 表面电荷
surface chelate 表面螯合[物]
surface chemical behavior 表面化学行为
surface chemical composition 表面化学组成
surface chemical property 表面化学性质
surface chemistry 表面化学
surface complex 表面络合物
surface condition 表面状态
surface conductivity 表面导电[性]率
surface coverage 表面覆盖度
surface covering 表面覆盖(物)
surface cupric sulfide 表面硫化铜
surface drainage 地表水
surface drilling 地表钻探
surface electrokinetic property 表面电动性
surface electron 表面电子
surface energy 表面能
surface energy value 表面能值
surface enthalpy 表面焓
surface entropy 表面熵
surface estimation 表面估算
surface excess 表面超量
surface excess concentration 表面超量浓度
surface factor(SF) 表面系数(缩写 SF)
surface force 表面力
surface free energy 表面自由能
surface freshening 表面新鲜[淡化]（程度）
surface group 表面基团

surface hand sample 手采地表样
surface hydration 表面水合
surface hydration force field 表面水合作用力场
surface hydrophobic 表面疏水的
surface hydrophobic layer 表面疏水层
surface hydrophobic species 表面疏水产物
surface hydrophobicity 表面疏水性
surface hydroxyl 表面羟基
surface ion 表面离子
surface ionic force 表面离子力
surface iron carboxylate 表面羧酸铁
surface lattice 表面晶格
surface layer of reagent 表面药剂层
surface liquid 表面水
surface Madlung constant 表面马德伦[隆]常数
surface metal soap 表面金属皂
surface modification 表面改变
surface oxidation 表面氧化(作用)
surface oxidation product 表面氧化产物
surface oxidation reduction reaction 表面氧化还原反应
surface phenomena 表面现象
surface photovoltage 表面光电压
surface plant 地面选矿厂
surface polishing 表面磨光(程度)
surface pressure 表面压力
surface property 表面性质
surface reaction product 表面反应产物
surface reaction product species 表面反应产物类别
surface relaxation 表面松弛
surface roughness 表面粗糙度
surface sample 地表[表面]样品[试样]
surface species 表面产物
surface state energy 表面状态能
surface structure 表面结构(结晶面的)
surface sulphur 表面硫
surface tensiometer 表面张力计
surface tension 表面张力
surface tension depressant 表面张力抑制剂
surface tension-concentration gradient 表面张力-浓度梯度
surface treatment method 表面处理方法
surface variance 表面方差
surface water 地表水
surface water system 地表水系
surface-active 表面活性[化]
surface-active counter ion 表面活性配衡离子
surface-inactive 表面非活性[化]
surface-nucleated precipitation 表面成核沉淀
surfactant 表面活性剂
surfactant additive 表面活化添加剂
surfaction 表面改性(质)

surfaction ion 表面活化离子
Surfonic 非离子型表面活性剂品牌名
Surfonic N-120 壬基酚聚乙氧烯基(12)醚 化学结构式：

R—⟨⟩—O(CH_2CH_2O)$_n$H,
R=C_9烷基, n=12

Surfonic N-40 壬基酚聚乙氧烯基(4)醚 化学结构式：

R—⟨⟩—O(CH_2CH_2O)$_n$H,
R=C_9烷基, n=4

Surfonic N-60 壬基酚聚乙氧烯基(6)醚 化学结构式：

R—⟨⟩—O(CH_2CH_2O)$_n$H,
R=C_9烷基, n=6

Surfonic N-95 壬基酚聚乙氧烯基(9)醚 化学结构式：

R—⟨⟩—O(CH_2CH_2O)$_n$H,
R=C_9烷基, n=9~10

surge chamber 调节[缓冲]室
surge facilities 缓冲设施
surge plate 涌浪挡板
surge pump 涌浪[膜式,隔膜]泵
surge tank 均[恒]压箱,缓冲槽
surplus power 剩余功率[动力]
survey report 调查报告
surveying instrument 测绘仪器
susceptor 感受[接受,衬托]器,基座
suspended dust 悬浮粉尘
suspended state 悬浮状态
suspended velocity 悬浮速度

suspension condition 悬浮条件[状态]
suspension density 悬浮浓[密]度
suspension effectiveness 悬浮效率
SUS Unit (SUS=Saybolt Universal Second) 赛波特通用黏度秒数(单位)
sweep behavior 扫描特性
sweep peak 扫描峰值
sweep rate 扫描率
sweep voltammetry curve 扫描伏安曲线
sweet ore 无[脱]硫矿石
swimming pool filter 游泳池[水隔离]型过滤机
swing jaw 动颚板
swing valve 旋转阀
swinging cut-off gate 摆动断流闸门
swirl cyclone 重介质旋涡旋流器
switch unit 转换开关,开关装置
switchable pneumatic fines handling system 可换向风力[压气]细粒输送系统
switching system 转接[开关]系统
swivel adjustment cable 旋转环调整钢绳
swivel belt 旋转[臂]胶带
symbiotic relationship 共生关系
symbolism 符号体系
symmetric axis 对称轴
symmetric stretching mode 对称伸缩模式
symmetrical electrolyte 对称电解质
Symons cone crusher 西蒙斯圆锥破碎

机(是美国密尔沃基西蒙斯兄弟二人设计的,故得名西蒙斯圆锥破碎机)
Symons short-head crusher 西蒙斯短头圆锥破碎机
Symons short-head crusher application information 西蒙斯短头圆锥破碎机使用资料
Symons single deck vibrating screen 西蒙斯单层振动筛
Symons standard cone crusher 西蒙斯型标准圆锥破碎机
symposium (专题)讨论会,(专题)论文(讨论)集,(专题)论丛
symposium co-chairman (专题)讨论会[座谈会]联席主席
symposium on mill design 选厂设计会议
synchro-drive variable speed drive 同步传动的变速驱动装置
synchro-generator 同步发电机
synchronizer 同步器
synchronous 同步
synchronous direct drive motor 同步直接驱动电机
synchro-receiver 同步接收器
synchro-transformer 同步控制变压器
synchrotransmitter 同步发送器
synfuel division 合成燃料处[部]
Syntex 辛太克斯[译](美国 Syntex 公司生产的表面活性剂品牌名)
Syntex L 辛太克斯 L(椰子油脂肪酸的磺化单甘油酯钠盐),可用作选矿药剂(辉钼矿捕收剂)
Syntex mobility determination 辛太克斯淌度测定
synthetic cassiterite 合成[人造]锡石
synthetic cloth 合成滤布
synthetic copper sulfide 合成硫化铜
synthetic cryolite 合成冰晶石
synthetic ferric oxide 合成氧化铁
synthetic fiber 合成纤维
synthetic filter bag fabric 合成滤袋布
synthetic mixture 人工混合矿
synthetic organic phosphate 合成有机磷酸盐
synthetic rubber 合成橡胶
synthetic salty mineral 合成[人造]盐类矿物
synthetic solution 合成溶液
Syntron F88 vibrating feeder Syntron F88 振动给矿机(美国 Syntron 公司产品)
Syntron feeder Syntron 给矿机(美国 Syntron 公司产品)
system 制度,体制
system acceptability 系统可接收[合格]性
system availability 系统的[可]利用率
system behavior 系统性能[特性]
system configuration 系统配置
system curve 系统曲线
system design data 系统设计数据
system energy loss 系统能量损失
system engineering 系统工程

system head 系统压头
system interaction 系统互相作用[影响]
system management capacity 系统管理能力
system of enclosure and dust collection equipment 封闭[密封]集尘设备系统,封闭[密封]罩和集尘设备系统
system of equation 方程组
system pressure curve 系统压力曲线
system resources 系统资源(计算机)
system software 系统软件(计算机)
system stability 系统稳定性
System's behavior 系统(工作)情况,系统(运转)状态
systematic analytic means 系统分析法
systematic error 系统误差
systematic fashion 系统方式
systematic manner 系统方式
systematic research 系统研究

T

T.D.A. 亚硫酸盐废液(品名,可用作氧化铁矿的抑制剂)
T matrix T矩阵
table 用摇床选矿(动词),床选
table below 下表
table of cost versus equipment parameter 投资费用与设备参数对照表
tabular method 制表方法
tabular vibrating conveyor 平板振动输送机
tacho 速(度),转速计[器]
tachometer 转[流]速计
tack welding 平头焊接
tactical stage 战术阶段[步骤]
tactite deposit 接触岩矿床
Tafel slope 塔费尔斜率
tail assay 尾矿品位[金属含量]
tail brake 尾部制动(胶带运输机的)
tail sprocket 尾部链轮
tailing assay 尾矿品位[金属含量]
tailing discharge line 尾矿排矿[出]管
tailing impoundment system 尾矿蓄水系统
tailing reclaim water pump 尾矿回水泵
tailing retreatment plant 尾矿再处理厂
tailing sampling box 尾矿取样盒
tailing slurry distribution system 尾矿浆分配系统
tailings 尾矿,筛余物
tailings backfill plant 尾矿回填车间

tailings disposal area 尾矿场[处理区]
tailings disposal facilities 尾矿处理设施
tailings disposal system 尾矿处理系统
tailings grade 尾矿品位
tailings impoundment 尾矿(坝)蓄水,尾矿水
tailings impoundment area 尾矿蓄水区
tailings pond 尾矿池
tailings pump 尾矿泵,浸渣泵
tailings pump house 尾矿泵站[房]
tailings sump 尾砂泵池
tailings system 尾矿系统
tailings thickener 尾矿浓缩机
tailings treatment plant 尾矿处理厂
tailings vent fraction 尾矿出口部分
tailings-dam area 尾矿坝区
tails line 尾矿管线
tails site 尾矿场
take-away conveyor system 排矿运输机系统
take-up assembly 拉紧装置组合
take-up movement 拉紧装置移动距离[行程]
talk on telephone 电话会谈
tall oil crude 粗塔尔油(可用作捕收剂)
tall oil distilled 蒸馏塔尔油(可用作捕收剂)
tall oil fatty acid 塔尔油脂肪酸
tallow 牛脂,动物油脂

tallow amine acetate 醋酸胺脂,牛脂胺醋酸盐
tallow fatty acid 牛[骨]脂酸,动物脂肪酸
tallow-based amine 硬脂胺
Tallso 粗制塔尔油皂(品名,可用作捕收剂)
tandem car dumper 双联翻车机,串联翻斗车
tandem tunnel 串联的地[隧]道
tangent 正切(的),切线(的)
tangent rule 切线法则
tangential entrance 切线进口
tangential feed 切线给矿(旋流器)
tangential feed design 切线给矿设计[设计图,图样,结构,形状]
tangential inlet 切线入口
tangible asset 有形资产
tank wall 池[罐,槽]壁
tanker 罐车,油船
tannin 单宁(鞣质),单宁酸,鞣酸(可用作抑制剂、絮凝剂、分散剂、调整剂、湿润剂、锰矿浮选时作为硅酸盐脉石的抑制剂)
tannin reagents 丹宁类药剂
tannin-oleate clathrate 丹宁(酸)油酸包合物
taoered slot 锥形截槽
tap water 自来水
tape length 胶带长度(胶带运输机的)
taper key 楔键,斜键,锥形键
tapered cavity 锥形破碎腔

tapered flight 锥形螺旋叶片
tapered profile design 锥形截面设计
tapering main shaft 锥形[带梯度]主轴
tapioca powder 木薯(淀)粉
tapioca starch 木薯淀粉(可用作抑制剂絮凝剂)
tar-coated vat pit 有焦[柏]油层的槽浸坑
tare weight 皮重,[包装]空车重量
target 指标,目标,靶子
target grade 计划[目标]品位
target investigation stage 目标勘测[探查]阶段,地质勘探阶段
target mineral 目的矿物
target recovery 回收率指标
target set 目标给定
tariff 关税
tarmac 煤焦油(可用作捕收剂、起泡剂)
tarnished film 锈[氧化]膜
tarnished lead ore 生锈[表面氧化]铅矿石
tarnished non-floatable gold 生锈[带污垢](失去光辉)的不能浮起的金
tarnished oxidized molybdenite 生锈[带污垢]的氧化辉钼矿
tarnished sulfide 表面氧化[生锈](失去光泽)的硫化矿
tarnished sulfide ore 表面氧化[生锈]的硫化矿石
tartaric acid 酒石酸(可用作石英抑制剂及 pH 值调和剂) $(COOH)_2(CHOH)_2$
task force 工作组,任务班子,特遣部队
task force personnel responsibility 特别工作组人员职责
task forces 特别工作组,特别任务班子,特遣部队
task scheduler 任务调度程序
taste threshold 味觉阈
tax authorities 税务机关[管理局]
tax deduction 减免税款,税收扣除(额)
tax rate 税率
tax return 所得税申报(单)
tax statement 纳税申报
taxable profit 应税利润
taxation 征税
taxation law 税收[征税]法
taxation liability 税款债务
Taylor series 泰勒级数
TBE(tetrabromoethylene) 四溴乙烷(重液)$Br_2CH-CHBr_2$
TDPB(tetradecyl pyridinum bromide) 十四烷基溴化吡啶(可用作捕收剂)
technical and economic feasibility 技术经济可行性
technical and economic objective 技术经济目标
technical and economic summary 技术经济要览
technical and operating personnel 技术操作人员

technical annual meeting 技术年会
technical assistance 技术支援
technical assistance representative office 技术服务办事处
technical base 技术基础
technical characteristics 技术特性
technical contact 技术接触
technical development 技术发展[开发]
technical information 技术资料
technical input 技术输入[进口,引进]
technical language 技术语言
technical limitation 技术限制
technical literature 技术文献
technical manual 技术手册
technical operating personnel 技术操作[生产]人员
technical paper 技术论文
technical people 技术人员
technical performance 技术性能
technical publication 技术刊物
technical quality 技术质量
technical risk 技术风险
technical service 技术性服务
technical service department 技术服务部门
technical specification 技术规格[说明书]
technical support 技术援助[支援]
technical term 技术术语
technically advanced mathematic model 技术先进的数学模型

technologic(al) 工艺(学)的,(科学)技术的,因工业技术发展而引起的
technological advancement 技术进步
technological base 技术基础
technological control system 技术控制系统
technological innovation 技术革新[改革,创新]
technological limitation 技术限制
technological process 工艺过程
technological progress 技术进步
technology 技术,工艺(学),(工业,生产)制造学
technology agreement 技术协议
technology evolution 技术进[发]展
technology import tax 技术进口税
technology of metals 金属工艺学
Teepol(=Tepol) 仲烷基磺酸钠(阴离子表面活性剂品名,中文名称:涕波尔,可用作起泡剂、捕收剂、抑制剂)
teeter chamber 搅拌室
teeter column 摇摆柱[室,塔](分级用)
teeter layer 搅拌层
teeter zone 摇摆化流态区
Teflon 聚四氟乙烯(塑料,绝缘材料),特氟隆
Teflon coated magnet 聚四氟乙烯覆盖磁铁
Teflon cup 聚四氟乙烯杯
Teflon stopcock 特氟隆活栓
Teflon stopcock manifold 特氟隆活栓

歧管

teleconference 电话会议

telephone conversation 电话通话

telephoto lens system 远距照相镜头[物镜]系统

teletype terminal 电传打字机终端

television monitor 电视监控器

television monitoring 电视监控

telex 电传

telinite 结构凝胶体

Telsmith cone crusher 特尔史密斯型圆锥破碎机(美国 Telsmith 公司)

Telsmith gyrosphere crusher 特尔史密斯型回转球面式破碎机(美国 Telsmith 公司)

Telsmith heavy-duty vibration grizzly 特尔史密斯重型振动格筛(美国 Telsmith 公司)

Telsmith type cyclone spherical surface crusher 特尔史密斯型旋回球面破碎机(美国 Telsmith 公司)

temperature adjustment 温度调整

temperature change 变温,温度变化

temperature dependence 温度依赖性

temperature of conditioning 调整温度

temperature probe 温度探针

temperature related test 与温度有关的试验

temperature reversibility 温度可逆性

temperature variation 变温,温度变化

temporal response 时间[瞬态]响应

temporary control panel 临时性控制屏

temporary power supply 临时供电

temporary water supply 临时供水

tender 看管人,招投标,提议

tender document 投标文件

tender price 投标价格

ten-fold range 十倍范围[幅度]

Tensatil DA120 辛(烷)基硫酸酯[盐](品名) $C_8H_{17}OSO_3M$

tension plate 拉[张]紧板

tension rail 横拉杆[条]

tension rating 张力额定值

tension utilization 张力利用率

tentative crusher selection 破碎机试探性[暂时性]选择

tentative installation angle 试用[暂时]安装角

tentative size 试用规格

terebene oil 松节油

terebenthene 松节油

Tergitol 表面活性剂品牌名

Tergitol 08 2-乙基己基硫酸钠(可用作湿润剂)
$CH_3(CH_2)_3CH(C_2H_5)CH_2OSO_3Na$

Tergitol 4 仲—十四烷硫酸钠(可用作湿润剂)

Tergitol 7 仲—十七烷硫酸钠(可用作湿润剂)

Tergitol NP-14 壬基酚聚氧乙烯醚(可用作润湿剂)化学结构式:

R—⟨⟩—$O(CH_2CH_2O)_nH$,

Tergitol NP-27 壬基酚聚氧乙烯醚（可用作润湿剂）化学结构式：

R—⟨⟩—O(CH$_2$CH$_2$O)$_n$H,

R=C$_9$烷基, n=7

Tergitol NP-35 壬基酚聚氧乙烯醚（可用作润湿剂）化学结构式：

R—⟨⟩—O(CH$_2$CH$_2$O)$_n$H,

R=C$_9$烷基, n=15

Tergitol P-25 2-乙基己基膦酸钠（可用作捕收剂）

[C$_4$H$_9$CH(C$_2$H$_5$)CH$_2$O]$_2$P(O)ONa

term 术语,专门名词;期间,限期;任[学]期,条款

term bank loan 定期银行贷款

term loan agreement 限期贷款协议

terminal 末端,终点,终端机

terminal application 终端应用

terminal assembly 端部组合[装置]

terminal free falling velocity of particle 颗[矿]粒自由沉降末端速度

terminal pulley 端轮,两端滚筒（胶带机首轮,尾轮）

terminal pulleys 首尾轮

terminal settling velocity 沉降末端速度

terminal usage 终端应用

terms of reference 职责[权]范围

ternary mixture 三元混合矿

terpenol 萜烯醇

terpineol 萜烯醇,松油[萜品]醇(可用作起泡剂) C$_{10}$H$_{18}$O

terpineol thiol 萜烯硫醇(可用作起泡剂)

terrace 台地,平台

tertiary amine salt 叔胺盐

tertiary ammonium salt 叔胺盐

tertiary amyl alcohol 叔戊醇,特戊醇（可用作起泡剂）C$_2$H$_5$·C(CH$_3$)$_2$OH

tertiary concave 三段破碎机固定锥体

tertiary crushing and screening circuit 第三段破碎筛分流程[回路]

tertiary crushing circulating loading conveyor 细碎循环负荷运输机

tertiary mantle 三段破碎机可动锥体

tertiary sampling machine 第三次取样机

test accuracy 试验准[精]确性

test and analysis method 测试分析方法

test and measurement 测试

test and study 试验研究

test condition 试验条件[情况]

test data 试验数据

test data sheet 试验数据表

test facilities 试验设施

test heap 试堆

test liner 试验衬板

test material 试验物料

test mill 试验磨机

test method 试验方法

test mill power 试验磨机功率

test number 试验号

test of significance 显著性检验
test of statistical significance 统计显著性检验
test period 试验周期
test piece 试件,试样,制取试样的金属块
test plan 试验计划
test procedure 试验程序[步骤,方法]
test program(me) 试验项目[计划,程序]
test result 试验结果
test sample 试验样品
test sample lots 试验用样批次
test scheme 试验方案
test sensitivity 试验灵敏度
test series 试验组[批,系列]
test sieve size 试验筛孔尺寸
test unit 试验装置
test variable 试验变量
test work 试验工作
tested deck 试验筛(面)
testing equipment 试验设备
testing facilities 试验设施
testing of composite 组合样试验
testing of ore 矿石试验
testing program 试验程序方案[大纲,计划]
testing purpose 试验目的
tetrabromophenol sulfonphthalein (= bromophenol blue) 四溴苯酚磺酞,溴酚蓝
tetragonal prism 正[四]方柱,四方[角]棱柱
tetragonal structure 正[四]方晶体结构,四方[角]结构
tetrahedral coordination 四面体配位
tetrahedrally bonded 四面键合的
tetrahedrally coordinated 四面体配位的
tetrahedrite 黝铜矿
tetrahedrite production 黝铜矿生产
tetrahydroabietyl amine 四氢松香胺(可用作捕收剂)
tetrahydronaphalene 四氢萘
tetrasodium ethylenediamine tetraacetate 乙烯二胺四醋酸四钠,乙二胺四乙酸四钠
tetra sodium pyrophosphate 焦磷酸(四)钠(可用作分散剂)$Na_4P_2O_7$
Texafor Cl 油酸与环氧乙烷缩合物(品名,可用作起泡剂)
Texapon 表面活性剂品牌名
Texapon extract A 十二烷基硫酸铵(可用作湿润剂)$C_{12}H_{25}OSO_3NH_4$
Texapon extract T 十二烷基硫酸三乙醇胺盐(可用作湿润剂) $C_{12}H_{25}OSO_3H \cdot N(CH_2CH_2OH)_3$
Texapon W 十二烷基硫酸盐(可用作湿润剂) $C_{12}H_{25}OSO_3M$
Texapon Z 伯—烷基硫酸盐(可用作湿润剂)
text 文本,正[原]文
textile industry 纺织工业
textile mill waste 纺织厂废弃物

textural data 结构数据
textural description 结构描述
texture 结构,构造,组织,质地,纹理
thallium 铊 Tl
thallium butyl dithiophosphate 丁基二硫代磷酸铊
thallium ethyl dithiophosphate 乙基二硫代磷酸铊
thallium isoamyl dithiophosphate 异戊基二硫代磷酸铊
thallium m-cresyl dithiophosphate m-甲苯基二硫代磷酸铊
thallium methyl dithiophosphate 甲基二硫代磷酸铊
thallium o-cresyl dithiophosphate o-甲苯基二硫代磷酸铊
thallium p-cresyl dithiophosphate p-甲苯基二硫代磷酸铊
thallium phenyl dithiophosphate 苯基二硫代磷酸铊
thallium propyl dithiophosphate 丙基二硫代磷酸铊
thallium salt 铊盐
the best scheme 最好的方案
the requirement from design [task book] 设计[任务书]要求
theoretical analysis 理论分析
theoretical aspect 理论方面
theoretical attempt 理论尝试
theoretical basis 理论基础
theoretical calculation 理论计算
theoretical composition 理论组成

theoretical consideration 理论研究[见解,理由]
theoretical curve 理论曲线
theoretical force 理论力
theoretical formula 理论公式
theoretical investigation 理论研究[探索]
theoretical isotherm 理论等温线
theoretical laboratory result 实验室理论试验结果
theoretical model 理论模型
theoretical model of mineral processing 选矿理论模型
theoretical photovoltage 理论光电压
theoretical recovery 理论回收率
theoretical relationship 理论关系
theoretical research 理论研究
theoretical screen separation 理论筛分
theoretical study 理论研究
theoretical value 理论值
theoretical viewpoint 理论观点
theoretical worth 理论价值
theoretically available value 理论计算[有效]值
theoretically infinite particle size 理论上无穷大的粒度
thermal conductivity 导热系数,热导率
thermal device 热力装置
thermal drier 火[热]力干燥机
thermal drying 热[火]力干燥
thermal engineering 热工学

thermal gradient 热[温度]陡[梯]度,温度坡差
thermal meter 热工[电式]仪表,热能表
thermal power 火电,热能,热功率
thermal power generation equipment 火力发电设备
thermal power plant 热电厂
thermochemical study 热化学研究
thermocouple-actuated controller 热电偶驱动控制器
thermodynamic calculation 热力学计算
thermodynamic driving power 热力学驱动力[功率]
thermodynamic equilibrium calculation 热力学平衡计算
thermodynamic information 热力学信息
thermodynamic property 热力学性质
thermodynamic quantity 热力学量
thermodynamic study 热力学研究
thermodynamics of flotation 浮选热力学
thermodynamics of surface 表面热力学
thermodynamics of wetting 湿润热力学
thermoelectric instrument 热电式仪表
thermostatically controlled water bath 恒温控制水槽[浴器]
thick overflow 浓溢流

thick overflow slurry 浓溢流矿浆
thick wall rubber tube 厚壁胶管
thickened deslimed pulp 已浓缩脱泥的矿浆
thickened tailings 浓密[缩]尾矿
thickener area 浓缩机面积
thickener area requirement calculation 浓缩机面积需要量计算
thickener rake 浓缩机耙齿
thickener rake gear 浓缩机耙子齿轮(装置,机构)
thickener underflow 浓缩机底流
thickener underflow percent solids 浓缩机底流固体百分比[数]
thickening area 浓密[缩]面积
thickening device 浓缩装置
thickening operation 浓缩[密]作业
thickening test 浓缩试验
thickness deviation 厚度偏差[误差]
thickness variable 厚度变量
thickness variation 厚度变化[偏差]
thick-pulp conditioning 稠矿浆调整[搅拌]
thin anode 薄阳极
thin film phenomena 薄膜现象
thin line 细线
thin section 薄片[剖面,切片]
thin solid film 固体薄膜
thio- 硫,硫代(词头)
thioarsenic compound 硫代砷化合物
thiocarbamate 硫代氨基甲酸盐[酯] $NH_2C(S)OM$(或 R)

thiocarbamide(=thiourea) 硫脲 H_2NCSNH_2

thiocarbanilide 白药(二苯基硫脲,可用作捕收剂)

thiocarbanilide 130 一种易润湿及易分散的白药(可用作捕收剂)

thiocyanate 硫氰酸盐[酯]

thiol 硫醇(类)

thiol anion 硫醇阴离子

thiol collector 硫醇类捕收剂

thiol polar group 硫醇极性基团

thiol reagent 硫醇类药剂

thiolate 硫醇盐,烃硫基金属

thiol-type anion 硫醇类阴离子

thiol-type surfactant 硫醇类表面活性剂

thionate 硫代硫酸盐[酯]

thiono- 硫羰=CS,硫逐(硫代 CO 中之 O,当与"硫代"有区别之必要时用此词头)

thionocarbamate 硫逐[羰]氨基甲酸酯

thionocarbonate 硫代氨基甲酸酯(硫代碳酸酯,可用作捕收剂)

thiophosphorus compound 硫代磷化合物

thiosalt-producing potential 生成硫代盐的潜力[可能性]

thio succinamate 硫代琥珀酰胺酸盐

thiourea(=thiocarbamide) 硫脲 $(NH_2)_2CS$

third cleaner 第三段[次]精选

third generation analog device 第三代模拟设备

third power 三次幂

third shift 第三班

third theory energy calculation 第三理论破碎能量计算

third theory of comminution 第三破碎理论

third-phase 第三相

third-stage flotation 第三段浮选

Thomas J-36 pump 托马斯 J-36 泵

Thomas roll crusher 托马斯辊式破碎机(第一台辊式破碎机是英国人 A.Thomas 发明的)

Thompson procedure 汤普逊程序[方法]

Thompson theory 汤普逊理论

thorough and systematic manner 周密[缜密,详细]和系统的方式

thorough consideration 充分[严密,周到]考虑

thorough test program 详细[周到,严密]的试验程序

thread count 纱支[织物经纬]密度,线程数

three bladed impeller 三叶片叶轮

three digit facility code 三位数设备代号[编码]

three digit number 三位数号码

three dimensional data 三维数据

three dimensional model 三维模型

three dimensional sampling 三维取样

three lattice ion 三晶格离子
three stage counter-current decantation 三段逆流倾析
three stage counter-current washing 三段逆流洗涤
three stage ore reduction 三段碎矿
three stage reduction 三段碎矿
three way tap 三通旋塞
three-leaf clover 三叶草形
three-liter flotation cell 三升浮选机[池,槽]
three-mineral system 三种矿物组织,三矿物系统
three-pen strip chart recorder 三笔带式图表记录仪
three-phase 三相
three-phase contact 三相接触
three-phase frothing test 三相起泡试验
three-phase pulp 三相矿浆(气,液,固相)
three-port flotation cell 三端口浮选槽
three-position blend knob 三位置的混合按[旋]钮
three-roll idler set 三辊式托辊组
three-shift per day basis 每日三班制
three-shift system 三班制
three-stage approach 三(个阶)段法
three-stage counter-current carbon adsorption 三段逆流碳吸附
three-stage crushing plant 三段破碎厂
three-stage flotation method 三段浮选法
three-stage sampling system 三段取样装置[系统]
three-stage vertical turbo pump 三级立式透平泵
three-way flop gate 三通卸料闸门
three-way pipe 三通管
three-way stopcock 三通活栓[旋塞],三通阀
threshold hydrophilicity 临界亲水性
threshold of liberation 解离界限
throat liner 入口衬板
throughput 处理量,通过量[速度]
throughput method 通过量法
throughput rate 处理量[速度],吞吐率
throughput ratio 通过[生产]率,吞吐率,流量比
throughput vs. particle size curve 处理量与粒度关系[对比]曲线
throughs 筛下物
throw 投掷,偏心距(离)
throwaway product 废产品,废品
thrust 逆断层,止推,推力
thrust plate 止推[推力]板,冲断层板状体
Tibalene AM (一种法国产硫酸化或磺化脂肪酸,可用作捕收剂)
ticket pickup station 取票[牌]站
tie-in 捆成束,打结,相配,连接
tight restriction 苛刻限制

tight side 紧边
tight side drive tension 紧边驱动张力(带式运输机的)
tighter specific product 更严格特定产品
tightly closed container 密闭[封]容器
Tilden Concentrator 蒂尔登选矿厂(选氧化燧铁岩)
tilt switch 杠杆开关
tilting frame 翻转式跳汰盘
tilting vehicle 倾斜式载重[运输]车辆
time constant 时间常数
time consuming and expensive procedure 耗时且昂贵的办法
time curve 时间曲线
time cycle 时间周期
time delay 延时,时间延迟
time delay element 延时元件
time estimate 时间估计[值]
time factor 时间因素
time flow function 时间流(动)函数
time flow of profit 利润时间流量
time frequency 时间频率
time of production 生产期间
time period 时间周期
time sharing system 分时系统
time smoothing 时间平滑[修匀]
time study man 工时研究员,工序时间研究员
time value of money 金钱[货币]的时间价值
time-consuming 时间消耗

time-consuming procedure 耗时过程
timed sampler main frame 定时取样器主机架
timed solenoid valve 定时电磁阀
time-intensity relationship 时间-强度关系
timer 定[计]时器,跑表
timer controlled solenoid valve 计时器控制电磁阀
timer period 时标周期
time-recovery relationship 浮选时间与回收率的关系
time-sharing terminal system 分时终端系统
time-surface curve (磨矿)时间与(矿石)表面积关系曲线
timing function 定时功能
timing operation 定时操作
timing sampling 定时取样
tin fuming 锡烟化
tin yield 锡生产率
tine 耙齿
tin-salicylaldehyde chelate 锡-水杨醛螯合物
tin-salicylaldehyde compound 锡水杨醛化合物
tiny fleck of metal 金属微(细)粒(子)
tiny grain 极细颗粒
tipper vehicle 倾卸式载重[运输]车辆,翻斗车
tire driven 轮胎传动的
tire vulcanizing machine 轮胎热补机,

轮胎(橡胶)硫化机 控制
titanium dioxide pigment 钛白粉颜料
title 标题,题目,名称,头衔,所有权,权利
title of all managerial staff 全体管理人员职称
title page 扉[书名]页
titre(=titer) 脂酸冻点,滴定度[率]
titrimetic method 滴定法
to use(period) test 使用(期)试验
today's economic situation 当前[今天,现在]的经济形势
today's metal price 今天的金属价格
toe problem 坝趾问题
tolerance limit 允许限度,公差极限,耐受界限
tolerant plant 耐药力植物,耐阴[病]植物
toluene 甲苯 $CH_3C_6H_5$
toluene arsenic acid 甲苯亚砷酸
toluol 甲苯 $CH_3C_6H_5$
toluol arsonic acid 甲苯胂酸
toluyl(=toluoyl)(o,m 或 p) 甲苯酰基,甲苯甲酰(邻、间或对位),$CH_3C_6H_4CO-$
tolyl(o,m 或 p) 甲苯基(邻、间或对位),$CH_3C_6H_4-$
tolyl-triazole 甲苯三唑(可用作硅孔雀石的捕收剂)
tonnage calculation 吨位计算
tonnage control (给矿)吨位[吨数]

tonnage conversion factor 吨位换算系数
tonnage estimate 吨数估算
tonnage of liquid 液体吨位[数]
tonnage of pulp 矿浆吨位[数]
tonnage of solid 固体物料吨位[数]
tonne (公)吨(1吨=1000公斤)
tool crib 工具间
tool outfit 成套工具
tool-steel industry 工具钢工业
tooth-paste 牙膏
top cover 顶盖,(胶带)顶[面]层
top impeller 顶叶轮
top of pool 沉降区顶部,溢流池面
top screen 上层筛
top shell 顶罩
top size 最大粒度
top size feed 上限[最大]粒度给矿
top suspension 顶部悬挂[吊]
top unloader 顶部[上部]卸载机
topic 题目,课[论,主,标,话,专]题
topography 地形[势]
topological transformation 拓扑变换
topology 拓扑(学,结构),(集成电路元件)布局,地志学
topsoil 表土
torch ring 管形圈
torque arm 扭矩臂
total amount 总金额
total amount of funds 资金总额
total annual operating supply expense

年度生产用品总费用
total average 总平均数(值)
total cash outflows 现金总支出(额)
total cash receipts 现金总收入(额)
total circuit performance 整个回路指标
total current assets 共计流动资产
total dissolved solids 溶解固体总量
total dynamic head 总动压头
total dynamic head of water 水的总动压头
total estimated capital cost 总估算基建成本
total figure 总数值[字]
total installed horse power 总安装马力
total interaction energy 相互作用总能
total interaction potential energy 相互作用总位能
total maintenance labor cost 总维修劳动费
total material lot 材料总量
total net operating cost 总的净生产费用
total organic carbon 总有机碳量
total organic carbon analysis 有机碳总量分析
total physical plant cost 实际选矿厂总费用
total plant recovery 工厂总回收率
total process equipment cost 工艺设备费总额

total product cost 总产品费用
total project cost 总工程费用
total recovery 总回收率
total surface energy 总表面能
total system 总体系统
total tonnage 总吨位[数]
total value of industrial output 工业总产值
totalizer 加法器
tote bin 运输吊斗,搬运箱
tote box 吊桶,提桶,工具箱
tower mill 塔式磨机
toxic hydrogen sulfide gas 有毒的硫化氢气体
toxicity data 毒性资料[数据]
toxicity information 毒性资料
trace analysis technique 痕分析方法[技术]
trace element 痕[微]量元素
trace quantity 痕[微]量
tracer solution 示踪溶液
tracer test 示踪试验
tracer test procedure 示踪试验法
tracing 描图,跟踪
tracing machine 描图机,电子轨迹描绘器
track alignment 轨道方向
track scale 轨道衡[地秤]
tracking error (运输机)铺轨道误差
tracking sensitivity 轨道灵敏[敏感]性
trackless haulage equipment 无轨运输设备

trackless rubber tired vehicle 无轨橡胶轮胎车辆
trackless vehicle 无轨矿车(坑内装载和运输)
traction drive 牵引驱动
tractive power curve 牵引功率曲线
trade credit 贸易信贷
trade wind 信风
traditional approach 传统方法
traditional chemical reaction theory 传统化学反应理论
traditional firm 传统企业
traditional method 传统方法
traditional practice 传统[习惯,通常]作法
traffic light 交通灯
tragacanth 黄蓍胶,黄芪胶
Tragol-4 一种德国产煤焦油起泡剂
trailer camp (可用拖车牵引的矿工宿舍,活动房营地,汽车营地)
trained crew 训练有素的[操作]人员
trained mineralogist 训练有素的矿物学家
training course 培训课程
training facilities 培训[训练]设施
training operator 培训操作工
training staff 培训人员
training technique 训练技术
trajectory of discharge 排矿流轨[轨迹],排矿抛射线
tramming capacity 运载能力
tramming distance 运输距离

tramp 错配物(如精煤中的高比重物,废渣中的低比重物,筛上产品的细粒,筛下产品中的超粒等)
tramp iron magnet 除铁器,磁力吸铁器
tramp material 外来杂质,混入物
tramp material detection 异[错配]物探测
tramp oversize material 混入的超粒级物料
tramp protection screen 除屑筛
tramp screen 错配物筛
trampoline effect 蹦床效应
tranquil flow 缓流
transaction loan 交易贷款
transactions 会[学]报,会刊,报导,会议录
transcript 手[底,原]稿,抄本,副本
transducer 变[转]换器,转换机构,换能[流]器,变流[频]器,传感器,变送器,发射器
transfer belt 转载[运]胶带运输机,传送带
transfer belt conveyor 转载[运]胶带运输机
transfer box 转送箱,传动箱
transfer conveying device 转运装置(物料堆场装载输送系统中的一部分)
transfer energy 转移能,传递能量
transfer equipment 转运[传送]设备
transfer line 转运管线
transfer of charge 电荷转移

transfer point 转载点
transfer station component 转运站部[组]件
transfer tower 转运站[塔]
transferred equation 变[转]换方程式
transformation matrix 变换矩阵
transformed variable 折算变量
transformer oil 变压器油
transient operating condition 过渡运行情况
transient performance 过渡过程特性,瞬时[态]特性
transition hopper 过渡漏斗
transition radius 过渡半径,转位半径
transitional zone 过渡带
transloader 转载机,装运机
transmission case 变速箱,传动箱
transmission electron microscopy 透射电子显微术,透射电镜术
transmission equipment 传动[输]设备
transmission power 传动功率
transmission spectra 透射光谱
transmission type beta radiation gauge 透射式β(射线)辐射仪
transmissivity 透射率
transmitted radiation 透射的辐射线
transparent hutch 透明筛下室
transport characteristic 传输特性
transport cost 运输费
transport facilities 运输设施
transport lag 传输滞后
transport lag time 传输滞后时间

transport tonnage 运输吨位[数]
transportation cost 运输费用
transportation insurance premium 运输保险费
transversal belt conveyor 横向胶带运输机
Transweigh conveyor scale Transweigh运输机皮带秤
trapped gravity line 加闸自流管线
trapped mineral 捕获的桶集的矿物
trapping effect 俘获效应
trapping rate constant 俘获率常数
trash 碎屑,废料
trash box 废料槽[箱],垃圾[渣滓]箱
trash screen 隔碴筛
Traube's rule 特劳贝规则(有机物溶液的表面张力随溶液浓度的变化的经验规则)
travel 旅行,行程,移[运]动
travel arrangement 旅游安排
travel of material 物料移动
traveling baffle plate 移动式挡[阻]板
traveling belt conveyer 移动式胶带运输机
traveling carriage 行走车架,移动(运转机)支座
traveling rotary plough feeder 移动式回转刮板[犁]给矿机
traveling winged stacker 移动式翼臂堆料机
traversing system 横移装置[系统]
tray belt 槽型带

tray concentrator	槽[盘]式选矿机
tray dryer	盘式干燥机
Traylor gyratory crusher	特雷勒旋回破碎机(美国 Fuller 公司的设计制造技术)
treasury stock	库存股份
treated barren bleed	经处理的贫[废]液排放
treated effluent	经处理的废水
treatment method	处理方法
treatment option	处理方案(供选择的)
treatment procedure	处理[选矿]方法[程序]
treatment rate	处理量,处理速率[度]
treatment system	处理系统(水质处理)
treatment technique	处理方法[技术]
treatment tonnage rate	处理吨位(数量)
trench sample	探槽[槽沟]矿样
trench sampling	探槽[槽沟]取样
trend surface analysis	趋势表面分析法
Trent process	特伦特制团法(液中造粒法,全油浮选细末团煤法,美国)
trial and error procedure	尝试法,试错法
trial-and-error	试错[逐次逼近,反复试验]法
trial-and-error method	尝试法,试算法,试错[逐次逼近,反复试验]法
triangle belt drive	三角皮带传动
triangular block	三角形地[矿]块(法)
triangular potential sweep	三角(波)电位扫描
tricalcium phosphate	三钙磷酸盐
trichloroethylene	三氯乙烯(重液)
tried and proven chemical analysis	久经考验的化学分析法
triethoxybutane	三乙氧基丁烷(可用作起泡剂)$C_4H_7(OC_2H_5)_3$
triethoxybutanol	三乙氧基丁醇 $(C_2H_5O)_3C_4H_6OH$
trigger	起动[触发]器,雷管引发物,闸柄,制动器
trigonal coordination	三角配位
trigonometric formula	三角(学)的公式
tri-ionic crystal	三离子晶体
tri-ionic crystal mineral	三离子晶体矿物
tri-ionic solid	三离子固体
trimethyl	三甲基
trimethylchlorosilane	三甲基氯硅烷
trimethyl dodecyl ammonium hydroxide	三甲基十二烷氢氧化铵
trimming dozer	平整场地推土机
trioxane	三聚(甲)醛(三噁烷,三氧杂环己烷)$(CH_2O)_3$
trip report	出差[考察]报告
triple deck screen	三层筛
triple superphosphate	三元过磷酸钙
triple-deck wash screen	三层选矿筛

triply distilled water　三次蒸馏水
tripper unit　卸矿器[装置]
Triton　美国道公司(DOW)表面活性剂品牌名
Triton K-60　二甲基-正十六烷基-苄基季铵盐氯化物(可用作捕收剂、乳化剂、湿润剂)
Triton NE　烷基芳基聚乙二醇醚(可用作捕收剂、乳化剂、湿润剂)
Triton X-100　叔辛基苯基聚乙二醇醚(可用作湿润剂)
$C_8H_{17} \cdot C_6H_4O(CH_2CH_2O)_{9\sim10}H$
Triton X-102　叔辛基苯基聚乙二醇醚(可用作湿润剂)
$C_8H_{17} \cdot C_6H_4O(CH_2CH_2O)_{12\sim13}H$
Triton X-114　叔辛基苯基聚乙二醇醚(可用作湿润剂)
$C_8H_{17} \cdot C_6H_4O(CH_2CH_2O)_{7\sim8}H$
Triton X-200　烷基苯酚聚乙醇硫酸盐(可用作湿润剂、乳化剂)
$R \cdot C_6H_4(OCH_2CH_2)_nOSO_3M$
Triton X-400　二甲基-正十六烷基-苄基季铵盐氯化物(可用作捕收剂、乳化剂、湿润剂)
Triton X-45　叔辛基苯基聚乙二醇醚(可用作湿润剂)
$C_8H_{17} \cdot C_6H_4O(CH_2CH_2O)_5H$
trivalent anion　三价阴离子
trivalent cation　三价阳离子
trivalent metal cation　三价金属阳离子
Tropari compass　特罗帕里钻孔方位测量仪,罗显仪

tropical climate　热带气候
tropical countries　热带国家
tropical rain forest　热带雨林
tropical region　热带地区
Trostol　粗塔[妥]尔油(品名,可用作捕收剂)
trouble shooting　排除故障,故障分析
trouble-free　无故障的
trouble-free installation　无事故的装置
troughing angle　槽角(槽形胶带运输机的)
troughing idler　槽形(成槽)托辊
trough shape multiplier　槽形系数
trough type hopper　槽型漏斗
trout　鲑鱼,鳟鱼
trowel　泥铲,泥刀,馒刀,修平刀;用泥刀涂抹
trowel mix　泥刀[瓦刀]混合
troy　金衡,金衡制(英制)
troy weight　(英国)金衡,金衡制(用于衡量金、银、宝石的质量的单位制)
truck　卡[矿,货]车
truck crane　随车吊,汽车式起重机
truck disposal　汽车排弃(运往废石场)
truck door　卡车大门,载重汽车门
truck dump　载重汽车堆场
truck lead-out bin　汽[卡]车装载仓
truck lead-out scale　卡车装载用地磅
truck mounted　车载式
truck scale　载重汽车秤,卡车地磅
truck shift　汽车工作班

truck with crane 随车吊,汽车式起重机	tungsten value 钨的价值
truckload 卡车装载,货车荷载	tunnel conveyor 平洞[坑道]运输机
true compiled language 真正编译语言	turbid region 混浊区
true middlings 真中矿	turbidity 混浊度[性]
true power 实际[有效]功率	turbidness 混浊,混乱
true specific gravity 真[绝对]比重	turbine impeller 透平叶轮
true value 真值,真实价值	turbo blower 涡轮[透平]式鼓风机[增压器]
truly linear grinding system 真正的线性磨矿系统	turbo type propeller 透平式螺旋桨
trunnion type pivot 耳轴式枢轴	turbo-blower 离心式扇风机,涡轮吸风机
truss 桁[构]架	turbocyclone 透平[涡轮]旋流器
trust company deposit 信托公司存款	turbomachinery 透平[涡轮]机械
try and error search 尝试[试错,试验,逐次渐近]法探索[研究]	turbulence level 湍流水平(浮选柱内的)
T-T collector T-T捕收剂(15%白药加85%邻甲苯胺)	turbulent agitation 紊流搅动
	turbulent flow field 湍流场
T-T mixture T-T混合剂(15%白药加85%邻甲苯胺,可用作捕收剂)	turbulent region 湍流区域
	Turkey red oil 土耳其红油,硫酸化蓖麻油酸钠皂(含3.5%有机SO_2,可用作捕收剂、起泡剂、乳化剂和湿润剂)
tube digestor 管式蒸煮[浸煮,压煮]器	
tube feeder 管式给矿机	
tube sheet 制管薄钢板	Turkey red sulfonated oil 土耳其红磺化油
tube-type mixer 管式混料器[机]	
tubular filter bag 管式滤袋(织物收尘器)	turnkey contract 总承包合同,交钥匙合同
tumbling action 滚动作用	turn-key contractor 成套(设备)承包商
tumbling mill 滚(筒式)磨机	
tumbling mixer 转筒混合器	turn-key X-ray analyzer 交钥匙(直接交付使用的)X射线分析仪
tune-up expense 调整费	
tune-up period 调整期	turpentine 松节油
tune-up phase 调整阶段	twelve-fold coordination 12∶1配位
tungsten manufacturer 钨生产厂家	twice crystallized borate 二次结晶硼

twill 酸盐
twill 斜纹织物
twin cells 双槽
twin helical sluice 双螺旋流槽(螺旋选矿机)
twin mass system 双块体系统
twin pendulum test 双摆锤试验,用于破碎
twin-screw feed distributor 双螺旋给矿[料]分配器
two bearing circular motion inclined vibrating screen 双轴承圆周运动倾斜振动筛
two bearing screen 两轴承筛
two compartment H-cell 双室 H 形槽
two flight system 两(输送)机段系统
two product size 两种产品粒度
two sections of abundance level 两个富含量区间
two shift basis 按两班考虑
two shift operation 两班生产[运行]
two size product 两种粒度产品
two stage adsorption process 两段吸附过程
two stage classification 两段分级
two stage closed circuit rod-ball mill 两段闭路棒球磨
two stage desliming cyclone classifier 两段脱泥旋流分级机
two stage rod-ball mill 两段棒球磨(前开路后闭路)
two stage rod-ball mill with intermediate treatment 带中间选矿的两段棒球磨
two way pulp distributor 二流矿浆分配器
two-arm whirling sprinkler 双臂旋转喷洒器
two-dimensional aggregate 二维聚集[合]体,二维骨料
two-dimensional data 二维数据
two-dimensional projection 二维投影
two-level factorial design 二级析因设计
two-phase fluid vortex motion 两相流体涡漩[流]运动
two-phase frothing effect 两相起泡效应
two-phase frothing test 两相起泡试验
two-phase mixing region 两相混合区(浮选机的)
two-position flop gate 两位置的自动放水[导向]闸门
two-product operation 两种产品的生产
two-product size split 两种产品粒度分离
two-section secondary-tertiary oversize conveyor system 两段中细碎筛上产品(胶带)运输机系统
two-shift operation 两班工作[作业]制
two-shift-day five-day-week 每天两班每周五天工作日
two-spindle attrition machine 双轴擦

洗机[槽]

two-stage centrifugal pump 两级离心泵

two-stage counter current leach 两段逆流浸出

two-stage crusher control 两段碎矿控制

two-stage crushing and screening plant 两段破碎筛分厂

two-stage cyclone classification 两级旋流器分级

two-stage grinding section 两段磨矿工段

two-stage open circuit crushing 两段开路破碎

two-stage reduction 两段碎矿

two-stage reverse flotation process 两段反浮选法

two-stage sampling system 两段取样装置[系统]

two-stage system 两段系统

two-stage wet vacuum pump 两段湿式真空泵

TX-2043 石油磺酸盐(品名,可用作铍矿捕收剂)

TX-2044 石油磺酸盐(品名,可用作铍矿捕收剂)

Tygon tubing 泰贡塑管

Tyler double-deck F-900 Tyrock vibrating screen 泰勒双层 F-900 Tyrock 振动筛

Tyler scale 泰勒标准

Tylor double-deck scalping screen 泰勒式双层除大块筛

Tylor screen series 泰勒套筛

Tylor sieve 泰勒筛

Tylose 表面活性剂品牌名(德国 Hoeschst AG 公司生产)

Tylose CBR4000 羧甲基纤维素(可用作抑制剂、絮凝剂)

Tylose H. B. R. 羧甲基纤维素钠(可用作抑制剂、絮凝剂)

Tylose MH-200 甲基纤维素(可用作抑制剂、絮凝剂)

type and dimensions 类型和规格

type C cutter C 型切取器

type CX extra heavy duty thickener mechanism CX 型超重负载工作浓缩机机构

type D sand D 型砂

type H classifier H 型分级机(高堰式分级机)

type of adsorption 吸附类型

type of belt carcase 胶带层类型

type of capital fund 投资资金类型

type of cavity 破碎腔型式

type of crushing equipment 破碎机型式[类型]

type of drill 钻机类型

type of equipment 设备型号

type of estimation 估算类型

type of flotation 浮选类型

type of intergrowth 共[连,交]生体类型

type of mineralization 成矿[矿化]类型
type of motion 运动形式[类型,方式]
type of organizational 组织形式
type of research 研究类型
type of resin 树脂类型
type of secondary mill 二段磨机类型
type of service 操作形式,工作类型
type of solids 固体物料类型
type of test 试验类型
type of water 水的类型
type S classifier S型分级机(浸没式分级机)
types of work 工种
typical carrier concentration 典型载流子浓度
typical closed circuit flotation test 典型闭路浮选试验
typical construction material 典型的结构[制造]材料
typical deck preparation 典型筛面制备
typical dust 典型粉尘
typical fabric 标准织物
typical flotation test 典型的浮选试验
typical flow sheet 典型流程图
typical gravity processing 典型重力选矿
typical grinding circuit 典型磨矿流程
typical ground line 标准地平线(选厂地形断面线)
typical laboratory rod mill 典型实验室棒磨机
typical liquid hydrocarbon 典型液态烃
typical low-cost product 典型低成本产品
typical methodology 典型方法论[学]
typical milling cost 典型选矿费用
typical milling plant 典型选矿厂
typical nuclear level switch 典型的核料位开关
typical one stage crushing plant 典型的单段破碎车间
typical operating range 典型操作范围
typical operating strategy option 典型的生产策略方案
typical photoresponse curve 典型光响应曲线
typical plot of separation efficiency 典型分离效率曲线
typical procedure 典型步骤
typical product specification 典型产品规格
typical size analysis of product 典型的产品粒度分析
typical speed 典型速度
typical standard laboratory grinding 典型的标准实验室磨矿
typical test program 典型试验过程
typical tube-type mixer 标准管式混料器[机]
typical value 典型值
typical variation 常用变量

typical variogram 典型变异函数
Ty-Rod type woven wire cloth 泰罗德型金属丝编织筛布
α-terpineol α-萜烯醇(可用作起泡剂)
β-terpineol β-萜烯醇(可用作起泡剂)
γ-terpineol γ-萜烯醇(可用作起泡剂)

U

U.S. base rate 美国基准利率
U.S. exporter 美国出口商
U.S. goods 美国货
Ucon 联合碳化物公司(Union Carbide 公司)生产的聚合物品牌名
Ucon frother 190 高分子高级醇加聚丙二醇(可用作起泡剂)
Ucon LB-100X 聚丙二醇(可用作捕收剂、起泡剂)
ultimate design calculation 最后设计计算
ultimate evaluation 最终评价
ultimate flowsheet 最终流程图
ultimate loading capacity 最终载荷[金]能力
ultimate operation 最终生产
ultimate stress 极限应力
ultimate test 最后[终]试验
ultimate testing 最终试验
ultrafiltration 超细过滤(作用)
ultra-high purity nitrogen 高纯纯氮
Ultranat 1 油溶石油磺酸盐(品名,可用作氧化铁矿的捕收剂)
Ultranat 3 油溶石油磺酸盐(品名,可用作氧化铁矿的捕收剂)
Ultrapon S 脂肪酰胺(品名,可用作湿润剂)
ultra-small cell 超[特]小型浮选槽
ultra-small flotation cell 超小[微]型浮选槽
ultrasonic attenuation 超声波衰减
ultrasonic bin level control system 超声波矿仓料位控制系统
ultrasonic cleaning device 超声波清扫[洗]器
ultrasonic density gauge 超声波浓度计
ultrasonic energy 超声波能量
ultrasonic excitation 超声波振荡
ultrasonic oscillation 超声波振荡
ultrasonic probe 超声波探头
ultrasonic treatment 超声波处理
ultraviolet radiation 紫外(线)辐射
ultraviolet spectra 紫外(线)光谱

ultra-violet spectrometry 紫外(线)光谱法
ultraviolet spectrophotometric analysis 紫外(线)分光光度分析
ultraviolet spectrophotometry 紫外(线)分光光度测定法
ultra-violet spectrum 紫外线光谱
Ultravon K 湿润剂(十七烷基苯骈咪唑—磺酸盐)
Ultravon W 湿润剂(十七烷基苯骈咪唑二磺酸盐)
Ultrawet 表面活性剂品牌名
Ultrawet 烷基苯磺硫钠(可用作氧化铁矿的湿润剂) $RC_6H_4SO_3Na$
Ultrawet 30-DS 烷基苯磺酸盐(可用作湿润剂、起泡剂及滑石和蜡石的捕收剂)
Ultrawet 35KX 32%十二烷基苯磺酸钠加68%水(可用作湿润剂)
Ultrawet 40A 烷基芳基磺酸钠(可用作滑石的捕收剂及湿润剂)
Ultrawet DS 烷基芳基磺酸钠(含活性物质85%,可用作湿润剂、捕收剂、起泡剂)
Ultrawet K 烷基苯磺酸盐(可用作湿润剂)
Umix 塔尔油与一种中性油(即柴油或燃料油)混合成的乳剂,品名,与烷基芳基硫酸盐一起使用,从赤铁矿中分选磷矿时用作捕收剂
unabsorbed electrolyte 未被吸收的电解质

unbalance condition 不平衡条件
unbalance weight 不平衡配重
unbalanced force 不平衡力
unbiased fashion 无偏离方式
un-biased sample 无偏差的样品[试样]
unbound water 非结合水
uncalibrated instrument 未校准的仪表
unclassified ball mill discharge 未(经)分级的球磨机排矿
uncleaned raw coal 未精选的原煤
uncoded form 非编码形式
uncommon processing 特殊[不常用的]加工[处理]
uncompensated charge 未补偿电荷
unconfined compressive strength 无侧限抗压强度
uncrushable material 不能破碎的物料
undecyl 十一(烷)基 $C_{11}H_7$—
undecylic acid 十一(烷)酸 $C_{10}H_{21}COOH$
undercut gate 下截[下开]式闸门(扇形闸门)
undercut splitter gap 底部截取分流孔隙
underdesign 冒进[欠安全]的设计
underdrain system 暗沟(排水)系统
under-fed 加[给]料不足
underflow % solids 底流固体量%[百分数]
underflow blockage 底流堵塞
underflow dry solids 底流干固体

underflow dryness 底流[返砂]干燥度
underflow percent solids 底流[沉砂]固体百分率[数]
underflow percent solids by weight 底流固体重量百分数
underflow solids 底流固体量
underflow solids portion 底流固体部分
underflow stream 底流[沉砂](矿流)
underflow volumetric flowrate 底流[沉砂]体积流量
underflow water 底流水量
underground conveyor 地[井]下运输机
underground development work 地下开拓[采准]工作
underground drift 地下坑道
underground exploration 地下勘探
underground haulage level 地下运输平巷
underground mining operation 地下开采作业,坑采生产
underground operation 地下[坑内]生产[开采作业]
underground ore 地下矿石
underground service 地下服务[工作,操作]
underground train 井下[地下,坑内]列车
underground workings 地下[采区]工作面
underhand cut and fill mining method 下向分层充填采矿法
underhung crane 悬挂式起重机
underloading 欠载,欠装载
underlying random component 基底随机分量
undersaturation 不饱和
undersize 筛下产品
undersize equipment 尺寸[规格]过小的设备
undertaking 事业,任务
under-utilization 不能利用
undesirable component 无用成分
undesirable impurity 不需要的[无用]杂质
undissociated 不[未]离解的
undue stock buildup 过量库存的形成
uneconomical 不经济的
unemployment tax 失业税
uneven distribution 不均匀分布
unfamiliar equipment 不熟悉的设备
unfinished tailings 非最终尾矿
unforeseen problem 未[不]可预见的问题
unheated slurry 未加热的矿浆
unheating building 不采暖的厂房
uni- 单,一
unidirectional fan 单向风扇
Unifloc 絮凝剂(品名,土豆淀粉与$CaCl_2$及$ZnCl_2$制成的产品)
uniform dissolution 均一分解
uniform flow 均匀流动
uniform flow pattern 均匀流动型[流

uniform ore grade 均匀矿石品位
uniform outflow 均匀流出量
uniform quantity 均匀数量
uniform range 均匀变程
uniform rate of depreciation 相同折旧率
uniform residence 均匀停留(期间)
uniform sample length 统一的试样长度
uniform spacing 均匀间距(法)
uniform surface 均匀表面
uniform top size 均匀上限粒度
uniformly distributed 均匀分布的
Union Carbide MIBC frother MIBC 起泡剂(美国联合碳化物公司生产)
Union Carbide PP425 PP425 起泡剂(美国联合碳化物公司生产)
Union Carbide R-133 frother R-133 起泡剂(美国联合碳化物公司生产)
union officer 工会干部
union-company meeting 联合公司会议
unique combination 唯一组合(形式)
unique function 单值函数,独特功能
unique value 独特价值,唯一值
unissued share 未发行的股份
unit 单位[元],装置,部件
unit A flotation A 段浮选
unit area 单位面积
unit cell 晶[单]胞
unit charge 单位电荷,每千瓦·小时的电费
unit construction 整套组合体,构件组合,独立装置[结构]
unit flotation cell 单槽浮选机
unit matrix 单位矩阵
unit mechanism 单一机构
unit of construction 建筑结构单元,构件
unit of production 生产单位
unit of production depreciation rate 生产单位折旧率
unit of structure 构件,结构单元
unit operation 单元运行[设备作业]
unit plant 单元设备[机组]
unit power 单位功率
unit power consumption 单位功(率消)耗
unit power cost 单位功率价格,单位电费
unit power efficiency 单位功率效应
unit power plant 成套动力[发电]装置
unit power requirement 单位功率需要量
unit process 单元作业[操作](过程)
unit pulp volume 单位矿浆体积
unit sampling cell length 单位取样单元长度
unit thickening area 单位浓缩面积
unit time 单位时间
unit tonnage rate 单位吨位流量
unit volume of pulp 单位矿浆体积

unitary matrix equation 单一矩阵方程	unsaturated hydrocarbon 非饱和烃
unity 单一,个体,团结,一致,联合	unscheduled downtime 计划外[不定期间断的]停车[工,机]时间
uni-univalent salt 1-1价盐	unscheduled plant shutdown 无计划的选矿厂停产
univalent ion 一价离子	unsecured loan 无担保贷款
univalent ionic solid 一价离子型固体	unsecured promissory note 无担保期票
universal adaptor base 万向[能]连接器底座	unsized feed 未分级的给矿
universal jaw crusher 万向[能]型颚式破碎机(可动颚板装)	unsized flotation 不分级浮选
	unstable effect 不稳定效应
universal simulator 通用模拟器	unstable form 不稳定形态
university laboratory 大学实验室	unstable froth 不稳定泡沫
unknown interference 未知干扰	unstable influence 不稳定影响
unliberated 未单体解离的	unstable sulfide 不稳定硫化矿
unliberated mineral particle 未单体解离矿物颗粒	unsteady flow 不稳定流
	untested procedure 未经试验的方法
unloading ramp 卸车台	untrained operator 未经训练的操作人员
unmanageable froth 难控制的泡沫	
unoxidized clean galena 未[非]氧化纯净方铅矿	untreated galena 未处理方铅矿
	untried procedure 未经尝试的方法
unpleasant odor 令人不愉快的[不良]气味	unused portion 未动用部分
	unusual physical factor 异常的物理因素
unpleasant taste 令人不快[不好]的味道	unusual screen surface 特殊筛面
	unusual screen surface area 特殊筛面面积
unpolished particle 非磨光颗[矿]粒	
unprotonated amine 非质子化胺	unusual selectivity 异常选择性
unproven high technology 未经改进的高级技术	upflow exchange column 上流型交换柱
unsaturated chemical bond 不饱和化学键	up-front cost 预先[先期]费用[成本]
unsaturated compound 不饱和化合物	upgrade 提高(质量,等级,品位)
unsaturated fatty acid 不饱和脂肪酸	

up-graded bulk product 提高品位的批量产品
upgraded concentrate 最终精矿
upgraded zinc concentrate 最终[提高品位的]锌精矿
upgrading 精选
upgrading circuit 精选[提高品位,升级]回路
upper bearing 上部轴承
upper constraint 约束域上限
upper curve 上部曲线
upper flotation edge 上部浮选边界
upper holder 上部夹子,上托架
upper level 上层,上能级
upper limit 上限
upper mantle 上部动锥
upper mantle liner 上部动锥衬板,动锥上部衬板
upper screen cloth oversize 上层筛网筛上物
upper step bearing plate pin 上部立式止推轴承垫板销钉
upper stop bearing plate pin 上部承重板销钉
uppernatant solution 上部[层]清溶液,清液层溶液,上清液,悬浮溶液
upright convergent jaw crusher 立式收缩破碎腔颚式破碎机
upset condition 翻花状态
upstream method 上游[逆流]法(筑坝)

Uramin 乌洛托品(品名,可用作湿润剂)
uranium extraction plant 铀萃取厂
urban sewage 城市污水
urethane disperser 尿烷分散器
urotropine (=hexamethylene tetramine) 乌洛托品(六亚甲基四胺,可用作湿润剂)
Ursol 低温煤焦油(品名,可用作起泡剂、捕收剂)
usage 使用(量),用法
usage factor 利用系数,利用率
use rate 使用率
use tax 使用税
useful component 有用成分
user manual 用户手册
user's personnel 用户人员
utilization 利用率,运转率(磨机的)
utilization of available time 可用[有效]时间利用率
utilization of capacity (处理)能力利用率
utilization of total time 总时间运转率
utilization of waste heat 废热利用率
utilization rate 利用率
Utinal 德国产十二烷基硫酸钠(品名,可用作(重晶石、钾盐)捕收剂) $C_{12}H_{25}OSO_3Na$
Uvermene 38 (=Alamine 26) 牛脂胺(品名,87%伯胺,可用作(磷矿)捕收剂)

V

vacation allowance 假日津贴[补助]
vacuum belt filter 真空带式过滤机
vacuum desiccator 真空干燥器
vacuum disc filter 真空圆盘[盘式]过滤机
vacuum disk filter 真空圆盘过滤机
vacuum drier 真空干燥器
vacuum drum filter 真空鼓式[滚筒,圆筒]过滤机
vacuum equipment 真空设备(过滤系统的)
vacuum filtering 真空过滤
vacuum flotation technique 真空浮选法
vacuum machine 真空浮选机
vacuum micro balance 真空微量天平
vacuum pressure gauge 真空压力表
vacuum rotary dryer 真空回转干燥机
vacuum tower 真空(蒸馏)塔
vacuum-pneumatic delivery system 真空风力输送系统
valence band edge 价(电)带限界
valeron(e) 二丁基甲酮或壬酮-[5](可用作起泡剂)$(C_4H_9)_2CO$
valid model 有效模型
valid request 有效请求
valid tool 有效工具
valonia extract 柞壳浸液,橡椀栲胶,橪树富萃取液(含单宁,可用作抑制剂)
valuable constituent 有价成分
valuable element 有价元素
valuable heavy mineral 有用重矿物
valuable heavy minerals 有用重矿物类
valuable insight 有价值的见解
valuable metal 贵重[有价值,有用]金属
valuable mineral 有用矿物
valuable portion 有用部分
valuable reclaimed material 有价值的回收物料
valuable species 有用矿料[种]
valuable sulfide 有用硫化矿
value 价值,面值,面额,有用成分
value determination 价值确定
value engineering 工程经济学,价值工程(学)
value of ASARCO's ownership 美国熔炼[冶金]公司拥有的投资值
value of concentrator output 选矿厂产值

valve actuator pump 闸阀致[驱]动器泵
valve station 阀门站
valving system 阀门系统
Van der Waals 范德华[范德瓦耳斯]（荷兰物理学家）
Van der Waals attractive force 范德华吸引力
Van der Waals attractive potential energy 范德华吸引位[势]能
Van der Waals bond 范德华键
Van der Waals cohesive energy 范德华内聚能
Van der Waals dispersion force 范德华色散力
Van der Waals force 范德华力（又称分子作用力，产生于分子或原子之间的静电相互作用）
Van der Waals interaction 范德华相互作用
Van der Waals radius 范德华半径
Vandeveer diagram 范迪维尔图解法
Vanport limestone Vanport 石灰石（一种含有海洋动物海百合形成的石灰质的石灰石）
vapo(u)rizer 汽化[蒸发，蒸馏，喷雾]器
vapor adsorption 水蒸气吸附
vapor oil 蒸气油
vapor pressure 水蒸气压力，水汽压
variability 变化率
variability analysis 变化分析

variability of operation 操作[生产]可变性
variable abrasiveness 不同磨蚀性
variable cost 非固定费用，可变费用
variable element 可变参[元]数
variable flow 变流量
variable frequency AC system 变频交流系统
variable gear 变速齿轮
variable hardness ore 不同硬度的矿石
variable inflation rate 可变的通货膨胀率
variable interval 变量间距
variable mineralogy 变化的矿物学
variable motion 变速[不稳定]运动
variable motor 变速电机
variable nozzle 可变截面喷管[嘴]，可调喷嘴
variable ore 变异矿石
variable pump 变量泵
variable speed conveyor drive 变速运输机传动装置
variable speed drive 变速传动[装置]
variable speed drive unit 变速传动装置
variable speed feeder 变速给矿[料]机
variable speed fluid drive 变速液压驱动装置
variable speed slurry pump 变速泥浆泵，变速砂泵
variable unit 变量单位
variable-speed drive 变速传[驱]动

variable-speed eddy current drive belt conveyer 变速涡流电流驱动胶带运输机
variable-speed transmission 变速传送[动](装置)
variable-speed unit 无级变速器[装置]
variance 方差,变异(数),偏差,差异
variance analysis 方差[偏差,差异]分析
variation 变化[动,更,异,种],改变
variation interval 变化间距[区间]
variogram 变异函数,变量图
variogram analysis 变异函数分析
variogram curve 变异函数曲线
variogram model 变异函数模型
various ceramics 各种陶瓷制品
various combination 不同[各种]组合
various government agencies 各政府机构[关]
various options of operation philosophy 各种运行基本原理方案
various scale 不同尺度[规模]
various time period 不同[各个]时期
various types of ore 不同类型的矿石
vari-speed fluid drive (无级)变速液压传动装置
varying mineralogy 矿物性质变化
varying time constant 时变常数
varying weight 不同的重量
vat (大)桶,(大)槽,(大)缸,瓮
vat cyanide leaching 槽内氰化浸出
V-belt drive V形[三角]皮带传动装置

vector 向量,矢(量)
vector of flow rate 流速向量
vector of kinetic parameter 动力参数向量
vector of recovery 回收率向量
Vee box V形分级箱
Vee box sizer V形分级箱[箱式分级机]
vegetable fatty acid 植物脂肪酸
vegetable tall oil 植物塔尔油
vegetation 植被
vegetative cover 植被,植物覆盖层
vehicle shop 车辆修理车间
velocity component 分速度,速度分量
velocity constant 速度常数
velocity distribution 速度分配
velocity effect 速度效应
velocity field 速度场
velocity gradient 速度梯度
velocity head loss 速度压头损失
velocity of efflux 流出速度
velocity ratio 流速比
vendee 买方,买主
vender 供应商,供方,自动售货机
vender quotation 供方报价单
vender's standard design 供方的标准设计
vender's project engineer 供方的项目工程师
vending machine 自动售货机
ventilation criteria 通风标准

ventilation hood 通风罩
ventilation requirement 通风需要量
venturi scrubber 文丘里式洗涤器
venturi throat section 文丘里管喉部[喷嘴段]
verification test 验证试验
vertical auto-scale feeder 立式自动称量给料机
vertical bed 直立层
vertical column 竖栏,立柱
vertical conditioner 立式调整槽
vertical development 垂直开拓
vertical gate 竖[立]式闸门
vertical gravity take-up 垂直重力拉紧装置
vertical height 垂直高度
vertical running load 垂直动[运行]负荷
vertical slurry pump 立式矿[泥]浆泵
vertical submersible pump 立式潜水泵
vertical turbine pump 立式涡轮泵
vertically adjustable spout 垂直可调斜槽
VHF radio 甚高频无线电话[无线电广播]
VHF repeater station 甚高频中继站
viable project 可行的[有生存力的,富有生命力的]工程项目
Vibra screw feeder 维勃拉[振动]螺旋给料机
vibra screw gyrator 振动螺旋回旋器
[回转螺旋]
vibrated hopper 振动漏斗
vibrating conveyor feeder 振动输送给矿机(槽式或管式)
vibrating conveyor trough 振动输送槽(槽式给矿机)
vibrating conveyor tube 振动输送管(管式给矿机)
vibrating Derrick screen 德瑞克型振动筛
vibrating discharge 振动排料[矿]
vibrating feeder-grizzly 振动格筛式给矿[料]机
vibrating grizzly feeder 振动(棒条)筛给矿机
vibrating packer 振动装填[包装]器
vibrating pan ore discharger 振动槽排矿机
vibrating reclaim feeder 振动取[出]料给矿机
vibrating roller compactor 振动辊式压实机
vibrating screen bar pitch 振动筛棒距
vibration exciter 振子,振动激励器
vibration isolation 振动隔离
vibration type 振动型
vibrator housing 振动器外壳
vibrator shaft 振动轴
Viburnum weigh launder Viburnum型称量流槽
vice-chairman 副主席
vice-president 副董事长[总裁,总经

理,会长]

Victamin 月桂胺磷酸乙酯(品名,可用作(锌矿)捕收剂)
$C_{12}H_{25}NH \cdot P(O)(OC_2H_5)(ONH_3C_{12}H_{25})$

Victamin D 硬胺磷酸乙酯(品名,可用作捕收剂)
$C_{18}H_{37}NH \cdot P(O)(OC_2H_5)(ONH_3C_{18}H_{37})$

Victawet 表面活性剂品牌名

Victawet 12 辛基磷酸酯(非离子型,可用作黏土分散剂)

Victawet 14 辛基磷酸酯(非离子型,可用作黏土分散剂)

Victawet 35B 2-乙基己基三磷酸钠
$Na_5R_5(P_3O_{10})_2, R=2$-乙基己基

Victawet 58B 己基三磷酸钠
$Na_5R_5(P_3O_{10})_2, R=$己基C_6H_{13}—

violamine 紫胺(可用作染料及脉石抑制剂)

violet red 紫红色

virgin surface 原始[自然,新鲜,未污染]表面

viscosity inhibitor 黏度抑制剂

viscosity reducing reagent 降低黏度剂

viscous drag force 黏滞阻力

viscous slurry 黏性矿浆

visible water 可见水

visual colorimetry 目视比色法

visual comparison 直观比较,视觉比较

visual display 可见显示,直观显示

visual display terminal 可见显示终端,视屏终端

visual examination 外部[外表]检查,表观检查,目测

visual inspection 目测检查

visual means 视觉手段

viton 氟化橡胶

vitrified-clay product 陶土产[制]品

vitrinite 镜质组[体]

vitrinite class 镜质型(煤岩)

vivid green color 鲜绿色

VK 3 machine VK 3型浮选机

voice communication 语音通信

void volume 孔隙体积

volatile matter 挥发性物质

volatile organics 挥发性有机物

volcanic genesis 火山成因

volcanic origin 火山成因

voltage current curve 伏安曲线

voltage sweep 电压扫描

voltammetry sweep 伏安测量扫描

voltammogram 伏安(测量曲线)图

volume conversion constant 体积换算常数

volume element 体积元[单元,元素]

volume estimate 体积估算

volume estimating procedure 体积估算过程

volume flow rate 体积流量[速]

volume flowrate ratio 体积流速比

volume throughput 体积通过量

volume-tonnage conversion factor 体积吨量换算系数

volumetric capacity 体积能力

volumetric feeder 体积[量]给矿

[料]机
volumetric flask 量瓶
volumetric flow 体积流量
volumetric flowrate 体积流量[速]
volumetric froth withdrawal rate 体积泡沫回收速度
volumetric gas or vapor adsorption 体积气体或蒸汽吸附(作用)
volumetric percent 体积分数
volumetric percent solids 固体体积分数
volumetric proportion 体积比
volumetric throughput information 体积通过[处理]能力资料

volute discharge pump 蜗壳式排出泵
vortex action 漩涡[涡流]作用
vortex center 漩涡[涡流]中心
vortex core 漩涡[涡流]中心
vortex diameter 旋流器溢流管直径
vortex finder 导流管,涡流探测器
vortex screen 漩流筛
vortex shedding meter 漩涡放射仪表
vortex size 漩涡导流管规格
vortexing action 漩涡[涡流]作用
vulcanized rubber 硬化[加硫]橡胶
vulcanized splice separation 硫化拼接脱离
vulkollan 氨基甲酸乙酯橡胶

W

Waelz(fuming) process 威尔兹回转窑烟化法(冶锌)
Waelz furnace 威尔兹炉
Waelz method 威尔兹(回转窑烟化)法
wagon loading 装车(矿车,铁路货车)
walk-in plenum 可(步)入式强制通风
wall friction angle 仓壁摩擦角
wall material 壁面材料
wall panel (护)墙板
wall slope (露天矿场)边坡
Warco 表面活性剂品牌名

Warco A-266 椰子油酰-N-甲氨基乙酸钠(可用作捕收剂、洗涤剂)
R—CO—N(CH$_3$)CH$_2$COONa,R—CO— 为椰子油酰
warehouse expense 仓库保管费
Warman pump 瓦尔曼泵
warranty 保证(书),担保(书),保单
wash circuit 洗涤回路
wash down 冲洗[刷,掉,净]
wash efficiency 洗涤效率
wash filtration test 过滤洗涤试验
wash ratio 洗涤比率(涤涤用水量与物

料之比)
wash water 洗涤水
washed concentrate salt level 洗涤后精矿含盐量
washing section 冲洗区
washing solution 洗涤液
washing spray 洗矿喷水
washing thickener 洗涤浓缩机
waste control system 废料[尾矿]控制系统
waste dam 废石坝
waste disposal area 废石处理场,废石场
waste gas treatment 废气处理
waste heat recovery 废热回收
waste heat system 废热系统
waste pond 尾矿池
waste portion 废石部分
waste product 废品
waste product analysis 废石分析
waste rock 废石
waste stream 废液流
waste wash solution 废渣洗涤液
waste water treatment 废水处理
wasting asset 消耗性资产
water addition 水的添加(量)
water addition indicator 给[加]水指示器
water addition rate 加水量[率],水的添加量[率]
water availability 水可获量,水的可利用性

water capacity (处理)清水能力
water capacity for standard cyclone 标准旋流器(处理)清水能力
water catchment 集水,汇水,汇集水
water chamber and jet 水室及喷头
water chemistry 水化学性质
water conservation 水资源保护
water cooled bearing 水冷却轴承
water disposal system 水处理系统
water eductor 水(力)喷射器
water emission 排放污水
water flea 水蚤
water flow meter 给水流量计
water flow rate 水流速度[流量]
water glass 水玻璃
water head 水头
water jacket 水套
water pollution situation 水污染情况
water proportioning 按比例配水,水的配比
water quality 水质
water quality data 水质数据[资料]
water quality demand 水质要求
water quality parameter 水质参数
water quality status 水质状态
water reclaiming system 回水系统
water reclamation (废)水回收
water reclamation circuit 回水流程图
water recovery 水回收率
water refrigerator 水冷箱
water repellant film 水排斥膜,防水膜
water reuse practice 水的再利用生产

W

实践

water ring vacuum pump　水环式真空泵

water softening agent　水软化剂

water source　水源

water split equation　水分流方程式

water spray for dust suppression　喷水抑制粉尘

water sprinkler truck　洒水汽车,喷水汽车

water storage　储水池

water structure　水结构

water supply source　供水水源

water tank　储水池[罐]

water temperature　水温

water test data　水[清水]试验数据

water treatment plant　水处理厂[车间]

water treatment process　水处理过程

water treatment system　水处理系统

water treatment technique　水处理技术

water truck　洒水车,水罐车

water vapor adsorption method　水汽吸附法

water-avid part　亲水部分

water-butyl alcohol　水-丁醇体系

water-deficient area　缺水地区

waterfowl　水禽,水鸟

water-gas tar　水煤气焦油(可用作锌矿的捕收剂,起泡剂)

water-insoluble collector　不溶于水的捕收剂

water-softening method　水(质)软化法

water-solid pulp ratio　矿浆液[水]固比

water-soluble constituent　水溶性组分

water-soluble reagent　水溶性药剂

watertight　防水

water-to-solid ratio　液固比

watt transducer　功率变送器

wattle bark extract　荆树皮萃取液(类似南美产白坚木产生的单宁)

wave length detector　波长探测器

wave number　波数

weak allergen　弱变态反应原,弱变应[过敏]原

weak alteration　弱蚀变

weak base　弱碱

weak base resin　弱碱树脂

weak electrolyte surfactant　弱电解质表面活性剂

weak hydrofluoric acid　弱氢氟酸

weak link　薄弱环节

weak market　滞销市场

weak solution　稀溶液

weak zone　薄弱带[面]

weak-base collector　弱碱性捕收剂

weakly bonded ore(s)　弱结合[键合,黏着]矿石(类),软矿石(类)

weakly ionizable　弱电离

weakly magnetic separator　弱磁选机

wear characteristic　磨损[耐磨]特性

[性质]
wear element 耐磨[磨损]元件
wear life 耐磨寿命
wear metal cost 磨损金属费用[成本]
wear parts 磨损件
wear ring 耐磨[磨损]环
wear study 磨损(试验)研究
wear-resistant plate 耐磨板材
wear-resistant polyurethane 耐磨聚胺酯
weather shutdown 天气影响的停产
weathered ore 风化矿石
weatherproof 防晒防雨,耐气候
Weber number 韦伯数(韦伯数是流体力学中的一个无量纲数,代表惯性力和表面张力效应之比)
Wedag laboratory cell 德国维达格型实验用浮选槽
Wedag machine 德国产维达格型浮选机(实验室用)
wedge bar screen 楔形棒条筛,楔形格筛
wedge bar stationary screen 固定楔形棒条筛
wedge blinding 楔形孔堵塞
wedge hopper 楔形漏斗
wedge-shaped deflector liner 楔形导流衬板
wedge-shaped liner 楔形衬板[里]
weekly activity report 周活动报告
weigh beam 秤杆
weigh bin 计量仓

weigh bridge 称量桥(胶带称的),桥式天秤
weigh idler 称量托辊(胶带称的)
weigh launder 称(重)量槽
weigh platform 称量台,称料台
weigh scale 重量秤,磅秤
weigh span 称量跨度[范围](胶带称的)
weigh tank 称重槽,称量桶
weighbatcher 称量配料器,重量配料斗
weigh-belt feeder 计量胶带[带式]给矿机
weighbridge 桥[台,地]秤,地磅,地中衡
weighing batcher 称量配料器,重量配料斗
weighing cell 重量传感器
weighing feeder 计量给矿器[机]
weighing table 称量台
weighing unit 称量装置
weigh-launder 计量流槽
weighmatster panel 司秤[过磅]员控制盘[操作台]
weight conversion constant 重量换算常数
weight display 重量显示(器)
weight fraction 重量百分率
weight of charge 装料重量(磨机的)
weight of pulp hold-up 存储[滞留]矿浆重量,矿浆容重
weight of the largest piece 最大件重量
weight percent 重量百分比

weight percentage 重量百分比
weight sensing element 重量感应元件
weight signal 重量信号
weighted assay 加权平均成分,加权值[法,分析]
weighted average 加权平均值
weighted noise level 加权噪声级
weighted sum of squares objective function 加权目标函数平方和
weighting factor 加权因数[子]
Weinig flotation machine 威涅格型浮选机
weir box 溢流堰箱(浮选机的)
weir crest 堰口[顶]
weir end 溢流堰端(螺旋分级机的)
weir gate 闸门堰(浮选机的)
weir gate splitter 溢流堰闸门分流器,堰板分流器
welded steel scoop box 大杓焊接钢槽
well defined process 精确限定的过程
well mineralized quartz-porphyry stockwork 矿化良好的石英-斑岩网状脉
well stratified 层状良好的
well stratified zone 良好层状地带
well water 井水
well-defined grain structure 界限分明[清楚]的晶粒构造[结构]
well-defined pattern 意义明确[轮廓分明,周密布置]的形式
well-stratified zone 层状良好的矿带,明显的层状矿带
Wemco depurator flotation machine 威姆科型净化(器)浮选机
Wemco flotation cell 威姆科浮选机(改进后成为机械搅拌式自吸空气浮选机)
Wemco flotation machine 威姆科浮选机
Wemco HMS cone 威姆科型重介质圆锥分选机[分选圆锥]
Wemco hydrocleaner flotation machine 威姆科型水净化浮选机
Wemco model 120 flotation machine 威姆科120型浮选机
Wemco Ni-hard pump 威姆科型硬镍泵
Wemco-Fagergren (1+1) mechanism flotation machine 威姆科-法格古伦(1+1)机构型浮选机
Wemco-Fagergren 1+1 design 威姆科-法格古伦1+1设计(美国威姆科公司将法格古伦浮选机的笼形转子改为星形转子,笼形定子改为带肋条的多孔圆筒,新机型称为威姆科1+1)
Wemco-Fagergren 1+1 rotor 威姆科-法格古伦1+1转子
Weserhütte primary screen Weserhütte第一段分级筛
WESTATES coconut carbon WESTATES椰子壳碳
western hemisphere 西半球
Western Uzbekestan 乌兹别克西部
wet analysis 湿法分析
wet autogenous mill 湿式自磨机

wet classification step 湿式分级作业[工序,方法,步骤]	wet semi-autogenous mill 湿式半自磨机
wet clay 潮湿黏土	wet shaking table 湿式摇床
wet grate ball mill 湿式格子球磨机	wet slurry dust collector 湿式泥浆集尘器
wet gravity concentrating table 湿式重力选矿摇床	wet sticky clay 湿黏土
wet gravity separation 湿式重选	wet table 湿式摇床
wet grinding alternative 湿磨方案	wet-overflow closed circuit 湿式溢流型闭路(磨矿的)
wet injection-type exhaust scrubber 湿式喷射型乏气洗涤器	wet-overflow open circuit 湿式溢流型开路(磨矿的)
wet mineral concentration plant 湿式选矿厂	wet-residence time of pulp 矿浆湿润停留时间
wet ore 湿矿石	wettability of surface 表面湿润性
wet overflow ball mill 湿式溢流型球磨机	wetting behavior 湿润状态
wet overflow rod mill 湿式溢流型棒磨机	wet-type scrubber 湿式除尘器
	whale oil 鲸油(可用作铬铁矿捕收剂)
wet peripheral rod mill 湿式周边排矿棒磨机	whale oil sulfonate 磺化鲸油(可用作氧化铁矿捕收剂)
wet plant 湿式重选厂	wharf storage yard 码头储[堆]料场
wet preparation 湿式预备,湿式处理(筛分)	white filler 白色填料
	white medical oil 白色医用油
wet preparation screen 湿式预筛	white spirit 白精油(石油产品,沸点 145~205℃,含芳烃小于10%)
wet process grinding mill 湿式[法]磨机	whole coal 全煤
wet processing 湿式工艺[加工,选矿]	wide band differential amplifier 宽带差动放大器
wet screening duty 湿式筛分任务[工作]	wide gap 大间隙
wet screening factor 湿(式)筛(分)系数	wide throat opening 宽入口孔
	wide tracked dozer 宽履带推土机
wet scrubber collection efficiency 湿式洗涤器收尘效率	wide-angle scanning photosedimentometer 宽角扫描沉淀摄影计,光透式粒度测

定仪
wiggler 摇摆式胶管洒水器
wildlife 野生生物
wild-life area 野生动物区
wildlife mammal 野生哺乳动物
Wilfley concentrating table 威尔弗利选矿摇床(美国)
Wilfley table 威尔弗利摇床(美国)
Williams' impact crusher 威廉姆斯冲击式破碎机
Windarra type sparger 温达拉型空气喷[雾]器
wind-driven ram 风动撞击机[撞锤]
winding 绕组,线圈
windrow method 长(条)形堆积法(分层掺和堆积法之一)
wiper brush 刮料刷
wiping seal 擦抹式密封,柔性密封
wire bar casting 丝棒铸造
wire-rope conveyor 钢丝绳(胶带)运输机
wire-rope-hung belt conveyer 钢丝绳吊挂胶带运输机
wiring diagram 接线图
Wisprofloc 英国生产的改性淀粉絮凝剂品牌名
Wisprofloc 20 絮凝剂(阴离子型水溶性淀粉)
Wisprofloc P 絮凝剂(阳离子型水溶性淀粉)
withdrawal hole 排出[矿,料]孔
wobbler feeder 摇摆给矿(料)机

wood and steel construction with galvanized roofing 钢木结构镀锌板屋顶
wood ash 木灰,可用作黄铁矿的抑制剂
wood baffle 木挡板
wood chip screen (隔,除)木屑筛
wood tank 木槽
wood weir block 木闸板(浮选机的)
woody taste 木头味
word digit 字码数字
work 作用,功,工作,劳动,加工,事业,业务,制[作,产]品,著作,机[配,工,构]件,工作量,工厂,工程
work classification 工作分类
work environment 工作环境[场合,场所]
work force requirement 劳动力需要量
work index analysis 功指数分析
work index calculation 功指数计算
work index determination 功指数测定
work index figure 功指数数值[字]
work index guide 功指数指南
work input 输入功
work of adhesion 黏附功
work of cohesion 黏结[合]功,内聚功
work sheet 工作图表[卡片,单,表]
workable circuit 可采用的回路,可用[行]回路
workable control system 可行[使用,运转]的控制系统
workable flowsheet 可采用的流程图
workable froth condition 适合的泡沫条件[状态]

workable idea 切实可行的主意[打算,想法]
worked-out area 采空区
working bench 工作[开采]台阶,工作台
working capital 流动[周转]资金
working capital estimate 流动资金估算
working capital maintenance contract 周转资金维持协议
working capital portion 流动资金[资本]部分
working condition 工作条件
working cost 工作费用,使用[经营]费用
working day 工作日
working electrode 工作电极
working flowsheet 作业[工作]流程图
working memory 工作存储器
working stope 采(矿)场
workload 工作负担
workman's compensation 工人补偿费
workmate 同事
works 著作,作品

workshop 车间,研讨会
World Bank 世界银行
world economy 世界经济
world production 世界产量
world supply-demand position 世界供需状态[形势]
World War I 第一次世界大战
World War II 第二次世界大战
worldwide competition 国际竞争
wormwood oil 苦艾油(可用作起泡剂)
worn parts 磨损件
wrap angle 包角
wrap-around motor 卷[环]绕式电动机
written application 书面申请
written approval 书面批准
written notice 书面通知
written off 冲销,注销,一笔勾销
wrong concentrate 不当精矿
wrong fraction 错误部分

X

X-2610(=Separan 2610) 聚丙烯酰胺(品名,相对分子质量 150 万~170 万,与 Separan2610 等同,可用作絮凝剂、分散剂、抑制剂、捕收剂)
X_{50C} equation X_{50C}方程
xanthan gum 黄原胶

xanthate 黄原酸盐,黄药 R-OC(S)SM
xanthate band position 黄原酸盐[黄药]谱带位置
xanthate coating 黄原酸盐[黄药]覆盖层
xanthate-dixanthogen couple 黄原酸盐[黄药]—双黄药电偶
xanthate-dixanthogen redox couple 黄原酸盐[黄药]—双黄药氧化还原电对
xanthate-dixanthogen redox potential 黄原酸盐[黄药]—双黄药氧化还原电位
xanthate group 黄酸基
xanthate ion 黄原酸离子
xanthate polar group 黄药极性基
xanthate radical 黄原酸根
xanthogen 黄原酸,双黄药,植物黄素(可用作捕收剂)
xanthogen formate 黄原酸甲酸盐[脂]
xenon arc lamp 氙弧灯
X-kuchen 捕收剂(萘胺)
X-ray analysis X射线分析
X-ray analyzer X射线分析仪
X-ray assay X射线分析
X-ray beam X射线束
X-ray calibration X射线标定[校准,定标],标度
X-ray console X射线控制台
X-ray density X射线密度
X-ray diffraction analysis X射线衍射分析
X-ray diffraction data X射线衍射资料
X-ray diffraction machine X射线衍射设备
X-ray diffraction technique X射线衍射技术[方法]
X-ray emission spectroscopy X射线发射光谱法[学]
X-ray energy dispersive system X射线能量色散系统
X-ray fluorescence analysis method X射线荧光分析法
X-ray fluorescent spectrometer X射线荧光分光计[光谱仪]
X-ray head X射线探头
X-ray installation X射线(流程)分析仪,X射线设备
X-ray intensity measurement data X射线强度测定数据
X-ray irradiation X射线照射[辐照]
X-ray measurement X射线测定[量]
X-ray monitor X射线控制测量仪,X射线监视器
X-ray on-line analysis X射线在线[联机]分析
X-ray on-stream analyzer X射线载流分析仪
X-ray photoelectron spectroscopy X射线光电子光谱法[学]
X-ray room X射线室
X-ray sedimentation X射线沉降[积,淀](测定,分析用)
X-ray unit X射线装置
X-Y mixture X-Y混合剂(α-萘胺溶于60%二甲基苯胺溶液中的混合物,可用

作捕收剂)
xylenol 二甲酚(二甲苯酚,可用作捕收剂和起泡剂)

Y

15 yard shovel 15立方码电铲
Yang's equation 杨氏方程式(界面化学的基本方程之一,它是描述固气、固液、液气界面自由能与接触角之间的关系式,也称润湿方程)
yard lighting 场地照明
yard maintenance 场地维修
yard piping design 场地管道设计
yarder 集材绞盘机,木材堆垛机,集材机
Yarmer F 松油(品名,可用作捕收剂、起泡剂)
Yate's method 耶特方法(造表方法)
yellow cake 黄色团块[块状物]
yellow dextrin 黄色糊精(经热处理的一种糊精,可用作发散剂、抑制剂)
yield 产率[量],收益(率),屈服

Z

Z3-xanthate Z3-黄药(乙基钾黄药,乙基黄原酸钾,可用作捕收剂)
$C_2H_5OC(S)SK$
Z4-xanthate Z4-黄药(乙基钠黄药,乙基黄原酸钠,可用作捕收剂)
$C_2H_5OC(S)SNa$
Z5-xanthate Z5-黄药(戊基钾黄药,戊基黄原酸钾,可用作捕收剂)
$C_5H_{11}OC(S)SK$
Z6-xanthate Z6-黄药(异戊基钾黄药,异戊基黄原酸钾,可用作捕收剂)
$(CH_3)_2CHCH_2CH_2OC(S)SK$
Z7-xanthate Z7-黄药(丁基钾黄药,丁基黄原酸钾,可用作捕收剂)
$C_4H_9OC(S)SK$
Z8-xanthate Z8-黄药(仲丁基钾黄药,仲丁基黄原酸钾,可用作捕收剂)
$CH_3CH_2CH(CH_3)OC(S)SK$
Z9-xanthate Z9-黄药(异丙基钾黄药,异丙基黄原酸钾,可用作捕收剂)
$(CH_3)_2CHOC(S)SK$
Z10-xanthate Z10-黄药(己基钾黄药,

己基黄原酸钾,可用作捕收剂)
$C_6H_{13}OC(S)SK$

Z11-xanthate Z11-黄药(异丙基钠黄药,异丙基黄原酸钠,可用作捕收剂)
$(CH_3)_2CHOC(S)SNa$

Z12-xanthate Z12-黄药(仲丁基钠黄药,仲丁基黄原酸钠,可用作捕收剂)
$CH_3CH_2CH(CH_3)·OC(S)SNa$

Z-200 丙乙硫氨酯(品名,O-异丙基-N-乙基硫代氨基甲酸酯,可用作捕收剂)

Zadra cell 扎德拉槽(回收金、银的电解槽)

Zeiss PMQII type spectrophotometer 蔡司[Zeiss]PMQII型分光光度计(德国)

Zephiramine 十四烷基二甲基苄基氯化铵[氯化十四烷基二甲基苄基铵](可用作捕收剂)
$C_{14}H_{29}-N^+(CH_3)_2(CH_2-C_6H_5), Cl^-$

zero contact angle 零接触角

zero current condition 零电流条件[状态]

zero order kinetics 零阶[级]动力学

zero vector 零向[矢]量

Zerolit FF 一种碱性离子交换树脂(品名,可用于萃取铀)

Zeromist 75%$NaHCO_3$和25%氟代戊基磺酸盐的混合物(品名,可用作抑制剂、湿润剂)

zeta coefficient Z[ζ]系数

zeta potential 电动电势,界面动电势,动电位,Z[ζ]电位

zeta potential curve 动电位曲线,Z[ζ]电位曲线

zeta reversal 动电位逆转,Z[ζ]电位逆转

Zetag 22[51,49] 阳离子型聚胺(液体)絮凝剂

Zetag 32[57,76,63,92] 阳离子型聚丙烯酰胺(固体)絮凝剂(英国)

Zetag 88 阴离子型高分子量聚电解质(浓乳剂)絮凝剂

zeta-potential 动电势[位],Z[ζ]电位[电势]

Zeten 一种动物胶品牌名(可用作絮凝剂)

Zetex 一种动物胶品牌名(可用作絮凝剂)

Zetol-A[B,C] 一种动物胶(品名,可用作絮凝剂)

zigzag path 曲折[之字形]路径

zinc 锌 Zn

zinc area 锌区(选锌工段)

zincate ion 锌酸盐离子

zinc butyl dithiophosphate 丁基二硫代磷酸锌

zinc chromate primer 铬酸锌底漆,锌黄底漆

zinc cone 锌粉给料斗

zinc cyanide 氰化锌(可用作闪锌矿、黄铁矿抑制剂) $Zn(CN)_2$

zinc cyanide chelate 锌氰化物螯合剂

zinc cyanide complex 氰化锌络合物

zinc cyanide compound 氰化锌化合物

zinc dross retort furnace　锌渣蒸馏罐炉
zinc dust precipitation method　锌粉置换法
zinc electrolysis plant　锌电解厂
zinc ethyl dithiophosphate　乙基二硫代磷酸锌
zinc flotation scavenger concentrate　锌浮选扫选精矿
zinc fuming furnace　锌烟化炉
zinc hexyl xanthate　己基黄原酸锌
zinc inhibiting reagent　锌抑制剂
zinc isoamyl dithiophosphate　异戊基二硫代磷酸锌
zinc m-cresyl dithiophosphate　m-甲苯基二硫代磷酸锌
zinc-mercaptan key　锌硫醇键
zinc methyl dithiophosphate　甲基二硫代磷酸锌
zinc mixing cone　锌混合漏斗
zinc o-cresyl dithiophosphate　o-甲苯基二硫代磷酸锌
zinc oxide　氧化锌(铜—铅—锌矿浮选时作为锌矿抑制剂)
zinc p-cresyl dithiophosphate　p-甲苯基二硫代磷酸锌
zinc phenyl dithiophosphate　苯基二硫代磷酸锌
zinc precipitate　锌沉淀物(金泥)
zinc precipitation technique　锌粉置换技术

zinc primary concentrate　最初级[第一阶段]锌精矿
zinc propyl dithiophosphate　丙基二硫代磷酸锌
zinc-silver-gold precipitate　锌-银-金泥
zinc sulfate base　硫酸锌基
zinc sulfate-cyanide complex　硫酸锌氰化物络合物
zinc sulfide　硫化锌(矿)
zinc sulfite　亚硫酸锌
zinc sulphate base compound　硫酸锌基化合物
zinc xanthate　黄原酸锌
zircon magnetics　锆磁石(电气石在工业上的称号)
zonal pattern　带状[分区]模式
Zn amyl xanthogenate　戊基黄原酸锌
Zn butyl xanthogenate　丁基黄原酸锌
Zn ethyl xanthogenate　乙基黄原酸锌
Zn heptyl xanthogenate　庚基黄原酸锌
Zn hexyl xanthogenate　己基黄原酸锌
Zn isoamyl xanthogenate　异戊基黄原酸锌
Zn isobutyl xanthogenate　异丁基黄原酸锌
Zn isopropyl xanthogenate　异丙基黄原酸锌
Zn octyl xanthogenate　辛基黄原酸锌
Zn propyl xanthogenate　丙基黄原酸锌
Zn thiocyanate　硫氰酸锌

附录

附录1 缩略语

A	altitude	海拔高度,标高
	ampere	安,安培(电流强度单位)
	Architecture	建筑(工程专业简称)
	argon	氩(第18号元素)
a	annum	年
	anthracite	无烟煤
	axial	轴向的,轴流式
A. P. I.	accurate position indicator	精确位置指示器
	American Petroleum Institute	美国石油学会
	application program interface	应用程序接口
A. S. T. M.	American Society for Testing Materials	美国材料试验学会
A. W. W. A.	American Water Works Association	美国水行业协会
AA	acrylic acid	丙烯酸
	activation analysis	活化分析,放射性分析
	atomic absorption	原子吸收
AAA	Aluminum Association of America	美国铝业协会
AASHO	American Association of State Highway Officials	美国各州公路与运输工作者协会
ACFM	actual cubic feet per minute	实际立方英尺/分(每分钟流量或产量)
	alternating current field measurement	交流电磁场检测
ADR	address	地址
	address register	地址寄存器
	adsorption-desorption-refining	吸附解吸精炼

Adv.	advance	推进,进尺
	advanced	先进的
adv.	advertisement	广告
	advice	建议,忠告,劝告,通知
AE	absolute error	绝对误差
	aeronautical engineering	航空工程
	aminoethyl	氨基乙基
AFR	acceptable failure rate	允许故障率
	air/fuel ratio	空气燃油混合比
AIME	American Institute of Mining Engineers	美国采矿工程师协会
	Association of the Institute of Mechanical Engineers	美国机械工程师学会
AL	Albania	阿尔巴尼亚
	algorithmic language	算法语言
	aluminum	铝
	amendment list	订正表,勘误表
AM	activity model	活动模式
	airlock module	气密舱
	ammeter	安培计
	amplitude modulation	调幅
	amylase	淀粉酶
	Armenia	亚美尼亚
AMAX Inc.	American Metal Climax Inc.	美国克利马克斯金属公司
AMSL	above mean sea level	平均海平面以上(的高度)
AN	access network	接入网
	account number	账号
	acid number	酸值
	acrylonitrile	丙烯腈,氰乙烯
	ammonium nitrate	硝酸铵,硝铵炸药
	Antarctica	南极洲
AO	amplifier output	放大器输出
	anaerobic aerobic process	厌氧好氧工艺法
	Angola	安哥拉

	auditor office	审计师办公室
APT	advanced passenger train	高级特快列车
	ammonium paratungstate	仲钨酸铵
aq.	aqueous	水的
AR	acceptance requirement	验收条件
	account receivable	应收款
	American Standard	美国标准
	Argentina	阿根廷
	aspect ratio	纵横比,屏幕宽高比
AS	acrylonitrile-styrene	丙烯腈-苯乙烯
	application system	应用系统
	Australian Standards	澳大利亚标准
	automatic synchronizer	自动同步器
ASARCO	American Smelting and Refining Company	美国熔炼[冶金]公司
ASME	American Society of Mechanical Engineers	美国机械工程师学会
AT	ambient temperature	环境温度,室温
	assistive technologies	辅助技术
	Austria	奥地利
	automatic transmission	自动变速,自动换挡
	auxiliary transformer	辅助变压器
ATS	absolute temperature scale	绝对温标
	American Technical Society	美国技术协会
	automatic transfer switch	自动转换开关
	automatic transfer system	自动切换系统
AU	alarm unit	报警装置
	atomic unit	原子单位
	Australia	澳大利亚
AX	amyl xanthate	戊基黄药
AZ	aluminum zinc	铝锌
	azaguanine	氮鸟嘌呤
	Azerbaijan	阿塞拜疆

	azimuth	方位
B. E. T. equation	Brunauer-Emmett-Teller equation	布鲁诺尔·埃梅特·泰勒公式
B. P. L.	bone phosphate of lime	骨质磷酸钙
BC	Before Christ	公元前
	benzoyl choride	苯酰氯
	binary code	二进制码
	bronze casting	青铜铸件
BD	Backward Diode	反向二极管
	Bangladesh	孟加拉
	basic design	基础设计
	benzidine	联苯胺,对二氨基联苯
	bidding document	招标文件
	break down	分解,出故障
BE	base excess	碱过剩
	Belgium	比利时
	bill of exchange	汇票
BF	bachelor of finance	金融学学士
	basis function	基函数
	blast furnace	高炉
	blind flange	盲板法兰
BG	back gain	背齿轮
	background	背景
	battened grating	板条格栅
	Bulgaria	保加利亚
	butylene glycol	丁二醇,保湿剂
BH	Bahrein	巴林
	boiler house	锅炉房
	Brinell hardness	布氏硬度
BI	built-in	内置,嵌入的,内装的
	Burundi	布隆迪
	bus interface	总线接口
	business intelligence	商业智能

bit.	bitumen		沥青
	bituminous coal		烟煤
BJ	bachelor of journalism		新闻学士
	ball joint		球形接头, 球窝接头
	barrage jamming		阻塞干扰, 抑制干扰, 全波段干扰
BL	bachelor of law		法学士
	base line		基准线
	bend line		弯曲线
	bill of lading		提货单
	butyrolactone		丁内酯
blk	black		黑
	blank		坯料, 空白的
	block		闭塞, 封锁, 阻碍, 块, 滑车, 块料
BM	Babbitt metal		巴氏合金
	ball mill		球磨机
	bench mark		水平点, 基准点
	bill of material		材料表, 材料单
	bi-motor		双发动机
BN	bachelor of nursing		护理学士
	balancing network		平衡网络
	boron nitride		氮化硼, 一氮化硼
	Brunei Darussalam		文莱
BO	base oil		基础油, 原油
	blow out		停炉, 爆裂
	Bolivia		玻利维亚
	branch office		分支机构
	butyl oleate		油酸丁酯
BR	blast rate		风速
	boiler room		锅炉房
	border router		边界路由器

	Brazil	巴西
BRD	building research division	建筑研究部门
	business requirement document	商业需求文档
BSCL	British Smelter Construction Limited	英国冶炼厂建设有限公司
BW	band width	频带宽度,带宽
	basic weight	基本质量,自重
	black and white	黑白
	Botswana	博茨瓦纳
BY	billion years	十亿年(美),万亿年(英)
	Byelorussia	白俄罗斯
Bz	benzene	苯
	benzoyl	苯(甲)酰
	bronze	青铜
C	carbon	碳
	Civil	土建(工程专业简称)
	concentrate	精矿,精煤
	electrostatic capacity	电容
C. A. C.	Canadian Alice Chalmers Company	加拿大阿里斯·查尔默斯公司
c. f. c.	critical flotation concentration	临界浮选浓度
C. I. P.	carbon-in-pulp	炭浆,炭浆法,发浆法
C. I. P. plant	carbon-in-pulp plant	炭浆厂
C. O. B.	coarse ore bin	粗矿仓
C. P. S.	cyclo-pneumatic separator	回旋风力分选器
C. S.	carbon steel	碳钢,素钢
	cast steel	铸钢
CA	Canada	加拿大
	chloraniline	氯苯胺
	cold air	冷空气,冷风
	compressed air	压缩空气
	controlled atmosphere	控制气氛
CAC	carbon-arc cutting	碳弧切割
	cascade	瀑落式

CAS		Chinese Academy of Sciences	中国科学院
CC		control console	控制台
		crushing control	破碎控制
		cyanogen chloride	氯化氰
CF		carrier-free	无载体的
		Central Africa	中非
		cresol-formaldehyde	甲酚,甲醛
		cubic feet	立方英尺
CG		center of gravity	重力中心,重心
		cholyglycine	甘胆酸
		coal gas	煤气
		Congo	刚果
CH		compass heading	罗盘航向
		conductor head	排水斗
chl		chloride	氯化物
		chlorination	氯化(作用,处理)
		Switzerland	瑞士
CI		cast iron	铸铁
		concentration index	浓度指数,集中指数
		Cote d'Ivoire	科特迪瓦
CIL		carbon in leach	浸出碳吸附(法),浸碳(法)
CIM		the Canadian Institute of Mining and Metallurgy	加拿大采矿冶金协会
CL		car load	车辆荷载
		center line	中心线,轴线
		Chile	智利
clnr.		cleaner	精选,精选机
CM		Cameroon	喀麦隆
		control memory	控制存储器
		core memory	磁心存储器
CMC		carboxymethyl cellulose	羧甲基纤维素
		critical micelle concentration	临界胶束[团]浓度

CMIEC	China Metallurgical Import and Export Company	中国冶金进出口公司
CN	cellulose nitrate	硝酸纤维素
	China	中国
	chloroacetophenone	氯乙酰苯,氯苯乙酮
	communicating net	通信网络
CNT	carbon nanotube	碳纳米管
	contrast	对比度
CNY	China Yuan	人民币
CO	carbon monoxide	一氧化碳
	classifier overflow	分级机溢流
	Colombia	哥伦比亚
COD	chemical oxygen demand	化学需氧量
COD value	chemical oxygen demand value	化学需氧量
COMBO vessel	combination container/break-bulk vessel	集装箱/杂货船
comp.	comparative	相对的
	compare	比较
	composite	复合物,合成物,化合物
conv.	converter	转化器,换流器
	conveyor	运输机,传送机
COS	coarse ore stockpile	粗矿堆场
CPB	cetyl pyridinium bromide	十六烷基溴化吡啶
CPC	cetyl pyridinium chloride	十六烷基氯化吡啶
	Corn Products Company	谷物产品公司
CPD(cpd)	compound	化合[混合,复合]物
CPM	critical path method	关键路线法(工程管理用语),临界运行图
cps	centipoise	厘泊(动力黏度单位,等于百分之一泊)
	cycles per second	每秒周数
CPS	central processing system	中央处理系统

	control power supply	控制电源
CR	chlorination residue	氯化残渣
	conditioned reflex	条件反射
	Costa Rica	哥斯达黎加
CROFT	cyanide retreatment of flotation tails	浮选尾矿的氰化再处理
CROFT surge tank	cyanide-retreatment-of-flotation-tailings surge tank	浮选尾矿氰化再处理缓冲槽[罐]
CS	common steel	普通钢
	control system	控制系统
	crystal structure	晶体结构
	Czech Republic	捷克
CSIRO(=C.S.I.R.O.)	Commonwealth Scientific and Industrial Research Organization	联邦科学与工业研究组织（澳大利亚）
CSM	Chinese Society for Metals	中国金属学会
CSS	cast semi-steel	半铸钢,钢性铸铁
	close side setting	最小排矿口宽度设置
CT	captive test	静态试验
	concentrate thickener	精矿浓缩机
	current transformer	电流互感器
CTAB	cetyltrimethyl ammonium bromide	十六烷基三甲基溴化铵
CTAC	cetyltrimethyl ammonium chloride	十六烷基三甲基氯化铵
CU	close up	闭合,接通,接通电流,特写镜头
	coefficient of utilization	利用系数
	control unit	控制器,控制单元,控制装置
	Cuba	古巴
cum.	cumulative	累计的,累计积的
CY	calendar year	日历年度
	cyclohexane	环乙烷
	cylinder	圆柱体,气缸
	Cyprus	塞浦路斯
CYN	cyanide	氰化物
D	density	浓度,矿浆浓度

		diameter	直径
		drain	排水,排水管,排水沟
		draw	拉,牵,拔制,拉制
		dust	细粉状物质,尘灰,尘埃
DAA		dodecyl ammonium acetate	十二烷基醋酸铵
DAC		dodecyl ammonium chloride	十二烷基氯化铵
DAP		diammonium phosphate	磷酸氢二铵
DCF		direct centrifugal flotation	直接离心浮选
		discounted cash flow	贴现现金流
DDC		digital data converter	数字数据变换器
		direct digital control	直接数字控制
DE		decision element	判定元件
		driving end	驱动端,传动端
		the internet domain name for Germany	德国
DECO		decreasing consumption of oxygen	降低氧消耗
		Denver Equipment Company	丹佛设备公司(美国)
		direct energy conversion operation	直接能量转换操作
DF		direction finder	测向器
		Dow frother	陶氏起泡剂(美国)
DK		Denmark	丹麦
		diketene	二乙烯酮
		distance kilometer	里程
DMMF		dry mineral mater free	无水干矿物质
DO		data output	数据输出
		diesel oil	柴油
		dissolved oxygen	溶解氧
		Dominican Republic	多米尼加共和国
DOC		direct operating cost	直接成本,直接运行费用
		document	文件,文献
DS		data sheet	设计技术要求数据表
		design specification	设计任务书,设计技术要求说明书

	disconnecting switch	隔离开关
DSM	design standard manual	设计标准手册
DSM screen	Dutch State Mine screen	荷兰国家矿业公司产弧形筛
DWP	Dyna whirl pool separator	Dyna 漩涡分选机
DWT	dead weight	自重,静重,静载荷
	deadweight tonnage	载重吨位,实载吨数
DZ	Algeria	阿尔及利亚
	dozen	一打(12个)
	drop zone	空投区
E	eccentricity	偏心率,偏心距,离心率
	Electrical	电气(工程专业简称)
	ethylene	乙烯
E.P.P.A.(=EPPA)	ethyl phenyl phosphonic acid	乙苯基膦酸
E/MJ	Engineering and Mining Journal	工程与采矿杂志
EA	each	每个,各
	elemental analysis	元素分析
	ethyl acrylate	丙烯酸乙酯
EC	Ecuador	厄瓜多尔
	effective concentration	有效浓度
	elution conditioner	洗提搅拌槽
	ethyl cellulose	乙基纤维素(塑料的一种成分)
EDMJ	Engineering and Mining Journal	工程与采矿杂志
EE	electrical eye	电眼
	electronic engineer	电子工程师
	engineering economic	工程经济
	Estonia	爱沙尼亚
EG	Egypt	埃及
	ethylene glycol	乙二醇
	grid voltage	栅压
ENFI	China ENFI Engineering Corp.	中国恩菲工程技术有限公司(原中国有色冶金设计研究总院)
EP	epoxide	环氧化物

	equilibrium potential	平衡电位
	error promise	可能偏差
	ethylene propylene	乙烯丙烯
EPA	Environmental Protection Agency	美国环境保护局
EQPT	equipment	设备
ES	earth switch	接地开关
	echo sounding	回声测深
	electro slag	电渣
	Spain	西班牙
ET	effective temperature	实际温度,有效温度,实感温度
	electrolytic[al]	电解的,电解质的
	Ethiopia	埃塞俄比亚
	express transportation	特快运输
Et.	Ethyl	乙烷基
EU	energy unit	能量单位
	enriched uranium	浓缩铀
	equivalent unit	等效单位
	Europe	欧洲
	European Union	欧盟
	polyether urethane	聚醚型聚氨酯,聚醚氨酯
EVOP	evolutionary operation	调优运算,改良操作法,渐变操作(选矿厂在生产过程中采用的一种统计技术)
F.O.B.	fine ore bin	粉矿仓
	free on board	离岸价格
F.R.	flocculation ratio	絮凝比,絮凝率
fdr	feeder	给料机,给矿机,馈电线,馈电回路
FF	feedforward	前馈
	fire fighting	防火,消防
	fuel flow	燃料流量
FI	field intensity	磁场强度

	field ionization	场电离
	Finland	芬兰
FJ	Fiji	斐济
	fixed jack	固定插孔,固定塞孔
	flush joint	平头接合
fl	flash flushing	闪光,闪蒸
	flow line	流体输送管线
	flush	齐平的,埋头的(螺钉等)
FR	failure rate	故障率
	field rheostat	激磁变阻器
	flow recorder	流量记录器
	France	法国
FRP	fibre reinforced plastic	纤维增强塑料
FRR	Fibre Reinforced Rubber	纤维增强橡胶
	froth removal rate	除沫率
G	General Arrangement	总图(工程专业简称)
g.t.	gross ton	长吨,英吨(等于1.01615公吨)
ga	general arrangement	总体布置,安装图
	grate area	筛棒面积,炉算面积
GA	Gabon	加蓬
	general arrangement	总装配
	glyoxal	乙二醛
GAC	Geological Association of Canada	加拿大地质协会
	Guangxi Aluminum Corporation	广西铝公司
GB	GigaByte	十亿字节(计算机)
	Great Britain	英国
	grounded base	接地底盘
GD	Gas Detector	气体检漏器
	gravimetric density	重量密度
	Grenada	格林纳达
	ground detector	接地探测器(电)
GE	gas ejection	瓦斯爆发,煤气喷射,注气

	General Electric Co.	美国通用电气公司
	Georgia	格鲁吉亚
	gross energy	总能量
GH	gaseous hydrogen	气态氢,氢气
	Ghana	加纳
	grid heading	网格航向
	growth hormone	生长激素
GN	gaseous nitrogen	气态氮,氮气
GPG	grams per gallon	克/加仑
GPL	general precision laboratory	通用精密实验室
gpl	grams per liter	克/升
GR	grade resistance	坡度阻力
	Greece	希腊
gr.	graphite	石墨
	grind	磨碎,研磨,磨削
grav.	gravel	砾石
	gravity	重力,引力
GSA	Geological Society of America	美国地质学会
	the General Service Administration	(美国)国家服务管理处
GT	gasoline tight	汽油密封的,不漏汽油的
	Grease Trap	滑脂分离器
HMS	heavy media separator	重介质分选机
	heavy medium separation	重介质分选
HU	Hungary	匈牙利
	hydroxy urea	羟基脲
HV	Heating and Ventilation	暖通(工程专业简称)
	high voltage	高压(电)
I	input	输入
	Instrumentation	仪表(工程专业简称)
	iodine	碘
i. e. p.	isoelectric point	等电点,等电位点

I. M. M. E.	Institute of Mining and Mechanical Engineers	采矿与机械工程师协会
ID	identity document	身份证件
	indicating device	指示装置
	Indonesia	印度尼西亚
	internal diameter	内径
IE	index error	指标误差
	industrial engineer	工业工程师
	industrial engineering	工业管理学
	Ireland	爱尔兰
IEEE	Institute of Electrical and Electronic Engineers	美国电气与电子工程师协会
IGS	inertial guidance system	惯性制导系统
	Interactive Graphic Services	电子计算机绘图仪,简称IGS系统
IL	initial line	开始线,起始线
	insertion loss	插入损失
	Israel	以色列
IM&C	International Minerals & Chemicals Corporation	国际矿物化学公司
IMM	Institution of Minerals and Metals	矿产与金属学会(英)
	Institution of Mining and Metallurgy	采矿冶金学会(英)
IMPC(I. M. P. C.)	International Mineral Processing Congress	国际选矿会议
IN	India	印度
	input	输入,进料
	insoluble nitrogen	不溶氮
INCO	International Nickel Corporation	国际镍业公司
ind.	index	索引
	individual	个别的
	industry	工业
Info.	information	信息,资料

IOM(I.O.M.)	Institute of Metals	金属学会
	Institution of Metallurgists	冶金工作者协会(英)
IOS	intermediate ore storage	中间矿石储存堆场,中间矿堆
	International Organization for Standardization	国际标准化组织
IQ	import quota	进口限额
	intelligence quotient	智力商数
	Iraq	伊拉克
IR	infrared radiation	红外辐射
	interrupt register	中间存储器
	Iran	伊朗
IRR	infrared rays	红外线
	internal rate of return	内部收益
IS	Iceland	冰岛
	input signal	输入信号
	interior surface	内表面
ISA	International Standardization Association	国际标准化协会
ISO	International Standardization Organization	国际标准化组织
IT	information technology	信息技术
	insulating transformer	隔离变压器
	Italy	意大利
IX	ion exchange	离子交换
JKMRC	Julius Kruttschmitt Mineral Research Center, university of Queensland	朱利叶斯·克鲁特施密特矿物研究中心(澳大利亚昆士兰大学)
JM	Jamaica	牙买加
	master of law(Juris Master)	法律硕士
JP	Japan	日本
	jet propulsion	喷气推进器
KE	Kaiser Engineering	凯萨工程公司

	key error	键入错误
	kinetic energy	动能
KG	kilogauss	千高斯(电磁场强度单位)
	Kyrgyzstan	吉尔吉斯斯坦
KHA	potassium hydroxamate	氧肟酸钾
Kt(KT)	kiloton	千吨
L	length	长度
	level	料位,水平面,标高
LA	lactic acid	乳酸
	Laos	老挝
	lightning arrestor	避雷器
	low altitude	低空
lav	lavatory	盥洗室
LB	Lebanon	黎巴嫩
	life boat	救生船
	light bin	轻型斗仓
	line busy	占线
LC	landing craft	登陆艇
	lethal concentration	致死浓度
	liquid chromatography	液体色谱分析法,液相色谱法
	logic circuit	逻辑电路
LC50	50% lethal concentration	50%致死浓度
LI	level indicator	液面指示器
	Liechtenstein	列支敦士登
	liquidity index	流动性指数,液性指数,流化指数
	longitudinal interval	纵向间隔
LK	link	环节,耦合线,连接线
	Sri Lanka	斯里兰卡
LMB	Laboratory of Molecular Biology	分子生物实验室(英国医学研究委员会)
logical I/O	logical input and output	逻辑输入输出

LOI(L.O.I.)	loss on ignition	烧失量,灼减
LPF	leaching-precipitation-flotation	浸出沉淀浮选(法)
	lowest operating frequency	最低操作频率
LR	laboratory reactor	实验室反应堆
	Liberia	利比里亚
	loading room	矿山装料室,装药车间
	long range	远距离,远程的
	low resistance	低阻力,低电阻
LT	letter message	书信电报
	Lithuania	立陶宛
	lost time accidents	有工时损失的事故
LU	logical unit	逻辑单元
	loudness unit	响度单位
	Luxembourg	卢森堡
LV	laser vision	激光视盘,激光录像
	Latvia	拉脱维亚
	leaching vessel	浸出槽[罐]
	linear velocity	线速度
	low voltage	低压(电)
M	Mechanical	机械(工程专业简称)
	mega-	百万,兆
	mine	矿山
	moisture	水分
M.	mill	选矿厂,工厂,轧钢厂
M.C. precipitation	Merill Crowe precipitation	金氢化池沉淀
M.E.	manufacturing engineer	制造工程师
	mining engineer	采矿工程师
M.S.	master switch	主(总)开关
	mild steel	软钢,低碳钢
	mine superintendent	矿长
	minerals separation	选矿,矿物分选

M. T. E.	Mining and Transport Equipment Engineering		采矿及运输设备工程公司
M/M	man/month		人/月(一个人一个月完成的工作量)
MA	magnetic amplifier		磁放大器
	methyl acrylate		丙烯酸甲酯
	microanalysis		微量分析,微区分析
MC	methyl cellulose		甲基纤维素
	mode control		模式控制
	Monaco		摩纳哥
MCC	motor control center		电机控制中心
mct	millions of cubic feet		百万立方英尺
MD	manual data		手控数据
	maximum demand		最大需要量,最大电负荷
ME	microelectrophoresis		微电泳
med.	medicine		医学,药
	medium		介质,中间的
MES	Mechanical Engineering Society		(美国)机械工程学学会
MG	machine gun		机关枪
	Madagascar		马达加斯加
	marginal		极限的,边缘的,临界的
	motion graphic		动态图像,动态影像
MHD	magnetohydrodynamic		磁流体动力(的)
MHS	magnetohydrostatic		磁流体静力(的)
	magnetohydrostatic separator		磁流体静力分离器
MIBC	methyl isobutyl carbinol		甲基异丁基甲醇
MIM	Mount Isa Mines Limited		芒特艾萨矿业股份有限公司(澳大利亚)
MIN(Min.)	minimum		最小值
	mining, mine		矿业,矿山
ML	machine language		机器语言
	Mali		马里

	maximum likelihood	最大相似性
	mean level	平均位面,平均高度
MM	mineralogical museum	矿物博物馆
	mischmetal	混合稀土,稀土金属混合物,铈镧稀土合金,米氏合金
	molten metal	熔融金属,熔态金属,金属熔液
M-M	man-month	人月(一个人一个月完成的工作量)
MMS(M.M.S.)	Manchester Metallurgical Society	曼彻斯特冶金学会(英)
	methylmercuric sulfate	甲基汞化硫酸酯
MMSA	Mining and Metallurgical Society of America	美国采矿与冶金学会
MMTC	Minerals and Metals Trading Corporation	矿产和金属贸易公司
MMU	memory management unit	存储器管理单元
	million monetary unit	百万货币单位
MN	magnetic north	磁北
	meganeuton	兆牛顿(力的单位)
	Mongolia	蒙古
MO	master oscillator	主振荡器,主控振荡器(电子)
	method of operation	操作方法
moly	molybdenum	钼
Moly-Cop	molybdenum-copper	钼铜合金
MPL	message processing language	信息处理语言
	mole per liter	摩尔/升
MS	medium steel	中碳钢,中硬钢
	memory system	存储系统
	mineral separation machine	浮选机
MSA	Mechanical Signature Analysis	机械特征分析
	mine safety appliance	矿山安全设备
	Mineralogical Society of America	美国矿物学会
MT	machine translation	机器翻译
	Malta	马耳他

		maximum torque	最大扭矩,最大转矩
		mean time	平均时间
MU		marginal utility	边际效用
		Mauritius	毛里求斯
		memory unit	储存设备
		methylene unit	亚甲基单位
MW		Malawi	马拉维
		medium wave	中波
		megawatt	兆瓦
		molecular weight	分子量
MX		Mexico	墨西哥
		multiplex	多路传输
		m-xylene	间二甲苯
MY		Malaysia	马来西亚
		methyl yellow	甲基黄
		million years	百万年
		motor yacht	机动游艇
MZ		mean grain size	平均粒度
		Mesozoic	中生代
		middle zone	中间地带
		Mozambique	莫桑比克
N. A.		not available	不可利用的,不能提供,不能用
		nucleic acid	核酸
n. d.		no date	日期不详,未注明日期
		no delivery	未交货
		not detected	未查获的
N. V. F.		nuclear vane feeder	核子叶片式给料器(机)
NA(N. A.)		Namibia	纳米比亚
		North America	北美洲
		not announced	未宣布
		not applicable	不适用的
NaAF		sodium aerofloat	钠黑药

NaAX	sodium amyl xanthate	戊基钠黄药
NaIX	sodium isopropyl xanthate	异丙基钠黄药
nat.	native	本国的,土著的,天生的
	natural	天然的
NBHC	New Broken Hill Consolidated	新布罗肯希尔联合公司（澳大利亚）
NCF	Net Cash Flow	净现金流动量
ND	national debt	国债
	natural draught ventilation	自然通风
	non-delay	瞬时动作,无迟发的
	nuclear device	核装置
NE	net energy	净能量
	Niger	尼日尔
	nitroethylene	硝基乙酸
	not exceeding	不超过
NEMA	National Electrical Manufactures Association	(美国)国家电气制造商协会
NG	narrow gauge	窄轨距
	Nigeria	尼日利亚
	nitroglycerin	硝酸甘油
NI	national insurance	国民保险制度
	Nicaragua	尼加拉瓜
	noise index	噪声指数
NIM	South African National Institute for Metallurgy	南非国立冶金研究所
NIM	National Institute for Metallurgy	国家冶金学会
	network installation management	网络安装管理
	nuclear instrument module	核仪器插件标准
NL	Netherlands	荷兰
	non-linear	非线性的
	north latitude	北纬
NLT	night letter telegram	夜间电报

	no lost time accidents	没有工时损失的事故
	not later than	不晚于
Nm^3	nominal cubic meter	标称立方米
NO	nobelium	元素锘的符号
	normally open	正常断开的(触点)
	Norway	挪威
	number	数字
NP	name plate	铭牌
	Nepal	尼泊尔
	neutral point	中性点
	nickel-plated	镀镍的
	nitro-paraffin	硝基烷
NST	new serial titles	新刊题录(期刊名)
	non stress test	无负荷试验
	normal starting torque	正常起动转矩
NTIS	National Technical Information Service	美国国家技术情报局
NY(N.Y.)	naphthol yellow	萘酚黄
	New York	纽约
nzls	nozzles	喷嘴
o.p.t.	oz per ton	盎司/吨
O/F(O'F)	overflow	溢流
	oxidizer to fuel	氧化剂与燃料比
O/S(o/s)	oversize	筛上产品,过大粒度,尺寸过大
OA	Oceania	大洋洲
	office automatic	办公自动化
	operation analysis	操作分析
ODAA	octyl decyl amine acetate	辛基癸基醋酸胺
OM	Oman	阿曼
	operation maintenance	操作维护,日常维修
	optical microscope	光学显微镜
OSA	on-stream analysis	在线分析,流程分析
OSS	open side setting	松边排口设定

	orbital space station	轨道空间站
OUT	output	输出,出料,输出量,生产量,输出功率
P	absolute pressure	绝对压力
	pipeline	管道,管线,输油管线
	Piping	配管(工程专业简称)
	propylene	丙烯
P.H.(PH)	power house	动力车间,发电厂
	precipitation hardening	析出硬化,沉淀硬化
	preheating	预热
p.ct(pct.)	per cent, percentage	百分数(%),百分率
P.D.	potential determination	电位测定,电位测定法
	potential difference	电位差,势差
	polymerization degree	聚合度
P.E.I.C.	periodic error integrating controller	周期误差累积控制器
P.I.F. analysis	portable isotope X-ray fluorescence analysis	便携式同位素X射线荧光分析(仪)
	program information file	程序信息文件
p.m.	pebble mill	砾磨机,球磨机,碎石磨机
	per minute	每秒
	post meridiem	下午
P.O. box	post office box	邮箱
P.T.A.A.(=PTAA)	para-toluene arsonic acid	对甲苯胂酸
PA	Panama	巴拿马
	polyacrolein	聚丙烯醛
	power amplifier	功率放大器
PAX	potassium amyl xanthate	戊基钾黄药
	private automatic (telephone) exchange	专用自动交换机
pcf	pound per cubic foot	磅/立方英尺
PE	Peru	秘鲁
	polyelectrolytes	聚合电解质,聚合高分子电解质
	polyethylene	聚乙烯

	power equipment	动力设备
PG	Papua New Guinea	巴布亚新几内亚
	program guidance	程序制导,项目指南
	propyl gallate	没食子酸丙酯,酸丙酯,五倍子酸丙酯
	protective ground	保护接地
pH	potential of hydrogen	酸碱度,酸碱值,氢离子浓度指数
pHc	pH critical	临界pH值,临界氢离子浓度
PK	Pakistan	巴基斯坦
	private key	内线电键,专用电键
	psychokinesis	意志力
PL	parting line	分界线
	pay load	有效负载,有效载重量
	phospholipid	磷脂
	pilot lamp	信号灯
	Poland	波兰
PLC	programmable logic control	可编程逻辑控制
	programmable logic controller	可编程逻辑控制器
	polycaprolactone	聚己酸内酯
PMS	particle measuring system	粒度监测仪
	pressure measuring source	测压源
	purchase management system	采购管理系统
ppb	parts per billion	十亿分率
	pounds per barrel	磅/桶
PRC	the People's Republic of China	中华人民共和国
preg'n	pregnant solution	贵液(含贵重矿物溶液)
PRI	pressure-ratio indicator	压力比率指示器
	primary	最初[原始,初步,基本,初级]的
PS	metric horse power	公制马力
	poly styrene	聚苯乙烯

		proof stress	弹性极限应力,最大容许应力,屈服点
PSM		particle-size monitor	粒度监测[控]器
		Pulse Slope Modulation	脉冲斜度调制
PT		platinum	铂金
		Portugal	葡萄牙
		potential transformer	电压互感器,变压器
PTY		party	当事人
		property	财产
Pty. Ltd.		private limited company	私人有限公司
PVC		polyvinyl chloride	聚氯乙烯
Py		pyridine nucleus	吡啶核
		pyridine ring	氮苯环,吡啶环
		pyrope	镁铝榴石
PZC		piperazine chloride	盐酸哌嗪
		point-of-zero charge	零点电荷
QA		Qatar	卡塔尔
		quality assurance	品质保证
R. I. P.		resin-in-pulp	矿浆树脂离子交换
RADA		random access discrete address	随机存取离散地址
		rosin amine denatured acetate	变性松酸胺醋酸盐
rclnr.		recleaner	再精选机,二次洗涤机
rec.		received	收到,接受
		recovery	回收率,回采率,恢复,再生
recip.		recipient	容器,受体
		reciprocating	往复的
regr.		regrinding	再磨
rel.		relative	相对的
		relative to	关于,对于
		reluctance	磁阻
ret.		retention	停(滞)留
		retire	退休

		return	返料,返回,回路
rghr.		rougher	粗选,粗选槽,粗切机
RM		radar mapper	雷达测绘器
		reaction mass	反应质量
		rod mill	棒磨机
RMB		Renminbi	人民币
RO		Read Only	只读
		Receive Only	只接收
		recovery operation	回收作业,收回作业
		reverse osmosis	反渗透,逆渗透
		Romania	罗马尼亚
ro. conc.		rougher concentrate	粗选精矿
ROS		reactive oxygen species	活性氧
		read-only-storage	只读存储器
		run off switch	出轨[跑偏]开关
rot.		rotation, rotor	旋转,转子
RU		Resource Unit	资源单元
		response unit	回应单位
		Russia	俄罗斯
rubr		rubber	橡胶
S		sample	样品
		silicate	硅酸盐
		Structure	结构(工程专业简称)
S. P. A.		strong phosphoric acid	浓磷酸
		styrene phosphoric acid	苯乙烯磷酸
S. R. plant		stacker and reclaimer plant	堆取料机车间
S. R. L.		soft rubber lined	(软)橡胶衬里的
s. t.		short ton	短吨,美吨(等于0.9072吨)
SA		Saudi Arabia	沙特阿拉伯
		semi-automatic	半自动的
		South Africa	南非
		South America	南美洲

	sulfamic acid	氨基磺酸
	surface area	表面面积
SAG	semi-autogenous	半自磨(的)
SAG feed box	semi-autogenous feed box	半自磨给矿盒
SAG mill	semi-autogenous mill	半自磨机
SAR	specific absorption rate	特定吸收率
	sodium absorption ratio	钠的吸附比值
SC	Seychelles	塞舌尔
	silicon control	可控硅控制
	standard conductivity	标准电导率
	system controller	系统控制器
scav.	scavenger	扫选,扫选机,净化剂,清除机
	scavenging	扫选,清除
scav. conc.	scavenger concentrate	扫选精矿
SCFM(scfm)	standard cubic feet per minute	标准立方英尺/分(每分钟流量或处理量)
	subcarrier frequency modulation	副载波调频
SCR	semiconductor controlled rectifier	半导体可控整流器
	silicon controlled rectifier	可控硅整流器
	solids contact reactor	固体接触反应器
SD	sight draft	侧视图
	slow down	减速,减慢
	standard deviation	标准偏差
	Sudan	苏丹
SDS	scientific data system	科学数据系统
	sodium dodecyl sulfate	十二烷基磺酸钠
SE	southeast	东南方
	sulfoethyl	硫代乙基
	Sweden	瑞典
	system engineer	系统工程
SEM	scanning electron microscope	扫描式电子显微镜
	standard electronic modules	标准电子组件

SF330		Superfloc 330	330絮凝剂
SG		Screen grid	屏栅极
		Singapore	新加坡
		soluble gelatin	溶性明胶
		spheroidal graphite	球状石墨
SI		silicone	聚硅酮,硅酮树脂
		Slovenia	斯洛文尼亚
		spark ignition	火花塞点火
Siemens AG		Siemens Company	西门子公司
SK		segerkegel	塞氏测温熔堆
		service kit	维修工具
		Slovakia	斯洛伐克
SM		master of science	科学硕士
		San Marino	圣马力诺
		sheet metal	金属薄板,金属片
		synchronous motor	同步电机
SME		Society of Manufacturing Engineers	(美国)制造工程师学会
		Society of Mining Engineers	(美国)采矿工程师学会
SN		Senegal	塞内加尔
		serial number	顺序号
		shipping note	装货通知
		succinonitrile	琥珀腈,丁二腈
		synchronizer	同步器,同步机
SO		Sales Order	销售订单
		shift out	移出
		Shipping Order	装货单
		solid	固体
		Somalia	索马里
		special order	特种订货,特别指令
		Stop Order	停机指令
		Sub-Office	分局
SP		self propelled	自行驱动的

	single pole	单极
	streaming potential	流动电位
sp	sample	样品,试样
	solubility product	溶度积
sprg.	sparger	起泡装置,喷雾器,喷洒器
SPS	safety pull switch	安全拉线开关
	samples per second	每秒样品数
	standard pipe size	标准管径,标准管子尺寸
SRL	Scientific Research Laboratory	科学研究实验室
SSDEVOP	simplex self-directing evolutionary operation	单纯形自定向调优运算
ST	shipping ticket	装货单
	standard time	标准时间
	storage tank	贮槽
STCF	slime-tailing cyclone feed	矿泥尾矿旋流器给矿
STPD	short ton per day	短吨/日,美吨/日
	standard temperature and pressure, dry	干态标准温度与压力
subbit.	sub-bituminous coal	副烟煤,次烟煤
surf.	surface	地面,表面,路面
	surfactant	表面活化剂
SY	square yard	平方码
	synchronize	使同步,整同步
	Syria	叙利亚
Syn. (syn.)	synchronization	同步,整步
	synthetic	合成的
SZ	size	粒度,尺寸,大小
	Swaziland	斯威士兰
TD	task description	任务说明书
	time delay	时间延迟
	turbine driver	涡轮机传动装置
TDS	total dissolved solid	溶解固体总量,溶解固体总浓度
TEB	tri-ethoxy butanol	三乙氧基丁醇

TEFC	totally enclosed fan cooled	全封闭风扇冷却
Tel. (tel.)	telephone	电话
TG	tachometer generator	测速发电机
	thermogravimetry	热解重量分析法
	Togo	多哥
	triglyceride	甘油三酸酯
TH	tap hole	出铁口,出钢口
	tape handler	磁带处理机
	Thailand	泰国
	thermometer	温度计
TJ	Tajikistan	塔吉克斯坦
	turbo-Jet	涡轮喷气发动机
TM	technical manual	技术手册
	temperature meter	温度计
	true mean value	真平均值
	Turkmenistan	土库曼
TMD	temperature of maximum density	最大密度时的温度
	tramp metal detector	金属杂质探测器
TN	total nitrogen	全氮量,总含氮量
	trade name	商品名
	Tunisia	突尼斯
toz	troy ounce	金衡盎司,金衡制盎司,金衡唡(英两=480格令)
TOC	total organic carbon	总有机碳
TOC value	total organic carbon value	总有机碳量[值]
tpa	tons per annum	吨/年(每年吨数)
TR	temperature recorder	温度记录器
	thio rubber	聚硫橡胶
	transmitter-receive	发送-接收机
	Turkey	土耳其
TW	tail water	尾水,废水,下游水

	terawatt	兆兆瓦,万亿瓦特
	tool wagon	工具车
	twin wire	双心导线
U/F(U'F)	underflow	底流
u/s(us.)	undersize	细粒,筛下产品,粒度过小,尺寸不足,过小
	unsatisfactory	不满足的,不充分的
UA	Ukraine	乌克兰
	ultra-audible	超声,超可听见的
	uric acid	尿酸
UG	Uganda	乌干达
	underground	地下,井下,地铁
UK	United Kingdom	英国
	urokinase	尿激酶
UN	United Nations	联合国
	uranyl nitrate	硝酸铀铣
UNCTAD	United Nations Conference on Trade and Development	联合国贸易与发展会议
UNIDO	United Nations Industrial Development Organization	联合国工业发展组织
US	ultrasound	超声,超声波
	underseal	底封
	United States	美国
USBM	United States Bureau of Mines	美国矿务局
UY	uranium Y	铀 Y
	Uruguay	乌拉圭
UZ	uranium Z	铀 Z
	Uzbekistan	乌兹别克斯坦
VC	Saint Vincent	圣文森特岛
	venture capital	风险投资
	vinyl chloride	氯乙烯

		virtual circuit	虚拟电路
		volume control	音量控制
VE		valence electron	价电子
		value engineering	价值工程,价值工程学
		velocity error	速度误差
		Venezuela	委内瑞拉
vol.		volatile	挥发(性)的,挥发物(质)
		volatilize	使挥发,使发散
		voltage	伏特,电压
		volume	体积,容积,卷
W		Watt	瓦,瓦特(功率单位)
		weigh scale	计量秤
		West	西,西方,西部
W/M		weight of measurement	测定重量
W. M.		whit metal	白合金,轴承合金
w/w		weight in weight	重量/重量比
		weight to weight ratio,% percentage by weight	重量百分比
WC		water circulation	水循环
		water closet	洗手间
		water cooled	水冷却
		weight concentration	重量浓度
WP		water pump	水泵
		water-proof	防水的
		white phosphorus	白磷
		working pressure	工作压力
XRF		X-ray fluorescence	X射线荧光(光谱)
ZA		Z Axis	Z轴
		zone alarm	防火墙(反病毒软件)
		zone of alarm	报警区
ZM		Zambia	赞比亚
		zone marker	区域指点标,区域标志

ZR	Zaire	扎伊尔
	zero resistance	零电阻
	zero restriction	零约束
μm	micrometer	千分尺,测微计
	micron	微米(等于百万分之一米)

附录2　机构、企业、刊物名称

A. M. Gaudin Flotation Symposium	A. M. 高登浮选论丛[美国论文集]
Academy of Science of Romania	罗马尼亚科学院
Agrico Chemical Company	阿格里科化学公司(产磷酸盐)
AIME Centennial Meeting	美国矿冶[采矿]工程协会百年纪念会议
Air Pollution Control Association(APCA)	空气污染控制协会(缩写 APCA)
Algoma Steel Corp., Ltd.	阿尔戈马钢铁公司(加拿大)
Allied Colloids	美国联合胶体公司
Allis-Chalmers family	阿利斯·查尔默斯家族(美国)
Allis-Chalmers Manufacturing Company	阿利斯·查尔默斯制造公司(美国)
Aluminum Company of America	美国铝业公司
AMAX Extractive Metallurgy Laboratory Colorado	美国科罗拉多州阿马克斯(AMAX)萃取冶金实验室
American Association of State Highway Officials(AASHO)	美国各州公路与运输工作者协会(缩写 AASHO)
American Cyanamid Company	美国氰胺公司
American Geological Institute	美国地质学会
American Institute of Mining Engineers(AIME)	美国采矿工程师协会(缩写 AIME)
American Institute of Mining, Metallurgical and Petroleum Engineers	美国采矿、冶金和石油工程师协会
American Metal Climax Inc.	美国克利马克斯金属公司
American Metal Company, Ltd.	美国金属股份有限公司
American Metal Market	美国金属市场(美刊)
American Meteorological Society	美国气象学会
American Mineralogist	美国矿物学家(美刊)
American Mining Congress Journal	美国采矿大会会志
American Mining Congress Meeting	美国采矿(专业)会议

American Mining Congress(AMC)	美国采矿会议[联合会](缩写 AMC)
American Nuclear Society	美国核学会
American Society for Testing and Materials	美国材料与试验协会
American Society of Heating, Refrigerating and Air Conditioning Engineers	美国供热制冷和空气调节工程师协会
American Society of Mechanical Engineers	美国机械工程师学会
American Society of Testing Materials	美国材料实验协会
American Society of Testing Materials (ASTM)	美国材料试验学会(缩写 ASTM)
American Technology Corporation(AMTECH)	美国技术公司(缩写 AMTECH)
Anaconda Copper Mining Company	阿纳康达铜矿公司[蟒蛇铜矿公司](美国)
Anglo American Corporation of South Africa Limited	南非英美资源集团有限公司
Applied Science Research	应用科学研究(荷刊)
Arizona Chemical Company	亚利桑那化学公司(美国)
Armco Automatics	阿姆科自动计量公司(美国)
Army Corp. of Engineers	美国陆军工程兵团
Arthur Concentrator	阿瑟选厂(美国)
ASARCO Central Engineering Department	美国熔炼公司中央工程处
Ashland Oil Refining Co.	阿什兰炼油公司(美国)
Association of Tungsten Producers	钨生产者协会(私人企业间国际组织)
ASTM Standardization News	美国材料试验学会标准化新闻[ASTM 标准化新闻](期刊)
Atomic Energy Commission(AEC)	(美国)原子能委员会(缩写 AEC)
Aufbereitungs-Technik(AT)	选矿技术(德刊,缩写 AT)
Australia Mineral Industries Review(quarterly)	澳大利亚矿物产业(季刊)
Australia Mining Industry Council	澳大利亚采矿工业委员会[协会]
Australian Institute of Mining and Metallurgy (AIMM), Melbourne	澳大利亚采矿冶金学会(缩写 AIMM, 墨尔本)
Australian Institute of Mining and Metallurgy Proceedings	澳大利亚采矿与冶金学会会刊

Australian Minerals Industry Research Association	澳大利亚矿物工业研究协会
Australian Mining	澳大利亚采矿(杂志)
Australian Standard	澳大利亚标准
Bechtel Corp.	柏克德[贝克特尔]公司(美国,综合性工程公司)
Bechtel Inc.	柏克德工程公司(美国)
Bechtel International service Inc.	美国柏克德国际服务公司
Bechtel's Hydro & Community Facilities Division(H & CF)	柏克德公司的水电和公共设施部(美国)
Bishop Scheelite Concentrator	比索白钨选矿厂
Boan Consolidated, Zambia	赞比亚博安联合公司
Boliden Group	波利顿集团公司(瑞典)
Bor mines, Yugoslavia	南斯拉夫博尔矿
Bougainville Copper Limited	布干维尔铜有限公司(位于巴布亚新几内亚)
Bougainville Copper Limited Concentrator	布干维尔铜有限公司选矿厂(位于巴布亚新几内亚)
Bougainville Island	布干维尔岛(位于巴布亚新几内亚)
Brewster Phosphates	布鲁斯特磷酸盐公司(美国)
British Book News	英国书讯
British Patents Abstracts	英国专利文摘(英刊)
British Smelter Construction Limited	英国冶炼工程建设有限公司
Broken Hill Proprietary Company Ltd.	布罗肯希尔控股有限公司(澳大利亚)
Building Research Division	建筑研究部(美国)
Bull and Ball Co. Inc. Mineral Processing Consultant, Colorado	美国科罗拉多州布尔和鲍尔公司选矿顾问
Bulletin of American Geological Institute	美国地质学会[研究所](AGI)通报
Bulletin of American Mathematical Society	美国数学协会通报[会刊]
Bulletin of American Meteorological Society	美国气象学会通报[会刊]
Bureau of Mines	(美国)矿务局

C. S. I. R. O Division of Mineral Chemistry	(澳大利亚)联邦科学工业研究组织矿物化学部
Canada-China Trade Council	加中贸易理事会
Canadian Alice-Chalmers(CAC)	加拿大阿里斯-查尔默斯公司(缩写CAC)
Canadian Concentrator	加拿大选矿厂
Canadian Institute of Mining and Metallurgy (CIMM), Montreal	加拿大采矿冶金学会(缩写CIMM),蒙特利尔
Canadian Mining & Metallurgical Bulletin (=CMM Bulletin)	加拿大采矿冶金通报[会刊]
Canadian Mining and Metallurgical Bulletin	加拿大采矿冶金会报[刊]
Canadian Mining Journal	加拿大采矿杂志
Canadian Superior Oil Co.	加拿大高级油品公司
Canex Placer Limited, Endako Mines Division	Canex砂矿公司,恩达科[英达科]矿业分公司(加拿大)
Caterpillar Far East Limited	卡特彼勒远东有限公司(美国)
Central Research Laboratory Stamford	斯坦福德中央研究实验室(美国)
Central South Institute of Mining and Metallurgy	中南矿冶学院
Chemical Abstract	化学文摘(美刊)
Chicago Branch Office	芝加哥分支机构(美国)
Chile Copper	智利铜业公司
China Commercial and Industrial Press	中国工商出版社
China Commodity Inspection Bureau	中国商品检验局
China Coordination office	中国联络处
China Metallurgical Import and Export Corporation	中国冶金进出口公司
China National Nonferrous Metals Import & Export Company	中国有色金属进出口总公司
China National Technical Import and Export Corporation(CNTIEC)	中国技术进出口总公司(缩写CNTIEC)
Chinese Government	中国政府
CIM Bulletin	加拿大采矿冶金通报

Cleveland-Cliffs Iron Co.	克利夫兰·克利夫斯钢铁公司(美国)
Climax Mine	克利马克斯矿(美国)
Climax Molybdenum Company	克利马克斯钼公司(美国)
Climax Molybdenum Company's Conversion Plant	克利马克斯铜公司制酸厂(美国)
Climax Plant	克利马克斯选矿厂(美国)
Coal Age	煤时代(美刊)
Coal Mining and Processing	煤矿开采与加工(美刊)
Coba Mines Ltd.	科巴矿业有限公司(澳大利亚)
College Park Metallurgy Research Center, Bureau of Mines	矿山局学院园区冶金研究中心
Colloid and Surface Chemistry Group, Department of Physical Chemistry, University of Melbourne	墨尔本大学物理化学系胶体和表面化学教研组
Colombia University Laboratory	哥伦比亚大学实验室(美国)
Colorado School of Mines	科罗拉多矿业学院(美国)
Colorado School of Mines Quarterly	科罗拉多矿业学院季刊(美刊)
Cominco American Incorporated Company, Spokan, Washington	华盛顿州斯波坎市美国科明科公司
Computer Journal	计算机杂志(英刊)
COMSAT Technical Review	通讯卫星公司技术评论(美刊)
Concentrator of Lake Dufault Mines, Noranda	杜福尔特湖矿选矿厂,诺兰达(加拿大)
Consolidated Gold Field Limited in South Africa	南非联合黄金股份有限公司
Conveyor Components Company	输送机配件公司(美国)
Conveyor Equipment Manufacturers Association	运输机设备制造商协会
Copper Inc.	铜业有限责任公司
Council for Mutual Economic Assistance (CMEA)	经济互助委员会(简称经互会),是由苏联组织建立的一个由社会主义国家组成的政治经济合作组织,1991年6月28日,该组织在布达佩斯正式宣布解散(缩写CMEA)
CPC International Inc.	CPC(谷物产品)国际公司(美国)
Credit Lyonnais Representative Office in China	法国里昂信贷银行中国代表处

Current Science	现代科学(印度,期刊)
Current Technology Index	当代工艺索引(英刊)
Custom Controls Company(CCC)	(客户)自定制控制公司(提供工业防爆空调通风设备,美国,缩写 CCC)
Cyanamid(Far East)Ltd.	氰胺(远东)有限公司
Cyanamid Australian Pty.,Ltd.	澳大利亚氰胺有限公司
Cyanamid De Mexico,SA De CV	墨西哥氰胺有限公司
Cyanamid GmbH Germany	德国氰胺股份有限公司
Cyanamid International Division	氰胺国际部
Cyanamid Japan Ltd.	日本氰胺有限公司
Cyanamid Mining Chemicals Handbook	氰胺公司矿用化学药剂手册
Cyanamid of Canada Ltd.	加拿大氰胺有限公司
Cyanamid of Great Britain Ltd.	英国氰胺有限公司
Cyanamid Taiwan Corporation	台湾氰胺公司
Cyprus Bagdad Copper Co.	塞浦路斯巴格达铜业有限公司
Cyprus Pima Concentrator	塞浦路斯皮马选矿厂
Cyprus Pima Mining Co.	塞浦路斯皮马矿业有限公司
Cyprus Pima Mining Company,Tucson,Arizona	亚利桑那州图森市塞浦路斯皮马采矿公司(美国)
Data Processing	数据处理(英刊)
Denver Equipment Co.	丹佛设备公司(美国)
Department of Materials Science and Engineering,University of California	加利福尼亚大学材料科学工程系(美国)
Department of Metallurgical Engineering,South Dakota School of Mines and Technology	南达科塔矿业技术学院冶金工程系(美国)
Department of Mineral Engineering,University of British Colombia,Canada	加拿大不列颠哥伦比亚大学矿物工程系
Department of Mineral Processing,Pennsylvania State University	宾夕法尼亚州立大学选矿系(美国)
Department of Mining & Mineral Technology,Imperial College,London	伦敦皇家学院采矿和矿物工艺系

Department of Mining and Mineral Engineering, Faculty of Technology, Tohoku University, Aramaki, Sendai, Japan	日本仙台东北大学工学院矿冶工程系
Dept. Chem. Eng., Univ. of Natal, Durban	纳塔尔学院化学工程系,德班(南非)
Dept. of Chemistry, Univ. of Lethbridge, Alberta	阿尔伯塔省莱斯布里奇大学化学系(加拿大)
Dow Corning	道康宁公司(道康宁公司是康宁玻璃制造公司和陶氏化学公司于1943年合资建立的,是有机硅产品行业的全球领先企业)
Dutch State Mine	荷兰国家矿业公司
Duval Sierrita Concentrator	杜瓦尔·西雅里塔选矿厂(美国)
Duval Sierrita Corporation	杜瓦尔·西雅里塔公司(美国)
E & MJ(Engineering & Mining Journal)	工程与采矿杂志(美刊)
Earth Resources Company	地球资源公司(美国)
Earth Science Inc. Colorado	科罗拉多州地球科学公司(美国)
Ecstall Concentrator of Texas Gulf	德克萨斯海湾埃克斯塔尔选矿厂(美国)
Ecstall Mining Co.	埃克斯塔尔矿业公司(美国)
Eimco Envirotech	埃姆科[艾默克]环境技术公司(美国)
Eimco Process Equipment Company	埃姆科[艾默克]工艺设备公司(美国)
Elandsrand Gold Mining Company	依兰茨兰德矿业公司(南非)
Electric Engineering University of Toronto	多伦多电力工程学院(加拿大)
Electronic/Electric Product News	电子与电气产品新闻(美刊)
Electro-Sensors	美国伊莱克森公司(提供转速,振动,温度监测及传感器和电机控制产品)
Energy Conversion & Management	能源转换与管理(英刊)
Energy Digest	能源辑要[文摘](美刊)
Engineering & Mining Journal(E & MJ)	工程与采矿杂志(缩写E & MJ,美刊)
Engineering Division, Outokumpu Oy, Finland	芬兰奥托昆普公司工程部
Engineering Economist	工业经济学家(英刊)
Engineering Index(Annual)	工程索引(年刊,美刊)

Engineers' Digest	工程师摘要(英刊)
Envirotech Canada Ltd.	加拿大环境技术有限公司
Envirotech Corporation	环境保护技术公司(美国)
Erzmetall	金属学(德刊)
European Chemical News	欧洲化学药品新闻(刊物)
European Communities	欧洲共同体
Falcon-Bridge Mines Ltd.	鹰桥矿业有限公司(加拿大)
Falcon-Bridge Nickel Mines	鹰桥镍矿业公司(加拿大)
Federal Agency	联邦机构(美国)
Federal and State Environmental Regulations	联邦和州的环保条例(美国)
Federal Highway Administration	联邦公路管理局(美国)
Federal Water Pollution Control Act Amendment	联邦水污染控制法修正案(美国)
Federal Water Pollution Control Law Amendments	联邦水质污染控制法修正案(美国)
Float Ore Ltd.	浮选矿有限公司
Flotation	浮选(日刊)
Fluor China Inc.	福禄中国公司(美国)
Franklin Institute	富兰克林研究院(美国)
Frontino Gold Mines Limited	弗龙蒂诺金矿有限股份公司(哥伦比亚)
Fuel Science Section, Pennsylvania State University	宾夕法尼亚州立大学燃料科学部(美国)
Galigher Company	加利格公司(美国)
General Service Administration (GSA)	通用服务管理局(缩写GSA)(美国)
Geological Society of America Bulletin	美国地质学会(GSA)通报[学报]
Glückauf-Forschungsheffe	矿业研究(德刊)
Gold Fields of South African	南非金矿
Government Report Annual Index	美国政府报告年度索引
Ground Water Monitoring Review	地下水监测综论(美刊)
Guangxi Aluminium Corporation	广西铝业公司(即广西平果铝矿)
Heating/Piping/Air Conditioning	供暖,配管与空调(美刊)
Henry Krumb School of Mines, Colombia University	美国哥伦比亚大学亨利·克鲁姆矿业院

Hercules Incoporated (a global manufacturer of chemical specialties)	美国赫克力士公司(是一家跨国企业集团,专门生产特种化学品,成立于1913年)
Heyl & Patterson Inc.	美国海尔&帕特森公司(专业工程公司,成立于1887年)
Hitachi Ltd.	日立株式会社(日本)
Honeywell International	霍尼韦尔国际公司(是一家国际性从事自控产品开发及生产及在多元化技术和制造业方面占世界领导地位的跨国公司,成立于1885年,总部在美国新泽西州莫里斯镇)
Houston International Minerals	休斯敦国际矿物公司(美国)
IEEE Transactions of Automatic Control	IEEE自动控制汇[会]报(美刊)
Imperial Chemical Industries (ICI)	英国帝国化学工业集团(缩写ICI,是一个全球性的化工集团,成立于1926年)
Import and Export Administration Committee	进出口管理委员会
Import and Export Control Commission of Guangxi Zhuang Autonomous Region	广西壮族自治区进出口控制委员会
Import Duty and Customs Commission	进口关税与海关委员会
Import-Export Bank	进出口银行
Industrial Minerals	工业矿物(英刊)
Industrial Research Assistance Program (IRAP)	工业(科学)研究协助计划(加拿大,缩写IRAP)
Industrie Minerale	采矿工业(法刊)
Institute of Geology and Geophysics, CAS	中国科学院地质与地球物理研究所
Institution of Mining and Metallurgy, London	英国采矿冶金学会(伦敦)
Instruments & Control Systems	仪表与控制系统(美刊)
International Glossary of Coal Petrography	煤岩学国际术语
International Handbook of Coal Petrography	煤岩学国际手册
International Journal of Applied Radioaction and Isotope	国际应用放射和同位素杂志(英刊)

International Journal of Energy Research	国际能源研究杂志(英刊)
International Journal of Mineral Processing	国际选矿杂志(荷兰刊)
International Mineral Processing Congress	国际选矿会议
International Minerals & Chemicals Corporation(IMC)	美国国际矿物化学公司(缩写 IMC)(产磷酸盐)
International Nickel Company of Canada, Ltd.,Thomson,Manitoba	加拿大国际镍公司,曼尼托巴省汤姆森厂
International Nickel Company of Canada, Ontario	加拿大安大略省国际镍公司
International Nickel Company,Sudbury, Canada	加拿大萨德柏里国际镍公司
International Nickel Corporation(INCO)	加拿大国际镍公司(缩写 INCO)
International Nickel's Clarabelle Concentrator	国际镍公司克拉拉贝尔选矿厂(加拿大)
International Organization for Standardization	国际标准化组织
International Standardization Association(ISA)	国际标准化协会(缩写 ISA)
International Standardization Organization(ISO)	国际标准化组织(缩写 ISO)
Japanese Industrial Standard	日本工业标准
Joint Industry Conference Electrical Standards	工业通用电器标准联合会
Joint Industry Council	联合工业委员会(美国)
Journal of Environmental Science	环境科学杂志(美刊)
Journal of Geology	地质学杂志(美刊)
Journal of Geophysical Research	地球物理研究杂志(美刊)
Journal of Mining and Metallurgical Institute of Japan	日本矿业会志(日刊)
Journal of Nonferrous Metal	有色金属杂志
Journal of Scientific and Industrial Research	科学与工业研究杂志(印度刊)
Journal of Solar Energy Engineering	太阳能工程杂志(美刊)
Journal of South Africa Institute of Mining & Metallurgy(SAIMM)	南非采矿与冶金学会会刊(缩写 SAIMM)
Journal of Structural Mechanics	结构力学杂志(美刊)
Journal Water Pollution Control Federation (=Journal WPCF)	污水处理联合会会刊(美刊)

Julius Kruttschnitt Mineral Research Center (JKMRC), University of Queensland, Australia	澳大利亚昆士兰大学朱利叶斯·克鲁特施尼特矿物研究中心(缩写JKMRC)
Kaiser Engineers, Inc. (KE)	凯萨工程(咨询)公司(美国,缩写KE)
Kawasaki Heavy Industries Ltd.	川崎重工业株式会社(日本)
Kennecott Copper Corporation	肯尼科特铜业公司(美国)
Kennecott Copper Corporation Bonneville Concentrator	肯尼科特铜公司博纳维尔选矿厂(美国)
Kennecott Nevada Mines Division	肯尼科特内华达州矿业分公司(美国)
Kobe Steel Ltd.	神户制钢所(公司,日本)
Komatsu	株式会社小松制作所(即小松集团,是全球最大的工程机械及矿山机械制造企业之一,成立于1921年,小松集团总部位于日本东京)
Laboratory Management	实验室管理(杂志)
Laboratory Practice	实验室技术(杂志)
Lakefield Research	湖地研究所(加拿大)
Lakefield Research (A Division of Falcon-Bridge Limited)	鹰桥有限公司湖地研究所(加拿大)
Larox Oy	芬兰拉罗克斯公司(生产固液分离设备,净化过滤机和阀门等)
Lead-Zinc Update	铅锌进展(美刊)
Linatex	耐纳特耐磨胶业有限公司(诞生地:澳大利亚)
Linatex Corp. of America	美国耐纳特公司(生产耐磨耐腐蚀的橡胶)
Liquid Solid Separation Ltd.	英国液-固分离公司
Lurgi (Germany)	鲁奇公司(德国,化工领域的工程公司)
Mackay School of mines, University of Nevada	内华达大学麦凯矿业学院(美国)
Magmont Concentrator	马格蒙特选矿厂(铜、铅、锌选矿厂,位于美国密苏里州)

Magna Concentrator	马格纳选矿厂(位于美国犹他州)
Marine Sciences	海洋科学(日刊)
Material Controls Inc.	物[材]料控制公司(美国)
McDonald Douglas China Inc.	麦克唐纳·道格拉斯中国分公司(美国)
Mckesson Chemical Company	麦克森化学药品公司(美国)
Measurement & Control	测量与控制(英刊)
Metals Bulletin (London)	伦敦金属导报(英国)
Mine & Quarry	地底与露天采矿(英刊)
Mine and Smelter Division, Barber-Green Company	巴伯-格林公司矿山和冶炼部(美国)
Mine Safety Appliances Co.	矿山安全用品公司(美国)
Mineral Deposit Ltd. (The Manufacture of Reichert Cones)	矿产有限公司(澳大利亚,生产赖克特圆锥选矿机)
Mineral Engineering Department of Britain Colombia University	不列颠哥伦比亚大学矿业工程系(加拿大)
Mineral Processing & Extractive Metallurgy	矿物加工和提炼冶金学(英刊)
Mineral Processing Division of the Royal Institute of Technology, Stockholm	斯德哥尔摩皇家技术协会选矿事业部(瑞典)
Mineral Science and Engineering	矿物科学与工程杂志
Minerals Division of Dresser Industries, Houston, Texas	德莱赛(Dresser)选矿工业分公司,位于美国德克萨斯州休斯敦市
Mines Magazine	矿业杂志(美刊)
Mining Congress Journal	采矿大会会刊(美国)
Mining Engineer	采矿工程师(英刊)
Mining Engineering	采矿工程(美刊)
Mining Industry	采矿工业(英刊)
Mining Journal	采矿周刊(英刊)
Mining Journal Books Ltd.	矿业书刊出版公司(英国)
Mining Journal Ltd.	采矿杂志(出版)有限公司
Mining Magazine	采矿杂志(英刊)
Ministry of Foreign Trade and Economic Cooperation	外经贸部

Ministry of Metallurgy	冶金部
Minmet Scientific Limited	明美科技有限公司
Mitsubishi Corporation	三菱商事株式会社(日本)
Mitsui	三井(集团)(日本)
Mitsui & Co.,Ltd.	三井物产株式会社(日本)
Mitsui Mining and Metals	三井采矿冶金公司(日本)
Molybdenum Products Corporation	钼产品公司(美国)
Molycorp,Inc.,Questa Division	钼有限公司奎斯塔分公司(美国)
Monthly Weather Review	气象月报(美刊)
Mount Isa Mines Limited	芒特艾萨矿业有限公司(澳大利亚)
Muruntau open pit	穆伦陶露天金矿(乌兹别克斯坦)
NALCO	美国纳尔科化学公司(是世界上最大的水处理化学品供应商)
National Bureau of Standards(NBS)	国家标准局(美国,缩写 NBS)
National Crushed Stone Association(NCSA)	国家轧石协会(美国,缩写 NCSA)
National Electrical Manufacturers Association	美国电气制造商协会
National Fire Protection Association Codes	国家防火协会标准(美国)
National Institute for Metallurgy in Johannesburg	约翰内斯堡国家冶金研究所(南非)
National Institute for Metallurgy,Johannesburg, South Africa	南非约翰内斯堡国立冶金学院
National Iranian Copper Industries Co.	伊朗国家铜业公司(缩写 NICICO)
National Research and Development Corporation (NRDC)(UK)	国家研究和发展公司(英国,缩写 NRDC)
National Research Corporation	国家研究中心(美国,致力于改善医疗保健行业提供的服务质量的组织)
National Research Council Canada(NRC)	加拿大国家研究委员会(缩写 NRC)
National Sand & Gravel Association	国家沙石协会(美国)
National Slag Association	国家矿渣协会(美国)
National Technical Information Service(NTIS)	美国国家技术情报局(缩写 NTIS)
Navoi Mining Combine	纳沃伊采矿联合公司(乌兹别克斯坦)
Nchanga Consolidated,Zambia (Anglo-American)	赞比亚恩昌加公司(英美资源集团)

New Broken Hill Consolidated Limited	新布罗肯希尔联合有限公司(澳大利亚)
New Scientist	新科学家(国际化科学杂志)
Newmont Proprietary Limited	纽蒙特控股有限公司(美国)
Nippon Electric Co. ,Ltd.	日本电气公司
Noise Control Engineering Nordberg	噪声控制工程(美刊)
Nordberg Bulletin	诺德伯格产品通报
Nordberg Process Machinery Test Center	诺德伯格工艺机械试验中心
North American Mill	北美选矿厂
North Broken Hill Limited	北布罗肯希尔有限公司(澳大利亚)
Northeast Institute of Technology	东北工学院
NTIS Tech. Notes	美国国家技术情报局简编
Nuclear Science & Engineering	核子科学与工程(美刊)
Occupational Safety and Health Administration(OSHA)	职业安全和健康署(美国,缩写 OSHA)
Ocean Exploitation	海洋开发(日刊)
Ocean Industry	海洋工业(美刊)
Operating Committee of United State War Production Board	美国战时生产局营运委员会
Outokumpu Oy	奥托昆普公司(芬兰,矿业公司)
Outokumpu Oy Technical Export Division, Finland	芬兰奥托昆普公司技术输出部
Ozark Lead Company	奥扎克铅公司(美国)
Pacific Science	太平洋科学(美国太平洋科学协会官方杂志)
Panasonic Corporation	松下电器株式会社(日本)
Patino Mining Corp.	帕蒂诺采矿公司(加拿大)
Perkin-Elmer	美国珀金-埃尔默公司(是世界上最大的分析仪器设计、咨询公司及生产制造商,成立于1937年)
Phelps Dodge Corporation Tyron, Branch	费尔普斯·道奇公司蒂龙分公司(美国)
Phelps Dodge Tyrone Concentrator	费尔普斯·道奇公司蒂龙选矿厂(美国)
Phelps,Dodge & Company	费尔普斯·道奇公司(美国)

Philips Brothers China Inc.	菲利浦兄弟中国有限公司(荷兰)
Physical Review	物理评论(美刊)
Pingguo Aluminum Company	平果铝业公司(中国)
Pingguo Aluminum Industrial Base Project	平果铝工业基地工程(中国广西)
Pingguo Aluminum Plant	平果铝厂(中国广西)
Pollution Engineering	污染工程(美刊)
Popular Science	大众科学(美国月刊杂志)
Power Transmission Design	电力传输设计(美刊)
Primary Tungsten Association(PTA)	原生钨协会(缩写PTA)
Proceedings of American Mathematical Society	美国数学协会会报(美刊)
Proceedings of Central South Institute of Mining and Metallurgy	中南矿冶学院学报(中国)
Proceedings of National Academy of Sciences of USA	美国国家科学院学报(综合性期刊)
Progress in Material Science	材料科学进展(英刊)
Quarterly Journal of Mechanics and Applied Mathematics	力学与应用数学季刊(英刊)
Ramsey Engineering Company(REC)	拉姆齐工程公司(美国,缩写REC)
Randol International Ltd.	兰多尔国际公司(美国)
Rasa Trading Co.,Ltd.	拉萨商事株式会社(日本公司)
Rautaruukki Oy	劳塔鲁基钢铁公司(芬兰)
Representative Office	代[办]表处
Representative Office in China	驻华代[办]表处
Research Institute of Mineral Dressing and Metallurgy,Tohoku University,Aramaki, Sendai,Japan	日本仙台东北大学矿冶研究所
Rexnord Inc.	莱克斯诺公司(美国)
Rexnord Process Machinery Division	莱克斯诺工艺[选矿]机械分公司(美国)
Reynolds International Inc.	雷诺(铝业)国际公司(美国)
Rosario Dominica SA	南美罗萨里奥多米尼加公司
Rosario Resources	罗萨里奥资源公司(阿根廷)
Rosemount Inc.	罗斯蒙特公司(美国)

Royal Institute of Technology	皇家理工学院(英国)
Royal Institute of Technology in Stockholm	斯德哥尔摩皇家工学院(瑞典)
Royal School of Mines	皇家矿业学院(英刊)
Royal School of mines, London	伦敦皇家矿业学院
Sanyo Electric Trading Co., Ltd.	三洋电机贸易株式会社(日本)
Sar Cheshmeh, Iran	伊朗萨尔切什迈(铜,钼)矿(属伊朗国家铜业公司)
Science	科学(美刊,自然科学综合类期刊)
Science Digest	科学文摘(美刊)
Science News	科学新闻(美刊)
Science of Today	现代科学(美刊)
Science Progress	科学进展(美刊)
Science World	科学世界(美刊)
Scientific World	科学世界(英刊)
Selected Water Resources Abstracts	水资源文献选摘(美刊)
Silver State Mining Corp.	银州采矿公司(美国)
Simplified Practice Recommendation(SPR)	简化应用推荐协会(美国,缩写 SPR)
Society of Mining Engineers of AIME	AIME 采矿工程师协会(美刊)
Society of Women Engineers(SWE)	女工程师协会(美国,缩写 SWE)
South Africa Mining & Engineering Journal	南非采矿与工程杂志
South African Cyanamid Pty., Ltd.	南非氰胺有限公司
South African Institute of Mining and Metallurgy, Johannesburg	南非采矿冶金学会(约翰内斯堡)
Spaulding Equipment Company(SEC)	斯波尔丁设备公司(美国,缩写 SEC,生产农机及矿机等设备)
Stamford Research Institute	斯坦福研究所[院](美国)
State and Federal Agency	州和联邦机构[总署](美环保机构)
State and Federal Government	州政府和联邦政府(美国)
State and Federal Laws	州法和联邦法(美国)
State Council	国务院
State Council of the People's Republic of China	中华人民共和国国务院
State Department	美国国务院

State Mining Company	国家采矿公司
State Technical Research Centre of Finland	芬兰国家技术研究中心
State Water Pollution Board	州水污染(管理)局[委员会](美国)
Steel Co. of Canada, Ltd.	加拿大钢铁有限公司
Sumitomo Corporation	住友商事株式会社(日本)
Sweden's Boliden Group	瑞典 Boliden 集团(采矿和冶炼企业)
Swift Chemical Company	斯威夫特化学公司(美国,产磷酸盐)
Symons Co.	西蒙斯公司(美国)
Taishu Mine, Nagasaki Prefecture, Japan	日本长崎县对洲矿山
Taiyo Bussan Kaisha Ltd.	太阳物产株式会社(日本)
Technical Services Laboratory	技术(服务)研究所
Technology Update	现代技术杂志
Texas Gulf Inc.	得克萨斯海湾公司(产磷酸盐)(美国)
The Metallurgical Society of AIME	AIME 冶金学[协]会(美国)
Thin Solid Films	《固体薄膜》(瑞士半月期刊,属于工程技术行业,材料科学"膜"子行业的优秀级杂志)
Transactions of American Mathematical Society	美国数学协会汇刊
Transmission & Distribution	输电与配电(美刊)
Twin Cities Metallurgy Research Center, Minneapolis, Minnesota	明尼苏达州明尼阿波利斯双子城冶金研究中心(美国)
U.S. Government V-Loan Guarantees	美国政府五项贷款协定[保证书]
U.S. Public Health Service	美国公共卫生署
U.S. Soil Salinity Laboratory	美国[联邦]土壤盐度实验室
Union Carbide Corp.	碳化物股份有限公司(美国)
Union Carbide Corporation Technological Center, Tarrytown, New York	美国纽约州联合碳化物公司技术中心
United State Bureau of Mines	美国矿山局
United States Bureau of Mine Publications	美国矿业局出版社
United States Environmental Protection Agency(EPA)	美国环境保护署[美国国家环境保护局](缩写 EPA)
United States Government	美国政府

United States National Academy of Sciences	美国国家科学院
University of Leeds	英国利兹大学
Urals	乌拉尔(山脉地区)
US Bureau of Mines Center in Salt Lake City	美国盐湖城矿山中心局
Utah Copper Company	犹他州铜公司(美国)
Utah Copper Division	犹他州铜分公司(美国)
Utah Copper Division of Kennecott Copper Corporation	肯纳可铜业公司犹他州铜分公司
W. R. Grace & Co.	W. R. 格雷斯公司(美国,产磷酸盐)
Ward's National Science Establishment, Inc., Rochester, New York	纽约州罗契斯特市华德自然科学中心(美国)
Washington Academy of Sciences	华盛顿科学院(美国)
Water & Waste Treatment	水与废水处理(英刊)
Water Pollution Control	水污染控制(英刊)
Weed Concentrator	威德选矿厂(美国)
Weekly Record	新书周报(美刊)
Wemco Co.	威姆科公司(美国)
Wemco Division, Envirotech Corporation	环境技术公司维姆科分公司(美国)
World Mining	世界采矿(美刊)
World Mining Equipment	世界采矿设备(美刊)
World Patents Abstracts Journal	世界专利文摘杂志(英刊)
World Pump	世界泵业(英刊)
WRC Information (Publication & Meetings) (WRC=Water Research Centre)	水研究中心(WRC)通报(英刊)
Wright Engineers Limited	莱特工程公司(加拿大)

参 考 文 献

[1] 英汉金属矿业词典 1984 年版.
[2] 英汉冶金工业词典 1988 年版、1999 年版.
[3] 英汉科学技术词典 1991 年版.
[4] 英汉科技缩写词汇 1981 年版.

注：1. 除以上文献外编者陈贤书自己参考的其他中外文资料文献已无从查到；
　　2. 1999 年版的英汉冶金工业词典是现在整理人员参考的书目。

后 记

在此附上本书作者、我的父亲生前为《英汉选矿工业词典》撰写的"内容简介和特点"手迹，根据其中内容，这本词典应该共收编技术词汇约 50000 余条、缩略语约 1500 条。遗憾的是在我们整理父亲的手稿时，只找到了大约原著词条的一半，所以现在出版的这本词典的技术价值、完整性和实用性都与父亲的原著有一定距离。也由于原作是手写稿件，在后期计算机录入、整理和排版中难免会出现漏录或错误，敬请读者在从中获取知识的同时给予谅解。

陈继贤

2017 年 11 月

作者手稿

英汉选矿工业词典
English-Chinese Mineral Processing
Industry Dictionary
内容简介和特点

本书是为适应我国书荒和送书下乡以及分家和满足我国选矿工业界的技术工作者的需要,而首次创造性地以选矿专业为主所编写的。

本书收集了技术词语50000余条,缩略语约1500条。其中以选矿为主,并充分收入了矿物、选矿机械、设备配置、水冶、土建、药剂、总图运输、水尾、环保、自动控制、水暖通风、技术经济、计算技术中也有关词语,也适量兼收些亲矿、地质、岩石技术名词。

本书特点是新词多、收词全面、实用价值。在5万余条词语中约有两万余条是现今国内外刊行的采选译合工具书中尚付缺如的新词,缩略词中也有相当部分属全新的。这便提供读此查找它们的间潜有处;所收集的词条,从上面所列的专业数看,是比较全面的,在阅读国外技术文献时,手此一册,即可减轻另查他书之劳。本书正文与缩略语之别，文后附些常用的便览表以供使查之效。

　　本书可供送矿工业界的科研、设计、生产建设的工作者、翻译人员及大专院校师生等同行参考。

　　　　　　　　　　　　　1989.7.30.